Heath
ALGEBRA 1

Clyde A. Dilley

Steven P. Meiring

John E. Tarr

Ross Taylor

D.C. Heath and Company
Lexington, Massachusetts Toronto

Authors

Clyde A. Dilley
Author in residence; formerly Professor of Education,
University of Toledo, Ohio; and formerly mathematics teacher
in Iowa and Illinois

Steven P. Meiring
Supervisor of Mathematics, State of Ohio;
formerly mathematics teacher in Indiana

John E. Tarr
Professor, University of Northern Iowa;
and Mathematics Teacher, Malcolm Price Laboratory School

Ross Taylor
Supervisor of Mathematics, Minneapolis Public Schools;
formerly mathematics teacher in Illinois

Published simultaneously in Canada.

Printed in the United States of America.

International Standard Book Number: 0-669-06161-1

1 2 3 4 5 6 7 8 9 0

Contents

1 Algebraic Expressions

2 Real Numbers

3 Solving Equations

4 Polynomials

5 Graphing Linear Equations and Functions

6 Systems of Linear Equations

7 Multiplying and Factoring Polynomials

8 Algebraic Fractions and Applications

9 Inequalities and Their Graphs

10 Rational and Irrational Numbers

11 Solving Quadratic Equations

Supplementary Topics

A Letter to the Student

Algebra 1 is the first in a series of high-school mathematics courses that can prepare you for promising opportunities in higher education and in future careers. Algebra is like arithmetic because it deals with numbers; but it is a more general way (and a more powerful way) to think about relationships among numbers. It takes time and practice to become used to this new way of thinking, but the time and effort are well spent. Algebra provides a deeper undrstanding not only of mathematics but also of the world in which we live.

You will now be expected to take greater responsibility for your own learning. You will need to learn how to read mathematics, how to study mathematics, and how to use a textbook effectively. To help you develop these new learning skills we have provided study hints called *Strategies for Success.*

Expect to be successful in this algebra course. Believe in yourself, apply yourself, enjoy the work, and in all likelihood you will be successful.

Clyde A. Dilley   *John E. Tarr*

Steven P. Meiring   *Ross Taylor*

Strategy for Success Reading Mathematics

To read mathematics with understanding you need to use skills different from those you use when reading a novel, a poem, or even a science lesson. Part of the power of mathematics lies in using a very abbreviated notation to express ideas that in ordinary English might require long sentences. Since big ideas are expressed very compactly, mathematics must be read very slowly, and frequently must be reread many times to get the full meaning. Keep paper and pencil handy as you read mathematics and try to illustrate or write an example for each new idea.

1 Algebraic Expressions

In skydiving, algebra is used to calculate information the skydiver needs to know before attempting a jump. The jumper must determine the altitude of the plane, the wind velocity, and a safe time interval in which to freefall.

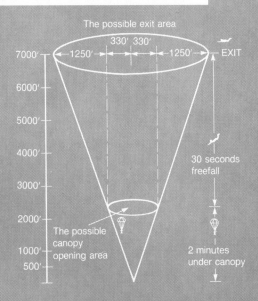

The possible exit area

7000′ — EXIT

1250′ — 330′ 330′ — 1250′

6000′

5000′

4000′ — 30 seconds freefall

3000′

2000′ — The possible canopy opening area

1000′ — 2 minutes under canopy

500′

1-1 Variables and Expressions

Preview

Diana Briggs is a skydiver. This table shows how far she traveled during the first few seconds of freefall.

Number of seconds	Distance fallen
1	16 ft
3	144 ft
5	400 ft
7	784 ft
9	1296 ft

- How many feet did she fall in the first three seconds?
- How many feet did she fall during the second and third seconds?
- Estimate how many feet she fell in the first two seconds.

Most people estimate that she fell 80 feet in the first two seconds because 80 is halfway between 16 and 144. The actual distance fallen is 64 feet. The table does not provide enough information for you to give an accurate estimate. By the end of this lesson you will be able to use algebra skills to answer these kinds of questions.

■ LESSON

Letters such as x, y, A, and g are used in expressions in place of numbers. When letters are used in this way, they are called **variables.** Here are some expressions with variables:

$$9 + x \qquad y - 17 \qquad A \times \frac{3}{4} \qquad \frac{g}{0.7}$$

Variables can be replaced by numbers. This process is called **substitution.** The set of numbers to be substituted is called the **domain of the variable. Braces** { } are used to indicate a set of numbers.

Example 1 Let the domain of x be $\{3, 4\frac{1}{2}, 5.2\}$. Substitute each number of the domain for x in this expression:

$$9 + x$$

Solution

Original expression	$9 + x$
Substitute 3 for x.	$9 + 3$
Substitute $4\frac{1}{2}$ for x.	$9 + 4\frac{1}{2}$
Substitute 5.2 for x.	$9 + 5.2$

These numerical expressions represent the same number.

$$12 - 4 \qquad \frac{6 \times 4}{3} \qquad 5 + 2 + 1 \qquad 8$$

The **simplest expression** is 8. Whenever you are asked to **simplify** a numerical expression, you are to write the simplest expression that represents the number.

Example 2 Simplify each of the following.

           a. $\dfrac{12}{24}$ **b.** $\dfrac{8 \times 3}{2}$ **c.** $3.4 - 1.6$

Solution   **a.** $\dfrac{12}{24} = \dfrac{1}{2}$ **b.** $\dfrac{8 \times 3}{2} = 12$ **c.** $3.4 - 1.6 = 1.8$

Answer   **a.** $\dfrac{1}{2}$ or 0.5   **b.** 12   **c.** 1.8

A common algebra activity, especially when checking your work, is to substitute for a variable and then simplify the resulting numerical expression. This process is called **evaluating** an expression for a given number.

Example 3 Evaluate $3 \times a$ for $a = 7$.

Solution   Original expression $3 \times a$
            Substitute 7 for a. 3×7
            Simplify.   21

Answer   When $a = 7$, the value of $3 \times a$ is 21.

■ CLASSROOM EXERCISES

Identify the variable in each expression.

1. $2 \times t$ **2.** $5 \times b$ **3.** $A - 6$ **4.** $x - 17$

Simplify.

5. 6×2 **6.** $26 - 12$ **7.** $\dfrac{3 \times 4}{2}$ **8.** $\dfrac{3}{4} - \dfrac{1}{2}$ **9.** 1.5×2.3

a	A	n	x
2	7	3	5

Substitute and simplify.

10. $a + A$ **11.** $2 \times x$ **12.** $10 - x$ **13.** $n \div 3$

14. $10 - n$ **15.** $3 \times n$ **16.** My dog is x years old.

Substitute each number of the domain for n and simplify. Domain $= \{0, 2, 4\}$

17. $n + 3$ **18.** $10 - n$ **19.** $3 \times n$ **20.** $\dfrac{n}{2}$

■ WRITTEN EXERCISES

The domain of each variable is {1, 4, 3.7}. Substitute each number of the domain for the variable. Do *not* simplify the numerical expression.

A
1. $17 - x$ **2.** $11 - t$ **3.** $\dfrac{2}{a}$ **4.** $\dfrac{b}{2}$

5. $1 + y$ **6.** $c + 3$ **7.** $6 \times N$ **8.** $10 \times P$

Substitute and simplify.

a	b	m	T	x
3	5	6	3	2

9. $a + 7$ **10.** $9 + m$ **11.** $12 - b$ **12.** $6 - m$

13. $4 \times T$ **14.** $5 \times x$ **15.** $T \div a$ **16.** $m \div a$

17. $m \times b$ **18.** $b \times x$ **19.** $m + m$ **20.** $a + a$

Substitute and simplify.

B	c	n	x	y
4	2	1	$\frac{2}{3}$	0

21. $\dfrac{24}{c}$ **22.** $\dfrac{44}{B}$ **23.** $n \times x$ **24.** $x \times y$ **25.** $\dfrac{B}{n}$

26. $\dfrac{c}{n}$ **27.** $c + n + x$ **28.** $x + c + n$ **29.** $y \times c \times 5$ **30.** $B \times c \times y$

Copy the expression and substitute the given values. Do *not* simplify.

31. Substitue 10 for l and 5 for w.
     a. The area of a rectangle is $l \times w$.
     b. The perimeter of a rectangle is $l + w + l + w$.

32. Substitute your age in years for A.
     a. I am A years old.
     b. In 5 years I will be $(A + 5)$ years old.
     c. Six years ago I was $(A - 6)$ years old.
     d. In A years I will be $(2 \times A)$ years old.

33. Substitute your height in inches for h.
     a. I am h inches tall.
     b. If I grow 3 inches taller, I will be $(h + 3)$ inches tall.

34. Substitute your weight in pounds for w.
     a. I weigh w pounds.
     b. If I lose 10 pounds, I will weigh $(w - 10)$ pounds.
     c. My brother weighs $\dfrac{3}{4}$ as much as I do. He weighs $\dfrac{3}{4} \times w$ pounds.
     d. Last year I weighed $(w - 12)$ pounds.

35. Substitute the number of girls in your algebra class for g and the number of boys for b. Simplify.
  a. There are g girls in my class.
  b. There are $(g + b)$ students in my class.
  c. If one girl moves away, there will be $(g - 1)$ girls and $(g + b - 1)$ students in my class.
  d. If two classmates move away, there will be $(g + b - 2)$ students in my class.

36. Substitute the number of hours of class you have in the morning for a and the number of hours of class you have in the afternoon for b. Simplify.
  a. I have b hours of class in the afternoon.
  b. I have $(a + b)$ hours of class each day.
  c. If I get out of school an hour early, I will have $(b - 1)$ hours of class in the afternoon and $(a + b - 1)$ hours of class for the day.
  d. If school is dismissed at noon after I arrive an hour late, I will have $(a - 1)$ hours of class for the day.

37. In the Preview, we can find the number of feet fallen by the skydiver by the expression $16 \times t \times t$.
  a. How many feet did the skydiver fall in the first 8 seconds?
    [*Hint:* Evaluate $16 \times t \times t$ for $t = 8$.]
  b. How many feet did she fall in the first 2 seconds?
  c. How many feet did she fall in the first 6 seconds?
  d. How many feet did she fall by the time she pulled the rip cord, assuming a freefall of 10 seconds?

Copy and complete the following tables.

38.

a	b	$a + b$	$a - 7$
7	4	?	?

39.

x	y	$x \times y$	$y \times x$
0.5	7	?	?

40.

m	n	$\dfrac{m}{n}$	$\dfrac{n}{m}$
12	4	?	?

41.

c	d	$c + d$	$c - d$
7	?	10	?

42.

r	s	$r + s$	$r - s$	$\dfrac{r}{s}$
?	5	?	5	?

43.

p	q	$p + q$	$p - q$	$\dfrac{p}{q}$
?	?	10	6	?

■ REVIEW EXERCISES

Keep your arithmetic skills sharp!

Simplify.

1. $\dfrac{3}{8} + \dfrac{1}{4}$ **2.** $\dfrac{2}{5} + \dfrac{1}{3}$ **3.** $\dfrac{5}{6} - \dfrac{1}{2}$ **4.** $\dfrac{7}{8} - \dfrac{2}{5}$

5. $\dfrac{1}{2} \times \dfrac{1}{3}$ **6.** $\dfrac{1}{6} \times \dfrac{3}{5}$ **7.** $\dfrac{2}{3} \div 3$ **8.** $\dfrac{4}{5} \div \dfrac{1}{3}$

1–2 Using the Order of Operations to Simplify Expressions

Preview   **Multiplication on a computer**

When Greg Mason was learning about a computer at school, he tried to use it to add 6 and 4 and then multiply the result by 9.

He typed:

PRINT 4 + 6 * 9   " * " is the computer symbol for multiplication.

The computer screen displayed this number:

58

That was not the answer that Greg expected. What do you think happened?

In this lesson you will learn some rules mathematicians have agreed on to be sure that an expression has just one meaning.

■ LESSON

In the Preview you saw that the expression

$$4 + 6 \times 9$$

could be considered to represent either of two numbers, depending on the order in which the operations (addition and multiplication) were done.

Multiplication first	*Addition first*
$4 + 6 \times 9 = 58$	$4 + 6 \times 9 = 90$
↑ ↑	↑ ↑
2nd 1st	1st 2nd

To avoid this confusing situation, mathematicians have agreed on the following rules for the **order of operations.**

Order of Operations

1. Simplify inside parentheses and other grouping symbols first.
2. Do multiplications and divisions next, in order from left to right.
3. Do additions and subtractions next, in order from left to right.

Example 1   Simplify. $3 \times (4 + 1)$

   Solution

                                                     2nd 1st

First: Simplify inside the parentheses. $3 \times (4 + 1) = 3 \times 5$
Second: Multiply 3 and 5. $= 15$

   Answer   15

Example 2.   Simplify. $3 \times 4 + 1$

   Solution

                                        1st 2nd

First: Multiply 3 and 4. $3 \times 4 + 1 = 12 + 1$
Second: Add 12 and 1. $= 13$

   Answer   13

Example 3   Simplify. $12 - (8 - 3)$

   Solution

                                            2nd 1st

First: Simplify inside the parentheses. $12 - (8 - 3) = 12 - 5$
Second: Subtract 5 from 12. $= 7$

   Answer   7

Example 4   Simplify. $12 - 8 - 3$

   Solution

                                        1st 2nd

First: Subtract 8 from 12. $12 - 8 - 3 = 4 - 3$
Second: Subtract 3 from 4. $= 1$

   Answer   1

Brackets [ ], like parentheses, are also grouping symbols used to enclose an expression. If there are grouping symbols within grouping symbols, simplify inside the innermost symbols first.

Example 5   Simplify. $26 - [(2 \times 3) \times 3]$

   Solution

   First: Multiply 2 and 3.
   Second: Multiply 6 and 3.
   Third: Subtract 18 from 26.

$$\overset{\text{3rd}\quad\text{1st}\quad\text{2nd}}{\downarrow\quad\downarrow\quad\downarrow}$$
$$26 - [(2 \times 3) \times 3] = 26 - [6 \times 3]$$
$$= 26 - 18$$
$$= 8$$

   Answer   8

Example 6   Simplify. $7 \times 5 + 9 \times 4 \div 3$

   Solution

   First: Multiply 7 and 5.
   Second: Multiply 9 and 4.
   Third: Divide 36 by 3.
   Fourth: Add 35 and 12.

$$\overset{\text{1st}\ \ \text{4th}\ \ \text{2nd}\ \ \text{3rd}}{\downarrow\ \ \ \downarrow\ \ \ \downarrow\ \ \ \downarrow}$$
$$7 \times 5 + 9 \times 4 \div 3 = 35 + 9 \times 4 \div 3$$
$$= 35 + 36 \div 3$$
$$= 35 + 12$$
$$= 47$$

   Answer   47

Example 7   Simplify. $\dfrac{8 + 5 \times 2}{12}$

   Solution   The "fraction bar" in this example does two jobs. It indicates the operation division and it is also a grouping symbol. It indicates that $8 + 5 \times 2$ is divided by 12.

   First: Multiply 5 and 2.

   Second: Add 8 and 10.

   Third: Divide 18 by 12.

$$\overset{\text{2nd}\ \ \text{1st}\qquad\text{3rd}}{\downarrow\quad\ \downarrow\qquad\ \big|}$$
$$\frac{8 + 5 \times 2}{12} = \frac{8 + 10}{12}$$
$$= \frac{18}{12}$$
$$= 1.5$$

   Answer   1.5

■ CLASSROOM EXERCISES

1. State the rules for order of operations.

Indicate the order of operations. Then simplify.

$$\overset{?\quad\ ?}{\downarrow\quad\ \downarrow}$$
2. $2 + 3 \times 7$

$$\overset{?\quad\ ?}{\downarrow\quad\ \downarrow}$$
3. $(2 + 3) \times 7$

$$\overset{?\quad\ ?\quad\ ?}{\downarrow\quad\ \downarrow\quad\ \downarrow}$$
4. $3 \times (2 + 2) \times 4$

$$\begin{array}{ccc} ? & ? & ? \\ \downarrow & \downarrow & \downarrow \end{array}$$
5. $4 \times [(3 + 4) \div 2]$

$$\begin{array}{cc} ? & ? \\ \downarrow & \end{array}$$
6. $\dfrac{20 + 4}{3 \times 2}$
$$\begin{array}{c} \uparrow \\ ? \end{array}$$

$$\begin{array}{cccc} ? & ? & ? & ? \\ \downarrow & \downarrow & \downarrow & \downarrow \end{array}$$
7. $12 \times 6 \div 3 \times 2 \div 8$

■ WRITTEN EXERCISES

Write $+$, $-$, $\times$, or $\div$ to identify the operation to be done first.

A
1. $(8 + 4) \div 2$
2. $(6 + 12) \div 3$
3. $8 + 2 \times 5$

4. $4 + 3 \times 6$
5. $8 + 4 \div 2$
6. $6 + 12 \div 3$

7. $(10 \div 2) \times 4$
8. $(15 - 2) \times 5$
9. $13 - 1 \times 6$

10. $4 + 7 \times 2 - 2$
11. $12 \div 4 + 15 \times 3$
12. $8 \times [2 \div (2 - 1)]$

Simplify.

13. $4 + 5 \times 6$
14. $5 + 6 \times 4$
15. $20 - 8 \div 4$

16. $30 - 10 \div 5$
17. $4 \times 5 + 6$
18. $6 \times [5 + 4]$

19. $(20 - 8) \div 4$
20. $(30 \div 10) + 5$
21. $3 \times 4 + 6 \times 4$

22. $5 \times [(2 + 5) \times 3]$
23. $(3 + 6) \times 4$
24. $5 \times (2 + 3)$

25. $\dfrac{70}{2 + 5}$
26. $\dfrac{18}{3 + 6}$

h	B	p	c
3	30	5	100

Substitute and simplify.

27. $4 \times p + 2$
28. $5 \times h + 4$
29. $4 \times (p + 2)$

30. $5 \times (B + 1)$
31. $2 \times B \times h$
32. $3 \times B \times p$

33. $B \times (h + 4)$
34. $c \times (p + 5)$
35. $5 + B \times p$

36. $200 + 3 \times B$
37. $c - (70 - B)$
38. $(B - h) + (c - h)$

Simplify.

B
39. $8 \times 2 - 3 \times 4 + 5$
40. $7 \times 3 - 5 \times 4 + 6$

41. $18 \times (12 - 9) + 2$
42. $15 \times (10 - 8) + 1$

43. $27 + 18 \div 9 - 3 + 1$
44. $40 + 24 \div 8 - 3 + 1$

45. $(27 + 18) \div 9 - (3 + 1)$
46. $(40 + 24) \div 8 - (3 + 1)$

47. $\dfrac{27 + 18}{9 - (3 + 1)}$
48. $\dfrac{40 + 24}{8 - (3 + 1)}$

Indicate whether the equation is true or false.

49. $10 + (12.5 - 7.5) = (10 + 12.5) - 7.5$

50. $23 - (10 + 7) = (23 - 10) + 7$

51. $30 - (17 - 13) = (30 - 17) - 13$

52. $40 - (23 + 12) = (40 - 23) - 12$

53. $7 \times 8 + 7 \times 2 = 7 \times (8 + 2)$

54. $\frac{40 - 8}{4} = 40 \div 4 - 8 \div 4$

55. $\frac{54 - 27}{9} = 54 - 27 \div 9$

56. $10 \times 3 + 10 \times 5 = (10 \times 3 + 10) \times 5$

Use each symbol once to write an expression for the number in the box.

Sample	$\boxed{\frac{1}{2}}$	$4, 5, 6, 8, -, \times, \div, (\,)$
Solution		$(6 - 5) \times 4 \div 8$

C **57.** $\boxed{6}$ $3, 4, 7, +, -$

58. $\boxed{8}$ $1, 2, 3, +, \times, (\,)$

59. $\boxed{10}$ $1, 2, 3, 15, +, -, \times$

60. $\boxed{0}$ $3, 4, 5, 7, 8, +, -, \times, \div, (\,)$

61. Puzzle. Pick a whole number less than 60. Do these steps.
   a. Divide the number by 3. Call the remainder a.
   b. Divide the number by 4. Call the remainder b.
   c. Divide the number by 5. Call the remainder c.
   d. Substitute the values of a, b, and c in this expression:

$$40 \times a + 45 \times b + 36 \times c$$

   e. Simplify the expression and divide by 60.
   Is the remainder the number you picked?

(This puzzle uses the "Chinese Remainder Theorem," which appeared in the writings of the Chinese mathematician Sun Tzu who lived in the 4th or 5th century A.D.)

62. Puzzle. I am thinking of a whole number that is less than 60. If you divide the number by 3, you get a remainder of 2. If you divide it by 4, you get a remainder of 1. If you divide it by 5, you get a remainder of 4. What is my number?

■ REVIEW EXERCISES

Maintain your skills!

Simplify.

1. $\frac{5}{6} + \frac{2}{3}$

2. $\frac{1}{4} + \frac{2}{5}$

3. $\frac{2}{3} - \frac{1}{6}$

4. $\frac{1}{2} - \frac{1}{5}$

5. $\frac{1}{4} \times \frac{1}{4}$

6. $\frac{5}{6} \times \frac{6}{5}$

7. $\frac{7}{8} \div 3$

8. $\frac{5}{9} \div \frac{2}{3}$

1–3 Symbols for Multiplication in Expressions

Preview **Historical note: Multiplication signs**

The use of "×" for multiplication may have come from an early printing custom of using crossed lines to indicate multiplication steps. An early book might have set out the problem of multiplying 52 by 38 as

By the year 1600, the symbol × was employed in England but did not come into wide use until the late 1800's.

In this lesson you will learn ways of expressing multiplication in algebra.

■ LESSON

Mathematicians use several different ways to indicate multiplication. Each of the following expressions means 7 times m.

$$7 \times m \qquad 7 * m \qquad 7 \cdot m \qquad 7m \qquad 7(m)$$

We avoid using "×" to show multiplication in algebra because it is easily confused with the variable "x". We use "$7 * m$" only with computers.

The following expressions are used in algebra:

7 times m	m times n	5 times a times b	2 times 6
$7 \cdot m$	$m \cdot n$	$5 \cdot a \cdot b$	$2 \cdot 6$
$7m$	mn	$5 \cdot ab$	$2(6)$
$7(m)$		$5a \cdot b$	
		$5ab$	

In a multiplication expression, numbers that are multiplied are called **factors.** The result is called the **product.**

$$\underbrace{3 \cdot a \cdot b}_{\text{product}}$$

factors
↓ ↓ ↓

For a product that includes a number and variables, the number is called the **coefficient** (or **numerical coefficient**) of the remaining expression.

For $3a$, 3 is the coefficient of a.

For $\left(\dfrac{1}{5}\right)xy$, $\left(\dfrac{1}{5}\right)$ is the coefficient of xy.

For $7(a + b)$, 7 is the coefficient of $(a + b)$.

Mathematicians invent simpler ways to write expressions. Omitting multiplication signs in expressions like $3ab$ is an example of using shortened notation. Another shortened notation allows us to write $2 \cdot 2 \cdot 2 \cdot 2 \cdot 2$ more simply:

$$\text{Long way} \qquad \text{Short way}$$
$$2 \cdot 2 \cdot 2 \cdot 2 \cdot 2 \qquad 2^5 \leftarrow \text{exponent}$$
$$\qquad\qquad\qquad\quad \uparrow\text{——base}$$

The **exponent** (5) indicates how many times the **base** (2) is used as a factor.

Read as:

$4 \cdot 4 \cdot 4 = 4^3$ "4 cubed"

$7 \cdot 7 \cdot 7 \cdot 7 = 7^4$ "7 to the 4th power"

$m \cdot m \cdot m \cdot m \cdot m = m^5$ "m to the 5th power"

$(a + 1)(a + 1) = (a + 1)^2$ "the quantity $(a + 1)$ squared"

If expressions such as $3 \cdot 4^2$ and $(1 + 2)^2$ are to have a clearly understood meaning, we must agree on an *order of operations* that includes **raising to a power.**

Order of Operations

After simplifying inside parentheses, evaluate powers before doing the multiplication and division.

Example 1 Simplify. $7 \cdot 2^2$

Solution

$$\overset{\text{2nd} \quad \text{1st}}{\downarrow \quad \swarrow}$$

First: Evaluate the power. $7 \cdot 2^2 = 7 \cdot 4$
Second: Multiply 7 and 4. $= 28$

Answer 28

Example 2 Simplify. $(7 \cdot 2)^2$

Solution

$$\overset{\text{1st} \quad \text{2nd}}{\downarrow \quad \swarrow}$$

First: Simplify inside the parentheses. $(7 \cdot 2)^2 = (14)^2$
Second: Evaluate the power. $= 196$

Answer 196

Example 3 Simplify. $2^3 + (4 + 1)^2$

Solution

$$\overset{\text{2nd 4th 1st 3rd}}{2^3 + (4 + 1)^2} = 2^3 + 5^2$$
$$= 8 + 25$$
$$= 33$$

First: Simplify inside the parentheses.
Second and third: Evaluate the powers.
Fourth: Add 8 and 25.

Answer 33

■ CLASSROOM EXERCISES

Give two factors of each product.

1. $2xy$ **2.** $5a^2$ **3.** $4(a + 1)$

Identify the coefficient in each expression.

4. For $7mn$, what is the coefficient of mn?   **5.** For $3y^2$, what is the coefficient of y^2?

Read.

6. m^2 **7.** 6^4 **8.** y^3 **9.** $4 + a^2$ **10.** $(4 + a)^2$

Simplify.

11. 4^2 **12.** 5^3 **13.** 10^5

14. $3 + 2^3$ **15.** $3^2 + 4^2$ **16.** $(3 + 4)^2$

Strategy for Success **Doing assignments**

> When you have time in class to begin an assignment, complete as many problems as possible. Then if you get "stuck" on a problem you can ask for help.

■ WRITTEN EXERCISES

A Simplify.

1. 3^2 **2.** 2^3 **3.** 10^6 **4.** 10^5

5. $10 \cdot 3^2$ **6.** $10 \cdot 4^2$ **7.** $6 + 2^3$ **8.** $2 + 3^3$

9. $(2 + 3)^2$ **10.** $(5 - 2)^2$ **11.** 3.5^2 **12.** 4.5^2

Evaluate for $n = 3$.

13. $2n$ **14.** $4n$ **15.** n^2 **16.** n^4

17. 2^n **18.** 4^n **19.** $5n^2$ **20.** $4n^2$

Evaluate for $n = 10$.

21. $6n$ **22.** $3n$ **23.** n^6 **24.** n^3

25. $n + 2^3$ **26.** $n + 3^2$ **27.** $(n + 2)^2$ **28.** $(n - 2)^2$

a	b	t	r	s
0	1	4	$\frac{2}{3}$	3

Substitute and simplify.

29. $3(2 + t)^2$ **30.** $10(1 + t)^2$ **31.** b^8 **32.** b^{10}

33. $(rs)^2$ **34.** rs^2 **35.** $(t - b)^2$ **36.** $(b - r)^2$

The expression e^3 is used to find the volume of a cube when the length of an edge e is known. Find the volumes of cubes with these edges.

37. 5 cm   **38.** 4 cm   **39.** 10 m   **40.** 15 m

41. 1.5 cm   **42.** 1.2 m   **43.** $\frac{1}{2}$ in.   **44.** $\frac{1}{3}$ in.

The expression $16t^2$ is used to find the distance in feet that an object falls in t seconds. Find the distance an object falls in these times.

45. $\frac{1}{2}$ second   **46.** 1 second   **47.** 5 seconds   **48.** 10 seconds

The expression $1000(1.1)^t$ is used to find the current value of a $1000 investment that earns 10% interest per year, compounded annually for t years. Find the current value of a $1000 investment after the following periods:

B **49.** 2 years   **50.** 3 years   **51.** 4 years   **52.** 5 years

The expression $0.01E + 0.003E^2$ gives the required thickness in millimeters for an electrical insulation material. If E is the number of kilovolts required to puncture the insulation, find the thickness of insulation material for these voltages:

53. 1 kilovolt   **54.** 2 kilovolts   **55.** 10 kilovolts   **56.** 20 kilovolts

x	y	z	R	S
2	4	6	8	10

Substitute and simplify.

57. $z^x + R^x - S^x$ **58.** $y^x + R^x$ **59.** $z^3 - y^3$ **60.** $R^x - x^y$

61. $(z + R - S)^x$ **62.** $(x + R - z)^x$ **63.** $(z - y)^3$ **64.** $(S - R)^3$

C Copy and complete each table.

65.

a	n	a^n
4	3	?

66.

a	n	a^n
2	?	64

67.

a	n	a^n
?	4	81

■ REVIEW EXERCISES

Area represents the number of 1-unit by 1-unit squares needed to cover the interior of a figure. The perimeter (peRIMeter) is the distance around the "rim" of the figure.

Find the area and perimeter of each figure. Include cm or cm² in your answers.

1.

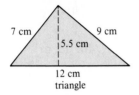

15 cm
rectangle

2.

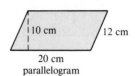

20 cm
parallelogram

3.

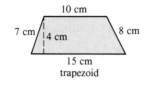

12 cm
triangle

4.

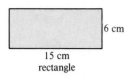

15 cm
trapezoid

Self-Quiz 1

a	A	d	D
3	8	2	4

1–1 Substitute and simplify.

    1. $(A + D) \div a$ **2.** D^a **3.** $D + ad$ **4.** $a(A - d)$

1–2 Use the order of operations to simplify.

    5. $3 + 4 \cdot 5$ **6.** $(3 + 4) \cdot 5$

    7. $8 + 4 \cdot 2^3$ **8.** $3 + 14 \div 2 - 1 \cdot 5$

1–3 Evaluate for $n = 2$.

    9. $4n$ **10.** n^4 **11.** $5(n - 1)$ **12.** $n + 6^2$

Strategy for Success **Using the self-quizzes**

Use the self-quizzes in the book to check your progress. If you miss any questions, find out why. First go back to the lesson. If you still don't understand, seek help.

1–4 Equivalent Expressions

Here are four expressions involving the variable a:

$$7a - 3a \qquad 4a \qquad \frac{1}{2}(8a) \qquad \frac{12a}{3}$$

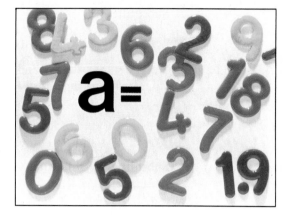

- Evaluate each expression for $a = 3$.

- Evaluate for $a = 5$.

- Evaluate for $a = 0.5$.

Expressions that have the same value for each substitution are very important in algebra. In this lesson you will begin the study of such expressions.

■ LESSON

Suppose that 7 is substituted for x in these two expressions:

$$3x + 2x \qquad 5x$$
$$3(7) + 2(7) \qquad 5(7)$$
$$35 \qquad\qquad 35$$

In both cases you get 35.

If 2 is substituted for x, you get 10 in both cases.

If $\frac{1}{3}$ is substituted for x, you get $\frac{5}{3}$ in both cases.

No matter what number is substituted for x, both expressions have the same value. For this reason, the two expressions $5x$ and $3x + 2x$ are called **equivalent expressions.** In the Preview, the four expressions are equivalent.

Definition: Equivalent Expressions

Two expressions are equivalent if the first has the same value as the second for every possible substitution.

Two expressions are **not equivalent** if they have different values for at least one substitution. You can prove that two expressions are not equivalent by finding one substitution that gives different values.

Example 1 Are $4a + 1$ and $5a$ equivalent?

Solution Substitute 1:

$$4a + 1 \qquad\qquad 5a$$
$$4 \cdot 1 + 1 \qquad\quad 5 \cdot 1$$
$$5 \;\;\leftarrow \text{same} \rightarrow\;\; 5$$

Substitute 3:

$$4a + 1 \qquad\qquad 5a$$
$$4 \cdot 3 + 1 \qquad\quad 5 \cdot 3$$
$$13 \;\leftarrow \text{different} \rightarrow 15$$

Answer We know that $4a + 1$ and $5a$ are not equivalent because they have different values when 3 is substituted for a.

Example 2 Are $4a + a$ and $5a$ equivalent?

Solution Substitute the following values: 2, 7, 12, and $\frac{1}{2}$.

a	$4a + a$	$5a$
2	$4 \cdot 2 + 2 = 10$	$5 \cdot 2 = 10$
7	$4 \cdot 7 + 7 = 35$	$5 \cdot 7 = 35$
12	$4 \cdot 12 + 12 = 60$	$5 \cdot 12 = 60$
$\frac{1}{2}$	$4 \cdot \frac{1}{2} + \frac{1}{2} = 2\frac{1}{2}$	$5 \cdot \frac{1}{2} = 2\frac{1}{2}$

In each case the values are the same.

Answer This information strongly suggests that the expressions are equivalent but does *not prove* it.

Example 3 Are aa and a^2 equivalent?

Answer No substitutions are necessary. We know that these two expressions are equivalent by the definition of exponents.

Example 4 Are $2a + b$ and $a + 2b$ equivalent?

Solution Substitute 0 for a and 0 for b.

$$2a + b \qquad\qquad a + 2b$$
$$2 \cdot 0 + 0 \qquad\quad 0 + 2 \cdot 0$$
$$0 \;\;\leftarrow \text{same} \rightarrow\;\; 0$$

Substitute 3 for a and 4 for b.

$$2a + b \qquad\qquad a + 2b$$
$$2 \cdot 3 + 4 \qquad\quad 3 + 2 \cdot 4$$
$$10 \;\leftarrow \text{different} \rightarrow 11$$

Answer The expressions are not equivalent.

CLASSROOM EXERCISES

1. How can you prove that two expressions are not equivalent?

Copy and complete the tables by substituting and simplifying.

2.

	x	$3x - 1$	$2x$
a.	1	?	?
b.	3	?	?
c.	4	?	?

d. Do you think that $3x - 1$ and $2x$ are equivalent?

3.

	a	b	a^2b^2	$(ab)^2$
a.	2	3	?	?
b.	4	1	?	?
c.	3	3	?	?

d. Do you think that a^2b^2 and $(ab)^2$ are equivalent?

WRITTEN EXERCISES

Copy and complete the tables by substituting and simplifying. Then state whether you think the two expressions are equivalent.

1.

	x	$x + 3x$	$2x + 2x$
a.	0	?	?
b.	1	?	?
c.	10	?	?

2.

	a	$3 + a$	$4a$
a.	0	?	?
b.	1	?	?
c.	3	?	?

For each pair of expressions, substitute 0, 2, and 5 for the variable. Simplify the expressions and state whether you think they are equivalent.

3. $3(a + 5)$
$3a + 5$

4. $4(a + 10)$
$4a + 10$

5. $x^2 + 10$
$7x$

6. $2x$
x^2

State whether you think the expressions in each pair are equivalent. If the expressions are not equivalent, find one number that produces different values when substituted for the variable.

7. $5 + a$
$a + 5$

8. $10 + z$
$z + 10$

9. $2x + 3$
$5x$

10. $x + 4x$
$3x + 2x$

11. x^3
$x \cdot 3$

12. $3x - 1$
$2x$

13. $2x^2$
$(2x)^2$

14. $2 + b$
$3b$

15. Steve thinks that $2x + 1$ and $3x$ are equivalent. Show that he is wrong.

16. Al thinks that $3x + 2$ and $10x$ are equivalent. Show that he is wrong.

17. Jo believes that $(5x)^2$ and $5x^2$ are equivalent. Do you agree with her?

18. Ed believes that $(3x)^2$ and $9x^2$ are equivalent. Do you agree with him?

Substitute the values of a and b in each expression. State whether you think the expressions are equivalent.

19.

a	b	$(2a)b$	$2(ab)$
5	10	?	?
1	1	?	?
3	0	?	?

20.

a	b	$5(a + b)$	$(5a) + b$
4	0	?	?
10	0	?	?
1	1	?	?

State whether you think the two expressions are equivalent. If the expressions are not equivalent, find one pair of numbers that produces different values when substituted for the variables.

B **21.** $\left(\frac{1}{2}b\right)h$ $\qquad$ **22.** $\left(\frac{1}{2}h\right)(a + b)$ $\qquad$ **23.** $6(ab)$ $\qquad$ **24.** $\left(\frac{1}{2}a\right)(2b)$

$\dfrac{bh}{2}$ $\qquad\qquad\qquad$ $\dfrac{h(a + b)}{2}$ $\qquad\qquad$ $(2a)(3b)$ $\qquad\qquad$ $(2a)\left(\frac{1}{2}b\right)$

25. $\dfrac{2}{3} + \dfrac{a}{b}$ $\qquad$ **26.** $\dfrac{a}{b} + \dfrac{1}{2}$ $\qquad$ **27.** $(a + b)(a + b)$ $\qquad$ **28.** $(p + 2)q + 3$

$\dfrac{2 + a}{3 + b}$ $\qquad\qquad$ $\dfrac{a + 1}{b + 2}$ $\qquad\qquad$ $a^2 + b^2$ $\qquad\qquad$ $pq + 6$

In each exercise, two of the three expressions are equivalent. Write the equivalent expressions.

29. $5x + 10y$ $\qquad$ **30.** $x^2 \cdot x^3$ $\qquad$ **31.** $10 - a - b,$ $\qquad$ **32.** $a - (b - 1)$
$\quad\;\; 15xy$ $\qquad\qquad\quad\; x^6$ $\qquad\qquad\quad\;\; 10 - (a - b)$ $\qquad\quad (a - b) - 1$
$\quad\;\; 5(x + 2y)$ $\qquad\quad\; x \cdot x^4$ $\qquad\qquad\;\; (10 - a) + b$ $\qquad\quad (a - b) + 1$

C **33.** These expressions can be put into three groups of equivalent expressions. List the expressions in each group.

$\quad 2x + 2y \qquad x + 2y \qquad\quad x + y + x \qquad\quad 2(x + y) \qquad x + y + y$

$\quad 2y + 2x \qquad y + y + x + x \qquad x + y + x + y \qquad 2y + x \qquad\quad y + x + x$

34. Show that these expressions are not equivalent.

$$\frac{x^2 - 4}{x - 2} \quad \text{and} \quad x + 2$$

■ REVIEW EXERCISES

Simplify each numerical expression. $\qquad\qquad\qquad\qquad\qquad\qquad\qquad$ [1–2, 1–3]

1. $3 + 4 \cdot 5$ $\qquad$ **2.** $7 - 1 \cdot 4$ $\qquad$ **3.** $5(7 - 4)$ $\qquad$ **4.** $5 \cdot 7 - 5 \cdot 4$

5. $2 \cdot 3^2$ $\qquad$ **6.** $2 + 2^3$ $\qquad$ **7.** $(2 + 3)^2$ $\qquad$ **8.** $2^2 + 3^2$

1–5 Basic Properties of Numbers

Preview

Pick any 3-digit number. Multiply it by 7. Multiply the product by 11. Multiply the new product by 13. Look at the number you picked and the final product. You should see something unexpected.

Can you explain why the two numbers are related in this way?

This lesson will help you understand what happened.

■ LESSON

In earlier grades, you discovered that the order of two addends does not affect their sum. You also discovered that the order of two factors does not affect their product.

$$6 + 5 = 5 + 6 \qquad 7 \cdot 2 = 2 \cdot 7$$

$$3\frac{1}{2} + \frac{1}{4} = \frac{1}{4} + 3\frac{1}{2} \qquad 9.5 \cdot 3 = 3 \cdot 9.5$$

These are examples of general properties of numbers.

> ## Commutative Property of Addition
>
> For all numbers a and b,
> $$a + b = b + a$$
>
> ## Commutative Property of Multiplication
>
> For all numbers a and b,
> $$ab = ba$$

Commutative property of addition

$$2x + 3 = 3 + 2x$$
$$m + n^2 = n^2 + m$$
$$4x + (3 + 2x) = (3 + 2x) + 4x$$

Commutative property of multiplication

$$(5a)4 = 4(5a)$$
$$rs^2 = s^2r$$
$$5(3y + 1) = (3y + 1)5$$

You also learned that the way three addends are grouped or associated does not affect their sum, and that the way three factors are grouped does not affect their product.

$$(4 + 6) + 3 = 4 + (6 + 3) \qquad (2 \cdot 5) \cdot 4 = 2 \cdot (5 \cdot 4)$$

$$3.2 + (5.7 + 2.1) = (3.2 + 5.7) + 2.1 \qquad \frac{1}{2} \cdot (2 \cdot 3) = \left(\frac{1}{2} \cdot 2\right) \cdot 3$$

These properties may be stated generally for numbers.

The Associative Property of Addition

For all numbers a, b, and c,

$$(a + b) + c = a + (b + c)$$

The Associative Property of Multiplication

For all numbers a, b, and c,

$$(ab)c = a(bc)$$

Associative property of addition

$$3a + (4a + 1) = (3a + 4a) + 1$$

$$(5x + 2) + 7 = 5x + (2 + 7)$$

$$2 + [3x + (4x + 1)] = (2 + 3x) + (4x + 1)$$

Associative property of multiplication

$$(2x^2)x = 2(x^2 x)$$

$$\frac{1}{2}(4a) = \left(\frac{1}{2} \cdot 4\right)a$$

$$3[6(4y - 5)] = (3 \cdot 6)(4y - 5)$$

Applied together, the commutative and associative properties of addition show that a collection of addends can be rearranged in any way without affecting the sum. The commutative and associative properties of multiplication show that a collection of factors can be rearranged in any way without affecting the product.

$$2x + 5 + 7x + 2 = (2x + 7x) + (5 + 2)$$

$$\left(\frac{1}{3} a\right)(6)(b) = \left(\frac{1}{3} \cdot 6\right)(ab)$$

The operations of addition and multiplication are combined in the following property.

The Distributive Property of Multiplication over Addition

For all numbers a, b, and c,

$$ab + ac = a(b + c) \qquad \text{and} \qquad ba + ca = (b + c)a$$

These equations are examples of the distributive property of multiplication over addition.

$$7 \cdot 2 + 7 \cdot 3 = 7(2 + 3)$$

$$8x + 5x = (8 + 5)x$$

Here are two other important properties.

Identity Property for Addition

For all numbers a,
$$a + 0 = a \qquad \text{and} \qquad 0 + a = a$$

Identity Property for Multiplication

For all numbers a,
$$1a = a \qquad \text{and} \qquad a \cdot 1 = a$$

Identity property for addition

$$7 + 0 = 7$$
$$0 + y = y$$

Identity property for multiplication

$$1 \cdot 9 = 9$$
$$x \cdot 1 = x$$

Some algebraic expressions are written without a numerical coefficient. In those cases, we say the coefficient is 1. Each of these expressions has a coefficient of 1:

$$b \qquad xy \qquad z^2 \qquad (c - d)$$

■ CLASSROOM EXERCISES

Give an example of each property.

1. The associative property of multiplication.

2. The commutative property of addition.

3. The distributive property of multiplication over addition.

4. The identity property for addition.

5. The identity property for multiplication.

Each equation illustrates one property. Name the property.

6. $(7a + 3) + 2a = 7a + (3 + 2a)$

7. $y \cdot 7 + y \cdot 3 = y(7 + 3)$

8. $(8b)\frac{1}{2} = \frac{1}{2}(8b)$

9. $x(3 + 4) = (3 + 4)x$

■ WRITTEN EXERCISES

Each equation illustrates the commutative property of addition or the associative property of addition. Name the property.

A **1.** $17 + (30 + 15) = (17 + 30) + 15$

2. $(4 + 5x) + x = 4 + (5x + x)$

Each equation illustrates the commutative property of addition or the associative property of addition. Name the property.

3. $6 + 3x = 3x + 6$

4. $14 + 0 = 0 + 14$

5. $12 + (4x + 2x) = (4x + 2x) + 12$

6. $(13 + 22) + 5x = 5x + (13 + 22)$

7. $(2 + 3x) + 7x = 2 + (3x + 7x)$

8. $23 + (47 + 2x) = (23 + 47) + 2x$

Each equation illustrates the identity property for addition or multiplication. Name the property.

9. $17 + 0 = 17$

10. $13 + (1 - 1) = 13$

11. $(5 - 5) + 6 = 6$

12. $0 + 11 = 11$

13. $6 \cdot \dfrac{3}{3} = 6$

14. $1 \cdot 4\dfrac{2}{3} = 4\dfrac{2}{3}$

15. $(5.3)1 = 5.3$

16. $\dfrac{5}{5} \cdot \dfrac{3}{8} = \dfrac{3}{8}$

Name the property illustrated in each equation.

17. $5 \cdot 7 = 7 \cdot 5$

18. $6000 \cdot 438 = 438 \cdot 6000$

19. $(17 \cdot 25) \cdot 4 = 17 \cdot (25 \cdot 4)$

20. $\left(\dfrac{4}{5} \cdot \dfrac{2}{3}\right) \cdot \dfrac{3}{1} = \dfrac{4}{5} \cdot \left(\dfrac{2}{3} \cdot \dfrac{3}{1}\right)$

21. $5 \cdot (10 + 3) = 5 \cdot 10 + 5 \cdot 3$

22. $(3 + 8) \cdot 10 = 3 \cdot 10 + 8 \cdot 10$

23. $(6 + 4.7) + 5.3 = 6 + (4.7 + 5.3)$

24. $(17.2 + 5.4) + 4.6 = 17.2 + (5.4 + 4.6)$

Use the property listed to write an expression equivalent to $5x + 2x$.

25. Commutative property of addition.

26. Distributive property of multiplication over addition.

Use the property listed to write an expression equivalent to $5(x + 2)$.

27. Distributive property of multiplication over addition.

28. Commutative property of addition.

29. Commutative property of multiplication.

Name the property illustrated in each equation.

30. $5x + (7 + x \cdot 2) = 5x + (7 + 2x)$

31. $5x + (7 + 2x) = 5x + (2x + 7)$

32. $5x + (2x + 7) = (5x + 2x) + 7$

33. $(5x + 2x) + 7 = (5 + 2)x + 7$

Use the distributive property of multiplication over addition to write an equivalent expression.

Sample: $5(100 + 3) = 5 \cdot 100 + 5 \cdot 3$

B **34.** $8\left(1 + \frac{1}{2}\right)$ **35.** $6\left(3 + \frac{1}{3}\right)$ **36.** $52(100 + 1)$

37. $43(100 + 1)$ **38.** $67(1000 + 1)$ **39.** $29(1000 + 1)$

Multiply in the order indicated. Simplify within parentheses first.

40. $[(123 \cdot 7) \cdot 11] \cdot 13$ **41.** $(123 \cdot 7) \cdot (11 \cdot 13)$

42. $123 \cdot (7 \cdot 11) \cdot 13$ **43.** $123 \cdot [(7 \cdot 11) \cdot 13]$

44. Explain why the puzzle in the Preview works. [*Hint:* What is $7 \cdot 11 \cdot 13$?]

Evaluate the following expressions for $a = 2$, $b = 60$, and $c = 10$. Decide if there are other "distributive" properties.

C **45.** $a(b - c)$ $ab - ac$

     Is multiplication distributive over subtraction?

46. $(b + c) \div a$ $b \div a + c \div a$ $\dfrac{a + b}{c}$ $\dfrac{a}{c} + \dfrac{b}{c}$

     Is division distributive over addition?

47. $a(b \div c)$ $ab \div ac$

     Is multiplication distributive over division?

48. $(b + c)^a$ $b^a + c^a$

     Is raising to a power distributive over addition?

■ REVIEW EXERCISES

Simplify.

1. $4.23 + 6.7$ **2.** $5 - 1.39$ **3.** $0.05 \cdot 7.3$

4. $72 - 0.6$ **5.** $\dfrac{1}{4} + \dfrac{2}{3}$ **6.** $\dfrac{1}{4} \cdot \dfrac{2}{3}$

7. a. Copy and complete the table by substituting and simplifying.

a	b	$a^2 + b^2$	$(a + b)^2$	$b^2 - a^2$	$(b - a)^2$
2	3	?	?	?	?
1	4	?	?	?	?
5	5	?	?	?	?

     b. Do you think that $a^2 + b^2$ and $(a + b)^2$ are equivalent?
     c. Do you think that $b^2 - a^2$ and $(b - a)^2$ are equivalent?

1-6 Using Basic Properties to Simplify Expressions

Preview

To explore a number trick, a student substituted different numbers for n into the expression:

$$[(6n + 30) \div 3 - 12] \div 2 + 1$$

The student kept getting the substituted number as a result. It seemed that the expression was equivalent to n, but the student didn't know how to show it.

In this lesson you will learn how to change complicated algebraic expressions into simpler ones.

■ LESSON

The expression shown below has four **terms**. Terms are parts of an expression "connected" by addition or subtraction signs.

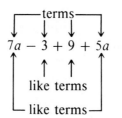

The terms $7a$ and $5a$ are called **like terms** because they contain the same variable factors. The terms 3 and 9 are also like terms because they contain no variable factors.

$$
\begin{array}{cc}
5x \quad 3x & 4a \quad 3ab & 2x \quad 3x^2 \\
\text{like terms} & \text{unlike terms} & \text{unlike terms}
\end{array}
$$

Exercise 45 in Section 1–5 showed that there is a distributive property of multiplication over subtraction.

Distributive Property of Multiplication over Subtraction

For all numbers a, b, and c,

$$ca - cb = c(a - b) \qquad \text{and} \qquad ac - bc = (a - b)c$$

The basic properties can be used to change an algebraic expression to a simpler *equivalent* expression.

Study these examples.

Example 1 Simplify. $4x + 3x$

   Solution $4x + 3x = (4 + 3)x$ *Distributive property*
   $\qquad\qquad\quad = 7x$ *Computation (4 + 3 = 7)*

Example 2 Simplify. $7y - 2y$

   Solution $7y - 2y = (7 - 2)y$ *Distributive property*
   $\qquad\qquad\quad = 5y$ *Computation*

Example 3 Simplify. $\frac{1}{2}(6y)$

   Solution $\frac{1}{2}(6y) = \left(\frac{1}{2} \cdot 6\right)y$ *Associative property of multiplication*

   $\qquad\qquad\quad = 3y$ *Computation*

Example 4 Simplify. $7a + a$

   Solution $7a + a = 7a + 1a$ *Identity property for multiplication*
   $\qquad\qquad\quad = (7 + 1)a$ *Distributive property*
   $\qquad\qquad\quad = 8a$ *Computation*

Example 5 Simplify. $2x + 3y + 4x + y$

   Solution $2x + 3y + 4x + y$
   $= 2x + 3y + 4x + 1y$ *Identity property for multiplication*
   $= (2x + 4x) + (3y + 1y)$ *Commutative and associative properties of addition*

   $= 6x + 4y$ *Distributive property and computation*

Example 6 Simplify. $2(4x + 1) + 3x$

   Solution $2(4x + 1) + 3x$
   $= 2(4x) + 2(1) + 3x$ *Distributive property*
   $= (2 \cdot 4)x + (2 \cdot 1) + 3x$ *Associative property of multiplication*
   $= 8x + 2 + 3x$ *Computation*
   $= (8x + 3x) + 2$ *Commutative and associative properties of addition*

   $= 11x + 2$ *Distributive property and computation*

When simplifying expressions, you often combine steps as in Examples 5 and 6.

■ CLASSROOM EXERCISES

1. What are the terms in this expression?

$2x - 2y + 7 + 4y$

2. Are $2x$ and 7 like terms?

3. Are $3x$ and $2yx$ like terms?

4. Are $2ab$ and $5ba$ like terms?

5. What are the like terms in this expression?

$$3a \ + \ 2b \ + \ 5a \ + 9 - 7 + \ 3b$$

Simplify.

6. $3(4x)$

7. $\frac{1}{2}(8c)$

8. $2x + 3x$

9. $5y + 4y$

10. $2ab + 5ba$

11. $2a + 6 + 3a$

12. $3x + 2y + 4y$

13. $3 + 2(x + 5)$

14. $6m + 3(4 + 2m)$

15. $2a + 5$

■ WRITTEN EXERCISES

State the number of terms in each expression.

A

1. $5x + 3$

2. $7 + 4x$

3. $5 + x + 3$

4. $7 + 4 + x$

5. $2xy + x$

6. $2x + xy$

7. $x - y - xy$

8. $xy - x + y$

List the terms in each expression.

9. $2x + 3y - 4x + 5xy$

10. $5x^2 - 3y^2 + x + y$

11. $7a^2 + 8b^2 - 6 + 3ab$

12. $17 - a^2 + b^2 + 12ab$

State whether the terms are like or unlike.

13. $5a$ and a

14. $6b$ and 6

15. $8c$ and 8

16. d and $12d$

17. $3xy$ and $7x$

18. $10y$ and $20xy$

19. $3x^2$ and $3x$

20. $13x$ and $2x^2$

Simplify these expressions.

21. $3m + 2m$

22. $5n + 4n$

23. $6x - 2x$

24. $5y - 2y$

25. $2a + a$

26. $3a + a$

27. $5b - b$

28. $9c - c$

29. $5a + 3a + 4b$

30. $2a + 3b + 4b$

31. $3 + 4a + 5$

32. $8 + 2a + 4$

33. $8a + 3 + 6a + 2b$

34. $5a + 2b + 2a + b$

35. $2(4x) + 3x$

36. $3(5x) + 2x$

37. $5(x + 2) + x$

38. $3(x + 7) - 6$

39. $3(2x + 5) + x + 1$

40. $2(3x + 10) + x + 9$

41. List the properties used to simplify the expression.

$$5(2x + 7) + x + 4 = 5 \cdot 2x + 5 \cdot 7 + x + 4$$
$$= (5 \cdot 2)x + 5 \cdot 7 + x + 4$$
$$= 10x + 35 + x + 4$$
$$= 10x + x + 35 + 4$$
$$= 10x + 1 \cdot x + 35 + 4$$
$$= (10 + 1)x + 35 + 4$$
$$= 11x + 39$$

a. ___?___
b. ___?___
Computation
c. ___?___
d. ___?___
e. ___?___
Computation

42. List the reason for each step used to simplify the expression.

$$7x + 3 + 8x + 2 = 7x + 8x + 3 + 2$$
$$= (7 + 8)x + 3 + 2$$
$$= 15x + 5$$

a. ___?___
b. ___?___
c. ___?___

43. Carmen simplified $2x + 1$ to $3x$. Show that she was wrong.

44. John simplified $5x - x$ to 5. Show that he was wrong.

45. Leonard simplified $x^2 + x$ to $3x$. Show that he was wrong.

46. Estela simplified $x^2 + x$ to $2x^2$. Show that she was wrong.

47. Amy simplified $3x^2 - x$ to $2x^2$. Show that she was wrong.

48. William simplified $(x + 2)^2$ to $x^2 + 4$. Show that he was wrong.

Simplify.

49. $5(x + 2) + 3(x + 1)$ **50.** $x(x + 5x) + x^2$ **51.** $\frac{1}{3}(6x + 12) - x$

52. $\frac{1}{4}(x + 3) + \frac{3}{4}(x + 3)$ **53.** $x(3 + x) + 3(5 - x)$ **54.** $2(x + 7) + x(x + 5)$

Simplify. List the reason for each step.

55. $5(x + 3) + 10x$ **56.** $2(3x + 8) + 5$

■ REVIEW EXERCISES

Simplify.   [1–2]

1. $8 + 4 - 2$ **2.** $2 + 3 \cdot 10$ **3.** $\frac{3}{4} - \frac{1}{3} + \frac{1}{2}$ **4.** $\frac{3}{5} - \frac{2}{3} + \frac{1}{2}$

Prove that the following pairs of expressions are not equivalent.   [1–4]

5. $5(a + 2)$ **6.** $(a - b)^2$ **7.** $6x + 1$
  $5a + 2$ $a^2 - b^2$ $7x$

State the property illustrated in each equation.   [1–5]

8. $(3a + 4) + 5a = 3a + (4 + 5a)$ **9.** $2a + 3 = 3 + 2a$

10. $5a + a = 5a + 1a$ **11.** $5a + 1a = (5 + 1)a$

Self-Quiz 2

1–4 Evaluate the expressions.

1.

	x	y	$6x + 9y$	$15xy$
a.	1	1	?	?
b.	0	0	?	?
c.	3	1	?	?

    d. Do you think that the expressions $6x + 9y$ and $15xy$ are equivalent?

1–5 State the number property illustrated in these equations.

    2. $a(b + c) = ab + ac$ **3.** $a \cdot 1 = a$ and $1 \cdot a = a$

    4. $(a + b) + c = a + (b + c)$ **5.** $ab = ba$

1–6 Simplify.

    6. $3x + 4x$ **7.** $a + 5a$ **8.** $2(b + 3) + 5$

    9. $2a + 8 + 5a + 15$ **10.** $\frac{1}{2}(6xy)(4z)$ **11.** $3(m + 2) + 4(m + 5)$

EXTENSION   Multiplication on a computer

For some operations, different symbols are used with computers than are used in algebra. To raise to a power, use the symbol ↑.
On a computer, 5 ↑ 2 means 5^2.

Example 1   Predict the computer output.

    PRINT 4 ↑ 2 + 3

    Solution $4^2 + 3 = 19$

Example 2   Write a computer command to find $(3 \cdot 5)^2$.

    Solution   PRINT (3 * 5) ↑ 2

Write the output for each command.

  1. PRINT 3 ↑ 2 + 1   **2.** PRINT 9 − 2 ↑ 3

  3. PRINT 4 ↑ 2 * 3   **4.** PRINT 4 * 3 ↑ 2

Write a computer command to evaluate.

  5. 7^2 **6.** $(4 + 2)^3$ **7.** $5^2 \cdot 3^2$

  8. $5 \cdot 6^2$ **9.** $(4^2)^3$ **10.** $6 \cdot 4 - 2^3$

1–7 Applications — Writing Expressions for Related Quantities

Preview

Consider these dot figures.

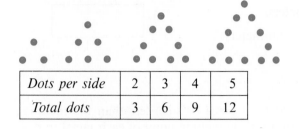

Dots per side	2	3	4	5
Total dots	3	6	9	12

What is the total number of dots for a "triangle" with:

- 10 dots per side?
- 100 dots per side?
- n dots per side?

In this lesson you will be given situations (in words) and asked to express the relationships using the language and symbols of algebra.

■ LESSON

Study the following examples to see how to select a variable and write an expression for a relationship stated in words. Note that two solutions are given for each example.

Example 1 Marcia is 2 years older than Jack. Write expressions using variables for their ages.

Solution 1 Select a variable for one number:
Let j be the number of years in Jack's age.

Write an expression for the other number:
Then $(j + 2)$ is the number of years in Marcia's age.

Solution 2 Let m years be Marcia's age.
Then $(m - 2)$ years is Jack's age.

At first, it is easy to make mistakes in writing expressions. You will make fewer mistakes if you check your work by substituting numbers that you *know* fit the sentence. For example, you know from reading the sentence that *if* Marcia is 16 years old, then Jack is 14 years old. Now check the expression.

Check Solution 1: Let 14 be the number of years in Jack's age. Then $(14 + 2)$ is the number of years in Marcia's age. Since $14 + 2 = 16$, the numbers check.

Check Solution 2: Let 16 years be Marcia's age. Then $(16 - 2)$ years is Jack's age. This also checks.

Get into the habit of checking!

Example 2 A rectangle is 3 times as long as it is wide. Write expressions using a variable for its dimensions.

Solution 1

Let w be the width in centimeters.
Then $3w$ is the length in centimeters.

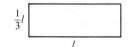

Solution 2

Let l be the length in centimeters.
Then $\dfrac{l}{3}$ or $\dfrac{1}{3}l$ is the width in centimeters.

Example 3 The melting point of lead is 90°C (degrees Celsius) less than the melting point of zinc. Write an expression for the melting point of each metal in degrees Celsius.

Solution 1

Let m be the melting point of lead in degrees Celsius.
Then $(m + 90)$ is the melting point of zinc in degrees Celsius.

Solution 2

Let n be the melting point of zinc in degrees Celsius.
Then $(n - 90)$ is the melting point of lead in degrees Celsius.

■ CLASSROOM EXERCISES

1. Let t be a number.
   a. What number is 3 greater than t?
   b. What number is 5 less than t?
   c. What number is 4 times as great as t?
   d. What number is $\dfrac{1}{2}$ as great as t?

2. A bicyclist pedals twice as fast as a marathoner runs.
   a. If r is the speed of the runner in miles per hour, then ___?___ is the speed of the bicyclist in miles per hour.
   b. If b is the speed of the bicyclist in miles per hour, then ___?___ is the speed of the runner in miles per hour.

3. Let n be a number.
   a. What number results if you increase n by 7?
   b. What number results if you decrease n by 10?
   c. What number results if n is tripled?
   d. What number results if n is halved?

■ WRITTEN EXERCISES

Complete.

A 1. Al is 3 years younger than Mary.
   a. If Al is 17 years old, then Mary is __?__ years old.
   b. Let A be Al's age in years. Then __?__ is Mary's age in years.

2. Juan is 4 cm taller than Jean.
   a. Let u be Juan's height in centimeters. Then __?__ is Jean's height in centimeters.
   b. Let e be Jean's height in centimeters. Then __?__ is Juan's height in centimeters.

3. John weighs 3 pounds more than Ted.
   a. Let d be John's weight in pounds. Then __?__ is Ted's weight in pounds.
   b. Let t be Ted's weight in pounds. Then __?__ is John's weight in pounds.

4. Jennifer scored 8 points more than Sarah.
   a. Let j be the number of points scored by Jennifer. Then Sarah scored __?__ points.
   b. Let s be Sarah's score. Then Jennifer scored __?__ points.

5. Susan lost 5 pounds more than Maria did.
   a. If Susan lost 12 pounds, then Maria lost __?__ pounds.
   b. Let s pounds be Susan's weight loss. Then Maria lost __?__ pounds.

6. The increase in Steve's height was 3 inches more than the increase in Barbara's height.
   a. If Barbara's height increased by 4 inches, then Steve grew __?__ inches.
   b. Let b inches be Barbara's increase in height. Then __?__ inches is Steve's increase in height.
   c. Let s inches be Steve's increase in height. Then __?__ inches is Barbara's increase in height.

7. **a.** If Mike is 17 years old, then 4 years ago he was __?__ years old.
   b. Let m be Mike's age in years. Then 4 years ago he was __?__ years old.
   c. In 5 years Mike will be __?__ years old.

8. **a.** If Kelly has 50 dollars and earns 10 dollars, then she has __?__ dollars.
   b. If Kelly has k dollars and earns 10 dollars, then she has __?__ dollars.
   c. If Kelly has y dollars and spends 10 dollars, then she has __?__ dollars.

9. A rectangle is 3 cm longer than it is wide.
   a. Let w be the width in centimeters. Then __?__ cm is the length.
   b. Let l be the length in centimeters. Then __?__ cm is the width.

10. The height of a parallelogram is 4 cm less than its base.

    a. Let h be the height in centimeters. Then the base is ___?___ cm.

    b. Let b be the base in centimeters. Then the height is ___?___ cm.

11. Lead weighs 154 times as much as balsa wood.

    a. Let x be the weight of a cubic foot of lead in pounds. Then a cubic foot of balsa wood weighs ___?___ pounds.

    b. Let y be the weight of a cubic inch of balsa wood in ounces. Then a cubic inch of lead weighs ___?___ ounces.

12. A jumbo jet weighs 946 times as much as the Wright brothers' first plane.

    a. Let p be the weight of a jumbo jet in tons. Then the Wright brothers' plane weighs ___?___ tons.

    b. Let m be the weight of the Wright brothers' plane in pounds. Then a jumbo jet weighs ___?___ pounds.

Choose a variable for the first number. Then write an expression for the second number using the same variable.

B **13.** Richard is 8 cm taller than Paula.

    a. Let ___?___ be Richard's height in centimeters. Then ___?___ is Paula's height in centimeters.

    b. Let ___?___ be Paula's height in centimeters. Then ___?___ is Richard's height in centimeters.

14. Jackie is 5 pounds lighter than Fred.

    a. Let ___?___ be Jackie's weight in pounds. Then ___?___ is Fred's weight in pounds.

    b. Let ___?___ be Fred's weight in pounds. Then ___?___ is Jackie's weight in pounds.

15. Arthur ran a mile 7.2 seconds faster than Walt.

    a. Let ___?___ seconds be Arthur's time. Then ___?___ seconds is Walt's time.

    b. Let ___?___ seconds be Walt's time. Then ___?___ seconds is Arthur's time.

16. Jill skied a slalom course 1.67 seconds slower than Jan.

    a. Let ___?___ seconds be Jill's time. Then ___?___ seconds is Jan's time.

    b. Let ___?___ seconds be Jan's time. Then ___?___ seconds is Jill's time.

17. Dana ran 3 times as many laps as Celina did.

    a. Let ___?___ be the number of laps that Dana ran. Then ___?___ is the number of laps that Celina ran.

    b. Let ___?___ be the number of laps that Celina ran. Then ___?___ is the number of laps that Dana ran.

Choose a variable for one number. Write an expression for the other number using the same variable.

18. Meaghan and Sue are skydivers. In one jump Meaghan was in freefall twice as long as Sue.

19. Morris and Ken ride motorcycles in hillclimbs. In one run Ken climbed three times as far as Morris.

20. Pam and Sharon were running laps after school. Pam ran 0.7 as many laps as Sharon.

21. Chris practiced gymnastics 3 hours longer than Frank did.

22. Pam and Sharon were running laps after school. Pam ran 0.8 as many minutes as Sharon.

23. Frank practiced 3 more piano pieces than Chris did.

Ⓒ 24. Steve and Martin swam a total of 60 lengths of the pool. [*Remember:* It can be helpful to think of numbers that satisfy the conditions. For example, suppose that Steve swam 25 lengths.]

25. Millie and Janell pitched a total of 48 minutes in a softball game.

26. Kevin and James spent a total of $15.20.

27. Phil and Sandy sold a total of 47 boxes of candy for the school band.

28. **a.** Jane scored twice as many points as Pat did.
    b. If Sharon had scored 1 more point, she would have scored 3 times as many points as Pat did.

29. **a.** Jack earned half as much as Bill earned.
    b. Howard earned 1 dollar more than half as much as Bill earned.

30. Half of the mathematics students in a university class wrote incorrect expressions for this sentence because they didn't bother to check by substituting numbers that they *knew* worked. Try it.
    There are 12 students for every professor at this university.

■ REVIEW EXERCISES

Substitute 1 for a and $\frac{1}{3}$ for b and simplify. [1–2, 1–3]

1. $(a + b)(a - b)$
2. $a^2 - b^2$

Show that the following pairs of expressions are not equivalent. [1–4, 1–6]

3. $6x + 1$ and $7x$
4. $3a - 1$ and $2a$
5. $3x + 4y$ and $(3 + 4)xy$
6. $(a + b)^2$ and $a^2 + b^2$

State the property illustrated in each equation. [1–5]

7. $a \cdot 2 + 3 = 2a + 3$
8. $1 \cdot a + 1 \cdot a = (1 + 1) \cdot a$

■ CHAPTER SUMMARY

● **Vocabulary**

variable	[page 1]	product	[page 10]
substitution	[page 1]	coefficient	[page 10]
domain of variable	[page 1]	numerical coefficient	[page 10]
braces	[page 1]	exponent	[page 11]
simplest expression	[page 2]	base (of an exponent)	[page 11]
simplify	[page 2]	raise to a power	[page 11]
evaluate	[page 2]	equivalent expressions	[page 15]
order of operations	[page 5]	terms	[page 24]
brackets	[page 6]	like terms	[page 24]
factor	[page 10]		

● Simplify numerical expressions in the following order: [1–2, 1–3]

1. Simplify inside the parentheses or other grouping symbols.

2. Evaluate all powers.

2. Do multiplications and divisions in order from left to right.

4. Do additions and subtractions in order from left to right.

The Basic Properties of Numbers

Property	Addition	Multiplication
Commutative Associative Identity	For all numbers a, b, and c $a + b = b + a$ $(a + b) + c = a + (b + c)$ $a + 0 = a$	$ab = ba$ $(ab)c = a(bc)$ $1 \cdot a = a$
Distributive	$ab + ac = a(b + c)$ $ba + ca = (b + c)a$	

■ CHAPTER REVIEW

1–1 Objective: To substitute numbers for variables and simplify numerical expressions.

r	s	t	u
3	6	12	2

Substitute and simplify.

1. $r + 10$ **2.** $t \cdot u$ **3.** $\dfrac{t}{u}$ **4.** $r + s + t$

34   Chapter 1 Algebraic Expressions

1–2 Objective: To simplify numerical expressions that have more than one operation.

Simplify.

5. $3 + 4 \cdot 5$ **6.** $2 \cdot 5 + 4 \cdot 5$ **7.** $(7 - 3) \cdot (7 + 3)$ **8.** $\dfrac{20 - 8}{4}$

1–3 Objective: To use algebraic notation for multiplication.

Simplify.

9. 3^4 **10.** $2 \cdot 5^2$ **11.** $(2 \cdot 5)^2$ **12.** $(10 - 5)^2$

1–4 Objective: To decide whether expressions are equivalent.

13. a. Copy and complete the table by substituting and simplifying.

x	$3(x + 2)$	$3x + 2$
0	?	?
1	?	?
5	?	?

b. Do you think that $3(x + 2)$ and $3x + 2$ are equivalent?

1–5 Objective: To identify and use basic properties of addition and multiplication.

Identify the property illustrated in each equation.

14. $3 + x = x + 3$ **15.** $3x + 4x = (3 + 4)x$

16. $4x + x = 4x + 1x$ **17.** $(67 \cdot 25) \cdot 4 = 67 \cdot (25 \cdot 4)$

1–6 Objective: To simplify algebraic expressions by combining like terms.

Simplify these expressions.

18. $x + 4x$ **19.** $4x + 5 + 3x$ **20.** $3a + 4b + 5a$ **21.** $5(x + 3) + 2x$

22. $4(x + 2) + 3$ **23.** $3(x + 6) + 2$ **24.** $3a + 6 + 4$ **25.** $4b + 3 + 2b$

1–7 Objective: To use algebraic expressions to express relationships that are stated in words.

Complete.

26. Mary has five dollars more than Carla.
   a. If Carla has c dollars, then Mary has __?__ dollars.
   b. If Mary has m dollars, then Carla has __?__ dollars.

27. A rectangle is three times as long as it is wide.
   a. If it is w cm wide, then it is __?__ cm long.
   b. If it is l cm long, then it is __?__ cm wide.

■ CHAPTER 1 SELF-TEST

a	A	B	x	y
2	0	4	1	3

1–1   Substitute and simplify.

1. $B(a + x)$ **2.** Ay **3.** ayB **4.** $3B + 4y$

1–2,
1–3   Simplify.

5. $3 + 6 \cdot 9$ **6.** $8 \cdot 2 + 6 \cdot 4$ **7.** $8 + 2 \cdot 6 + 4$

8. $3 \cdot 8^2 + 12 \div 4$ **9.** $8 - 4 + 6 \div 3$ **10.** $27 - 12 \div 3$

1–3   Let $n = 5$. Simplify.

11. $7n + 3$ **12.** $n - 2^2$ **13.** $(n - 2)^2$

14. $5n^2$ **15.** n^3 **16.** n^4

1–4   Copy and complete the tables by substituting and simplifying.

17.

x	$6x + 6$	$12x$
a. 0	?	?
b. 1	?	?
c. 2	?	?

d. Are the expressions $6x + 6$ and $12x$ equivalent?

18.

y	$4(y + 7)$	$4y + 7$
a. 0	?	?
b. 1	?	?
c. 2	?	?

d. Are the expressions $4(y + 7)$ and $4y + 7$ equivalent?

1–5   State the property illustrated in each equation.

19. $6 \cdot 12 + 6 \cdot 8 = 6(12 + 8)$ **20.** $(32 \cdot 14) \cdot 5 = 32 \cdot (14 \cdot 5)$

21. $8 + 42 = 42 + 8$ **22.** $1 + 0 = 1$

1–6   Simplify.

23. $3x + 8x$ **24.** $4a + 3b + 2a$

25. $5m + 7n + 2m + 3n$ **26.** $16 + 2y + 5$

27. $4m + 6 + 2m + 3$ **28.** $8m + 3n + 4n + m$

1–7   Write an algebraic expression.

29. Seven more than twice a number, n.

30. Sixteen less than the product of three and some number, k.

■ PRACTICE FOR COLLEGE ENTRANCE TESTS

Choose the best answer for each question.

1. If $y = 3^2$, then $y^2 =$ ___?___ .

   A. 3   **B.** 9   **C.** 27   **D.** 81   **E.** 243

2.

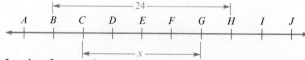

In the figure above, the lettered points divide line segment BH into segments of equal length. If $BH = 24$, then $x =$ ___?___ .

   A. 4   **B.** 8   **C.** 12   **D.** 16   **E.** 20

3. If $y = \dfrac{(a + b)^2}{a^2 - b}$, what is the value of y when $a = 2$ and $b = 1$?

   A. 1   **B.** 2   **C.** 3   **D.** 4   **E.** 6

4. If the average of two numbers is 9 and their product is 72, then the difference between the numbers is ___?___ .

   A. 2   **B.** 3   **C.** 4   **D.** 5   **E.** 6

5. If 2 dozen eggs cost $1.68, what is the cost of 7 dozen eggs?

   A. $2.52   **B.** $3.36   **C.** $4.20   **D.** $5.04   **E.** $5.88

6. If m and n are whole numbers and $m + n$ is divisible by 3, then which of the following must also be divisible by 3?

   A. $m^2 + n^2$ **B.** $m - n$ **C.** mn

   D. Either m or n, but not necessarily both   **E.** None of these

7. Juan, Henri, Erika, and Stasha each have a hamburger special lunch. Henri and Erika each also have a shake to drink. If the total bill is $15 and shakes cost $1.00, how much should Erika pay?

   A. $3.00   **B.** $3.25   **C.** $3.75   **D.** $4.00   **E.** $4.25

8. In the equation $5 + a + b = 17$, if a and b are positive 1-digit whole numbers, what is the smallest value that b can have?

   A. 1   **B.** 3   **C.** 5   **D.** 7   **E.** 9

9. If the height of the figure is 9 cm and the area of the square is 4 cm², what is the area of the triangle in square centimeters?

   A. 5   **B.** 7   **C.** 10

   D. 14   **E.** 28

10. In a class of 30 algebra students, 16 are boys and 24 are ninth graders. There are 2 boys who are not ninth graders. How many girls are ninth graders?

   A. 4   **B.** 6   **C.** 8   **D.** 10   **E.** 12

11. If the perimeter of a rectangle is 40 cm and the width is 8 cm, what is the area?

   A. 64 cm²   **B.** 72 cm²   **C.** 80 cm²   **D.** 88 cm²   **E.** 96 cm²

2 Real Numbers

An object experiences the weight of the air above it as a pressure. At sea level, this pressure is 1 atmosphere (14.7 pounds per square inch). In descending below the surface of water, the pressure due to the weight of the water increases at the rate of about 1 atmosphere for every 33 feet. The total pressure at a depth of 33 feet is 2 atmospheres, one atmosphere due to the weight of the air and one atmosphere due to the weight of the water.

We can use the algebraic equation $P = \dfrac{d}{33} + 1$ to compute the total pressure P (in atmospheres) for a given depth d (in feet).

2–1 Real Numbers

Preview

Mount Everest, on the border between Nepal and Tibet in southern Asia, is the highest mountain on earth. It extends 8848 meters above sea level. The Mariana Trench, in the West Pacific, is the deepest point of the oceans. It is 11,034 meters below sea level.

Since these two measurements have opposite directions, we can use positive numbers and negative numbers to describe them.

In this lesson you will begin to study the uses of positive and negative numbers in algebra.

■ LESSON

When **opposite directions** are involved in measurements, we use **positive** and **negative numbers** to describe those measurements.

For example: 7.3 meters to the right: $^+7.3$ Earning \$10: $^+10$

8 meters to the left: $^-8$ Spending \$15: $^-15$

15 miles to the east: $^+15$ \$50 profit: $^+50$

$9\frac{1}{2}$ miles to the west: $^-9\frac{1}{2}$ \$30 loss: $^-30$

0 is neither positive nor negative

The set consisting of positive numbers, negative numbers, and zero is called the **set of real numbers.** The real numbers can be represented as points on the **number line.**

The number line shows the **order** of the real numbers. For example, $^+6$ is to the *right* of $^+2$, so $^+6$ is *greater than* $^+2$. We write this relationship as $^+6 > {}^+2$. We can also state that $^+2$ is *less than* $^+6$. We write this relationship as $^+2 < {}^+6$.

We see by the number line that $^-2$ is to the *left* of 0, so $^-2$ is *less than* 0. That is, $^-2 < 0$. We can also write $0 > {}^-2$.

The number line below shows the order of some fractions and decimals:

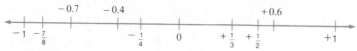

We see that $^+\frac{1}{3}$ is to the *right* of $^-\frac{7}{8}$, so $^+\frac{1}{3} > {}^-\frac{7}{8}$, and that $^-0.7$ is to the *left* of $^-0.4$, so $^-0.7 < {}^-0.4$.

Note that the point $^+3$ and the point $^-3$ are the same distance from zero. This distance is called the **absolute value** of $^+3$ and $^-3$.

The absolute value of any number (except zero) is a positive number. The absolute value of zero is zero. The absolute value is indicated by the symbol, $|\ |$. For example,

$|^-3| = {}^+3$ "The absolute value of negative three is positive three."

$|^+1.5| = {}^+1.5$ $\left|\dfrac{-1}{2}\right| = \dfrac{+1}{2}$ $|0| = 0$

Numbers that have the same absolute value and opposite signs are called **opposites.** For example, $^+3$ and $^-3$ are opposites, and $^+5$ is the opposite of $^-5$. It is also the case that zero is the opposite of zero.

The symbol "$-$" is used to express opposites. For example, to write an expression for the opposite of $^+6$, write $-\,^+6$. To write an expression for the opposite of the opposite of y, write $-(-y)$.

> **Example 1**   Simplify.   **a.** $-\,^+5$ **b.** $-\,^-3$ **c.** $-|^-3|$ **d.** -0
>
> *Solution*   **a.** $-\,^+5 = {}^-5$ **b.** $-\,^-3 = {}^+3$
> **c.** $-|^-3| = -(^+3) = {}^-3$ **d.** $-0 = 0$

To simplify notation, mathematicians do not use the raised $+$ and $-$ signs for positive and negative numbers. Instead, no sign is used with positive numbers and the symbol for opposites is used to indicate negative numbers.

> **Example 2**   Simplify, writing without raised $+$ and $-$ signs.
> **a.** $^+8$ **b.** $^-7$ **c.** $-(^+4)$ **d.** $-(-\,^+2)$
>
> *Solution*   **a.** $^+8 = 8$ **c.** $^-7 = -7$
> **c.** $-(^+4) = -4$ **d.** $-(-\,^+2) = -(-2) = 2$

■ CLASSROOM EXERCISES

Read the symbols.

1. $^+8$ **2.** $^-16$ **3.** $<$ **4.** $>$ **5.** $|^-3|$

Read. Then state whether the sentence is true or false.

6. $^+8 < {}^+10$ **7.** $^-8 > {}^-10$ **8.** $^-3 < 0$ **9.** $^-2 > {}^+1$ **10.** $^+7 > {}^-12$

11. $\dfrac{-1}{4} < \dfrac{-1}{2}$ **12.** $0.75 > 0.39$ **13.** $|-7| = -7$

14. $|-3| = 3$ **15.** $-(-5) = -5$ **16.** $-|-2| = -2$

Write using algebraic symbols.

17. The absolute value of negative three

18. The opposite of positive six

19. The opposite of negative three is equal to positive three.

20. The absolute value of negative two is equal to positive two.

■ WRITTEN EXERCISES

Use positive or negative numbers to represent each of the following:

A **1.** 30 feet below sea level

2. 520 feet above sea level

3. 23°C above freezing

4. 6°C below freezing

5. A loss of $380

6. A profit of $870

7. A gain of 18 yards in football

8. A loss of 3 yards in football

Write each statement in algebraic symbols.

9. Positive five is greater than negative three.

10. Negative two is less than zero.

11. One-fourth is greater than negative one.

12. The opposite of three is less than zero.

Replace ⑦ with $<$, $>$, or $=$ to make a true statement.

13. 2.1 ⑦ 7.3

14. 4.9 ⑦ 1.8

15. $3\frac{1}{2}$ ⑦ -1

16. $\frac{3}{4}$ ⑦ $6\frac{1}{4}$

17. 0 ⑦ -6

18. $-12\frac{7}{8}$ ⑦ 0

19. -3 ⑦ -5

20. -8 ⑦ -5

Simplify.

21. $|-8|$

22. $|-6|$

23. $|4|$

24. $|9|$

25. $|3\frac{2}{3}|$

26. $|-5\frac{2}{5}|$

27. $|-13.67|$

28. $|11.24|$

True or false?

29. $8 < 11$

30. $3.7 > 2.9$

31. $0 > -2$

32. $0 > -4$

33. $-4 < -9$

34. $-5.2 < -4.8$

B **35.** $|-8| < |3|$

36. $|2| > |-6|$

37. $|7| > |-8|$

38. $|-10| < |-6|$

39. The opposite of the opposite of y is y.

40. The value of the opposite of x can be positive, negative, or zero depending on the value of x.

41. The absolute value of zero is zero.

42. The opposite of zero is zero.

Write the numbers in order from least to greatest.

43. 0, 1, −2

44. −10, −3, 4, −1

45. 1.5, −$\frac{1}{2}$, 0, −1

46. −1.61, 1.61, −0.161, 16.1

A	a	B	M	p	r
−6	−4	−2	3	5	$\frac{1}{2}$

Substitute and simplify.

47. −B

48. −r

49. −$A + p$

50. −$a + M$

51. $|A|$

52. $|B|$

53. $|M|$

54. $|p|$

55. $|-r|$

56. $|-a|$

57. −$|r|$

58. −$|a|$

Solve.

C **59.** $|-6| = x$

60. $|x| = 6$

61. −$a = 2$

62. $|-a| = 2$

63. $|x| = 0$

64. −$x = 0$

■ REVIEW EXERCISES

Simplify.

1. 3.14 + 6.7

2. 4.34 + 17 + 3.9

3. $\frac{1}{2} + \frac{1}{3}$

4. $2\frac{3}{4} + 4\frac{1}{2}$

5. 6 − 1.27

6. 43.15 − 27.14

7. $\frac{2}{3} - \frac{1}{2}$

8. $2\frac{1}{2} - \frac{3}{4}$

EXTENSION   Checking accounts

Some bank statements for checking accounts use positive numbers for deposits and negative numbers for withdrawals. Deposits are called credits, and withdrawals are called debits. A summary of transactions might look like that shown below. Compute the new balance (the amount currently in the account).

Previous balance:	$845.32
Credits (2):	367.26
Debits (8):	−598.52
Service charge:	−.60
New balance:	?

Strategy for Success   Doing assignments

Be sure to complete your assignment every day. Each idea in algebra builds on ideas that come before. If you don't understand one idea, you may have difficulty with ideas that come later.

2–2 Adding Real Numbers

Preview A model for addition

One way to think about addition of real numbers is to use the following model. There are two types of electrical charge. One type is positive and the other negative. When a positive charge and a negative charge are combined the result is a 0 charge (neutral).

Positive	Negative	Neutral
⊕	⊖	⊕⊖

The charge on this set is 3.

Then 4 negative charges are added.

After pairing up the charges, the result is a charge of − 1.

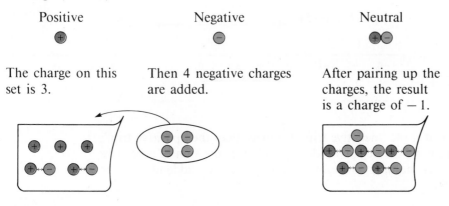

The example above shows that $3 + (-4) = -1$.

Write addition equations for these two examples.

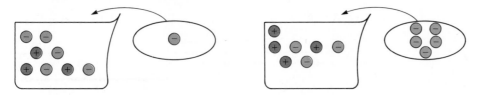

In this lesson, you will learn the rules for adding real numbers.

■ LESSON

We can use the number line to find the sums of real numbers. Think of positive numbers as indicating "trips" to the right and negative numbers as indicating "trips" to the left. For example, consider the sum $-1 + (-2)$.

Start at 0. Go 1 unit to the left. Then go 2 more units to the left. The single equivalent "trip" is − 3.

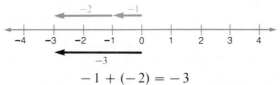

$$-1 + (-2) = -3$$

The example above illustrates that the sum of two negative numbers is a negative number. Similarly, the sum of two positive numbers is a positive number.

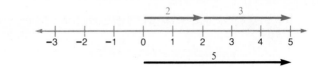

$$2 + 3 = 5$$

Cases in which a positive number is added to a negative number are more complex because one trip "cancels" all or part of another.

Example 1 Use the number line to compute the sum.

$$2 + (-5)$$

Solution

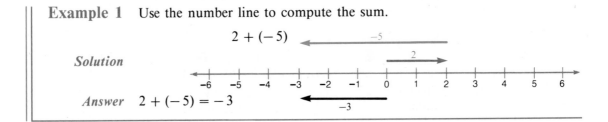

Answer $2 + (-5) = -3$

Example 1 shows that if the "negative trip" is longer than the "positive trip", then the ending point is on the negative side of 0. That is, the sum of a positive number and a negative number is negative if the negative addend has the greater absolute value.

Example 2 Use the number line to compute the sum.

$$-3 + 8$$

Solution

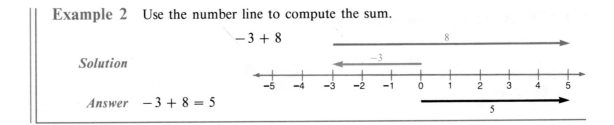

Answer $-3 + 8 = 5$

Example 2 shows that the sum of a positive number and a negative number is positive, if the absolute value of the positive addend is greater.

Example 3 Use the number line to compute the sum.

$$3 + (-3)$$

Solution

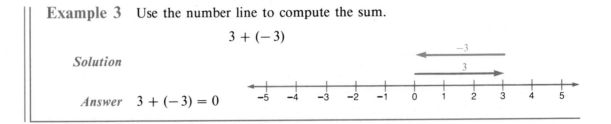

Answer $3 + (-3) = 0$

Example 3 shows that the sum of opposites is 0.

Here are the rules that can be used for adding real numbers.

Rule	Examples	
1. To add real numbers with the same sign:	$3 + 2$	$-4 + (-3)$
a. Find their absolute values.	$\|3\| = 3$ and $\|2\| = 2$	$\|-4\| = 4$ and $\|-3\| = 3$
b. Add their absolute values.	$3 + 2 = 5$	$4 + 3 = 7$
c. Give the result the same sign as the addends have.	5 $3 + 2 = 5$	-7 $-4 + (-3) = -7$
2. To add real numbers with different signs:	$5 + (-2)$	$1 + (-6)$
a. Find their absolute values.	$\|5\| = 5$ and $\|-2\| = 2$	$\|1\| = 1$ and $\|-6\| = 6$
b. Subtract the smaller absolute value from the larger.	$5 - 2 = 3$	$6 - 1 = 5$
c. Give the result the sign of the addend with the larger absolute value.	3 $5 + (-2) = 3$	-5 $1 + (-6) = -5$
3. If the addends are opposites, then the sum is 0.	$4 + (-4) = 0$	
4. The sum of a real number and 0 is that same real number.	$-5 + 0 = -5$	

■ CLASSROOM EXERCISES

Write an addition equation for each figure.

1.

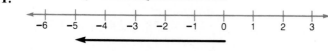

2.

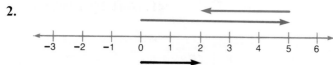

3.

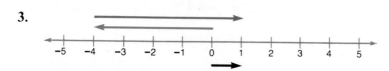

Draw a number line and state each sum.

4. $2 + (-7)$ **5.** $-3 + 1$ **6.** $5 + (-5)$ **7.** $0.2 + (-0.2)$ **8.** $-\frac{3}{5} + \left(-\frac{2}{5}\right)$

Simplify.

9. $4 + (-1)$ **10.** $4 + (-2)$ **11.** $4 + (-3)$ **12.** $4 + (-4)$ **13.** $4 + (-5)$

14. The sum of opposites is __?__.

15. In which of Exercises 4–13 are opposites added?

Simplify.

16. $(6 + (-3)) + -5$ **17.** $-(3 + (-5))$ **18.** $|-6 + 2|$ **19.** $|-6| + |2|$

■ WRITTEN EXERCISES

Simplify.

A
1. $3 + (-7)$ **2.** $6 + (-11)$ **3.** $-5 + (-9)$ **4.** $-6 + (-8)$

5. $-3 + 10$ **6.** $-2 + 6$ **7.** $7 + (-7)$ **8.** $9 + (-3)$

9. $6 + (-6)$ **10.** $8 + 9$ **11.** $-8 + 8$ **12.** $-6 + 5$

13. $9.3 + 8$ **14.** $7.6 + 5$ **15.** $5.31 + (-7.40)$ **16.** $4.95 + (-6.49)$

17. $-1.25 + 1.25$ **18.** $-1.35 + (-1.35)$ **19.** $-1.95 + 10$ **20.** $-3.45 + 20$

21. $-16 + 4$ **22.** $5 + (-20)$ **23.** $4 + (-16)$ **24.** $-20 + 5$

25. $(-8 + 4) + (-2)$ **26.** $(-11 + 8) + (-3)$ **27.** $-8 + (4 + (-2))$ **28.** $-11 + (8 + (-3))$

29. In which of Exercises 1–24 are opposites added?

Simplify.

B
30. $-(18 + (-12))$ **31.** $-(26 + (-32))$ **32.** $-18 + 12$

33. $-26 + 32$ **34.** $\left|7\frac{3}{4} + \left(-8\frac{1}{2}\right)\right|$ **35.** $\left(-2\frac{1}{2} + 3\frac{1}{4}\right)$

36. $\left|7\frac{3}{4}\right| + \left|-8\frac{1}{2}\right|$ **37.** $\left|-2\frac{1}{2}\right| + \left|3\frac{1}{4}\right|$ **38.** $|12.46 + (-3.51)|$

39. $|-41.28 + 84.35|$ **40.** $|12.46| + |-3.51|$ **41.** $|-41.28| + |84.35|$

a	b	c	x	y	z
-9	7	3	-2.5	3.6	-6.7

Substitute and simplify.

42. $(a + c) + y$ **43.** $(b + x) + z$ **44.** $a + (c + y)$ **45.** $b + (x + z)$

46. $|b + x|$ **47.** $|a + y|$ **48.** $|b| + |x|$ **49.** $|a| + |y|$

C **50.** Solvit Corporation stock was selling at $34\frac{3}{4}$ ($34.75) when trading began on Monday morning. Copy and complete the chart.

	Opening price	Net change	Closing price	
Monday	$34\frac{3}{4}$	$\frac{-1}{2}$	?	
Tuesday	$34\frac{1}{4}$	$+1\frac{1}{4}$	?	
Wednesday	?	$\frac{-3}{4}$	?	
Thursday	?	?	37	
Friday	?	$\frac{-1}{4}$	?	

51. The Central Tigers football team took possession of the ball on their 9 yard line. They ran a series of plays that showed the following changes before calling time out.

  gain of 3 yards, loss of 11 yards, gain of 17 yards, loss of 8 yards.

a. On what yard-line is the ball?
b. Over four plays, what was the team's net loss or gain?

52. Using the numbers $-4, -3, -2, -1, 0, 1, 2, 3,$ and 4 only once, make a magic square. (Each row, column, and diagonal must add to 0.)

?	?	?
?	?	?
?	?	?

53. Nancy Lopez had the following rounds on a Ladies Professional Golf Association (LPGA) tour event: 2 under par, par, 3 over par, 4 under par, and 1 over par. How did she finish with respect to par?

■ REVIEW EXERCISES

Substitute and simplify.

a	b	c
2	5	10

[1–1]

1. $\dfrac{b}{c}$ **2.** $c(b-a)$ **3.** $3(b+2)+b$

4. $10a^2$ **5.** $b^a - c$ **6.** $b - \dfrac{c}{a}$

2–3 Subtracting Real Numbers

The electrical charge model (see Preview 2–2) can also help us understand subtraction of real numbers. To subtract a number, think about removing charges. For example, $-2 - (-3)$

Start with -2.   Remove 3 negative charges.   The result is a charge of 1.

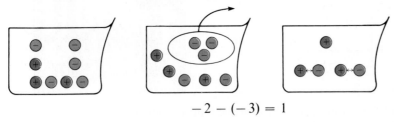

$$-2 - (-3) = 1$$

Write subtraction equations for each of the following examples.

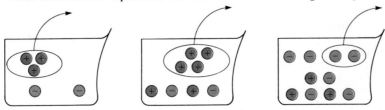

In this lesson you will learn how to subtract real numbers.

■ LESSON

The electrical charge model of subtraction leads to the conclusion that subtracting a real number has the same effect as adding the opposite of that real number. Number patterns such as the following support that conclusion.

A	B	C
$3 - 1 = 2$	$3 - 2 = 1$	$8 - 8 = 0$
$3 - 2 = 1$	$3 - 1 = 2$	$2 - 2 = 0$
$3 - 3 = 0$	$3 - 0 = 3$	$-3 - (-3) = ?$
$3 - 4 = ?$	$3 - (-1) = ?$	

The pattern in column A suggests that $3 - 4 = -1$. We also know that $3 + (-4) = -1$. In this case subtracting 4 and adding -4 give the same result.

The pattern in column B suggests that $3 - (-1) = 4$. Note also that since $3 + 1 = 4$, we can conclude that subtracting -1 and adding 1 give the same result.

Finally, the pattern in column C suggests that $-3 - (-3) = 0$. Since $-3 + 3 = 0$, we can conclude that subtracting -3 and adding 3 give the same result.

Definition: Subtraction of Real Numbers

Subtracting any real number is the equivalent of adding the opposite of
that real number.

For all real numbers a and b,
$$a - b = a + (-b).$$

According to this definition, any subtraction expression can be changed into
an addition expression and then simplified by using the rules for adding real
numbers.

Example 1   Simplify. $-3 - 7$

    Solution

	$-3 - 7$
Change to addition.	$-3 + (-7)$
Simplify.	-10

    Answer -10

Example 2   Simplify. $-2 - (-5)$

    Solution

	$-2 - (-5)$
Change to addition.	$-2 + 5$
Simplify.	3

    Answer 3

Example 3   Simplify. $(-6 - 7) - (-8)$

    Solution

	$(-6 - 7) - (-8)$
Change to addition.	$(-6 + (-7)) + 8$
Simplify.	$-13 + 8$
	-5

    Answer -5

■ CLASSROOM EXERCISES

Change these subtraction expressions to addition expressions.

1. $6 - (-3)$ **2.** $-4 - 5$ **3.** $-2 - (-7)$ **4.** $3 - 5$

Simplify.

5. $-3 - 4$ **6.** $2 - (-8)$ **7.** $4 - 7$ **8.** $-2 - (-8)$

9. $3 - (-2)$ **10.** $-5 - (-4)$ **11.** $8 - (-9)$ **12.** $-8 - (-10)$

■ WRITTEN EXERCISES

Write as an addition expression. Do *not* simplify.

A　**1.** $11 - 3$ 　　　　**2.** $14 - 8$ 　　　　**3.** $7 - (-2)$ 　　　　**4.** $16 - (-5)$

　　5. $6 - 9$ 　　　　**6.** $4 - 7$ 　　　　**7.** $10 - (-13)$ 　　　　**8.** $3 - (-5)$

Simplify.

　9. $42 - 21$ 　　　**10.** $26 - 13$ 　　　**11.** $-28 - 14$ 　　　**12.** $-31 - 17$

13. $61 - (-11)$ 　　**14.** $57 - (-19)$ 　　**15.** $-16 - (-25)$ 　　**16.** $-15 - (-32)$

17. $(13 - (-3)) - (-6)$ 　　　　　　　**18.** $(18 - (-4)) - (-5)$

19. $13 - (-3 - (-6))$ 　　　　　　　**20.** $18 - (-4) - (-5)$

B　**21.** $(13 - (-12)) + (-12)$ 　　　　**22.** $(82 - (-46)) + 46$

23. $(-22 - (-8)) + (-8)$ 　　　　　**24.** $(-36 - (-14)) + (-14)$

25. $-2\frac{1}{4} - \left(-1\frac{3}{4} + \frac{3}{4}\right)$ 　　　　　**26.** $-5\frac{2}{3} - \left(-2\frac{1}{3} + 1\frac{1}{3}\right)$

27. $\left(2\frac{1}{4} - \left(-1\frac{3}{4}\right)\right) - \frac{3}{4}$ 　　　　　**28.** $\left(-5\frac{2}{3} - \left(-2\frac{1}{3}\right)\right) - 1\frac{1}{3}$

29. $(-6.2 - 3.5) + (8.4 - (-2.8))$ 　　　**30.** $(9.7 - (-12.4)) + (-14.2 - 7)$

31. $(-6.2 + 8.4) - (3.5 + (-2.8))$ 　　　**32.** $(9.7 + (-14.2)) - (-12.4 + 7)$

33. In a football game, a kicker punted the ball from 15 yards behind the line of scrimmage. How far would he have to kick the ball in order to have a net gain of 40 yards? (*Net gain* is the distance from the line of scrimmage forward to the point where the kick is caught.)

34. One day the temperature in Billings, Montana, changed from $-4°C$ to $8°C$. How many degrees more or less, and in what direction was the temperature change?

a	b	m	n	x	y	z
7	-11	14	12	-13.8	26.4	0

Substitute and simplify.

C　**35.** $(m + n) - a$ 　　　　**36.** $m + (n - a)$ 　　　　**37.** $n - (m + b)$

　　38. $(n - m) - b$ 　　　　**39.** $(y + m) - m$ 　　　　**40.** $x - z$

　　41. $(b - x) + (z - y)$ 　　　**42.** $(b + a) - (x + y)$

43. Mr. and Mrs. Maxwell are planning to close out their checking account. Their September statement showed a balance of $200. After depositing $55, they wrote checks for $147, $23.50, $18.75, $62.25, and $15.50. The bank charges them $7.00 if their account is overdrawn, that is, if the total of the checks is more than the checking account balance. What size deposit must they make in order to show a balance of $0.00 on their closing statement?

44. Of the states listed, which had the greatest Fahrenheit temperature range for its record highest and lowest temperature? Which had the least range?

CA: 134°, −45°   FL: 109°, −2°   HI: 100°, 14°   ME: 105°, −48°

TX: 120°, −23°   MT: 117°, −70°   NY: 108°, −52°   CO: 118°, −60°

■ REVIEW EXERCISES

Simplify. [1–3]

1. $3^2 + 4^2$ **2.** $(3 + 4)^2$ **3.** $5^2 − 3^2$ **4.** $(5 − 3)^2$ **5.** $6.3 \cdot 0.4$

6. $\dfrac{2}{3} \cdot \dfrac{4}{5}$ **7.** $3.04 \cdot 1000$ **8.** $1\dfrac{2}{3} \cdot \dfrac{3}{4}$ **9.** $8.07 \div 100$ **10.** $\dfrac{2}{3} \div \dfrac{4}{5}$

Simplify.

11. $6.73 − 4.8$ **12.** $14 − 6.35$ **13.** $\dfrac{1}{2} − \dfrac{1}{3}$

14. $3\dfrac{1}{2} − 1\dfrac{3}{4}$ **15.** $1\dfrac{1}{3} − \dfrac{11}{12}$ **16.** $(1.6 + 3.5) + 2.2$

Self-Quiz 1

2–1   Simplify.

1. $−(−5)$ **2.** $−|−8|$ **3.** The opposite of $−x$

Here is a number line picture.

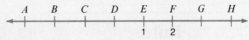

Write the number that corresponds to each of the following points.

4. A **5.** H **6.** Halfway between G and H

Use $<$ or $>$ to make the sentence true.

7. $−6 \; ? \; 0$ **8.** $2 \; ? \; |−3|$ **9.** $−\dfrac{1}{2} \; ? \; −1$

2–2   Simplify.

10. $8 + (−11)$ **11.** $14 + (−6)$ **12.** $−7 + (−2)$

13. $(−12 + 17) + 9$ **14.** $−20 + (3 + (−6))$ **15.** $−15 + (−4 + 18)$

2–3

16. $7 − (−4)$ **17.** $−19 − 12$ **18.** $−12 − (−14)$

19. $(14 − (−8)) − (−11)$ **20.** $16 − (9 − 15)$ **21.** $−12 − (6 − 16)$

2-4 Multiplying Real Numbers

Preview A model for multiplication

Think of a car traveling on an east–west road with east considered the positive direction. Traveling at a constant rate of speed, the car is *passing the house right now.* Think of time in the future as positive and time in the past as negative. Suppose that the car is traveling *east* at 50 mph. Two hours *from now* it will be 100 miles *east* of the house.

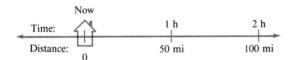

This is a distance–rate–time situation where rate × time = distance.

Equation: $50 \cdot 2 = 100$

If the car were traveling 50 mph to the *east*, 2 hours *ago* it was 100 miles *west* of the house.

Equation: $50 \cdot (-2) = -100$

If the car were traveling 50 mph to the *west*, 2 hours *ago* it was 100 miles *east* of the house.

Equation: $-50 \cdot (-2) = 100$

In this lesson you will learn the rules for multiplying real numbers.

■ LESSON

The examples in the Preview suggest the following rules for multiplication of real numbers.

Rules	**Examples**
To multiply two real numbers:	
1. Multiply the absolute values of the factors.	
2. Use these rules to find the sign of the product.	
a. If the factors have the same sign, the product is positive.	$-2 \cdot -3 = 6 \qquad 4 \cdot 3 = 12$
b. If the factors have different signs, the product is negative.	$3 \cdot -7 = -21 \qquad -5 \cdot 4 = -20$
c. If one of the factors is 0, the product is 0.	$-7 \cdot 0 = 0 \qquad 0 \cdot 8 = 0$

We can see that these rules make sense by studying the patterns in these equations.

$$2 \cdot 3 = 6$$
$$2 \cdot 2 = 4$$
$$2 \cdot 1 = 2$$
$$2 \cdot 0 = 0$$
$$2 \cdot -1 = ?$$

As the second factor decreases by 1, the product decreases by 2. This pattern suggests that the last product should be -2.

$$2 \cdot -1 = -2$$

This result agrees with the rule that the product of a positive number and a negative number is negative.

Look at these equations and find a pattern.

$$-2 \cdot 3 = -6$$
$$-2 \cdot 2 = -4$$
$$-2 \cdot 1 = -2$$
$$-2 \cdot 0 = 0$$
$$-2 \cdot -1 = ?$$

In this case, as the second factor decreases by 1, the product increases by 2. This pattern suggests that the last product should be 2.

$$-2 \cdot -1 = 2$$

This result agrees with the rule that the product of two negative numbers is positive.

Example 1   Simplify these products.

   a. $3 \cdot 4$ **b.** $-5 \cdot -3$ **c.** $5 \cdot -7$ **d.** $-25 \cdot 0$

Solution   **a.** $3 \cdot 4 = 12$ **b.** $-5 \cdot -3 = 15$
   c. $5 \cdot -7 = -35$ **d.** $-25 \cdot 0 = 0$

Example 2   Simplify. $-3 \cdot 6 - 8 \cdot -4$

Solution $$-3 \cdot 6 - 8 \cdot -4$$
   Simplify the products. $-18 - (-32)$
   Simplify the difference. 14

Answer   14

Example 3   Simplify. $-3(12 - 17)$

Solution $$-3(12 - 17)$$
   Simplify the difference. $-3(-5)$
   Simplify the product. 15

Answer   15

■ CLASSROOM EXERCISES

Simplify.

1. $-4 \cdot -6$
2. $(5) \cdot (3)$
3. $8 \cdot -4$
4. $-9 \cdot 6$
5. $-10 \cdot 5$
6. $-8 \cdot -5$
7. $(-6)^2$
8. $-12 \cdot 0$
9. $(-1 \cdot -1) \cdot -2$

■ WRITTEN EXERCISES

Simplify.

A 1. $8 \cdot 3$
2. $7 \cdot 4$
3. $9 \cdot -11$
4. $10 \cdot -12$
5. $25 \cdot 0$
6. $0 \cdot -20$
7. $0 \cdot -31$
8. $0 \cdot -13$
9. $-12 \cdot 4$
10. $-23 \cdot 6$
11. $-18 \cdot -3$
12. $-21 \cdot -5$
13. $4 \cdot -12$
14. $6 \cdot -23$
15. $-3 \cdot -18$
16. $-5 \cdot -21$
17. $(6 \cdot -3) \cdot -7$
18. $-5(-8 \cdot 2)$
19. $2(9 \cdot -4)$
20. $(-7 \cdot 3) \cdot -8$
21. $6 \cdot (-3 \cdot -7)$
22. $-8 \cdot (2 \cdot -5)$
23. $9 \cdot (-4 \cdot 2)$
24. $-7 \cdot (3 \cdot -8)$
25. $-8 \cdot (4 + (-6))$
26. $12 \cdot (-3 + 7)$
27. $(-8 \cdot 4) + (-8 \cdot -6)$
28. $(12 \cdot -3) + (12 \cdot 7)$
29. $9 \cdot (11 - (-4))$
30. $-6 \cdot (-3 - 5)$
31. $(9 \cdot 11) - (9 \cdot -4)$
32. $(-6 \cdot -3) - (-6 \cdot 5)$
33. $|-3| \cdot |-1|$
34. $|0| \cdot |-2|$
35. $|-3 \cdot -1|$
36. $|0 \cdot -2|$

a	b	c	d
$\dfrac{1}{2}$	$-\dfrac{1}{3}$	-7	5

Substitute and simplify.

B 37. $a \cdot b$
38. $b \cdot a$
39. $-2 \cdot b$
40. $-2 \cdot c$
41. $a \cdot |b|$
42. $a \cdot (b \cdot c)$
43. $(a \cdot b) \cdot c$
44. $a \cdot (c + d)$
45. $(a \cdot c) + d$
46. $(-c + d) \cdot \dfrac{1}{4}$
47. $(c + (-d)) \cdot \dfrac{1}{4}$
48. b^3

Use positive and negative numbers to write multiplication equations for these situations. Then answer each question with a real number.

49. The temperature falls 2° per hour. How many degrees more or less is the change in temperature after 3 hours?

50. A plane descends 100 meters per minute. How many meters more or less is the change in altitude after 5 minutes?

51. Erika has been spending $5 per day of her birthday gift money. How much more money or less money did she have a week ago?

52. A water tank has been leaking at the rate of 3 liters per day. How much more water or less water did the tank contain 5 days ago?

53. A water tank is being filled at the rate of 2 liters per minute. How much more water or less water did the tank contain 10 minutes ago?

54. Shawn earns $8.75 a day. How much more money or less money had Shawn earned 5 days ago?

Suppose that A, B, C, and D represent negative numbers. State whether these expressions are positive, negative, or zero.

Ⓒ **55.** $A \cdot B$ **56.** $A \cdot B \cdot C$ **57.** $A \cdot B \cdot C \cdot D$

58. $A + B$ **59.** $A + B + C$ **60.** $(A \cdot B) + (C \cdot D)$

61. $A \cdot (B + C)$ **62.** $A^3 B$ **63.** $A^3 + B$

64. $A^{10} B^9 C^8 D^7$ **65.** $A \cdot A \cdot A \cdot A \cdot B \cdot B \cdot B \cdot C \cdot C \cdot D$

■ REVIEW EXERCISES

Simplify.

1. $1.44 \div 2.4$ **2.** $1.38 \div 0.6$ **3.** $\dfrac{1}{2} \div \dfrac{2}{3}$

4. $4\dfrac{1}{2} \div 1\dfrac{1}{2}$ **5.** $1\dfrac{3}{4} \div 1\dfrac{3}{4}$ **6.** $0 \div \dfrac{1}{2}$

State whether the following pairs of expressions are equivalent.   [1–4]

7. $5a + 6 + a$ and $6(a + 1)$ **8.** $5(a + 2) + 4$ and $3a + 6 + 2a$

9. $x(x + 1)$ and $x^2 + 1$ **10.** $(a + 1)^2$ and $a^2 + 1$

Extension **Stock market reports** ─────────

Stock market reports show the closing price (price of the last sale of the day) and the change from the previous day's closing price in dollars. Positive numbers indicate increases and negative numbers indicate decreases in the prices. Compute the closing price of the previous day for each stock.

Company	Closing Price	Change
Chuckies Chicken	$17\dfrac{1}{2}$	$+\dfrac{5}{8}$
Hills Book Stores	$35\dfrac{7}{8}$	$-\dfrac{1}{2}$

2–5 Dividing Real Numbers

Preview Division by zero

Why is division by zero not permitted? It seems unusual that $\frac{12}{0}$ is undefined when 12×0, $12 + 0$, and $12 - 0$ have well-established meanings.

Look at the pattern in these division exercises:

$$\frac{12}{4} = 3 \qquad \frac{12}{2} = 6 \qquad \frac{12}{1} = 12 \qquad \frac{12}{\frac{1}{2}} = 24$$

What do these equal?

$$\frac{12}{\frac{1}{4}} \qquad \frac{12}{\frac{1}{8}} \qquad \frac{12}{\frac{1}{100}} \qquad \frac{12}{\frac{1}{100,000}}$$

Since the quotient becomes larger as the denominator gets closer to 0, does it seem likely that $\frac{12}{0}$ represents a number?

In this lesson you will learn how to divide by nonzero real numbers. Division by zero is an exception. It is *not* defined.

■ LESSON

In Section 2–3, we defined subtraction of real numbers in terms of addition of opposites. In a similar manner, we are going to define division of real numbers in terms of multiplication. Note that these expressions are equivalent.

$$4.8 \div 4 \qquad 4.8 \cdot \frac{1}{4}$$

Division by 4 is equivalent to multiplication by $\frac{1}{4}$. The numbers 4 and $\frac{1}{4}$ are **reciprocals** of one another.

Definition: Reciprocals

For all real numbers a and b (except 0),

if $ab = 1$, then a and b are reciprocals.

$\frac{1}{2} \cdot 2 = 1$ Therefore, $\frac{1}{2}$ and 2 are reciprocals.

$-\frac{1}{3} \cdot (-3) = 1$ Therefore, $-\frac{1}{3}$ is the reciprocal of -3, and -3 is the reciprocal of $-\frac{1}{3}$.

$\frac{2}{3} \cdot \frac{3}{2} = 1$ Therefore, $\frac{2}{3}$ and $\frac{3}{2}$ are reciprocals.

Since there is no number whose product with 0 is 1, 0 does not have a reciprocal.

The first example illustrated that division by 4 is equivalent to multiplication by $\frac{1}{4}$. This specific case illustrates the definition of division of real numbers.

Definition: Division of Real Numbers

Dividing by any real number (except 0) is equivalent to multiplying by the reciprocal of that real number.

For all real numbers a and b (b not equal to 0),

$$a \div b = a \cdot \frac{1}{b}$$

The reciprocal of a number a can be written $\frac{1}{a}$. Zero has no reciprocal.

Example 1 Simplify each quotient.

    a. $-49 \div 7$ **b.** $3 \div -\frac{1}{2}$ **c.** $-\frac{5}{9} \div -\frac{2}{3}$

Solution   Change divisions to multiplications. Then use the rules of multiplication.

    a. $-49 \div 7 = -49 \cdot \frac{1}{7} = -7$

    b. $3 \div -\frac{1}{2} = 3 \cdot -2 = -6$

    c. $-\frac{5}{9} \div -\frac{2}{3} = -\frac{5}{9} \cdot -\frac{3}{2} = \frac{5}{6}$

Example 2 Simplify. $0 \div -12$

Solution $0 \div -12 = 0 \cdot -\frac{1}{12} = 0$

Note these facts about the division of real numbers.

For all real numbers a and b (b not equal to 0):

$$\left|\frac{a}{b}\right| = \frac{|a|}{|b|}.$$

The quotient of two positive numbers or two negative numbers is positive.
The quotient of one positive number and one negative number is negative.

Example 3 Simplify each quotient.

    a. $\frac{12}{4}$ **b.** $\frac{-12}{4}$ **c.** $\frac{45}{-15}$ **d.** $\frac{-50}{-25}$

Solution   Use the facts about division given above.

    a. $\frac{12}{4} = 3$ **b.** $\frac{-12}{4} = -3$ **c.** $\frac{45}{-15} = -3$ **d.** $\frac{-50}{-25} = 2$

■ CLASSROOM EXERCISES

Give the reciprocal of each number in its simplest form.

1. 8   **2.** -8 **3.** $\dfrac{1}{2}$ **4.** $-\dfrac{1}{2}$

5. 0   **6.** 1   **7.** $-\dfrac{3}{4}$ **8.** 0.4

Complete.

9. Dividing by a real number is equivalent to multiplying by the ___?___.

10. Dividing by 6 is equivalent to multiplying by ___?___.

11. Dividing by $-\dfrac{2}{3}$ is equivalent to multiplying by ___?___.

12. The reciprocal of -1 is ___?___.

Simplify.

13. $-12 \div -6$ **14.** $20 \div -5$ **15.** $-20 \div 4$ **16.** $\dfrac{5}{-15}$

17. $\dfrac{0}{-5}$ **18.** $\dfrac{8}{-\dfrac{1}{2}}$ **19.** $-16 \div \dfrac{1}{4}$ **20.** $\dfrac{2}{3} \div -2$

■ WRITTEN EXERCISES

Write the reciprocal of each number in its simplest form.

A **1.** $-\dfrac{1}{6}$ **2.** $-\dfrac{1}{3}$ **3.** $\dfrac{2}{3}$ **4.** $\dfrac{3}{5}$

5. -6 **6.** -8 **7.** $|-5|$ **8.** $|-10|$

9. $-\dfrac{3}{8}$ **10.** $-\dfrac{5}{8}$ **11.** $-\left|-\dfrac{1}{5}\right|$ **12.** $-\left|-\dfrac{1}{7}\right|$

Write as a multiplication exercise. Do *not* simplify.

13. $\dfrac{2}{3} \div \dfrac{3}{8}$ **14.** $\dfrac{13}{5} \div \dfrac{3}{4}$ **15.** $-6 \div \dfrac{1}{4}$ **16.** $-5 \div \dfrac{1}{10}$

17. $\dfrac{3}{8} \div -\dfrac{3}{5}$ **18.** $\dfrac{5}{8} \div -\dfrac{5}{6}$ **19.** $-\dfrac{3}{4} \div -\dfrac{2}{3}$ **20.** $-\dfrac{2}{3} \div -\dfrac{5}{6}$

Write each division as a multiplication and simplify.

21. $-1\dfrac{2}{3} \div 4$ **22.** $-\dfrac{3}{5} \div 6$ **23.** $5 \div -3$ **24.** $6 \div -5$

25. $3 \div -1\dfrac{1}{3}$ **26.** $4 \div -2\dfrac{1}{4}$ **27.** $-\dfrac{3}{8} \div -\dfrac{5}{8}$ **28.** $-\dfrac{5}{6} \div -3\dfrac{1}{6}$

a	b	c	d
-48	12	-16	4

Substitute and simplify.

29. $\dfrac{a}{b}$ **30.** $\dfrac{a}{c}$ **31.** $\dfrac{b}{a}$ **32.** $\dfrac{c}{a}$

33. $\dfrac{a}{d}$ **34.** $\dfrac{b}{c}$ **35.** $\dfrac{a+b}{d}$ **36.** $\dfrac{b+c}{d}$

r	s	t	u
5.4	-6	36	-0.3

Substitute and simplify.

B **37.** $\dfrac{r}{-3} + 0.8$ **38.** $\dfrac{s}{t} + \dfrac{1}{3}$ **39.** $\dfrac{s}{u} + (-t)$ **40.** $\dfrac{r}{-u} + \dfrac{t}{s}$

41. $\dfrac{s+t}{u}$ **42.** $\dfrac{s}{u} + \dfrac{t}{u}$ **43.** $\dfrac{r+t}{s}$ **44.** $\dfrac{r}{s} + \dfrac{t}{u}$

Use positive and negative numbers to write division equations for these situations. Then answer each question with a real number.

45. Jon loses 10 pounds in 4 weeks. What is his average change in weight per week over the 4 weeks?

46. The temperature dropped 12 degrees in the past 4 hours. What was the average temperature change per hour over that time?

C **47.** Ten years ago, the population of Dry Gulch was 840 more than it is today. What was the average population change per month?

48. Jan was $10\frac{1}{2}$ pounds heavier 3 weeks ago than she is today. What was her average weight change per day over the 3 weeks?

Suppose A, B, C, and D represent negative numbers. State whether these are expressions for positive numbers, negative numbers, or zero.

49. $\dfrac{A}{B}$ **50.** $\dfrac{C \times D}{A}$ **51.** $\dfrac{A+B}{C}$ **52.** $\dfrac{A \times B}{C+D}$

53. $\dfrac{-A}{B}$ **54.** $\dfrac{-A+A}{C}$ **55.** $\dfrac{A^5 \cdot B^6}{C^7}$ **56.** $\dfrac{A^7 + B^9}{C^{16}}$

57. $\dfrac{-A}{-B}$ **58.** $\dfrac{A^{20}}{A^{10}}$ **59.** $(-A)^2$ **60.** $\dfrac{(-B)^2}{B}$

61. Describe what happens to $\dfrac{12}{n}$ when n is replaced with -4, -2, -1, $-\dfrac{1}{2}$, $-\dfrac{1}{4}$, $-\dfrac{1}{100}$, and $-\dfrac{1}{1,000,000}$.

■ REVIEW EXERCISES

State the property illustrated by each equation.   [1–5]

1. $(364 + 97) + 3 = 364 + (97 + 3)$ **2.** $2 \cdot 5 + 8 \cdot 5 = (2 + 8)5$

3. $(7 + 0) + 5 = 7 + 5$ **4.** $(7 + 0) + 5 = 5 + (7 + 0)$

Simplify.   [1–6]

5. $3x + 5 + x$ **6.** $10a + 3b + 5a + b$

7. $6(x + 3) + 4(2x + 1)$ **8.** $3(a + b) + 2b + 4b$

A rectangle is 5 cm longer than it is wide. Complete each statement.   [1–7]

9. If the rectangle is w cm wide, then it is __?__ cm long.

10. If the rectangle is l cm long, then it is __?__ cm wide.

EXTENSION   The absolute value function on a computer

Computers interpret multiple signs before a number as shown in these examples:

Computer command	Screen display
PRINT $--3$	3
PRINT $-5 + -6$	-11

A special function programmed into the computer is the *absolute value function.* In computer terminology, ABS (-2) means $|-2|$.

EXAMPLES

Computer command	Screen display
PRINT ABS (-9)	9
PRINT ABS $\left(-\dfrac{12}{4}\right)$	3

Write the display for each command. Check on a computer if one is available.

1. PRINT $-4--10$ **2.** PRINT $---8$

3. PRINT $-4 * -5$ **4.** PRINT ABS (-6)

5. PRINT ABS $(-5--3)$ **6.** PRINT ABS $(-5) -$ ABS (-3)

Write a computer command to evaluate each expression. Check on a computer if one is available.

7. $-9 - (-2)$ **8.** $(-2) \cdot (-3) \cdot (-4)$ **9.** $8 + (-12)$

10. $(-3)^2$ **11.** $|-12|$ **12.** $|-7 + 4|$

2–6 Addition and Subtraction in Expressions

Preview   Historical Note: Negative numbers

The notation used today to represent negative numbers is a rather recent invention, coming after North America was first visited by European explorers.

Hindu writers expressed negative numbers by placing a dot or a small circle over or beside each, as in $\overset{\circ}{6}$ or $\circ 6$ for -6. The Chinese either wrote positive numbers in red or negative numbers in black or they used a diagonal stroke through the rightmost-digit figure as in $=\text{Ⅲ}$ for -24.

Other early mathematicians used notations such as $m{:}3$, $m.3$, $0-3$, and $\tilde{m}3$ for "negative three."

In this lesson we will make a change in the way negative numbers are written and use the new notation to simplify expressions.

■ LESSON

Recall that "-4" can be read as "the opposite of four" or as "negative four." However, "$-x$" should be read only as "the opposite of x." "Negative x" might be misleading because it may suggest that $-x$ is a negative number. The expression $-x$ can be positive, negative, or zero depending on the value of x. For example, if x takes the value of -2, then $-x = -(-2) = 2$. Note that $-x$ has a positive value in this case.

Because of the relationship between addition and subtraction, we can write expressions in several equivalent forms. You should be able to change one form to another. Study these examples.

Example 1   Rewrite $a - (-2)$ as an addition expression.

Solution

Subtracting a number is the same as adding its opposite.

$$a - (-2) = a + 2$$

> **Example 2**   Rewrite $x + (-3)$ as a subtraction expression.
>
> *Solution*
>> Adding a number is the same as subtracting its opposite.
>>
>> $$x + (-3) = x - 3$$

> **Example 3**   Rewrite $y - 2y - 3$ so that all subtractions are additions.
>
> *Solution* $y - 2y - 3 = y + (-2y) + (-3)$

Expressions with more than one sign between numbers or variables can also be simplified. When an expression is in simplest form, only the first term can have a negative or opposite sign.

> **Example 4**   Simplify. $a - (-b)$
>
> *Solution* $a - (-b) = a + (--b)$ *The opposite of the opposite of b is b.*
> $ = a + b$

> **Example 5**   Simplify. $2x - (-4) + (-y)$
>
> *Solution* $2x - (-4) + (-y) = 2x + 4 - y$

■ CLASSROOM EXERCISES

True or false? If the statement is false, give a counterexample to show that it is false.

1. $-6 = {}^-6$

2. $-a$ is negative for all values of a.

3. $-b$ is positive for all values of b.

4. $-(-x)$ is positive for all values of x.

Change these subtractions to additions.

5. $a - 7$

6. $2x - (-8)$

7. $9 - y$

8. $-a - (-y)$

Change these additions to equivalent subtractions.

9. $b + (-3)$

10. $2a + (-y)$

11. $2a + y$

12. $-x + (-2)$

Simplify.

13. $a - (-4)$

14. $b + (-2)$

15. $2x - (-4)$

16. $3a + (-2b)$

17. $c - (+5)$

18. $3x + (-y)$

19. $2n - (-3m)$

20. $4x + (-3)$

■ WRITTEN EXERCISES

Write with the symbols described in this lesson.

A 1. The opposite of a 2. The opposite of b

3. The opposite of negative four   4. The opposite of negative three

True or false? If the statement is false, give a counterexample to show that it is false.

5. $-20 = {}^-20$ 6. $-30 = {}^-30$ 7. $-|-1| = -1$ 8. $-(-a) = a$

9. $a > -a$ for all values of a 10. $-b < b$ for all values of b

11. $-c < 0$ for all values of c 12. $d > 0$ for all values of d

Change these subtractions to equivalent additions.

13. $h - 5$ 14. $j - 15$ 15. $k - (-3)$ 16. $m - (-7)$

17. $x - y$ 18. $y - x$ 19. $-2 - p$ 20. $-3 - q$

Change these additions to equivalent subtractions.

21. $r + (-9)$ 22. $s + (-6)$ 23. $t + (-u)$ 24. $f + (-g)$

25. $k + n$ 26. $x + y$ 27. $2x + (-x)$ 28. $3y + (-y)$

Simplify.

29. $5 - (-6)$ 30. $7 - (-10)$ 31. $-5 + (-10)$ 32. $-10 + (-5)$

33. $x - (-2)$ 34. $y - (-1)$ 35. $2x + (-4)$ 36. $2y + (-3)$

B 37. $2x - (-8) + (2y)$ 38. $5x - (-10) + (-3w)$

39. $-12 + (-3w) - (-15)$ 40. $-15 + (-x) - (-10)$

Which is the larger number?

41. $5.6 + (-4.9) - (-3.6)$ or $5.6 - 4.9 - 3.6$

42. $\frac{3}{8} - \left(-\frac{3}{4}\right) - \left(-\frac{3}{2}\right)$ or $\frac{3}{8} + \left(-\frac{3}{4}\right) + \left(-\frac{3}{2}\right)$

43. $|17 + (-20) - (-1)|$ or $17 + (-20) - (-1)$

44. The opposite of 3 or the reciprocal of $\frac{1}{3}$.

a	A	n	x
-5	5	-10	$-\frac{1}{2}$

Substitute and simplify.

C 45. $A - (-a)$ 46. $A - (-n)$ 47. $x + (-A)$ 48. $A + (-x)$

49. $n + [-a + (-A)]$ 50. $[n + (-a)] + A$ 51. anx 52. $n \div (Ax)$

■ REVIEW EXERCISES

State the property illustrated by each equation. [1–5]

1. $(293 \cdot 25) \cdot 4 = 293 \cdot (25 \cdot 4)$

2. $(a + b) + 2a = 2a + (a + b)$

3. $4x + x = 4x + 1x$

4. $3a + 2a = (3 + 2)a$

Simplify. [1–6]

5. $3a + 4b + a$

6. $x + 3 + 6x + 5$

7. $x + 6 + 2(x + 6)$

8. $3(2a + 5b) + 2(b + 2a)$

Use an algebraic expression to answer each question. [1–7]

Randy has \$10 more than Matt.

9. How much does Matt have if Randy has R dollars?

10. How much does Randy have if Matt has M dollars?

Self-Quiz 2

Simplify.

2–4   **1.** $-6 \cdot -9$ **2.** $-1 \cdot -2 \cdot -3$ **3.** $(-5 \cdot 8) - (-5 \cdot 3)$

2–5   **4.** $-84 \div 7$ **5.** $-\dfrac{3}{8} \div -\dfrac{3}{4}$ **6.** $-12 \div |-2|$

2–4,
2–5

a	b	c	d
-1	6	9	-3

Substitute and simplify.

   7. $b \cdot d$ **8.** $a \div d$ **9.** $\dfrac{b + c}{c}$ **10.** $\dfrac{a \cdot |d|}{b}$

Write in symbols.

2–5   **11.** The reciprocal of negative seven

2–6   **12.** The opposite of negative one-half

2–6   Rewrite the additions as equivalent subtractions and the subtractions as equivalent additions.

   13. $x - 5$ **14.** $b + (-3)$

   15. $21 - m$ **16.** $-6p + (-1)$

EXTENSION The change sign key on a calculator

A special key on calculators is used in computations involving signed numbers. This key is called the **change sign key,** indicated on the calculator by $\boxed{\text{CHS}}$ or $\boxed{+/-}$. If we depress this key *after* entering a number, the opposite of the entered number is used by the calculator.

Examples

Compute $(-4) + (-3.2)$. $\dfrac{2 - (-4)}{-3}$.

Key Sequence	Display		Key Sequence	Display
$\boxed{4}$	4		$\boxed{2}$	2
$\boxed{\text{CHS}}$	-4		$\boxed{-}$	2
$\boxed{+}$	-4		$\boxed{4}$	4
$\boxed{3}\ \boxed{.}\ \boxed{2}$	3.2		$\boxed{\text{CHS}}$	-4
$\boxed{\text{CHS}}$	-3.2		$\boxed{=}$	6
$\boxed{=}$	-7.2		$\boxed{\div}$	6
			$\boxed{3}$	3
			$\boxed{\text{CHS}}$	-3
			$\boxed{=}$	-2

Use a calculator to simplify these expressions.

1. $1 - (-3)$ **2.** $(-9) + (-4.5)$ **3.** $-[-(-6)]$ **4.** $(-3.4)(-2)$

5. $(-15) \div 3$ **6.** $4(-6) + (-1)$ **7.** $\dfrac{-2}{5} + \dfrac{3}{-5}$ **8.** $\dfrac{14 + (-6)}{-4}$

Strategy for Success Preparing for class

Learning will be easier if you read the preview and skim over the lesson before class.

2–7 Using the Basic Properties to Simplify Expressions

Preview Biography

Major contributions to the development and understanding of the structure of modern algebra were made by the young Frenchman, Évariste Galois. At the age of 21 he spent an entire night recording his mathematical discoveries because he feared that he would be killed the following day in a duel. He did not want his findings to die with him. He was killed in that duel on May 30, 1832, but his contributions to mathematics live on.

In this lesson you will examine the fundamental principles of real numbers and use them to simplify algebraic expressions.

■ LESSON

We have discovered that the basic properties of positive numbers that we studied in Chapter 1 also apply to all numbers, including negative numbers. In this chapter we have also worked with inverse properties for addition and multiplication.

These basic properties are listed below.

Basic Properties of Real Numbers

	Addition	*Multiplication*
	For all real numbers *a, b,* and *c*	
Commutative properties	$a + b = b + a$	$ab = ba$
Associative properties	$a + (b + c) = (a + b) + c$	$a(bc) = (ab)c$
Identity properties	$a + 0 = 0$	$1a = a$
Inverse properties	$a + (-a) = 0$	$a\left(\dfrac{1}{a}\right) = 1 \quad (a \neq 0)$
Distributive properties	$ab + ac = a(b + c)$ $ba + ca = (b + c)a$	

We can derive additional properties from the basic properties listed on page 66.

	For all real numbers a and b,
-1 property of multiplication	$(-1)a = -a$
0 property of multiplication	$0a = 0$
Multiplying opposites property	$(-a)b = a(-b) = -(ab)$ $(-a)(-b) = ab$
Dividing opposites property	$\dfrac{-a}{b} = -\dfrac{a}{b} = \dfrac{a}{-b} \, (b \neq 0)$
Opposite of an opposite property	$-(-a) = a$
Distributive property of opposites	$-(a + b) = -a + (-b)$

We can use the real number properties to explain how algebraic expressions are simplified.

Example 1 Simplify. $3x - x$

Solution

$$
\begin{aligned}
3x - x &= 3x + (-x) && \textit{Subtracting is equivalent to adding the opposite.} \\
&= 3x + (-1)x && \textit{-1 property of multiplication} \\
&= (3 + (-1))x && \textit{Distributive property} \\
&= 2x && \textit{Computation}
\end{aligned}
$$

Example 2 Simplify. $3b - 7 + b$

Solution

$$
\begin{aligned}
3b - 7 + b &= 3b + (-7) + b && \textit{Change to addition.} \\
&= 3b + b + (-7) && \textit{Commutative property of addition} \\
&= 3b + 1b + (-7) && \textit{Identity property for multiplication} \\
&= (3 + 1)b + (-7) && \textit{Distributive property} \\
&= 4b + (-7) && \textit{Computation} \\
&= 4b - 7 && \textit{Change to subtraction.}
\end{aligned}
$$

There are three basic steps for simplifying expressions like those in Examples 1 and 2:

1. Change all subtractions to additions.

2. Use the commutative and associative properties of addition to rearrange terms so that like terms are together.

3. Use the distributive property to replace sums of like terms by single terms.

Sometimes other steps are necessary.

Example 3 Simplify. $2m - 3n + 4n - 7m$

Solution $2m - 3n + 4n - 7m$

$= 2m + (-3n) + 4n + (-7m)$ *Change to additions.*

$= 2m + (-7m) + (-3n) + 4n$ *Use the commutative and associative properties of addition to collect like terms.*

$= 2m + (-7)m + (-3)n + 4n$ *Multiplying opposites property*

$= (-5)m + 1n$ *Use the distributive property to combine like terms.*

$= (-5)m + n$ *Identity property for multiplication*

$= -5m + n$ *Multiplying opposites property*

As you practice the skill of simplifying expressions, look for shortcuts and do much of the work mentally.

Example 4 Simplify. $5x + 6y - 8x + 7y$

Solution $5x + 6y - 8x + 7y = -3x + 13y$ *Collect and combine like terms.*

■ CLASSROOM EXERCISES

Give an example of each property.

1. The associative property of multiplication

2. The distributive property

3. The commutative property of addition

4. The identity properties

Simplify.

5. $3x + 5x$

6. $5x + 3x$

7. $-(-4a)$

8. $1b$

9. $-4m + m$

10. $8n - 2n$

11. $2n - 8n$

12. $2a + 5 - a$

13. $-3c - c + 2a + a$

■ WRITTEN EXERCISES

Change to equivalent expressions involving addition but not subtraction.

A 1. $5a - a$ 2. $3r - r$ 3. $5 + 10s - 8s$

4. $10 + 3b - 2b$ 5. $-5a - 4a + b$ 6. $-7x - 6x + y$

Simplify.

7. $-8 + 5 + 2a$ 8. $-7 + 3 + 4b$ 9. $12a + 3a$ 10. $15b + 5b$

11. $-7c + c$ 12. $-9p + p$ 13. $10q - q$ 14. $30r - r$

15. $-6s - 5s$ 16. $-3t - 2t$ 17. $2a + 3b + 4a$ 18. $7x + 3y + 2x$

19. $6r + 3s + 2s + 4r$ 20. $7x + 3y + 7y + 5x$

21. $w + 3z - 5w + 2z$ 22. $6g + 5h - 15g + h$

State the property illustrated in each equation.

23. $8a + 5a + 3 = (8 + 5)a + 3$ 24. $10 + (4x + 5x) = 10 + (4 + 5)x$

25. $-3x + (5y + 2x) = -3x + (2x + 5y)$ 26. $4a + 3b + 2a = 3b + 4a + 2a$

27. $5c + c = 5c + 1c$ 28. $5r + 0 = 5r$

29. $7(10a) = (7 \cdot 10)a$ 30. $3(a + 2b) + 4b = 3a + 6b + 4b$

Write a simplified rule for each situation.

31. An orchard owner charges 15¢ per apple plus 10¢ per apple for picking, cleaning, and packaging. The cost of n apples is, therefore, $15n + 10n$.

32. The orchard owner charges 10¢ for a peach plus 8¢ per peach for picking, cleaning, and packaging. The cost of m peaches is, therefore, $10m + 8m$.

B 33. The total cost of n apples and m peaches is $15n + 10n + 10m + 8m$.

34. The total cost of g grapefruit and r oranges is $35g + 20r + 15g + 8r$.

35. After cooking, the total number of calories in a cups of asparagus and b cups of lima beans is $45a + 190b - 5a - 20b$.

36. After the waste has been removed, the total number of calories in c cantaloupes and p plums is $120c + 25p - 20c - 3p$.

Simplify these expressions.

37. $x^2 + x^2 + x$ 38. $y^2 + y^2 + 2y$

39. $xy + 2x + x$ 40. $xy + 2x + 2xy$

41. $0.5x - x$ 42. $-a^2 + 2b^2 + \frac{1}{4}a^2$

43. $3x - x - x^2$ 44. $5x^2 - x^2 - 3x$

45. $-5 - x^2 + 6x^2$

Simplify these expressions.

 46. $-8 + 8x - 8x^2 - 8 - 8x - 8x^2$

47. $8 - 8x + 8x^2 - 8 - 8x - 8x^2$

48. $x + \dfrac{x^2}{2} + 2x^2 - x$

49. $-(a^2 - 3a + 2) + 3(-a^2 + 3a - 2)$

Write the property listed in each step of these simplifications.

50. $12 - 5x + 3x = 12 + -(5x) + 3x$ *Subtracting is equivalent to adding the opposite.*

$= 12 + (-5)x + 3x$ **a.** _____?_____

$= 12 + [(-5)x + 3x]$ **b.** _____?_____

$= 12 + [(-5 + 3)x]$ **c.** _____?_____

$= 12 + (-2)x$ *Computation*

$= 12 + -(2x)$ **d.** _____?_____

$= 12 - 2x$ **e.** _____?_____

51. $5x + 2y - 10x = 2y + 5x - 10x$ **a.** _____?_____

$= 2y + 5x + -(10x)$ **b.** _____?_____

$= 2y + [5x + -(10x)]$ **c.** _____?_____

$= 2y + [5x + (-10)x]$ *Multiplying opposites property*

$= 2y + [(5 + (-10))x]$ *Distributive property*

$= 2y + (-5)x$ **d.** _____?_____

$= 2y + -(5x)$ *Multiplying opposites property*

$= 2y - 5x$ **e.** _____?_____

Write an expression for each perimeter and simplify.

52.

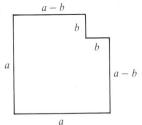

53.

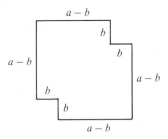

■ REVIEW EXERCISES

Use an algebraic expression to complete each statement.   [1–7]

1. Carl is driving 5 miles per hour under the speed limit. Let s be the speed limit in miles per hour. Then __?__ is Carl's speed in miles per hour.

2. An airplane is flying at twice the speed of sound. If S is the speed of sound, then __?__ is the speed of the airplane.

3. A house is 10 feet longer than twice its width. If w is the width in feet, then __?__ is the length in feet.

Simplify.   [2–13]

4. $-(-6)$ **5.** $--(x)$ **6.** $|-8|$ **7.** $-|9|$ **8.** $-\left|-\dfrac{2}{3}\right|$

2–8 Using the Distributive Properties to Simplify Expressions

Preview

An industrial designer submits a plan for blocks of various sizes, each depending on the lengths of a and b.

The perimeter of this figure is given by the following expression:

$$a + a + a + \frac{1}{2}(a - b) + b + b + b + \frac{1}{2}(a - b)$$

Suppose that $a = 25$ and $b = 5$. What is the perimeter?

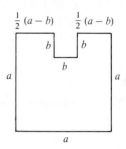

Another expression for the perimeter is $4a + 2b$. Substitute 25 for a and 5 for b in this expression to find the perimeter of the figure.

The two expressions for perimeter are equivalent, but the second is much simpler than the first. In this lesson we will continue to study methods of simplifying expressions.

■ LESSON

Sometimes part of an expression must be "expanded" before the whole expression can be simplified. Consider

$$2x + 3(x + 1)$$

We use the distributive property to obtain an expression without parentheses:

$$2x + 3x + 3$$

Then we simplify:

$$5x + 3$$

Study these examples.

> **Example 1** Simplify. $6 + 2(m - 3)$
>
> *Solution*
> $$\begin{aligned} 6 + 2(m - 3) &= 6 + 2[m + (-3)] &&\textit{Change to addition.}\\ &= 6 + [2m + 2(-3)] &&\textit{Distributive property}\\ &= 6 + 2m + (-6) &&\textit{Computation}\\ &= 2m + (-6 + 6) &&\textit{Rearrange terms.}\\ &= 2m + 0 &&\textit{Inverse property of addition}\\ &= 2m &&\textit{Identity property for addition} \end{aligned}$$

Example 2 Simplify. $2(a + 3) - 1(2a - 1)$

    Solution $2(a + 3) - 1(2a - 1)$

        $= 2(a + 3) + (-1)[2a + (-1)]$ *Subtracting is equivalent to adding the opposite.*

        $= 2a + 6 + (-2)a + 1$ *Distributive property*

        $= 2a + (-2)a + 6 + 1$ *Rearrange terms.*

        $= 0 + 7$ *Computation*

        $= 7$ *Identity property for addition*

Example 3 Simplify. $5a - \frac{1}{2}(2a + 4)$

    Solution $5a - \frac{1}{2}(2a + 4) = 5a + [-\frac{1}{2}(2a + 4)]$

$$= 5a + \left(-\frac{1}{2}\right)(2a) + \left(-\frac{1}{2}\right)(4)$$

$$= 5a + (-a) + (-2)$$

$$= 4a + (-2)$$

$$= 4a - 2$$

Here is a brief review of how to simplify expressions.

1. Change any subtractions to additions.

2. Use the distributive properties to expand terms.

3. Rearrange terms so that like terms are grouped, and then combine like terms.

As you practice, look for shortcuts.

■ CLASSROOM EXERCISES

Write an equivalent expression without parentheses.

1. $3(2a + 1)$ **2.** $7(a - 5)$ **3.** $-4(2y + 4)$ **4.** $-2(7 - 3m)$

5. $-1(4 - 3b)$ **6.** $-1(-6 - 2n)$ **7.** $-(8g + 2)$ **8.** $-(7a - 5)$

Simplify.

9. $3(2x + 1) - 8$ **10.** $4(3a - 2) + 5$ **11.** $2(5y - 2x) + x$ **12.** $13 + 5(2n - 2)$

13. $3y - (4y + 6x)$ **14.** $21 - (8 - 3c)$ **15.** $7b - (3a - 8b)$ **16.** $5 + (4g - 7)$

17. $4(3x - 1) + 2$ **18.** $5(6x - 1) + 4$ **19.** $2(3x - 4y) - x$ **20.** $3(2x + y) - y$

21. $2(x + 3) - 5$ **22.** $4(x + 1) - 4$ **23.** $4 - 3(x + 2)$ **24.** $6 - 2(x + 3)$

■ WRITTEN EXERCISES

Write an equivalent expression.

A
1. $3(x + 7)$ **2.** $5(x + 2)$ **3.** $7(a - 4)$ **4.** $4(a - 6)$

5. $-6(y + 2)$ **6.** $-2(y + 10)$ **7.** $-10(b - 8)$ **8.** $-8(b - 2)$

9. $5(3x - 6)$ **10.** $3(2x - 5)$ **11.** $-1(3a - 5)$ **12.** $-1(6a - 2)$

13. $-(3a - 5)$ **14.** $-(6a - 2)$ **15.** $3(2m + 4n)$ **16.** $3(4m + 2n)$

Simplify.

17. $8(b + 2) + 4$ **18.** $10(b + 3) + 7$ **19.** $6 + 2(x + 5)$ **20.** $2 + 6(x + 5)$

21. $3(2x - 6) + 8$ **22.** $4(5x - 2) + 3$ **23.** $3 - 2(x - 1)$ **24.** $10 - 3(x - 4)$

25. $5 - (x - 3)$ **26.** $6 - (x - 2)$ **27.** $5x + 2(x - 5)$ **28.** $10x + 3(x - 2)$

29. $3x - 5(x + 7)$ **30.** $2x - 6(x + 2)$ **31.** $2x - (x + 10)$ **32.** $3x - (x + 7)$

Write a simplified rule for each situation.

33. When Tom babysits for one child, he charges $1 plus $1.50 per hour. When he babysits for three children at once, he charges twice as much. Simplify $2(1 + 1.5\ h)$.

34. When Theresa babysits for one child, she charges $2 plus $1.00 per hour. When she babysits for three children at once, she charges twice as much. Simplify $2(2 + 1h)$.

B
35. An amount of 1 dollar invested at $r\%$ for one year equals $(1 + 0.01r)$ dollars. A $20 investment is worth 20 times as much as one for $1. Simplify $20(1 + 0.01r)$.

36. An amount of 1 dollar invested at $r\%$ for one year equals $(1 + 0.01r)$ dollars. A $100 investment is worth 100 times as much as for one for $1. Simplify $100(1 + 0.01r)$.

37. We can write 234 as $2h + 3t + 4n$ where h, t, and n represent 100, 10, and 1, respectively. Add $2h + 3t + 4n$ and $3h + 5t + 4n$. Does the result agree with $234 + 354$? Explain.

38. Add $1h + 4t + 8n$ and $5h + 3t$. Does the result agree with $148 + 53$? Explain.

Simplify.

39. $3(a + 2b) + 2(a + 5b)$ **40.** $6(a + b) - 5(a - b)$

41. $-7(a - b) + 3(2a + 3b)$ **42.** $5(2a + 6) - 2(5a - 4)$

43. $4(x - 3y) - 7(2x - y)$ **44.** $-3(2x - 3y) + 2(3x + 5y)$

45. $-5(x + 2y) - 6(x - 3y)$ **46.** $-3(3x - 5y) + 5(x + 4y)$

Consider the following example of a useful shortcut.

> **Sample** Simplify. $6(a - 2) + (a - 2) - 3(a - 2)$
>
> *Solution* Note that all the terms are like terms. So, we can add the coefficients of $(a - 2)$. Since $6 + 1 - 3 = 4$, the solution is $4(a - 2)$ or $4a - 8$.

C **47.** $4(2x - y) - 3(2x - y)$

48. $-10(3x - 4y) + 5(3x - 4y)$

49. $-(a + 2b) + 3(a + 2b) - 2(a + 2b)$

50. $\frac{1}{2}(3x - 2y) - 2(3x - 2y) + 3x - 2y$

51. $-3(a^2 + a - 1) - 2(-a^2 - a + 1)$

52. $a^2 - a + 1 - (-a^2 + a - 1)$

Write an expression for the perimeter of each figure. Simplify the expression.

53.

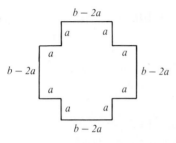

54.

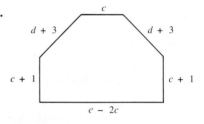

55.

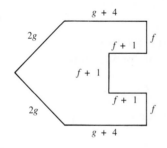

56.

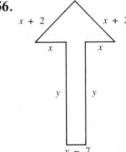

Strategy for Success Actively questioning

Formulate questions while you are learning mathematics. Think: Is there a different way of doing this problem? What would happen if ———— were changed in the problem? What other problems can be solved using this method?

■ REVIEW EXERCISES

Use an algebraic expression to complete each statement.   [1–7]

1. A rectange is 3 cm longer than it is wide.
   a. If its width is w cm, its length is ___?___ cm.

   b. If its length is l cm, its width is ___?___ cm.

2. Jean is $1\frac{1}{2}$ years older than her brother.

   a. If Jean is j years old, her brother is ___?___ years old.

   b. If her brother is b years old, Jean is ___?___ years old.

Simplify.   [2–2]

3. $7 + (-12)$ **4.** $-6 + (3) + (-10)$ **5.** $|-3 + 5|$ **6.** $|-3| + |5|$

Simplify.   [2–3]

7. $5 - 34$ **8.** $-2 - (-12)$ **9.** $19 - (-13)$ **10.** $|-13 - (-6)|$

Self-Quiz 3

2–7   Match the property name with the appropriate equation:

    1. Commutative property of addition   **a.** $3(s + t) = 3s + 3t$

    2. Associative property of multiplication   **b.** $5 \cdot \frac{1}{5} = 1$

    3. Identity property for multiplication   **c.** $-4 + x = x + (-4)$

    4. Inverse property of addition   **d.** $\frac{2}{3} + -\frac{2}{3} = 0$

    5. Inverse property of multiplication   **e.** $8 \cdot (2x) = (8 \cdot 2)x$

    6. Distributive property   **f.** $1 \cdot (xy) = xy$

    7. -1 property of multiplication   **g.** $(-1)(9a) = -9a$

**2–7,
2–8**   Simplify.

    8. $8x - 5x$ **9.** $-(-2y)$ **10.** $26 + 2(3x - 9)$

    11. $3(5x - 9) - 5$ **12.** $9 - 2(x + 3)$ **13.** $(3a + b) - (2a - 3b)$

    14. $4(2x - 1) + 3$ **15.** $3 - 2(x + 1)$ **16.** $(2x + 3) + 4x$

    17. $5(x + 3) + 2x$ **18.** $4 - 2(x + 3)$ **19.** $(2x + y) + (3x + 2y)$

Preview **Information from tables**

When at the beach, have you ever seen a small child shivering with cold while adult swimmers seemed to be comfortably warm? The table below will help you understand why the child is cold when the adults are not.

The table shows that as a cube becomes larger, the ratio of surface area to volume becomes smaller. That is, the surface area increases at a slower rate than the volume does. Although people are not shaped like cubes, the same relationship holds: Adults have a smaller surface area relative to their volume than children do.

The amount of heat produced by a body is relative to its volume. The amount of heat lost is related to its surface area. Therefore, a small body (child) has a larger ratio of heat loss to heat produced than a large body (adult) does.

The table shows the information in an organized way so that patterns and relationships are easy to see. In this lesson you will use tables in a variety of ways to solve problems.

Edge of cube	1	2	3	4	5	6	7	8	9	10
Surface area	6	24	54	96	150	216	294	384	486	600
Volume	1	8	27	64	125	216	343	512	729	1000
Ratio of surface area to volume	$\frac{6}{1}$	$\frac{3}{1}$	$\frac{2}{1}$	$\frac{3}{2}$	$\frac{6}{5}$	$\frac{1}{1}$	$\frac{6}{7}$	$\frac{3}{4}$	$\frac{2}{3}$	$\frac{3}{5}$

■ LESSON

When solving a problem, it may be possible to make a table that shows how the quantities (numbers) are related for some cases. Then, by extending the table we can find a particular case.

Example 1 Two printing presses were used to print a total of 19,500 copies of a booklet. Press *A* could print 2000 copies per hour, and Press *B* could print 1500 copies per hour. Press *A* started at 8:00 A.M. Press *B* started an hour later. At what time was the job completed?

Solution The quantities that vary are the time, the number of copies printed by each press, and the total number of copies. We can organize the information in the following table.

Time	8:00	9:00	10:00	11:00	12:00	1:00	2:00
Total number of copies printed by Press A	0	2,000	4,000	6,000	8,000	10,000	12,000
Press B	0	0	1,500	3,000	4,500	6,000	7,500
Total number of copies printed	0	2,000	5,500	9,000	12,500	16,000	19,500

Answer The job of printing 19,500 copies was completed by 2:00 P.M.

Example 2 Terri and Kim babysit after school. Terri is paid $1 plus $1.25 per hour while Kim is paid $2 plus $1 per hour. How many hours did they work if they were paid the same amount and worked the same number of hours?

Solution

Number of hours	1	2	3	4
Terri's pay (in dollars)	2.25	3.50	4.75	6
Kim's pay (in dollars)	3	4	5	6

Answer If they worked 4 hours, they earned the same amount.

Once we have used a table to solve a problem, we can use the same table to answer other questions.

Example 3 How many hours did Terri and Kim work if they worked the same number of hours but Kim was paid 50¢ more than Terri? (See Example 2.)

Solution Look for the column in which Kim's pay is 50¢ more than Terri's.

Answer They worked 2 hours.

Example 4 How many hours did Terri and Kim work if they worked the same number of hours but Terri was paid 50¢ more than Kim? (See Example 2.)

Solution Continue adding values to the table until Terri's pay is 50¢ more than Kim's.

Number of hours	4	5	6
Terri's pay (in dollars)	6	7.25	8.50
Kim's pay (in dollars)	6	7	8

Answer They worked 6 hours.

■ CLASSROOM EXERCISES

Mr. and Mrs. Helper were moving from one city to another. At 8:00 A.M., Mrs. Helper left their old home driving a truck containing their furniture. She drove at a steady rate of 50 mph. At 8:30 A.M. Mr. Helper left the old house in the family car. He drove the same route at a steady rate of 55 mph. At what time did he catch up with his wife?

Answer these questions.

1. At 9:00 A.M., how far had each driven? How far apart were they?

2. Between 9:00 A.M. and 10:00 A.M. how far did each drive? How far apart were they?

3. Make a table. [*Hint:* Use 30-minute intervals in your table.]

4. Solve the problem.

■ WRITTEN EXERCISES

Copy and complete the tables to solve the problems.

A 1. Kim had 10 coins, consisting of dimes and quarters. How many of the coins were quarters if the total value of the coins was $2.05?

Number of dimes	9	8	7	· · ·
Number of quarters	1	2	3	· · ·
Value of dimes (in dollars)	0.90	0.80	0.70	· · ·
Value of quarters (in dollars)	0.25	0.50	0.75	· · ·
Total value	1.15	1.30	1.45	· · ·

2. Machine *A* can produce 50 parts per hour, whereas Machine *B* can produce 30 parts per hour. At what time will the job of producing 500 parts be completed if Machine *A* starts at 8:00 A.M. and is joined by Machine *B* at 10:00 A.M.?

Time	8:00	9:00	10:00	11:00	· · ·
Number of parts from Machine A	0	50	100	150	· · ·
Number of parts from Machine B	0	0	0	30	· · ·
Total number of parts	0	50	100	180	· · ·

3. A painter could paint a given house by himself in 6 days. It would take his assistant 12 days to paint the same house. (This means that the painter could paint $\frac{1}{6}$ of the house each day and the assistant can paint $\frac{1}{12}$ of the house each day.) How long would it take them to paint the house if they worked together?

Day number	1	2	· · ·
Part painted by painter	$\frac{1}{6}$	$\frac{1}{6}$	· · ·
Part painted by assistant	$\frac{1}{12}$	$\frac{1}{12}$	· · ·
Part painted together in 1 day	$\frac{3}{12}$	$\frac{3}{12}$	· · ·
Total painted together	$\frac{3}{12}$	$\frac{6}{12}$	· · ·

4. One grain loader working alone can load a ship with wheat in 24 hours. A second loader working alone can load the same ship in 8 hours. (This means that the first loader can load $\frac{1}{24}$ of the ship in 1 hour and the other loader can load $\frac{1}{8}$ of the ship in 1 hour.) If both loaders are used, how long will it take to load the ship?

Hour	1	2	3	· · ·
Part loaded by first loader	$\frac{1}{24}$	$\frac{1}{24}$	$\frac{1}{24}$	· · ·
Part loaded by second loader	$\frac{1}{8}$	$\frac{1}{8}$	$\frac{1}{8}$	· · ·
Part loaded by both in 1 hour	$\frac{4}{24}$	$\frac{4}{24}$	$\frac{4}{24}$	· · ·
Total loaded by both	$\frac{4}{24}$	$\frac{8}{24}$	$\frac{12}{24}$	· · ·

5. What number added to the numerator and to the denominator of $\frac{3}{10}$ makes a new fraction equal to $\frac{1}{2}$?

Numerator	3	4	5	· · ·
Denominator	10	11	12	· · ·
Fraction	$\frac{3}{10}$	$\frac{4}{11}$	$\frac{5}{12}$	· · ·

6. A rumor was spread by the following process. Each person hearing the rumor told 3 additional people each hour. How many people had heard the rumor after 6 hours?

Hour	0	1	2	3	4	· · ·
Number of people who are told the rumor during the hour.	0	3	12	48	192	· · ·
Total number of people who know the rumor.	1	4	16	64	256	· · ·

Make a table to solve each problem.

7. Mrs. Anderson's sports car goes 12 kilometers on 1 liter of gasoline. She drives at a rate of 84 kilometers per hour. How many hours can she drive using 28 liters of gasoline?

8. Mr. Anderson's car goes 32 miles on 1 gallon of gasoline. He drives at a rate of 56 miles per hour. How many hours can he drive using 7 gallons of gasoline?

B 9. Jack wants his diet to be low in calories in relation to protein. Which of these five foods should Jack choose on the basis of the calorie/protein ratio?

	Broiled steak (85 g)	Lamb chop (135 g)	Oysters (225 mL)	Buttermilk (225 mL)	Liver (55 g)
Food energy (in calories)	330	400	160	90	130
Protein (in grams)	20	25	20	9	15

10. The ABC Electric Repair Shop charges $25 plus $20 per hour for house calls. The XYZ Fixit Service charges $40 plus $15 per hour for house calls.
   a. How many hours did a house call take if either company would have charged the same amount?
   b. How many hours did a house call take if the ABC shop would have charged $20 more than the XYZ service?

11. On the days of the week that start with the letter S or T, Pop's Grocery Store made a profit of $400 per day. On the days of the week that start with F, M, or W, it lost $300 per day.
   a. What is the total profit or loss for the store during a two-week period?
   b. What will be the total profit or loss in 50 weeks?

12. The Red-Rent-A-Car Company charges $15 per day plus $.12 per mile, the Yellow-Rent-A-Car Company charges $12 per day plus $.15 per mile, and the Green-Rent-A-Car Company charges $27 per day with unlimited mileage. After how many miles of driving would one day's total bill for each company be the same? [*Hint:* Use 20-mile intervals in your table.]

□ 13. In the classic "Frog-in-the-Well" problem, a frog is at the bottom of a well 10 meters deep. In the morning the frog climbs up 2 meters. But in the afternoon the frog slips back down 1 meter. How many days does it take the frog to escape from the well?

14. In a variation of the "Frog-in-the-Well" problem (see Exercise 13), the frog climbs up 3 meters in the morning and slips back 1 meter in the afternoon. How many days does it take the frog to escape from the well?

15. Sue plays a game in which she gains 1 point on the first play, loses 2 points on the second play, gains 4 points on the third play, loses 8 points on the fourth play, and so on. She has to stop when she is either ahead or behind by 100 points. When she has to stop will she be ahead or behind? On what play?

16. At what temperature is the Fahrenheit reading numerically equal to the Celsius reading? [*Hint:* Use the formula $F = \frac{9}{5}C + 32$; begin the table at $0°C$ and use $-10°$ intervals.]

■ REVIEW EXERCISES

Simplify.

1. $6.23 + 7 + 135.1$
2. $3.07 - 0.395$
3. $6.3(0.007)$
4. $45 - 0.03$
5. $1\frac{2}{3} + 2\frac{1}{2}$
6. $6\frac{1}{4} - 3\frac{1}{2}$
7. $\frac{3}{4}\left(1\frac{1}{2}\right)$
8. $2\frac{3}{4} - \frac{1}{2}$

■ CHAPTER SUMMARY

Strategy for Success   Using chapter summaries ────────

The summary at the end of the chapter lists the new words and symbols introduced in the chapter with page references. The chapter summary also collects the major ideas presented in the chapter. Use the summary when you prepare for a test.

- **Vocabulary**

opposite directions	[page 39]
positive numbers	[page 39]
negative numbers	[page 39]
set of real numbers	[page 39]
number line	[page 39]
order (of real numbers)	[page 39]
absolute value	[page 40]
opposites	[page 40]
reciprocals	[page 56]
inverse properties	[page 66]

- *Positive* and *negative* numbers are used to measure quantities that have opposite directions.   [2–1]
- The *real numbers* consist of the positive numbers, the negative numbers, and zero.
- The *absolute value* of a number is its distance from zero on the number line. The symbol "| |" is used to indicate the absolute value.
- Numbers that have the same absolute value and opposite signs are called *opposites*. Zero is the opposite of zero. The symbol "−" is used to indicate the opposite.
- Addition of Real Numbers   [2–2]

   1. To add real numbers with the same sign:
   a. Add their absolute values.
   b. Give the result the same sign as the two numbers.

   2. To add real numbers with different signs:
   a. Subtract their absolute values (the smaller from the larger).
   b. Give the result the same sign as the number with the greater absolute value.

- Subtraction of Real Numbers   [2–3]

   Subtracting any real number is equivalent to adding the opposite of that real number.

   For all real numbers a and b, $a - b = a + (-b)$.

- Multiplication of Real Numbers [2-4]

  To multiply two real numbers:

  1. Multiply the absolute values of the factors.

  2. Use these rules to find the sign of the product.
     a. If the factors have the same sign, the product is positive.
     b. If the factors have different signs, the product is negative.
     c. If one of the factors is 0, the product is 0.

- Two real numbers are *reciprocals* if their product is 1. [2-5]

  1. Dividing by any real number (except 0) is equivalent to multiplying by the reciprocal of that number.

  For all real numbers a and b (b not equal to 0), $a \div b = a \cdot \dfrac{1}{b}$.

  2. Division by 0 is not defined.

- Additional Properties of Real Numbers [2-7]

Property	For all real numbers a and b
-1 property of multiplication	$(-1)a = -a$
0 property of multiplication	$0a = 0$
Multiplying opposites property	$(-a)b = a(-b) = -(ab)$ $(-a)(-b) = ab$
Dividing opposites property	$\dfrac{-a}{b} = \dfrac{a}{-b} = -\dfrac{a}{b}$
Opposite of an opposite property	$-(-a) = a$
Distributive property of opposites	$-(a + b) = -a + (-b)$

Rules for simplifying expressions: [2-8]

1. Change any subtractions to additions.

2. Use the distributive properties to expand terms.

3. Rearrange terms so that like terms are together, and then combine like terms.

Strategy for Success Using chapter reviews

The review at the end of each chapter lists the objective for each lesson and includes several exercises for each lesson. Use the review when you prepare for a test.

■ CHAPTER REVIEW

2–1 **Objectives:** To relate real numbers to real-life situations.

To order real numbers.

To give the absolute value of a real number.

Use real numbers to represent each of the following.

1. A temperature 10°C below freezing **2.** A $200 profit

Simplify.

3. $|-2.4|$ **4.** $|4\frac{2}{3}|$

True or false?

5. $-10 < -6$ **6.** $|-10| < |-6|$

2–2 **Objective:** To add real numbers.

Simplify.

7. $8 + (-12)$ **8.** $-7 + (-4)$ **9.** $-6.5 + 4.25$

2–3 **Objective:** To subtract real numbers.

Simplify.

10. $5 - (-3)$ **11.** $-6 - 4$ **12.** $10 - (-5 - 7)$

2–4 **Objective:** To multiply real numbers.

Simplify.

13. $4 \cdot -5$ **14.** $-2 \cdot -10$ **15.** $-6 \cdot (-5 + 2)$

2–5 **Objective:** To divide a real number by a nonzero real number.

Write the reciprocal.

16. $-\frac{4}{9}$ **17.** -5

Write as a multiplication and simplify.

18. $-5 \div -4$ **19.** $-\frac{3}{5} \div \frac{4}{5}$

a	b	c
-36	6	-3

Substitute and simplify.

20. $\frac{a}{b}$ **21.** $\frac{a}{b} + \frac{b}{c}$

2–6 **Objective:** To use opposites to rewrite an addition or subtraction expression.

22. Change to an equivalent addition. $2x - (-3)$

23. Change to an equivalent subtraction. $a + b$

Simplify.

24. $12 - (-7)$ **25.** $-6 - (-x) - (-7)$

2–7 **Objective:** To simplify an expression involving addition and subtraction by using the properties of real numbers.

Simplify.

26. $7a - a$ **27.** $6r + 10 - 5r$ **28.** $5x + y - 6x + 2y$

2–8 **Objective:** To simplify an algebraic expression by using the distributive properties.

Simplify.

29. $6 + 4(x - 3)$ **30.** $3x - (10 + 2x)$ **31.** $3(x + 3y) + 5(x - y)$

2–9 **Objective:** To solve a problem by making a table.

Use a table to solve each problem.

32. What number added to the numerator and denominator of $\frac{1}{5}$ results in a new fraction equal to $\frac{1}{3}$?

33. A farmer has 100 feet of fencing to make a rectangular pen along the side of a barn, with the side of the barn being one side of the pen. What are the dimensions of the pen that will enclose the maximum area?

■ CHAPTER 2 SELF-TEST

Simplify.

2–1, **1.** $8 + -3$ **2.** $-9 - 26$ **3.** $-17 + (-14)$
2–2 **4.** $0 - (-7)$ **5.** $-|-8|$ **6.** $|-12 + 9|$

2–4, **7.** $3(-8)$ **8.** $-49 \div 7$ **9.** $(-5)(-12)$

2–5 **10.** $-10 \div \frac{-2}{5}$ **11.** $\frac{-4 + 9}{-10}$ **12.** $-6 \cdot (-3 - (-1))$

Simplify.

2–6,
2–7,
2–8
13. $-(-12)$ **14.** $x + (-3)$ **15.** $2y - (-2)$

16. $6x - 5x$ **17.** $4r + 3r$ **18.** $w + 3t - 2w$

19. $5(a + 1) - 3$ **20.** $y - (3 - y)$ **21.** $12 + 2(a - 6)$

22. $7(2b - 1)$ **23.** $(2a - 4) + (5a + 9)$ **24.** $(y + xy) - x$

State the property or definition used in each equation.

2–7
25. $5(a + b) = 5a + 5b$ **26.** $-(4xy) + 4xy = 0$

27. $3b - b = 3b + (-b)$ **28.** $(4y + 3x) + 7x = 4y + (3x + 7x)$

29. $-(4xy) \cdot 0 = 0$ **30.** $(6x + 9) - 2x = (9 + 6x) - 2x$

2–9 The length of a rectangle is twice its width. The width can be a whole number of centimeters from 5 to 10.

31. Make a table showing all possible whole number combinations of lengths and widths.

32. Compute the area for each case.

33. What are the dimensions of a rectangle with an area of 98 square centimeters?

■ PRACTICE FOR COLLEGE ENTRANCE TESTS

1. If $2 \cdot 5 = x$ and $2 \cdot 3 \cdot 5 = y$, then $x - y = 10 \cdot \underline{\ ?\ }$.
   A. -3 **B.** -2 **C.** 2   **D.** 3   **E.** 5

2. If $1, -1, 1, -1, 1, -1, \ldots$ is a sequence of numbers, what is the sum of the first 15 numbers of the sequence?
   A. -7 **B.** -1 **C.** 0   **D.** 1   **E.** 8

3. If for all real numbers

$$\ll a,\ b,\ c \Leftrightarrow d,\ e,\ f \gg = ad + be + cf,$$

then,

$$\ll -1,\ 2,\ -3 \Leftrightarrow 2,\ 0,\ -3 \gg = \underline{\ ?\ }.$$

   A. -11 **B.** -9 **C.** 7   **D.** 9   **E.** 11

4. If $y = -\dfrac{6}{8}$, then y is equivalent to

   A. $-\dfrac{3}{4}$ **B.** $\dfrac{-3}{-4}$ **C.** $-\dfrac{-3}{4}$ **D.** $-\dfrac{3}{-4}$ **E.** $\dfrac{-6}{-8}$

5.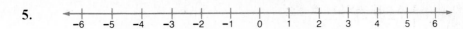

An object starts at 0 and moves 1 unit to the right, then from there it moves 2 units to the left; from there it moves 4 units to the right; from there it moves 8 units to the left; and so on. Where will the object be at the end of 6 moves?

A. -1 B. 3   C. -5 D. 11   E. -21

6. If for all numbers x, $\boxed{x}$ is defined by the equation $\boxed{x} = x(1 - x)$ then $\boxed{-4} = \underline{\ ?\ }$.

A. -20 B. -15 C. -12 D. 12   E. 20

7. If $a = -2$ and $b = -3$, then $3a - 2b - 1 = \underline{\ ?\ }$.

A. -13 B. -6 C. -4 D. -1 E. 5

8. For any number x, $[x]$ is defined as the greatest integer less than or equal to x. Then $[6.7] + [-3.3] = \underline{\ ?\ }$.

A. 2   B. 3   C. 3.4   D. 9   E. 10

9. If for all numbers a and b, $\bigotimes$ is defined by the equation $a \bigotimes b = a^2 - 2ab + b^2$, then $2 \bigotimes - 3 = \underline{\ ?\ }$.

A. -17 B. -11 C. 1   D. 5   E. 25

10. If xy is negative, which of the following is possible?

A. $x < y < 0$ B. $0 < x < y$ C. $y < x$ and $x = 0$

D. $x = y$ E. $x < 0 < y$

11. There are 3 piles that contain 6 sticks each. What is the least number of sticks that can be moved so that the second pile contains twice as many sticks as the first pile, and the third pile contains three times as many sticks as the first pile?

A. 2   B. 3   C. 4   D. 5   E. 6

12. A wooden cube is painted on all sides and then cut into 27 smaller cubes as shown in the figure. How many of the smaller cubes will be painted on exactly two sides?

A. 3   B. 4   C. 6

D. 8   E. 12

Strategy for Success **Beginning a chapter**

Skim over the Review at the end of the chapter to get an overview of what you will be studying in the chapter.

3 Solving Equations

Engineers apply the theories and principles of science and mathematics to practical technical problems. Their work is often the link between a scientific discovery and its useful application.

The engineer uses a principle such as centrifugal force in such diverse applications as the safe design of amusement rides, the separation of milk from cream, or the isolation of corpuscles from blood.

The engineer knows that, in order for a roller coaster car to complete a spiral loop, the centrifugal force must balance the weight of the car. Using this concept, the equation $v^2 = Rg$ can be obtained. In this equation, R equals the radius of the spiral loop; g equals the acceleration due to gravity; and v equals the minimum safe speed for the roller coaster to complete the loop.

3-1 Solving Equations by Repeated Trials

Preview

This item appears on a multiple-choice test:

Choose the one best answer to each question.

If $2(x + 3) = 5x - 6$, what does x equal?

A. -4 **B.** -1 **C.** 0 **D.** 3 **E.** 4

Five possible answers are given, and one of them must be correct. One way to find the correct answer is to substitute each number for x in the equation to determine whether it makes the equation true.

For example, we can substitute -4 for x in the equation.

$$2(x + 3) = 5x - 6$$
$$2(-4 + 3) = 5(-4) - 6$$
$$2(-1) = -20 - 6$$
$$-2 = -26$$

The last equation is obviously false. Therefore, -4 is not a solution. Try to find the correct answer.

■ LESSON

A **solution** of an equation with one variable is a number that changes the equation into a true statement when the number is substituted for the variable. The set of numbers that may be substituted for the variable is called the domain of the variable, or the **replacement set.** To **solve an equation** means to find all solutions of the equation. The set of numbers from the domain that are solutions of the equation is called the **solution set** of the equation.

Example 1 Find the solution set of the equation $x = \dfrac{4}{x}$ for the domain $\{-2, -1, 1, 2\}$.

Solution Substitute each member of the domain for x in the equation.

$-2 = \dfrac{4}{-2}$ True. -2 is a solution.

$-1 = \dfrac{4}{-1}$ False. -1 is not a solution.

$1 = \dfrac{4}{1}$ False.   1 is not a solution.

$2 = \dfrac{4}{2}$ True.   2 is a solution.

Answer The solution set is $\{-2, 2\}$.

For some equations, there are no solutions in the domain. In these cases the solution set for the equation is the **empty set.** A symbol for the empty set is $\emptyset$.

Example 2 Find the solution sets for these equations. The domain is $\{0, 1, 2\}$.

  a. $x + 1 = x$ **b.** $x - 2 = 0$ **c.** $x^2 = -4$

Solution Substitute each member of the domain in the equation.

  a. $x + 1 = x$ $0 + 1 = 0$ False.
  $1 + 1 = 1$ False.
  $2 + 1 = 2$ False.

Answer $\emptyset$

  b. $x - 2 = 0$ $0 - 2 = 0$ False.
  $1 - 2 = 0$ False.
  $2 - 2 = 0$ True.

Answer $\{2\}$

  c. $x^2 = -4$ $0^2 = -4$ False.
  $1^2 = -4$ False.
  $2^2 = -4$ False.

Answer $\emptyset$

■ CLASSROOM EXERCISES

1. Explain what is meant by the solution set of an equation.

2. What does the symbol $\emptyset$ represent?

Is -4 a solution for these equations?

3. $5 + x = 9$ **4.** $y - 1 = -3$ **5.** $a^2 = 16$ **6.** $5 \cdot |c| = 20$

Which of the numbers $\{-2, -1, 0, 1, 2\}$ are solutions of these equations?

7. $x^2 = 4$ **8.** $a + 1 = 3$ **9.** $1 - b = 2$ **10.** $|m| = m$

■ WRITTEN EXERCISES

Is 6 a solution for these equations?

A **1.** $x - 3 = -3$ **2.** $x - 4 = -2$ **3.** $3x = 18$ **4.** $5x = 30$

  5. $5x + 2 = 28$ **6.** $3x - 1 = 15$ **7.** $2x + 3 = x + 9$ **8.** $3x - 2 = x + 10$

  9. $10 - x = x - 2$ **10.** $20 - x = 2x + 14$ **11.** $-3\,|x| = 18$ **12.** $3|x| = -18$

Which of the numbers $\{1, 2, 3, 4, 5\}$ are solutions of these equations?

13. $x + 2 = 5$ **14.** $4 + x = 5$ **15.** $-2a = -8$

16. $-3x = -15$ **17.** $0x = 0$ **18.** $1 \cdot x = x$

Solve these equations for the domain $\{-2, -1, 0, 1, 2\}$.

19. $x - 2 = 0$

20. $2 - x = 2$

21. $\frac{1}{2}x = -1$

22. $\frac{1}{3}x = 0$

23. $28 + x = 27$

24. $-28 + x = -27$

25. $\frac{5}{x} = 2.5$

26. $\frac{3}{x} = -1.5$

27. $\frac{x}{2} = -0.5$

Which of the numbers $\{0, 1, 2, 3, \ldots, 100\}$ are solutions of these equations?
Write $\emptyset$ if there is no solution.

28. $x + 23 = 53$

29. $x + 17 = 57$

30. $x + 23 = x$

31. $x + 15 = x + 16$

32. $5x + x = 90$

33. $4x + x = 55$

34. $5x = x + 60$

35. $6x = x + 70$

36. $\frac{x}{5} = 20$

Solve these equations for the domain $\{-4, -2, 0, 2, 4\}$.

B   **37.** $x^2 + 2x = 8$

38. $x^2 - 2x = 8$

39. $5x = 3x - 8$

40. $5x = 3x + 8$

41. $x + x = 2$

42. $x = x \cdot 0$

43. $x = x + 0$

44. $a - a = 2$

45. $3x + 2x + x = 12$

Find the solution sets of these equations for the indicated domains.
Write $\emptyset$ if there is no solution.

C   **46.** $2^x = 8$

$\{3, 4, 6, 16\}$

47. $3x = 9$

$\{2, 3, 6\}$

48. $x^2 = -25$

$\{-12\frac{1}{2}, -5, 5\}$

49. $x^2 + 3 = 4x$

$\{-1, 1, 3\}$

50. $x^2 + x = 6$

$\{-3, -2, 2, 3\}$

51. $x^2 = 36$

$\{1, 2, 3, 4\}$

52. $(x + 1)^x = 64$

$\{1, 2, 3, 4\}$

53. $(x - 1)^x = 1$

$\{1, 2, 3, 4\}$

■ REVIEW EXERCISES

a	b	x	y
5	10	$\frac{1}{2}$	1.5

Substitute and simplify.   [1–1]

1. aby

2. $a - x$

3. $\frac{y}{b}$

Simplify.   [1–2]

4. $3 + 4 \cdot 5$

5. $(10 + 1)(10 - 1)$

6. $\frac{(9 + 18)}{3}$

Simplify.   [1–3]

7. $3^2 + 4^2$

8. $(3 + 4)^2$

Indicate whether the two expressions are equivalent.   [1–4]

9. $x(x + 2) + 3(x + 2)$ and $(x + 3)x + (x + 3)2$

10. $3(x + 5)$ and $5(x + 3)$

3-2 Equivalent Equations

Preview   A model for equations

Here is a *balance* with equal weights on the two pans.

What are some ways of changing the weights on the pans and still keeping the pans in balance? For example, if the weights on the left pan are doubled, what should be done to the weights on the right pan in order to keep the pans balanced?

■ LESSON

The two equations $2x = 8$ and $2x + 1 = 9$ have the same solution sets. For this reason they are called **equivalent equations.**

Equivalent equations	*Nonequivalent equations*
$2x = 8$ {4}	$3x = 12$ {4}
$2x + 1 = 9$ {4}	$x + 3 = 9$ {6}

It is not always necessary to find the solution sets of two equations in order to determine whether they are equivalent. The balanced pans in the Preview suggest other ways of knowing that equations are equivalent. Think of an equation as being a balance with equal quantities on the two pans. Just as some changes in the weights leave the pans balanced, some changes in equations give equivalent equations. Study the following properties.

> An equivalent expression may be substituted for one side of an equation without changing the solution set of the equation.

On a balance, we can rearrange the weights on one or both pans without adding or substracting weights. This idea is similar to using the basic properties of numbers in order to change one or both sides of an equation.

For example, consider the equation

$$2x + 3x = 30$$

We can apply the distributive property and computation to the expression $2x + 3x$ as follows:

$$(2 + 3)x = 30$$
$$5x = 30$$

These two equations are both equivalent to the given equation $2x + 3x = 30$.

The Symmetric Property of Equations

For all real numbers a and b, if $a = b$ then $b = a$.

On a balance, the weights in the two pans can be switched. The pans will still be in balance. This is similar to interchanging the expressions on the two sides of the equation. The new equation is equivalent to the original equation. For example, these two equations are equivalent to each other:

$$7 = 9 + x$$
$$9 + x = 7$$

On a balance, the pans stay in balance if the same weight is added to both sides or if the same weight is taken from both sides. Similarly, the same quantity can be added to, or subtracted from, both sides of an equation.

An equivalent equation is obtained when the same number is added to (or subtracted from) both sides of an equation.

Start with the equation	$x - 7 = 3$
Add 7 to both sides.	$x - 7 + 7 = 3 + 7$
Simplify.	$x = 10$

All three equations are equivalent.

Start with the equation	$a + 3 = 1$
Subtract 3 from both sides.	$a + 3 - 3 = 1 - 3$
Simplify.	$a = -2$

All three equations are equivalent.

On a balance, the pans stay in balance when the weights on both pans are doubled, or when the weights on both pans are halved. Similarly, both sides of an equation can be multiplied by, or divided by, the same nonzero quantity.

An equivalent equation is obtained when both sides of an equation are multiplied (or divided) by the same nonzero number.

Start with the equation	$\frac{1}{2}y = 6$
Multiply both sides by 2.	$2\left(\frac{1}{2}y\right) = 2(6)$
Simplify.	$y = 12$

All three equations are equivalent.

Start with the equation	$-3m = 12$
Divide both sides by -3.	$\frac{-3m}{-3} = \frac{12}{-3}$
Simplify.	$m = -4$

All three equations are equivalent.

Example 1 Write an equation equivalent to $a - 3 = 12$ by adding 3 to both sides and simplifying.

Solution

$$a - 3 = 12$$
Add 3 to both sides. $a - 3 + 3 = 12 + 3$
Simplify. $a = 15$

Example 2 Write two equations that are equivalent to $4x = 8$.

Solution 1

$$4x = 8$$
Subtract 3 from both sides. $4x - 3 = 8 - 3$
Simplify. $4x - 3 = 5$

Solution 2

$$4x = 8$$
Divide both sides by 4. $\frac{4x}{4} = \frac{8}{4}$
Simplify. $x = 2$

Answer $4x - 3 = 5$ and $x = 2$
There are many other equations equivalent to $4x = 8$.

Strategy for Success Using examples

When you have difficulty understanding an example, read it "backward." That is, start with the answer and read back through each step to the original problem.

Example 3 What equivalent equation results from following these directions in order? Simplify after each step.

Equation	Directions
$6x - 5 = 2x + 7$	**a.** Subtract $2x$ from each side.
	b. Add 5 to each side.
	c. Divide both sides by 4.

Solution

	$6x - 5 = 2x + 7$
a. *Subtract 2x from each side.*	$6x - 5 - 2x = 2x + 7 - 2x$
Simplify.	$4x - 5 = 7$
b. *Add 5 to both sides.*	$4x - 5 + 5 = 7 + 5$
Simplify.	$4x = 12$
c. *Divide both sides by 4.*	$\dfrac{4x}{4} = \dfrac{12}{4}$
Simplify.	$x = 3$

■ CLASSROOM EXERCISES

1. Explain what *equivalent equations* are.

2. State six ways of changing from one equation to an equivalent equation.

Write an equation equivalent to the given equation by adding 3 to both sides.

3. $2x = x + 2$

4. $2y - 3 = 17$

Write an equation equivalent to the given equation by multiplying both sides by 3.

5. $2a = 9$

6. $\frac{1}{3}t = 5$

Write two equations that are equivalent to the given equation.

7. $3x = 12$

8. $\frac{1}{2}d = 4$

State whether the two equations are equivalent. Explain your answer.

9. $5x = 15$

$-5x = -15$

10. $\frac{1}{2}c = 7$

$c = 3\frac{1}{2}$

What equations result from following these directions in order?

11. $2 - 3x = 10x + 7$
 a. Add $3x$ to both sides.
 b. Add -7 to both sides.
 c. Divide both sides by 13.

12. $-2x + 4 = 7 - 8x$
 a. Add $8x$ to both sides.
 b. Add -4 to both sides.
 c. Divide both sides by 6.

◼ WRITTEN EXERCISES

Write an equivalent equation by adding 4 to both sides of the given equation. Simplify both sides.

A **1.** $6x - 4 = 14$ **2.** $4x + 4 = 4$ **3.** $5x + 6 = 4$ **4.** $2x - 4 = 0$

Write an equivalent equation by subtracting n from both sides of the given equation. Simplify both sides.

5. $5n = n + 3$ **6.** $6n = n + 17$ **7.** $2 + n = 4n + 1$ **8.** $5 + n = 3n + 3$

Write an equivalent equation by multiplying both sides of the given equation by 3. Simplify.

9. $\frac{1}{3}n = 6$ **10.** $\frac{1}{3}n = -2$ **11.** $\frac{n}{3} = -4$ **12.** $\frac{n}{3} = 9$

Write an equivalent equation by multiplying both sides of the given equation by $\frac{1}{2}$. Simplify.

13. $2x = -6$ **14.** $2x = 20$ **15.** $4x + 6 = 10$ **16.** $8x - 6 = 30$

State whether the two equations are equivalent. Explain your answer.

17. $2x = 6$ **18.** $3x = 9$ **19.** $x = 7$ **20.** $x = 4$

     $6x = 18$ $12x = 36$ $5x = 35$ $\frac{1}{2}x = 2$

21. $3x + 7 + 5x = 20$ **22.** $4x + 6 + 3x = 27$ **23.** $2x + 1 = 3$ **24.** $x - 4 = 0$

     $3x + 5x + 7 = 20$ $4x + 3x + 6 = 27$ $x + 2 = 3$ $3x = -12$

What equations result from following these directions in order?

25. $4x - 5 = 15$
   a. Add 5 to both sides.
   b. Divide both sides by 4.

26. $7n + 14 = 35$
   a. Add -14 to both sides.
   b. Divide both sides by 7.

27. $\frac{1}{2}x - 12 = 18$
   a. Add 12 to both sides.
   b. Multiply both sides by 2.

28. $6 + x = 3x$
   a. Subtract x from both sides.
   b. Multiply both sides by $\frac{1}{2}$.

29. $5x + 3x + 2 = 20$
   a. Combine the x terms.
   b. Subtract 2 from both sides.
   c. Divide both sides by 8.

30. $6y + 4y + 5 = 25$
   a. Combine the y terms.
   b. Subtract 5 from both sides.
   c. Divide both sides by 10.

B **31.** $2(x + 3) = 25$
   a. Use the distributive property.
   b. Subtract 6 from both sides.
   c. Divide both sides by 2.

32. $\frac{1}{3}(x - 12) = 100$
   a. Use the distributive property.
   b. Add 4 to both sides.
   c. Multiply both sides by 3.

33. $\frac{1}{4}(x - 12) = 100$
   a. Use the distributive property.
   b. Add 3 to both sides.
   c. Multiply both sides by 4.

34. $10x + 6 - x = 8x$
   a. Simplify the left side by combining the x terms.
   b. Subtract $8x$ from both sides.
   c. Add -6 to both sides.

35. $5x + 15 = 50$
   a. Subtract 15 from both sides.
   b. Divide both sides by 5.

36. $8x + 13 = 41$
   a. Subtract 13 from both sides.
   b. Multiply both sides by $\frac{1}{8}$.

37. $5x + 15 = 50$
   a. Divide both sides by 5.
   b. Subtract 3 from both sides.

38. $8x + 13 = 41$
   a. Multiply both sides by $\frac{1}{8}$.
   b. Subtract $\frac{13}{8}$ from both sides.

39. True or false? If all the solutions of one equation are solutions of a second equation, then the equations are equivalent.

Fill in the missing reasons for these steps used in solving these equations.

40. $3(x + 5) = 10x - 3$
$3x + 15 = 10x - 3$ Use the distributive property.
$15 = 7x - 3$ **a.** Subtract __?__ from both sides and simplify.
$18 = 7x$ **b.** Add __?__ to both sides and simplify.
$\dfrac{18}{7} = x$ **c.** __?__ and simplify.

41. $3x + 5 = 10 - 2x$
$5x + 5 = 10$ **a.** _____?_____
$5x = 5$ **b.** _____?_____
$x = 1$ **c.** _____?_____

42. Puzzle. Study the two balance situations.

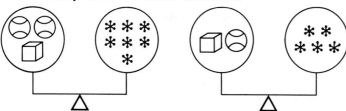

  a. How many $*$ are required to balance one ?

  b. How many $*$ balance one 🎲 ?

■ REVIEW EXERCISES

r	s	t	u
3	6	12	$\frac{2}{3}$

Substitute and simplify.

1. $r + s + t$

2. $\dfrac{u}{s}$ [1–1]

3. $r + st$

4. $\dfrac{t}{r + s}$ [1–2]

5. rs^2

6. $(rs)^2$

3–3 Solving Equations by Addition

Preview

Try to solve this equation by making repeated trials where the domain is the set of real numbers.

$$10a - 7 = 18$$

Organize your work by making a table of values you choose for a and the corresponding values of $10a - 7$.

Did you find that the solution to the equation is 2.5?

We can use repeated trials to solve equations like this one, but it may take a lot of time and effort. There are more efficient ways to solve equations, and you will begin to study them in this lesson.

■ LESSON

In Section 3–2, we learned that an equivalent equation is obtained when the same number is added to both sides of an equation. For example, starting with

$$x - 4 = 5$$

and then adding 4 to both sides and simplifying gives:

$$x = 9$$

The fact that the second equation has the same solution as the first is a consequence of the **addition property of equations.**

Addition Property of Equations

For all real numbers a, b, and c, if $a = b$, then $a + c = b + c$.

We also learned that an equivalent equation is obtained by subtracting the same number from both sides of an equation. However, we can think of subtraction of a number as addition of the opposite of the number. Therefore, subtracting the same quantity from both sides of an equation is also described by the addition property of equations. The addition property of equations can be used to solve complex equations by changing them to simpler equations whose solutions are obvious.

Strategy for Success   Using examples

The examples in a lesson are usually related to each other. Try to spot how an example differs from the preceding one.

Example 1 Solve. $3x - 7 = 2x + 5$

Solution

$$3x - 7 + = 2x + 5$$

Add −2x to both sides. $\quad 3x - 7 + (-2x) = 2x + 5 + (-2x)$

Simplify. $\qquad\qquad\qquad x - 7 = 5$

Add 7 to both sides. $\qquad\quad x - 7 + 7 = 5 + 7$

Simplify. $\qquad\qquad\qquad\qquad x = 12$

Answer $\{12\}$

The solution set of the last equation is $\{12\}$, so the solution set of the first equation is also $\{12\}$. Check by substitution.

Check
$$3x - 7 = 2x + 5$$
$$3 \cdot 12 - 7 \overset{?}{=} 2 \cdot 12 + 5$$
$$29 = 29 \quad \text{True!}$$

Note these things in Example 1.

1. We added $-2x$ to both sides of the equation in order to get an equivalent equation with variables on only one side.

2. We added 7 to both sides of the equation in order to get an equivalent equation in the simplest possible form: $x = 12$.

From now on, we will combine adding and simplifying into one step.

Example 2 Solve. $3 + a = -2$

Solution

$$3 + a = -2$$

Add −3 to both sides and simplify. $\qquad a = -5$

Answer $\{-5\}$

Check
$$3 + a = -2$$
$$3 + (-5) \overset{?}{=} -2$$
$$-2 = -2 \quad \text{True!}$$

Example 3 Solve. $8 - 5x = 2 - 6x$

Solution

$$8 - 5x = 2 - 6x$$

Add 6x to both sides and simplify. $\qquad 8 + x = 2$

Subtract 8 from both sides and simplify. $\qquad x = -6$

Answer $\{-6\}$

Check
$$8 - 5x = 2 - 6x$$
$$8 - 5(-6) \overset{?}{=} 2 - 6(-6)$$
$$8 + 30 \overset{?}{=} 2 + 36$$
$$38 = 38 \quad \text{True!}$$

Example 4   Solve. $5x + 3 = 6x + 7$

Solution

$$5x + 3 = 6x + 7$$

Subtract 5x from both sides and simplify. $3 = x + 7$

Subtract 7 from both sides and simplify. $-4 = x$

Some people like to end with the variable on the left side of the equation. They write the following extra step, using the symmetric property of equations to reverse the sides of the equation.

Reverse the sides of the equation. $x = -4$

Answer $\{-4\}$

Check   The check is left to the student.

■ CLASSROOM EXERCISES

1. State the addition property of equations.

2. Why is it unnecessary to write a subtraction property of equations?

State what was added to both sides of the first equation that resulted in the second equation.

3. $2x + 3 = x - 9$
$\quad\; x + 3 = -9$

4. $2x + 3 = x - 9$
$\qquad\quad 2x = x - 12$

5. $20x - 9 = 4x + 7$
$\qquad\; 20x = 4x + 16$

Solve by writing a series of equivalent equations.

6. $a + 3 = 9$

7. $b - 10 = 10$

8. $-3 = x - 3$

9. $x + 10 = 7$

10. $2n + 3 = n + 16$

11. $4y - 6 = 3y + 3$

■ WRITTEN EXERCISES

State what was added to both sides of the first equation that resulted in the second equation.

A   **1.** $8a + 3 = 7a + 20$
$\qquad\; a + 3 = 20$

2. $5a - 3 = 4a + 15$
$\qquad\; a - 3 = 15$

3. $8x + 3 = 7x + 20$
$\qquad\quad 8x = 7x + 17$

4. $5y - 3 = 4y + 15$
$\qquad\; 5y = 4y + 18$

5. $7a + 8 = -13$
$\qquad\; 7a = -21$

6. $5a - 5 = -9$
$\qquad\; 5a = -4$

Solve each equation by writing a series of equivalent equations.

7. $5t = 4t + 3$

8. $7t = 6t + 2$

9. $10x = 9x - 2$

10. $8x = 7x - 4$

11. $r - 6 = 10$

12. $r - 9 = 20$

13. $x + \dfrac{1}{2} = 2$

14. $w + \dfrac{1}{4} = 3$

15. $4a + 3 = 3a + 5$

16. $5a + 4 = 4a + 7$

17. $2x + 1 = 3x$

18. $4b + 8 = 5b$

100   Chapter 3   Solving Equations

19. $6x + 5 = 7x - 3$ **20.** $3x + 7 = 4x - 8$ **21.** $9x + 5\frac{2}{3} = 10x + 3\frac{2}{3}$

22. $7x + 5\frac{3}{4} = 8x + 2\frac{3}{4}$ **23.** $5z + 0.6 = 4z + 1$ **24.** $7z + 0.3 = 6z + 1$

B **25.** $x + \left(-2\frac{1}{2}\right) = 3\frac{1}{2}$ **26.** $x + \left(-4\frac{1}{4}\right) = 8\frac{1}{4}$ **27.** $x + 1.53 = 10.62$

28. $x + 4.02 = 9.48$ **29.** $1.2x = 0.2x + 3$ **30.** $1.7x = 0.7x + 2$

31. $5x + 1.8 = 4x - 3.2$ **32.** $3x + 4.7 = 2x - 5.3$ **33.** $1.6x + 4 = 2.6x$

34. $2.4x + 7 = 3.4x$ **35.** $2.3x - 2 = 3.3x + 2.5$ **36.** $7.5x - 3 = 8.5x + 3.5$

37. A student added $(-6x)$ to both sides of the equation $5x + 13 = 6x$. Then he became "stuck." Describe two different ways to complete the solution for him.

Identify the expression that can be added to both sides of each of these equations in order to solve them in one step.

> **Sample** $3x + 6 = 2x + 9$
>
> *Solution* $3x + 6 = 2x + 9$
> Add $(-2x + -6)$
> to both sides. $3x + 6 + (-2x + -6) = 2x + 9 + (-2x + -6)$
> $x = 3$
>
> *Answer* $(-2x + -6\}$

C **38.** $5x + 3 = 4x + 10$ **39.** $8x - 7 = 7x + 5$ **40.** $2x + 8 = 3x - 5$

Solve.

41. $5x + |-6| = 4x + |9|$ **42.** $3x - |-6| = 4x - |-8|$

43. This table shows the values of the left and right sides of the equation $2x - 4 = 10 - x$ for integral values of x from 0 to 8.

x	$2x - 4$	$10 - x$
0	-4	10
1	-2	9
2	0	8
3	?	?
4	?	?
5	?	?
6	?	?
7	?	?
8	?	?

  a. Copy and complete the table.

  b. Whenever x increases by 1, then $2x - 4$ __?__ (increases/decreases) by __?__.

  c. Whenever x increases by 1, then $10 - x$ __?__ (increases/decreases) by __?__.

  d. Between what integer values of x does the solution of $2x - 4 = 10 - x$ lie?

44. Between what integer values does the solution of the equation $5x - 3 = 2x + 10$ lie? Use a method similar to that used in Exercise 43.

■ REVIEW EXERCISES

State the property illustrated by each equation.   [1–5]

1. $(3x + 4) + 2x = 2x + (3x + 4)$ **2.** $2x + (3x + 4) = (2x + 3x) + 4$

3. $(2x + 3x) + 4 = (2 + 3)x + 4$

Simplify.   [1–6]

4. $5x - x$ **5.** $7a - 6a$ **6.** $5(2x + 3) + x + 4$

Use an algebraic expression to complete each statement.   [1–7]

Juan has \$25 more than Fernando.

7. If Fernando has f dollars, then Juan has __?__ dollars.

8. If Juan has j dollars, then Fernando has __?__ dollars.

The length of a rectangle is three times as long as the width.

9. If the width is w centimeters, then the length is __?__ centimeters.

10. If the length is l centimeters, then the width is __?__ centimeters.

Self-Quiz 1

3–1 Which of the numbers $\{-3, -1, 0, 1, 3\}$ is a solution of the equation?

    1. $2x + 4 = 10$ **2.** $5 - 3x = 8$

    3. $6|x| = 18$ **4.** $x^2 + 2x = 5x$

3–2 What equivalent equation results from performing the given operation?

    5. $4x + 8 = 22$ **6.** $7x - 6 = 39 - 2x$

        Subtract 8 from each side.   Add $2x$ to each side.

    7. $7x = -84$ **8.** $4(x + 1) - 2x = -6$

        Multiply each side by $\frac{1}{7}$.   Apply the distributive property.

3–3 Solve by writing a series of equivalent equations.

    9. $8x = 7x + 9$ **10.** $x + \frac{3}{4} = 2$

    11. $11x - 23 = 10x - 27$ **12.** $3(x - 1) = 4x$

3-4 Solving Equations by Multiplication

Preview

In the 1985 Boston Marathon, George Murray won the wheelchair division of the 26.2-mile race in an incredible time of 1 hour 45 minutes 34 seconds, or about 105.567 minutes. What was his average time for each mile of the race?

The answer to the question given above is the solution of this equation:

$$26.2x = 105.567$$

What operation could be used to solve the equation for x?

In this lesson you will learn how to use multiplication and division to solve equations.

■ LESSON

In Section 3–2 we learned that an equivalent equation is obtained when both sides of an equation are multiplied by the same nonzero number. For example, starting with

$$3a = 15$$

and then multiplying both sides by $\frac{1}{3}$ and simplifying gives:

$$a = 5$$

The fact that the second equation has the same solutions as the first is a consequence of the **multiplication property of equations.**

Multiplication Property of Equations

For all real numbers a, b, and c, if $a = b$, then $ac = bc$.

We also learned that an equivalent equation is obtained by dividing both sides of an equation by the same nonzero number. However, we can think of division by a number as multiplication by the reciprocal of the number. Therefore, dividing both sides by the same quantity is also described by the multiplication property of equations. We will use the multiplication property of equations to solve complex equations by changing them to simpler equations whose solutions are obvious.

Example 1   Solve. $5x = 37$

Solution $5x = 37$

Multiply both sides by $\frac{1}{5}$. $\frac{1}{5} \cdot 5x = \frac{1}{5} \cdot 37$

Simplify. $x = \frac{37}{5}$ or $7\frac{2}{5}$

Answer $\left\{\dfrac{37}{5}\right\}$

Check $5x = 37$

$5 \cdot \dfrac{37}{5} \stackrel{?}{=} 37$

$37 = 37$ True!

Notice in Example 1 that $\frac{1}{5}$ was used as the multiplier in the first step since it is the reciprocal of the coefficient of x. This choice of multiplier makes the coefficient of x equal to 1 in the next step.

Example 2   Solve. $-6a = 14$

Solution $-6a = 14$

Divide both sides by -6. $\dfrac{-6a}{-6} = \dfrac{14}{-6}$

Simplify. $a = -\dfrac{7}{3}$ or $-2\dfrac{1}{3}$

Answer $\left\{-\dfrac{7}{3}\right\}$

Check $-6a = 14$

$(-6)\left(-\dfrac{7}{3}\right) \stackrel{?}{=} 14$

$\dfrac{42}{3} \stackrel{?}{=} 14$

$14 = 14$ True!

Example 3   Solve. $\frac{2}{3}x = 16$

Solution $\dfrac{2}{3}x = 16$

Multiply both sides by $\frac{3}{2}$. $\dfrac{3}{2} \cdot \dfrac{2}{3}x = \dfrac{3}{2} \cdot 16$

Simplify. $x = 24$

Answer $\{24\}$

Check   The check is left to the student.

Example 4 Solve. $7 = \dfrac{m}{-3}$

Solution
$$7 = \dfrac{m}{-3}$$

Multiply both sides by −3. $-21 = m$

Answer $\{-21\}$

Check   The check is left to the student.

■ CLASSROOM EXERCISES

1. State the multiplication property of equations.

2. Why is it unnecessary to write a division property of equations?

State what both sides of the first equation were multiplied by or divided by that resulted in the second equation.

3. $2a = 7$

$a = \dfrac{7}{2}$

4. $\dfrac{m}{-4} = 8$

$m = -32$

5. $\dfrac{2}{5}y = -8$

$y = -20$

6. $0.2x = 1$

$x = 5$

Solve by writing simpler equivalent equations.

7. $-5n = 6$

8. $4a = 9$

9. $\dfrac{3}{4}y = 12$

10. $\dfrac{b}{5} = 10$

11. $-7 = -\dfrac{1}{2}z$

12. $8 = 0.4y$

■ WRITTEN EXERCISES

State what both sides of the first equation were multiplied by that resulted in the second equation.

A

1. $-3x = 8$

$x = -\dfrac{8}{3}$

2. $-2x = 7$

$x = -\dfrac{7}{2}$

3. $\dfrac{3}{8}x = 24$

$x = 64$

4. $\dfrac{4}{5}x = 20$

$x = 25$

5. $\dfrac{x}{7} = 2$

$x = 14$

6. $\dfrac{x}{6} = 12$

$x = 72$

7. $-a = 3$

$a = -3$

8. $-b = -4$

$b = 4$

State what both sides of the first equation were divided by that resulted in the second equation.

9. $7a = 18$

$a = 2\dfrac{4}{7}$

10. $9b = 16$

$b = 1\dfrac{7}{9}$

11. $-3c = 17$

$c = -5\dfrac{2}{3}$

12. $-5d = 12$

$d = -2\dfrac{2}{5}$

13. $-4p = -24$

$p = 6$

14. $-8q = -40$

$q = 5$

15. $\dfrac{1}{2}r = 3$

$r = 6$

16. $\dfrac{1}{3}s = 7$

$s = 21$

Check the solution listed for each equation. Write "OK" or "wrong."

17. $\frac{1}{5}x = 30$ **18.** $\frac{1}{4}x = 20$ **19.** $\frac{2}{3}x = -4$ **20.** $\frac{3}{4}x = -12$

{6} {5} {6} {−9}

Solve by writing simpler equivalent equations. Check your solutions.

21. $19 = 2x$ **22.** $13 = 2x$ **23.** $-5n = 31$ **24.** $-6x = 33$

25. $\frac{x}{-4} = 12$ **26.** $\frac{n}{-7} = 14$ **27.** $\frac{1}{6}p = -12$ **28.** $\frac{1}{5}q = -100$

29. $36 = \frac{3}{4}t$ **30.** $18 = \frac{2}{3}w$ **31.** $-\frac{5}{6}y = -3$ **32.** $-\frac{2}{5}z = -20$

B **33.** $\frac{4}{3}a = \frac{2}{3}$ **34.** $\frac{5}{3}b = \frac{1}{2}$ **35.** $-\frac{7}{8}c = \frac{3}{4}$ **36.** $-\frac{5}{8}d = -\frac{1}{3}$

37. $1\frac{1}{5} = \frac{3}{4}g$ **38.** $2\frac{1}{4} = \frac{2}{3}j$ **39.** $-\frac{5}{6}k = -1\frac{1}{3}$ **40.** $-\frac{3}{5}m = -2\frac{1}{3}$

41. $\frac{p}{1.3} = 2.1$ **42.** $\frac{q}{-5.2} = -7.1$ **43.** $-3.2r = 3.52$ **44.** $7.5s = -18$

C **45.** A student tried to solve the equation $x^2 = 4x$ by dividing both sides by x. The result was $x = 4$. The solution, 4, satisfies the first equation but is not the complete solution. What is the other solution? What property of numbers did the student violate when she divided by x?

46. Let k be an unknown constant. What value must k have if the equation $\frac{k}{3}x = 4$ has {10} as its solution set?

47. Let a, b, and c be constants (b not equal to 0). Show that $\frac{bc}{a}$ is a solution of $\frac{a}{b}x - c = 0$ if a is not equal to 0.

■ REVIEW EXERCISES

Simplify. [1–6]

1. $3x + 4 + 4x + 5$ **2.** $2(x + 3) + 3x + 4$

Use a variable expression to answer each question. [1–7]

3. What is the length (in meters) of a rectangle if the length is 4 meters greater than the width w?

4. What is Maria's age (in years) if she is half as old as her x-year-old brother?

5. Suppose that there are 17 dimes and nickels in a pile. How many of the coins are dimes if n of them are nickels?

6. The area of a rectangle is 24 cm². What is the length if the width is w?

3-5 Using Several Properties to Solve Equations

Preview

In business, one of the ways in which merchandise on the store shelves is managed is called the LIFO method. LIFO stands for "Last In, First Out." The merchandise last put on the shelves will be the first sold. In some stores, the packages far in the back may be quite old, while those in the front are fresh.

When you get dressed in the morning, you put on socks and then shoes. At the end of the day, you remove your shoes and then socks. The first thing done (in the morning) is the last thing undone (in the evening).

In this lesson you will learn a "last in, first out" technique for solving certain algebraic equations.

■ LESSON

The addition and multiplication properties of equations can be used together to solve equations. Notice how both properties are used.

$$
\begin{array}{ll}
\textit{Start with the equation} & 2m - 8 = 6 \\
\textit{Add 8 to both sides.} & 2m - 8 + 8 = 6 + 8 \\
\textit{Simplify.} & 2m = 14 \\
\textit{Multiply both sides by } \frac{1}{2}. & \frac{1}{2} \cdot 2m = \frac{1}{2} \cdot 14 \\
\textit{Simplify.} & m = 7 \\
\textit{Answer: } \{7\} &
\end{array}
$$

Notice that 8 was added to both sides, and then both sides were multiplied by $\frac{1}{2}$. That order was determined by the order of operations indicated in this expression:

$$
\begin{array}{cc}
\text{1st} & \text{2nd} \\
\downarrow & \swarrow \\
2m & - \ 8
\end{array}
$$

The order of operations indicates that the equation could have been "built" by starting with this simple equation:

$$m = 7$$

Multiply both sides by 2. $2m = 14$

Subtract 8 from both sides. $2m - 8 = 6$

To solve the equation, "undo" the operations in the reverse order:

Start. $2m - 8 = 6$

Undo the operation, subtracting 8, by adding 8 to both sides:

$$2m - 8 + 8 = 6 + 8$$

Simplify. $2m = 14$

Undo the operation, multiplying by 2, by dividing both sides by 2:

$$\frac{2m}{2} = \frac{14}{2}$$

Simplify. $m = 7$

Think: What is the order of operations?
Then undo the operations in the reverse order.

The operation that will "undo" an operation is called the **inverse** of that operation.

Operation	*Inverse operation*
Adding 3	Subtracting 3 (or adding -3)
Adding -3	Subtracting -3 (or adding 3)
Subtracting 4.3	Adding 4.3 (or subtracting -4.3)
Subtracting -4.3	Adding -4.3 (or subtracting 4.3)
Multiplying by 5	Multiplying by $\frac{1}{5}$ (or dividing by 5)
Multiplying by -5	Multiplying by $-\frac{1}{5}$ (or dividing by -5)
Dividing by 2	Multiplying by 2 $\left(\text{or dividing by } \frac{1}{2}\right)$
Dividing by -2	Multiplying by -2 $\left(\text{or dividing by } -\frac{1}{2}\right)$

Strategy for Success **Using examples**

You usually have to read an example several times. On the second reading, keep thinking about how each step gets closer to the answer.

Example 1   Solve. $\frac{x}{3} - 9 = 21$

Solution

Note the order of operations.   1st   2nd

$$\frac{x}{3} - 9 = 21$$

Perform the inverse operations in reverse order.

Add 9 to both sides. $\frac{x}{3} - 9 + 9 = 21 + 9$

Simplify. $\frac{x}{3} = 30$

Multiply both sides by 3. $3\left(\frac{x}{3}\right) = 3(30)$

Simplify. $x = 90$

Answer   {90}

Check $\frac{x}{3} - 9 = 21$

$$\frac{90}{3} - 9 \overset{?}{=} 21$$

$$30 - 9 = 21 \qquad \text{True!}$$

Example 2   Solve. $\frac{x - 9}{3} = 21$

Solution

Note the order of operations.   1st   2nd

$$\frac{x - 9}{3} = 21$$

Perform the inverse operations in reverse order.

Multiply both sides by 3. $3\left(\frac{x - 9}{3}\right) = 3(21)$

Simplify. $x - 9 = 63$
Add 9 to both sides. $x - 9 + 9 = 63 + 9$
Simplify. $x = 72$

Answer   {72}

Check   The check is left to the student.

Example 3 Solve. $-x - 6 = 5$

Solution

	$-x - 6 = 5$
Add 6 to both sides and simplify.	$-x = 11$
Multiply both sides by −1 and simplify.	$x = -11$

Answer $\{-11\}$

Check The check is left to the student.

Example 4 Solve. $3n - 2 = n + 9$

Solution

	$3n - 2 = n + 9$
Add −n to both sides and simplify.	$2n - 2 = 9$
Add 2 to both sides and simplify.	$2n = 11$
Divide both sides by 2 and simplify.	$n = \dfrac{11}{2}$

Answer $\left\{\dfrac{11}{2}\right\}$

Check The check is left to the student.

■ CLASSROOM EXERCISES

State the inverse operation.

1. Adding -5 **2.** Dividing by 3   **3.** Multiplying by -2

4. Subtracting 5   **5.** Multiplying by $-\dfrac{1}{3}$

$$\begin{array}{cc} ? & ? \\ \downarrow & \downarrow \end{array}$$

6. Indicate the order of operations: $6x + 8 = -5$

7. To solve $6x + 8 = -5$:
  a. What operation should be done first to both sides of the equation?
  b. What operation should be done second to both sides of the equation?

8. To solve $\dfrac{3x + 1}{2} = 8$:
  a. What should be done first?
  b. What should be done second?
  c. What should be done third?

Solve.

9. $2y + 7 = 23$ **10.** $5 + 3a = 14$ **11.** $\dfrac{b}{2} - 7 = 3$ **12.** $-x - 6 = 12$

Remember: When solving an equation, any operation performed on one side of the equation must be performed on the other side. Keep in mind your objectives: first to get the variable term on one side and real numbers on the other side, then to solve for the variable.

When you start your homework, review the examples in the lesson to re-
fresh your memory. If you get "stuck" on an exercise, find an example like
it and study the example.

■ WRITTEN EXERCISES

List the order in which the operations are performed.

$\overset{? \ ?}{\downarrow \downarrow}$

1. $3x - 2$

$\overset{? \ ?}{\downarrow \downarrow}$

2. $5x + 2$

$\overset{?}{\downarrow} \quad \overset{?}{\rceil}$

3. $\dfrac{x + 5}{3} \longleftarrow$

$\overset{?}{\downarrow} \quad \overset{?}{\rceil}$

4. $\dfrac{x - 4}{5} \longleftarrow$

State the operation to be performed first in solving the equation.

5. $5x - 2 = 10$

6. $6x - 8 = 9$

7. $7 + 8n = 20$

8. $6 + 4n = 24$

9. $\dfrac{r}{5} + 7 = 3$

10. $\dfrac{s}{3} + 8 = 5$

11. $\dfrac{x + 5}{3} = 10$

12. $\dfrac{x + 7}{4} = 5$

13. $8 - 2n = 40$

14. $7 - 3n = -2$

15. $3x + 8 = -10$

16. $4x + 4 = -12$

Solve.

17. $5 + 3x = 9$

18. $5 + 4x = 11$

19. $5x - 4 = 14$

20. $2x - 3 = 8$

21. $\dfrac{1}{2}x + 7 = 11$

22. $\dfrac{1}{3}x + 2 = 8$

23. $\dfrac{a}{3} - 5 = 11$

24. $\dfrac{a}{5} - 6 = 10$

25. $\dfrac{x - 7}{5} = 6$

26. $\dfrac{x - 2}{3} = 5$

27. $-2x + 7 = 15$

28. $-3x + 2 = 11$

29. $4a - 3 = 2a + 5$

30. $7a - 5 = 4a + 7$

31. $2a - 3 = 5a - 9$

32. $a + 7 = 3a + 1$

33. $5a + 3 = a - 9$

34. $4a - 2 = a - 14$

35. $7a + 8 = 3a - 6$

36. $6a - 3 = 4a - 10$

37. $4a + 4 = 6a - 8$

38. $2a - 6 = 3a - 4$

39. $4a - 8 = 5a - 6$

40. $2a - 4 = 5a + 8$

41. $6a - 4 = 3a + 8$

42. $2a - 3 = 4a - 7$

43. $6a - 3 = 5a + 2$

44. $2a + 3 = 4a - 1$

45. $5a + 1 = 2a + 3$

46. $7a + 2 = 3a - 1$

List the order in which the operations are performed in evaluating these
expressions.

$\overset{? \ ? \ ?}{\downarrow \downarrow \downarrow}$

B **47.** $3(5 \cdot 4 + 2)$

$\overset{?}{\downarrow} \quad \overset{?}{\downarrow} \quad \overset{?}{\downarrow}$

48. $3 + (10 - 4 \cdot 1)$

$\overset{? \ ? \ ?}{\downarrow \downarrow \downarrow}$

49. $(4 - 2 \cdot 1) \cdot 3$

Solve these equations by performing the inverse operations in the reverse order.

50. $3(x + 4) = 24$ **51.** $5(x - 2) = 10$ **52.** $18(3x + 4) = 18$

53. $4(5 - 2x) = 44$ **54.** $5(x + 3) = 23$ **55.** $3(x - 6) = 20$

56. Let $4(b + 2) = 52$ and $4b + 8 = 52$ be equivalent equations. List the order of operations for each equation, and then solve.

Solve these puzzles by starting with the result and performing the inverse operations in the reverse order.

C **57. Puzzle.** I am thinking of a number. If I add 2 to it, multiply the result by 3, subtract 1 from the product, and divide the difference by 2, I get 10. What number am I thinking of?

58. Puzzle. I am thinking of a number. If I multiply it by 3, add 2 to the product, divide the sum by 2, and then subtract 1, I get 6. What number am I thinking of?

59. Puzzle. I am thinking of a number. If I divide it by 2, subtract 1 from the quotient, add 2 to the difference, and multiply the sum by 3, I get 12. What number am I thinking of?

60. Write an equation for Exercise 57.

61. Write an equation for Exercise 58.

62. Write this equation as a puzzle similar to those in Exercises 57–59. "The number I am thinking of" is represented by n in this equation.

$$3 \cdot \left(\frac{n}{2} + 2\right) - 1 = 20$$

What number am I thinking of?

■ REVIEW EXERCISES

Change to expressions involving only addition.   [2–6]

1. $x - 4 + y$ **2.** $a + b - (c + d)$ **3.** $m - (n - 2)$ **4.** $3x - 5y - x$

Simplify.   [2–8]

5. $4a - 3b - 8b$ **6.** $12 - (2x - 4)$ **7.** $2x - 8 - (3x + 5)$

8. $5m - 6 - (4 - m)$ **9.** $3(2x - 6) - 5(3 - 3x)$ **10.** $-2(x + 5) + 3(2x - 6)$

Indicate whether the two expressions are equivalent.   [1–4]

11. $(a + b)^2$ and $a^2 + b^2$ **12.** $\dfrac{a+1}{b+1}$ and $\dfrac{a}{b}$

State the property illustrated by each equation.   [1–5]

13. $(5 + 0) + 6 = 5 + 6$ **14.** $(5 + 0) + 6 = 5 + (0 + 6)$

3–6 A Shortcut for Solving Some Equations

Preview

Consider the order of operations in this equation:

2nd 1st

↓ ↓

$$5 - 2x = 13$$

The last operation is subtracting $2x$. To solve the equation, first perform the inverse operation, adding $2x$, to both sides.

$$5 - 2x + 2x = 13 + 2x$$
$$5 = 13 + 2x$$

Finish solving the equation. Can you think of a shorter way to solve the equation?

In this lesson you will learn a more efficient way of solving equations like this one.

■ LESSON

If we strictly follow the equation-solving steps given in Section 3–5, we sometimes use more steps than are necessary. For example, follow the steps in solving this equation:

$$4 - 3x = 7$$

a. *Add 3x to both sides.* $4 - 3x + 3x = 7 + 3x$
   Simplify. $4 = 7 + 3x$
b. *Subtract 7 from both sides.* $4 - 7 = 7 + 3x - 7$
   Simplify. $-3 = 3x$

c. *Divide both sides by 3.* $\dfrac{-3}{3} = \dfrac{3x}{3}$

   Simplify. $-1 = x$

Answer: $\{-1\}$

If step (a) does not really seem like a step "forward" to you, you are correct. Here is another way of solving the equation.

$$4 - 3x = 7$$

Consider the equation to be $4 + (-3x) = 7$
a. *Subtract 4 from both sides and simplify.* $-3x = 3$
b. *Divide both sides by −3 and simplify.* $x = -1$

Answer: $\{-1\}$

Example 1 Solve. $9 - 5x = 7$

   Solution

Consider the equation to be	$9 + (-5x) = 7$
Add -9 to both sides and simplify.	$-5x = -2$
Divide both sides by -5 and simplify.	$x = \dfrac{2}{5}$

   Answer $\left\{\dfrac{2}{5}\right\}$

   Check $9 - 5x = 7$

$$9 - 5\left(\frac{2}{5}\right) \stackrel{?}{=} 7$$

$$9 - 2 = 7 \quad \text{True!}$$

Example 2 Solve. $7 - \dfrac{w}{2} = 10$

   Solution $\qquad\qquad\qquad\qquad\qquad\qquad 7 - \dfrac{w}{2} = 10$

   a. *Subtract 7 from both sides and simplify.* $\qquad -\left(\dfrac{w}{2}\right) = 3$

   b. *Multiply both sides by -1 and simplify.* $\qquad \dfrac{w}{2} = -3$

   c. *Multiply both sides by 2 and simplify.* $\qquad w = -6$

   Answer $\{-6\}$

   Check The check is left to the student.

In Example 2, if you remember that $-\left(\dfrac{w}{2}\right)$ is equivalent to $\dfrac{w}{-2}$, you can combine steps (b) and (c) into one step:

   b. *Multiply both sides by -2* $\qquad -2\left[-\left(\dfrac{w}{2}\right)\right] = -2(3)$
   and simplify. $\qquad\qquad\qquad\qquad\qquad w = -6$

Example 3 Solve. $-3 - \dfrac{y}{4} = 7$

   Solution $\qquad\qquad\qquad\qquad\qquad\qquad -3 - \dfrac{y}{4} = 7$

   Add 3 to both sides and simplify. $\qquad -\dfrac{y}{4} = 10$

   Multiply both sides by -4 and simplify. $\qquad y = -40$

   Answer $\{-40\}$

   Check The check is left to the student.

> If you have to miss algebra class, call a friend and find out what the assign-ment was. If possible, try to make up the assignment before you come back to class.

■ CLASSROOM EXERCISES

Rewrite each equation as an addition.

1. $2 - x = 3$

2. $5 - 2a = 4$

3. $8 - \dfrac{w}{4} = 12$

4. $7 - \dfrac{b}{5} = 8$

State the first step in solving each equation.

5. $3 - \dfrac{1}{4}w = 9$

6. $-1 - y = 5$

7. $\dfrac{2}{3} - 2a = 1$

8. $0.4 - 0.1x = 1.2$

Solve.

9. $5 - 3x = 14$

10. $-7 - 5b = 1$

11. $4 - \dfrac{x}{2} = 5$

12. $9 - \dfrac{a}{3} = 7$

■ WRITTEN EXERCISES

Solve. Check your solutions.

A

1. $2 - 3x = 8$

2. $5 - 3x = 8$

3. $4 - 2a = 2$

4. $6 - 2a = 2$

5. $3 - 5c = 2$

6. $4 - 2x = 1$

7. $5x - 3 = 7$

8. $4x - 2 = 5$

9. $3 + \dfrac{x}{2} = 4$

10. $5 - \dfrac{y}{2} = 7$

11. $8 - \dfrac{x}{3} = 4$

12. $7 - \dfrac{x}{3} = 5$

13. $\dfrac{a}{4} - 7 = 3$

14. $\dfrac{m}{3} - 5 = 2$

15. $-8 - \dfrac{n}{2} = 5$

16. $-3 - \dfrac{y}{3} = 5$

17. $-3 + 2x = 8$

18. $-5 + 3a = 7$

19. $-4 - \dfrac{2x}{3} = -2$

20. $5 - \dfrac{3x}{4} = -4$

Solve.

B

21. $2(3 - x) = 4$

22. $3(1 - y) = 6$

23. $2(6 - 2x) = 4$

24. $3(6 - 2y) = 6$

25. $\dfrac{1}{2}(4 - 6a) = 6$

26. $\dfrac{(8 - 2a)}{2} = 4$

27. $\dfrac{2}{3}(6 - y) = 8$

28. $\dfrac{3}{4}(8 - 4y) = 7$

29. $2x + 7 = 12$

30. $4a + 5 = 9$

31. $6b - 3 = 9$

32. $8b - 4 = -16$

33. $-2(2x + 7) = 16$

34. $3(2x + 2) = 12$

35. $5(3 - 5x) = -10$

36. $6(4 - 3y) = 18$

37. $5x - 2 = 4x + 5$

38. $6a - 5 = 4a + 2$

39. $7y + 4 = 3 - 2y$

40. $4 - 8a = 9 - 3a$

Solve these absolute value equations. If there is no solution, write ∅.

> **Sample**   Solve. $|x| + 2 = 12$
>
> *Solution* $|x| + 2 = 12$
> $|x| = 10$
>
> *Answer* $\{-10, 10\}$

C **41.** $|x| + 4 = 7$ **42.** $|y| + 5 = 3$ **43.** $5 - |z| = 1$ **44.** $2|x| - 4 = 18$

   45. $8 - |2a| = 3$ **46.** $8 - |3a| = 12$ **47.** $-|x| + 3 = -2$ **48.** $x + |x| = 0$

■ REVIEW EXERCISES

State the property illustrated by each equation.   [2–7]

1. $2 + (x - 3) = (x - 3) + 2$ **2.** $(a + b) = 1(a + b)$

3. $(a - b) + -(a - b) = 0$ **4.** $6(x + 3) = 6x + 18$

Solve by making a table.   [2–9]

5. Two teams had a contest to determine which one could collect the most newspapers in a paper drive. The Blue Team started with 2500 pounds and collected 100 pounds each day until the contest ended. The Red Team started with 1000 pounds, but they collected 200 pounds each day until the contest ended. How long did the contest last if the two teams were tied at the end?

Self-Quiz 2

3–4   Solve.

   1. $6x = 42$ **2.** $-2y = 18$

   3. $\dfrac{d}{5} = 17$ **4.** $\dfrac{2}{3}c = 66$

3–5   Solve.

   5. $3a - 4 = 17$ **6.** $2x + 3 = x + 12$

   7. $\dfrac{z + 6}{3} = 5$ **8.** $\dfrac{t}{4} - 9 = 11$

3–6   Solve.

   9. $12 - 3x = 8$ **10.** $4 - \dfrac{a}{2} = 7$

   11. $-5 - \dfrac{2y}{3} = 1$ **12.** $2(6 - t) = -10$

EXTENSION Checking the solution on a calculator _____

For many complicated equations, a calculator can be helpful in checking the solution. However, be careful to notice how your calculator carries out calculations.

Example 1. $\dfrac{x - 12}{3} = 16$

    Answer: {60}

      Check: Calculator Key Steps   Display

60 − 12 =	48.
÷ 3 =	16.

The key sequence 60 − 12 ÷ 3 = may give the result 56 on some calculators. Those calculators perform the operations using the algebraic order of operations.

Example 2. $4 - 3x = -71$

    Answer {25}

      Check Rewrite the equation. $4 + (-3)x = -71$
             Enter the answer and then "build" the equation.

25 × 3 $\boxed{\text{CHS}}$ =	−75.
+ 4 =	−71.

Check the answer to each equation with a calculator. If the answer is wrong, find the correct answer and check.

1. $5x + 12 = 87$; {15}

2. $4x + 19 = -17$; {−9}

3. $\dfrac{x}{3} - 12 = 16$; {81}

4. $\dfrac{2x}{5} + 10 = 26$; {45}

5. $7(x + 17) = 35$; {−12}

6. $\dfrac{x + 16}{-4} = -16$; {40}

7. $59 - 6x = 38$; {3.5}

8. $\dfrac{35 - 2x}{3} = 7$; {7}

9. $26 - 3x = 14$; {4}

10. $\dfrac{45 - 2x}{5} = 5$; {10}

Strategy for Success Using test results _____

Sometimes students keep repeating an error they made earlier. To help prevent this, make notes about any errors you have made when a quiz or a test is returned to you. Keep them and review them before taking another test.

3-7 Solving Equations

This is the epitaph of Diophantus, one of the greatest mathematicians of the Greek civilization, who lived in Alexandria, Egypt, around the 3rd century.

This tomb holds Diophantus. Ah, what a marvel! And the tomb tells scientifically the measure of his life. God vouchsafed that he should be a boy for the sixth part of his life; when the twelfth was added, his cheeks acquired a beard; He kindled for him the light of marriage after a seventh more, and in the fifth year after his marriage He granted him a son. Alas! late begotten and miserable child, when reached the measure of half his father's life, the chill grave took him. After consoling his grief by this science of numbers for four years, he reached the end of his life.

If x is the number of years of Diophantus's age when he died, we can write the equation from the epitaph:

$$\frac{1}{6}x + \frac{1}{12}x + \frac{1}{7}x + 5 + \frac{1}{2}x + 4 = x$$

After this lesson you can solve this equation and find out how many years Diophantus lived.

■ LESSON

When confronted with a more complicated equation, we must examine its structure and plan the method to use in solving it. We usually have more than one choice of method. We base our choice on the number of steps, the difficulty of the arithmetic, and the likelihood of making errors. Consider this equation, for example.

$$2(a + 4) = -3$$

There are two methods that most people would consider. Some would first expand the left side; others would first multiply both sides by $\frac{1}{2}$.

Expand first:

$$2(a + 4) = -3$$

Expand. $2a + 8 = -3$

Add -8. $2a = -11$

Multiply by $\frac{1}{2}$. $a = -\frac{11}{2}$

Use the multiplication property of equations first:

$$2(a + 4) = -3$$

Multiply by $\frac{1}{2}$. $a + 4 = -\frac{3}{2}$

Add -4 *and simplify.* $a = -\frac{11}{2}$

The method of expanding has the advantage of avoiding the use of fractions until the last step. The method of immediately reversing the order of operations has the advantage of taking one fewer step. The choice one makes is a personal one. The important points are to study the equation carefully, consider various possibilities, and choose the method that is easiest.

Example 1   Solve. $4(y - 3) = 6 + y$

Solution   Since there are variables in different terms and those terms must eventually be combined, it is probably easiest to expand first and then use the addition and multiplication properties of equations.

$$4(y - 3) = 6 + y$$

Expand. $4y - 12 = 6 + y$
Add $-y$ to both sides and simplify. $3y - 12 = 6$
Add 12 to both sides and simplify. $3y = 18$
Divide both sides by 3 and simplify. $y = 6$

Answer $\{6\}$

Check $4(y - 3) = 6 + y$
$4(6 - 3) \overset{?}{=} 6 + 6$
$4(3) = 12$ True!

Example 2   Solve. $4m - 3(m + 4) = m - 2 + 5m$

Solution   Since the variable occurs in more than one term on each side, expand and simplify each side of the equation first. Then use the properties of equations to solve.

$$4m - 3(m + 4) = m - 2 + 5m$$

Expand the left side. $4m - 3m - 12 = m - 2 + 5m$

Combine like terms on each side. $m - 12 = 6m - 2$

Add $-6m$ to each side and simplify. $-5m - 12 = -2$

Add 12 to each side and simplify. $-5m = 10$

Divide each side by -5 and simplify. $m = -2$

Answer $\{-2\}$

Check $4m - 3(m + 4) = m - 2 + 5m$
$4(-2) - 3(-2 + 4) \overset{?}{=} -2 - 2 + 5(-2)$
$-8 - 3(2) \overset{?}{=} -4 - 10$
$-8 - 6 \overset{?}{=} -4 - 10$
$-14 = -14$ True!

Some people want to avoid having to divide by a negative number in solving equations. They would prefer this alternative solution of Example 2:

$$4m - 3(m + 4) = m - 2 + 5m$$

Expand the left side.	$4m - 3m - 12 = m - 2 + 5m$
Combine like terms on each side.	$m - 12 = 6m - 2$
Add $-m$ *to each side and simplify.*	$-12 = 5m - 2$
Add 2 to each side and simplify.	$-10 = 5m$
Divide both sides by 5 and simplify.	$-2 = m$

Answer: $\{-2\}$

■ CLASSROOM EXERCISES

Describe the first step you would use in solving each equation. State the reason for your choice. Do not solve.

1. $5(x - 6) = 40$ **2.** $5(x - 6) = 41$ **3.** $2y + 6y = 15 - 2y + 8$

Solve.

4. $3(x + 1) = 36$ **5.** $5(a - 4) = 39$ **6.** $3x - 3(5 - x) = 40$

7. $2y + 8 - 3y = 24$ **8.** $5b = 2(3b - 8)$ **9.** $2(4x + 1) = 3(4 + 2x)$

■ WRITTEN EXERCISES

Solve.

A **1.** $5(a - 3) = 20$ **2.** $3(a - 6) = 15$ **3.** $2(b + 4) = 20$

4. $4(b + 7) = 12$ **5.** $3(c - 6) = 8$ **6.** $5(c - 4) = 12$

7. $6d + 3(d + 2) = 30$ **8.** $3d + 6(d + 2) = 30$ **9.** $5(k + 3) = 2(k + 10)$

10. $6(k + 3) = 4(k + 7)$ **11.** $6(n - 4) = 3n$ **12.** $5(n - 3) = 2n$

13. $5(2p + 3) = 50$ **14.** $2(5p + 3) = 50$ **15.** $2 - 3(x + 4) = 8$

16. $3 - 2(x + 4) = 8$ **17.** $10 = 3(5 - x)$ **18.** $20 = 5(3 - x)$

19. $2(3 - 2x) = 3(4 - x)$ **20.** $3(2 - 3x) = 4(3 - x)$

B **21.** $2x + 3x + 5x - 4x - x = 2x + 10$ **22.** $6x + 3x - 7x + 5x - 4x = x + 10$

23. $5x + \frac{1}{2}(x - 4) = 3(x + 2) - 6x + 3$ **24.** $7x + 4(x - 5) + 2(x + 5) = 10x + 1$

25. $x + 2(x - 1) = 3x$ **26.** $5x - x + 2(3 + x) = 0$

27. $(x - 2) - 2(x - 3) = 2x - 5$ **28.** $6x - 3(2 - x) = -6 + x$

29. $\frac{1}{2}(a + 4) = 5$ **30.** $\frac{1}{3}(a + 12) = 8$

31. Solve $2(1 - t) = 3$ two ways: (a) by first expanding the left side of the equation; (b) by first using the multiplication property of equations. Which method do you prefer and why?

Solve.

C **32.** $\dfrac{2(a + 4)}{3} = 5$ **33.** $\dfrac{3(x + 2)}{4} = 4$

34. $\left(x - \dfrac{2}{3}\right)6 = \left(x - \dfrac{3}{4}\right)8$ **35.** $\left(x + \dfrac{2}{3}\right)6 = \left(x + \dfrac{3}{4}\right)8$

36. $2[3 + 4(x + 5)] = 5[4 + 3(x + 2)]$ **37.** $4[3 - 2(x + 5)] = 5[2 + 4(x - 3)]$

38. $\dfrac{1}{3}(6 - 3a) = -12$ **39.** $2x = -\dfrac{1}{2}(4x - 8)$

40. $4(n - 7) - 2(1 - 3n) = 6n$ **41.** $\dfrac{1}{2}(4b + 1) + \dfrac{1}{4}(6b - 2) = 7$

42. $x - 5(x + 2) = x + 3(3 - 2x)$ **43.** $2(5a - 4) - 3(a - 5) = 8(2a - 7)$

44. Solve the equation in the Preview to determine how old Diophantus was at his death.

■ REVIEW EXERCISES

Write each of the following in algebraic symbols. [2-1]

1. The absolute value of negative three equals the absolute value of three.

2. The opposite of negative five equals the absolute value of negative five.

Simplify. [2-2]

3. $-3\dfrac{1}{2} + \left(-4\dfrac{1}{2}\right)$ **4.** $-4.1 + 5.7$

Simplify. [2-3]

5. $-12 - (-10)$ **6.** $45 - 63$

Simplify. [2-4]

7. $-5(25)$ **8.** $-12\left(-\dfrac{3}{4}\right)$

Simplify. [2-5]

9. $-27 \div 3$ **10.** $-18 \div -\dfrac{2}{3}$

Strategy for Success **Doing homework** _____

> Make a note in the margin of your homework paper of anything that you don't understand or cannot do. Then you can ask these questions during your next class.

3-8 Solving Literal Equations

Preview

The human brain has a remarkable capacity to store and recall information. For example, the average person uses approximately 25,000 words. However, most people prefer to memorize as few facts as necessary. Why try to remember two facts if one will do!

In this lesson we will see how one formula can be used to generate other formulas. Then, for example, you can memorize just one of the three equivalent distance-rate-time formulas and derive the other two whenever you need them: $d = rt \quad r = \dfrac{d}{t} \quad t = \dfrac{d}{r}$

■ LESSON

This formula relates temperature measurements on the Fahrenheit *(F)* scale to those on the Celsius *(C)* scale:

$$F = \frac{9}{5}C + 32$$

We can use the formula to change measurements from Celsius to Fahrenheit. For example, to change $10°C$ to Fahrenheit, substitute 10 for *C*:

$$F = \frac{9}{5}(10) + 32$$

$$F = 18 + 32$$
$$F = 50 \qquad \text{Thus, } 10°C \text{ is equivalent to } 50°F.$$

The same formula can be used to change $23°F$ to Celsius.

$$23 = \frac{9}{5}C + 32$$

$$-9 = \frac{9}{5}C$$

$$-5 = C \qquad \text{Thus, } 23°F \text{ is equivalent to } -5°C.$$

Notice that it is easier to use the formula $F = \dfrac{9}{5}C + 32$ to change from degrees Celsius to degrees Fahrenheit than to change from degrees Fahrenheit to degrees Celsius. That is because the formula is in the form "$F = \underline{\ \ ?\ \ }$." If we had many Fahrenheit temperatures to change to Celsius temperatures, we would want to have a formula in the form "$C = \underline{\ \ ?\ \ }$." We can change from one form to the other by using equation-solving skills.

Start with the equation	$F = \dfrac{9}{5}C + 32$
Add -32 to both sides and simplify.	$F - 32 = \dfrac{9}{5}C$
Multiply both sides by $\dfrac{5}{9}$ and simplify.	$\dfrac{5}{9}(F - 32) = C$
Use the symmetric property of equations.	$C = \dfrac{5}{9}(F - 32)$

Formulas and other equations containing more than one variable are called **literal equations.**

$$i = prt \qquad ab = c + d \qquad A = bh \qquad A = \frac{1}{2}h\,(b_1 + b_2)$$

A literal equation can be solved for any of its variables. For example, when the equation $i = prt$ is written in the form $p = \dfrac{i}{rt}$, it has been solved for p. Literal equations are solved using the sames properties that were used to solve other equations.

Example 1   Solve $V = lwh$ for w.

    Solution $V = lwh$

          Multiply both sides by $\dfrac{1}{lh}$. $\dfrac{V}{lh} = \dfrac{lwh}{lh}$

          Simplify. $\dfrac{V}{lh} = w$

          Use the symmetric property of equations. $w = \dfrac{V}{lh}$

Example 2

The formula for the area of a trapezoid is given below:

$$A = \frac{1}{2}h(b_1 + b_2)$$

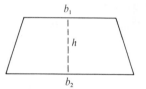

In this formula, h is the length of the altitude and b_1 and b_2 are the lengths of the two bases. The 1 and 2 in b_1 and b_2 are **subscripts** that indicate that b_1 and b_2 are related to each other (they are both lengths of bases) but are different variables. Subscripts should not be confused with exponents: b_2 means the length of the lower base of the trapezoid shown above, while b^2 means $b \cdot b$.

  Solve the formula for b_1.

    Solution $A = \dfrac{1}{2}h(b_1 + b_2)$

          Multiply both sides by $\dfrac{2}{h}$. $\dfrac{2}{h} \cdot A = \dfrac{2}{h} \cdot \dfrac{1}{2}h(b_1 + b_2)$

          Simplify the right side. $\dfrac{2}{h} \cdot A = b_1 + b_2$

          Simplify the left side. $\dfrac{2A}{h} = b_1 + b_2$

          Add $-b_2$ *to both sides.* $\dfrac{2A}{h} - b_2 = b_1$

          Use the symmetric property of equations. $b_1 = \dfrac{2A}{h} - b_2$

■ CLASSROOM EXERCISES

1. Explain why it is necessary to memorize only one of these four formulas.

$$i = prt \qquad p = \frac{i}{rt} \qquad r = \frac{i}{pt} \qquad t = \frac{i}{pr}$$

2. Solve for h. $\quad A = bh$ **3.** Solve for b. $\quad A = \frac{1}{2}bh$ **4.** Solve for n. $\quad C = 2n + 5$

■ WRITTEN EXERCISES

Solve the literal equation for n.

A   **1.** $an = 7$ **2.** $pn = -2$ **3.** $\frac{n}{r} = 0.2$ **4.** $\frac{n}{x} = \frac{3}{4}$

     5. $a + n = 15$ **6.** $c - n = 45$ **7.** $2n - j = 10$ **8.** $10n - t = 5$

Solve the literal equation for the indicated variable.

9. $ab + 2 = c$,   **10.** $ab - 5 = c$,   **11.** $pq = r + 3$,   **12.** $pq = 8 - s$,
     for a for b for p for q

13. $3q = r + s$,   **14.** $3q = r + s$,   **15.** $3x + 4y = 12$,   **16.** $3x + 4y = 12$,
     for r for s for y for x

17. $5x + 10 = 10y + 20$, for x **18.** $5x + 10 = 10y + 20$, for y

19. $y = \frac{1}{2}x + 3$, for x **20.** $y = \frac{1}{3}x + 4$, for x

21. $2x - 3y = 18$,   **22.** $3x - 2y = 18$,   **23.** $3x_1 + 4x_2 = 24$,   **24.** $3x_1 + 4x_2 = 24$,
     for y for y for x_2 for x_1

B   **25.** $rs + t = w$,   **26.** $a(b + c) = d$,   **27.** $ax + by = z$,   **28.** $x_1x_2 = x_3x_4$,
     for r for c for x for x_3

29. A formula for the area of a trapezoid is $A = \frac{1}{2}h(b_1 + b_2)$.

     a. Solve for h.

     b. Find h when $A = 240$, $b_1 = 11$, and $b_2 = 21$.

30. A formula for the expansion of a steel rod is $l = l_o(l + at)$.

     a. Solve for t.

     b. Find t when $l = 11$, $l_o = 10$, and $a = 2$.

Solve each formula for x_1 and x_2.

C   **31.** $x_1x_3 + x_2 = 10$ **32.** $x_1(x_2 - x_3) = 10$ **33.** $x_1x_3 = 10x_2$

34. $\frac{x_1}{x_3} = x_2 + 10$ **35.** $x_1 - x_2x_3 = 10$ **36.** $\frac{x_1x_3}{12} + \frac{x_2}{6} = 1$

■ REVIEW EXERCISES

Simplify.

1. $5x + (-10) - (-3x)$ **2.** $-3a + (-4b) - (-5a)$ [2–6]

3. $7x + 5y - 6x - y$ **4.** $4r - 5s - 6r + 12s$ [2–7]

Puzzle

I am thinking of a number. If 8 is multiplied by twice the number decreased by 4, the result is 16. What number am I thinking about?

You may find the number by solving the equation $8(2x - 4) = 16$. In this lesson you will write equations that fit word problems.

■ LESSON

Here is a "common sense" model for problem solving. The model has just three steps, but each step is very important.

1. *Study the problem until you understand it.*

Upon first reading a problem, you may not understand it. Reread it—perhaps several times. Sometimes reading softly aloud helps. You may have to find the meaning of key words that are unfamiliar to you. Perhaps the relationships in the problem may not be clear. Drawing pictures, diagrams, and figures can often help you organize your thoughts and understand what is being asked and what information is given.

2. *Decide what to do and then do it.*

You may decide to write and solve an equation. Perhaps the problem simply requires choosing the correct operation. Systematic trials can be used to solve some problems. Sometimes lists, tables, or examples from the text can be helpful. If the first method you try does not work, then try something else.

3. *Answer the question and check your answer.*

Think about your understanding of the problem, the method you used to solve it, and the calculations you made. Ask yourself, "How sure am I that my answer is correct?" Estimate to determine whether your answer is reasonable. If your answer seems wrong solve the problem in a different way. Check your arithmetic. Don't be satisfied with your first answer.

Thomas Edison once said: "Genius is 10% inspiration and 90% perspiration." The same can be said of problem solving. Hard work is required to solve real problems.

You may be able to solve some word problems without writing equations. However, it is important to develop the skill to translate word problems into algebraic equations. This skill is essential when you encounter problems that cannot be solved in any other way.

Example 1 Write an equation and solve the problem.
Eric is 1.5 times as old as Aretha.
If Eric is 21 years old, how old is Aretha?

Solution Since you want Aretha's age, it makes sense to do this:

Let a = Aretha's age in years.

Then $1.5a$ = Eric's age in years.

The problem indicates that Eric is 21 years old. There are two ways to write Eric's age in years:

$$1.5a \text{ and } 21$$

This gives the equation:

$$1.5a = 21$$
$$a = 14$$

Answer Aretha is 14 years old.

Example 2 Write an equation and solve the problem.
Nancy and Carmen are saving money to buy a motorcycle. Their combined savings is $180. Carmen has saved $10 more than Nancy. How much has each saved?

Solution 1 Let n = Nancy's savings in dollars.
Then $n + 10$ = Carmen's savings in dollars.

Nancy's savings	plus	Carmen's savings	is	180
n	$+$	$(n + 10)$	$=$	180
		$2n + 10$	$=$	180
		$2n$	$=$	170
		n	$=$	85

Answer Nancy saved $85 and Carmen saved $95.

Check Have they saved $180?
$95 + $85 = $180. Yes.
Did Carmen save $10 more than Nancy?
$95 = $10 + $85. Yes.

Solution 2 Let c = Carmen's savings in dollars.
Then $c - 10$ = Nancy's savings in dollars.

$$c + (c - 10) = 180$$
$$2c - 10 = 180$$
$$2c = 190$$
$$c = 95$$

Answer Carmen saved $95 and Nancy saved $85.

Note that the answers are checked by referring to the problem itself not just to the equation. The equation may have been solved correctly, but it may be the wrong equation.

Example 3 A rectangle is 2 times as long as it is wide. Its perimeter is 24 cm. What are the two dimensions?

Solution 1

Let w = width in centimeters.
Then $2w$ = length in centimeters.
Perimeter: 24 cm and $(w + 2w + w + 2w)$cm
Equation: $w + 2w + w + 2w = 24$
$$6w = 24$$
$$w = 4$$

Answer The width is 4 cm and the length is 8 cm.

Solution 2

Let l = length in centimeters.
Then $\frac{1}{2}l$ = width in centimeters.
Perimeter: 24 and $l + \frac{1}{2}l + l + \frac{1}{2}l$

Equation: $l + \frac{1}{2}l + l + \frac{1}{2}l = 24$
$$3l = 24$$
$$l = 8$$

Answer The length is 8 cm and the width is 4 cm.

Note in both Example 2 and Example 3 that the two solutions use different variables and different equations, and that the equations have different solutions. However, the answers to the original problems are the same.

■ CLASSROOM EXERCISES

Carry out these directions for each problem.

    a. Represent one number by a variable.
    b. Write expressions for other numbers using the variable.
    c. Give an equation.
    d. Solve the equation.
    e. Answer the question.
    f. Check the solution.

1. Sidney is 4 years younger than Marcia. Sidney is 17 years old. How old is Marcia?

2. Tracy scored 30 points. That was twice as many points as Kevin scored. How many points did Kevin score?

3. A rectangle is 4 times as long as it is wide. Its perimeter is 50 cm. What are the dimensions?

4. Joan and Carrie together scored 26 points. Joan scored 6 points more than Carrie. How many points did each score?

■ WRITTEN EXERCISES

Let w = width of the rectangle in centimeters. Write an equation that fits each problem. Do *not* solve the equation.

A **1.** Half the width of a rectangle is 30 cm. What is the width?

2. One third the width of a rectangle is 30 cm. What is the width?

3. If the width of a rectangle is increased by 5 cm, it will be 12.3 cm wide. What is the width?

4. If the width of a rectangle is decreased by 5 cm, it will be 12.3 cm wide. What is the width?

5. The length of a rectangle is 1.3 times the width. The sum of the length and width is 10.35 cm. What is the width?

6. The length of a rectangle is 2.8 times the width. The sum of the length and width is 133 cm. What is the width?

7. The length of a rectangle is 5 cm more than the width. If the perimeter is 78 cm, what is the width of the rectangle?

8. The length of a rectangle is 3 cm more than the width. If the perimeter is 96 cm, what is the width of the rectangle?

Let x = the number of dimes in Kari's collection. Write an equation that fits each problem. Do *not* solve.

9. Brad has 15 more dimes in his collection than Kari. Together they have 237 dimes. How many dimes does Kari have?

10. Kevin has 15 fewer dimes in his collection than Kari. Together they have 237 dimes. How many dimes does Kari have?

11. The value of the dimes in Kari's collection is $11.20. How many dimes does Kari have?

12. The value of the dimes in Kari's collection is $8.70. How many dimes does Kari have?

13. If Kari added 12 more dimes to her collection, she would have 200 dimes. How many dimes are in Kari's collection?

14. If Kari added 39 more dimes to her collection, she would have 400 dimes. How many dimes are in Kari's collection?

For each problem, write four things.

    a. What the variable represents (always a *number* of something).
    b. An equation that fits the problem.
    c. The solution to the equation.
    d. The answer to the question (with proper units).

15. James is 8 cm taller than Tom. James is 150 cm tall. How tall is Tom?

16. Sue is 3 cm shorter than Heidi who is 158 cm tall. How tall is Sue?

17. James is 32 pounds lighter than Carla. James weighs 111 pounds. How much does Carla weigh?

18. Lana is 25 pounds heavier than her brother. Her brother weighs 89 pounds. How much does Lana weigh?

B **19.** Mrs. Wilson has saved one-third the cost of the car she plans to buy. How much does the car cost if she saved $2715?

20. The combined weight of Terri and Keith is 230 pounds. Keith weighs 20 pounds more than Terri. How much does Terri weigh?

21. Gerry and Sandy bought a bicycle for $110. Sandy paid $10 more of the cost than Gerry. How much did each pay?

22. A rectangle is 3 m longer than it is wide. The perimeter is 34 m. What are the dimensions of the rectangle?

23. A rectangle is 3 times as long as it is wide. The perimeter is 48 m. What are the dimensions of the rectangle?

C **24.** Lyle bowled two games. His score in the first game was 18 higher than his second game score. The total was 226. What were his scores in the two games?

25. Connie has a record collection. Joel has half as many records as Connie. Together they have 54 records. How many records does Connie have?

26. Greg has some money in a savings account. After the bank added interest to his account equal to 0.05 times the amount he had in savings, he had $126. How much was in his account before the interest was added?

27. A rectangle is 1.4 times as long as it is wide. Its area is 315 cm². What is the width of the rectangle?

■ REVIEW EXERCISES

Simplify. [2-8]

  1. $15 - 2(2a - 4)$
                           2. $3(x - 5) - 2(2x + 5)$

  3. $2x - (5 - 2x) + (3 + 3x)$

Solve. [3-7]

  4. $2m - 4 = 5m + 12$
                           5. $2(3n - 5) = 14$

  6. $2(3x - 1) = 4(2 - x)$

Self-Quiz 3

3-7 Solve.

1. $4(n - 2) = 2n$
2. $32 - 3a = 17$
3. $3(x - 1) + 2x = 12$
4. $3n - 31 = 7 - n$
5. $9 - 2(x + 1) = 0$
6. $5(6 - x) = 2(7 - 2x)$

3-8 Solve the equation for the indicated variable.

7. $E = IR$, (Ohm's Law) for I
8. $A = 7 + 8t$, for t

3-9 Using the given variable, write an equation and solve.

9. The perimeter of a square is 84 cm. Find the length (l) of one side.

10. Drew is four years older than Mike (M). Their ages total 24 years. How old is each?

EXTENSION   Evaluating formulas on a computer

An equation in a computer program is a definition of a variable. For example, this equation

```
P = 2 * L + 2 * W
```

defines P in terms of L and W. Given values of L and W **(inputs)**, the computer will compute corresponding values of P **(output)**. If values of P and W are inputs, the computer cannot compute values of L using the above equation. The computer needs an equation such as:

```
L = P/2 - W
```

in order to find values of L.

When you write equations in a computer program, you should always be sure that the equation is solved for the desired output variable in terms of the given input variables.

Write equations for a computer to use given the following inputs and outputs.

1. Inputs: the length and width of a rectangle
   Output: the area of the rectangle

2. Inputs: A, B, C, and D for the equation AX + B = CX + D
   Output: X

3. Is there a Symmetric Property of Equations for equations in a computer program? (See page 93).

3–10 Problem Solving—Drawing Figures

Preview **History**

Archimedes (287 ?–212 B.C.) is considered one of the greatest mathematicians who ever lived. He lived in Syracuse, Sicily at the time the city was captured by the Romans. Archimedes was intently studying a geometric figure drawn in the sand when he was killed by a Roman soldier.

Mathematicians and mathematics students long ago discovered the value of pictures and diagrams in solving problems. This lesson will help you develop skill in using figures to solve problems.

■ LESSON

A good way to help yourself understand a problem, assign variables, and write equations is to draw pictures of the situation described in the problem. The pictures don't need to be "artistic." They merely should show the important relationships.

Example 1   A rectangle is 30 cm longer than it is wide. What are its dimensions if its perimeter is 88 cm?

Solution   A picture for this problem can help you set up an equation.

width $= w$

length $= w + 30$

           (... "30 cm longer than it is wide")

The perimeter is 88 cm. The sum of the lengths of the sides (2 lengths plus 2 widths) also represents the perimeter. Therefore,

$$88 = l + w + l + w$$
$$= (w + 30) + w + (w + 30) + w$$
$$88 = 4w + 60$$
$$28 = 4w$$
$$7 = w$$

Answer   The width of the rectangle is 7 cm and the length is 37 cm.

Check   The check is left to the student.

Example 2   Harry and Juan bought a hang glider for $900. Harry paid one half as much as Juan. How much did each pay?

Solution   You can represent the amount they each spent by segments.

Juan   ├─────────────────────────────┤
Harry ├──────────────┤

The amount they spent *together* can be represented by joining segments end-to-end.

```
            Juan                    Harry
├──────────────────────────┼──────────────────┤
├──────────────────────────────────────────────┤
            900 dollars
```

Now pick a variable for one dollar amount. Write a variable expression for the other dollar amount.

```
            Juan                    Harry
├──────────────────────────┼──────────────────┤
        x dollars          1/2 x dollars
├──────────────────────────────────────────────┤
            900 dollars
```

Two ways of writing the total length gives an equation.

$$x + \frac{1}{2}x = 900$$

$$\frac{3}{2}x = 900$$

$$x = 600$$

Answer   Juan spent $600, and Harry spent $300.

Check   The check is left to the student.

■ CLASSROOM EXERCISES

1. Draw a picture for this situation.
   You can drive straight from Cedar City to Nestone, a distance of 10 miles, or you can drive straight east for 8 miles and then straight north for 6 miles.

2. Draw a picture for this problem.
   James scored 4 points fewer than Wes. Together they scored 40 points. How many did each score?

■ WRITTEN EXERCISES

Draw and label a figure to organize the information in these problems. Do *not* solve the problems.

A

1. A rectangle is 20 cm longer than it is wide. What is its length if its perimeter is 92 cm?

2. The width of a rectangle is 10 cm less than its length. What is its width if its perimeter is 52 cm?

3. A recipe calls for 50 grams more flour than sugar. How much of each are in 380 grams of a flour-sugar mixture?

4. A recipe calls for 500 mL (milliliters) more apple juice than cranberry juice. How much of each are in 1000 mL of the apple-cranberry juice mixture?

5. Ted is 23 years older than Theresa. Vin is 2 years younger than Theresa. How old is Theresa if the sum of their ages is 60 years?

6. Heather is 3 years older than Steve. Vickie is 20 years older than Steve. How old is Steve if the sum of their three ages is 68 years?

7. One side of a triangle is 10 cm longer than another side. The third side is 12 cm longer than the shortest side. How long is the shortest side if the perimeter is 67 cm? (Draw a triangle. Label one side *x* cm. Label the other two sides in terms of *x*.)

8. One side of a triangle is 1.2 times as long as another side. The third side is 1.5 times as long as the shortest side. How long is the shortest side if the perimeter is 74 cm? (Draw a triangle. Label the shortest side *x* cm. Label the other two sides in terms of *x*.)

For each problem, write five things.

  a. What the variable represents (always a *number* of something).
  b. A figure that helps organize the information.
  c. An equation that fits the problem.
  d. The solution to the equation.
  e. The answer to the question.

9. The length of a rectangular cutting board is 7 cm more than its width. What is the width of the board if its perimeter is 94 cm?

10. The width of a rectangular desk is 8 cm less than its length. What is the length of the desk if its perimeter is 176 cm?

11. Robert and Jonathan bought a band saw for $250. Jonathan paid $80 more than Robert. How much did Robert pay?

12. Melanie and Shawn bought a chain saw for $420. Melanie paid $120 less than Shawn. How much did Melanie pay?

13. The Wildcat football team scored 7 more points in the second half of the game than it did in the first half. How many points did the Wildcats score in the second half if they scored 27 points in the game?

14. The Lynx football team scored 4 more points in the first half of the game than it did in the second half. How many points did the Lynx score in the first half if they scored 24 points in the game?

B 15. One side of a quadrilateral is 4 cm longer than the shortest side. Another side is 5 cm longer than the shortest side. The longest side is 10 cm longer than the shortest side. How long is the shortest side if the perimeter of the quadrilateral is 139 cm? (A quadrilateral is a 4-sided polygon.)

16. One side of a quadrilateral is 10 cm longer than the shortest side. Another side is 20 cm longer than the shortest side. The longest side is 30 cm longer than the shortest side. How long is the shortest side if the perimeter of the quadrilateral is 820 cm?

17. Cliff drove on a straight road from Town A to Town B, a distance of 35 miles. Darlene drove a route which was 7 miles longer. If one leg of her trip was 16 miles long, how long was the rest of her trip?

18. When Julius had driven from home to Town A, and then halfway back home he had covered 126 miles. How far did he live from Town A?

C 19. A carpenter cut a 12-foot board into three unequal lengths. Each of the shorter lengths was $\frac{1}{2}$ foot less than the next larger length. Find the three lengths.

20. Crystal, Sabrina, and Felipe divided $9 unevenly. Crystal received 50¢ less than the greatest amount, and Felipe received 25¢ more than the least amount. How much money did each receive?

21. The length of the sides of a triangle, in centimeters, are consecutive whole numbers. What is the length of the shortest side if the perimeter of the triangle is 96 cm?

22. The lengths of the sides of a quadrilateral, in centimeters, are consecutive whole numbers. What is the length of the longest side if the perimeter of the quadrilateral is 162 cm?

■ REVIEW EXERCISES

Complete.

1. 1 kilogram = __?__ grams

2. 1 liter = __?__ milliliters

3. 1.5 liters = __?__ milliliters

4. 27.3 centimeters = __?__ meters

Simplify.

[1–2]

5. $\dfrac{5 \cdot 2 - 1}{3}$

6. $8 - 4 \cdot 2 + 3$

■ CHAPTER SUMMARY

● **Vocabulary**

● The symmetric property of equations   [3–2]
  For all numbers a and b,
    if $a = b$ then $b = a$.

● Properties of equations
    For all real numbers a, b, and c:
  Addition: If $a = b$, then $a + c = b + c$.   [3–3]
  Multiplication: If $a = b$, then $ac = bc$.   [3–4]

● Steps in solving linear equations   [3–6]
  1. Simplify the expressions on both sides.
  2. Collect the variables on one side.
  3. Use the addition and multiplication properties of equations.

● To *solve* a literal equation for one of its variables, find an   [3–9]
  equivalent equation that has the variable on one side of the
  equation and does not contain the variable on the other side.

■ CHAPTER REVIEW

3–1   **Objective**   To solve an equation by repeated trials.

For each equation below, state which of the numbers $\{-2, -1, 0, 1, 2\}$
are solutions. Write "none" if there is no solution in the domain.

1. $x + 18 = 17$ **2.** $\dfrac{x}{-2} = -1$ **3.** $x^2 + 2 = 3x$

3–2   **Objective**   To write equations that are equivalent to a given equation.

State whether the two equations are equivalent.

4. $3x + 2 = 20$ **5.** $x = 6$

$3x = 22$ $\dfrac{x}{6} = 36$

3–3 Objective To solve equations by adding (or subtracting) the same quantity to (from) both sides.

Solve by writing a series of equivalent equations.

6. $7x = 6x - 5$ **7.** $6x + 5 = 5x + 10$ **8.** $3x - 2\frac{1}{2} = 4x + 3\frac{1}{2}$

3–4 Objective To solve equations by multiplying (or dividing) both sides by the same number.

Solve by writing simpler equivalent equations.

9. $3x = -10$ **10.** $\frac{3}{4}r = \frac{1}{2}$ **11.** $-4y = 13$

3–5 Objective To use several properties of equations to solve equations.

Solve.

12. $6z + 7 = 17$ **13.** $\frac{x}{4} - 10 = 8$

14. $\frac{x + 8}{3} = 12$ **15.** $6x - 3 = 2x + 11$

3–6 Objective To solve equations of the form $a - bx = c$.

Solve.

16. $7 - 2x = -3$ **17.** $5 - \frac{y}{3} = 8$ **18.** $-12 - 3x = 5$

3–7 Objective To solve linear equations in one variable.

Solve.

19. $6(a - 2) = 24$ **20.** $5y + 3(y - 4) = 20$

21. $5(x - 3) = 2(3 - x)$ **22.** $10 - 2(4 - x) = 15$

3–8 Objective To solve a literal equation for any one of its variables.

Solve each literal equation for the variable indicated.

23. $d = rt,$ **24.** $P = 2l + 2w,$ **25.** $y = 3x - 7,$
   for r for w for x

3–9 Objective To solve problems by writing equations.

For each problem define a variable and write an equation. Then solve the equation, and answer the question.

26. A rectangle is 3 times as long as it is wide. The perimeter of the rectangle is 52 m. What are the length and width?

27. Mark is 23 pounds heavier than Joseph. If Mark weighs 151 pounds, how much does Joseph weigh?

3–10 **Objective**   To use drawings to help solve problems.

Draw a figure. Then solve the problem by using an equation.

28. One side of a parallelogram is 5 cm longer than another. If the perimeter of the parallelogram is 70 cm, what are the lengths of the sides?

■ CHAPTER 3 SELF-TEST

Which of the numbers $\{-3, -2, -1, 0, 1\}$ are solutions of these equations?

3–1   **1.** $\dfrac{4.5}{x} - 5 = -50$ **2.** $x^2 = x$

Write the equation that results from following these directions.

3–2   **3.** $5x - 4 = x - 7$ **4.** $-6 - \dfrac{2}{3}y = 4$

a. Subtract x from both sides.   **a.** Add 6 to both sides.
b. Add 4 to both sides.   **b.** Multiply both sides by $-\dfrac{3}{2}$.
c. Divide both sides by 4.

Solve.

3–3   **5.** $12x - 7 = 13x + 4$ **6.** $4t - 2\dfrac{1}{3} = 1\dfrac{2}{3} + 3t$

3–4   **7.** $-9a = 12$ **8.** $\dfrac{b}{-4} = -8$

3–5   **9.** $2x + 15 = 19$ **10.** $\dfrac{1}{2}y - 3 = 1$

11. $\dfrac{w + 4}{2} = 3.3$ **12.** $6(r - 1) = 36$

State what is done to the sides of the equation at each step.

3–6   **13.** Solve. $\dfrac{6 - 2s}{3} = 4$

$$6 - 2s = 12 \qquad \textbf{a.} \underline{\qquad ? \qquad}$$
$$-2s = 6 \qquad \textbf{b.} \underline{\qquad ? \qquad}$$
$$s = -3 \qquad \textbf{c.} \underline{\qquad ? \qquad}$$

Solve.

3–6   **14.** $4 - \dfrac{x}{7} = 8$ **15.** $\dfrac{1}{4}(3 - 2w) = 5$

3–7   **16.** $2(p + 12) + p - 14 = 4$ **17.** $2(k - 2) = 4 - 2k$

3–8   Solve the equation for the indicated variable.

18. $A = 2\pi rh$, for r **19.** $y = 3 + 5x$, for x

3–9, Solve by writing and solving an equation.

3–10 **20.** One side of a triangle is 3 cm longer than the shortest side. The third side is 8 cm longer than the shortest side. What is the length of the shortest side if the perimeter of the triangle is 59 cm?

21. Kim and Karen bought a record for $5.50. Kim paid $5 more than Karen. How much did each girl pay?

■ PRACTICE FOR COLLEGE ENTRANCE TESTS

Choose the one best answer to each question.

1. If $2 \cdot x \cdot 3 = 9$, then $x = \underline{}$.

A. $\dfrac{2}{27}$ **B.** 3   **C.** $1\dfrac{1}{2}$ **D.** 6   **E.** $\dfrac{2}{3}$

2. What is the sum of five consecutive integers if the middle integer is 40?

A. 8   **B.** 40   **C.** 160   **D.** 200   **E.** 220

3. In a sequence of numbers, each number after the first is four more than three times the previous number. If the third number is 7, what is the first number?

A. -1 **B.** $\dfrac{-1}{3}$ **C.** 0   **D.** $\dfrac{1}{3}$ **E.** 1

4. If $4x - 2 = 10$, then $8x = \underline{}$.

A. 8   **B.** 16   **C.** 24   **D.** 32   **E.** 40

5. The operation $\boxed{\times}$ is defined for all numbers a and b as $a \boxed{\times} b = a^2 + ab$. If $3 \boxed{\times} n = 3$, then $n = \underline{}$.

A. -3 **B.** -2 **C.** -1 **D.** 1   **E.** 2

6. If $x - 3 = x + k$, then $k = \underline{}$.

A. $2x - 3$ **B.** $-2x + 3$ **C.** $-2x - 3$ **D.** -3 **E.** 3

7. Erik is two years older than Juanita, and Juanita is twice as old as Zelda. If Zelda is z years old, then an expression for Erik's age in terms of z is $\underline{}$.

A. $\dfrac{z+2}{2}$ **B.** $2(z + 2)$ **C.** $\dfrac{z-2}{2}$ **D.** $2(z + 2)$ **E.** $2z + 2$

8. If $a - b = 5$, then $a - b - 3 = \underline{}$.

A. -8 **B.** 1   **C.** 2   **D.** 4   **E.** 8

9. If a, b, and c are consecutive odd integers and $a < b < c$, then, in terms of c, $a = \underline{}$.

A. $c - 2$ **B.** $c - 3$ **C.** $c - 4$ **D.** $c + 2$ **E.** $c + 4$

10. If $1 - k^2 = m$ and $k = -2$, then $m = \underline{}$.

A. -3 **B.** -1 **C.** 1   **D.** 3   **E.** 5

11. The square piece of paper shown below is folded along the diagonal with point B placed on top of point D to form a triangle. Then the triangle is folded again along the dashed line shown with point C placed on top of point A to form a smaller triangle.

The shaded portion shown in the small section is cut from all four layers of the folded paper. When the paper is unfolded, which of the following diagrams would it look like?

A.   **B.**   **C.**

D.   **E.**

12. If $a + b = 0$, then which of the following must be true?

A. $a^2 = ab$ **B.** $ab = 1$ **C.** $ab = b^2$ **D.** $a - b > 0$ **E.** $a^2 = b^2$

■ CUMULATIVE REVIEW (Chapters 1–3)

1-1

a	B	r	t	y
3	0	9	18	$\frac{2}{3}$

Substitute and simplify.

1. $t - r$ **2.** $y \times r + a$ **3.** $B \div t$ **4.** $\frac{r}{t} + a \times y$

1-2   Simplify.

5. $4 \times 9 + 8$ **6.** $18 - 12 \div 2$ **7.** $9 \times (8 - 2)$ **8.** $\dfrac{30 - 6}{10 + 2}$

1-3   Evaluate for $x = 5$.

9. $6x$ **10.** $3x^2$ **11.** $x - 2^2$

1-4   State whether the expressions in each pair are equivalent.

12. $\dfrac{x + 2}{x + 4}, \dfrac{x + 1}{x + 2}$ **13.** $9a - a, 8a$

1–5 Identify the property illustrated in each equation.

14. $2x \cdot 1 = 2x$

15. $(36 + 4y) + 5 = (4y + 36) + 5$

16. $9 + 0 = 9$

17. $4 \cdot 20 + 4 \cdot x = 4 \cdot (20 + x)$

1–6 Simplify.

18. $12a + 6b - 3a$

19. $\frac{1}{2}(8x)$

20. $8(y + 2) - 7$

1–7 Let B be Bill's age in years. Write an expression for:

21. twice Bill's age

22. Bill's age in 3 years

23. Bill's age 4 years ago

24. one-fourth Bill's age 1 year from now

2–1 True or false?

25. $-14 < -13$

26. $|-9.2| = |9.2|$

27. No number is its own opposite.

2–2, 2–3, 2–4 Simplify.

28. $-14 + 8$

29. $-7 \cdot -5$

30. $9 - (-3\frac{1}{2})$

31. $5.6 + (-3.1)$

32. $-13 - (-7)$

33. $16 \cdot -4$

34. $(-9 + 14) + (-6)$

35. $-12 \cdot (-5 + 7)$

2–5 **36.** What is the reciprocal of -6?

a	b	c
-48	4	-16

Substitute and simplify.

37. $a \div b$

38. $\dfrac{b + c}{a}$

39. $\dfrac{c}{b} - \dfrac{a}{c}$

2–6, 2–7, 2–8 Simplify.

40. $4y - (-7)$

41. $13 + (-x) - 8.5$

42. $8b - 4b$

43. $9t + 3 - 8t$

44. $7x + y - 8x + 3y$

45. $5(a - 2) - 6$

46. $4y - (3 - 2y)$

47. $8(x - y) + 9(y + 2x)$

2–9 Use a table to solve.

48. What number added to the numerator and denominator of $\frac{1}{4}$ results in a new fraction equal to $\frac{3}{4}$?

Use a table to solve.

49. A bus and taxi leave Gotham City Airport traveling in the same direction. The taxi leaves at 8:00 A.M. traveling at 60 mph. The bus leaves at 8:30 A.M. traveling at 50 mph. At what time will the bus and taxi be 60 miles apart?

3–1 Solve the equation for the replacement set $\{-4, -2, -1, 2, 4\}$.

50. $2|x| = 4$ **51.** $\dfrac{4}{x} = -2$ **52.** $x^2 - 3x = 4$

3–2 What equation results from following these directions in order?

53. Original equation: $3y + 5 + 4y = 23$
    a. Combine the y-terms.
    b. Subtract 5 from both sides.

State whether each pair of equations is equivalent.

54. $x + 6 = 0, -4x = 24$ **55.** $2x + 3 = 4, x + 2 = 3$

3–3,
3–4, Solve.
3–5, **56.** $7x - 3 = 8x$ **57.** $\dfrac{n}{5} = -7$ **58.** $\dfrac{3}{8}p = 24$
3–6,
3–7 **59.** $\dfrac{y}{2} + 4 = 9$ **60.** $11x + 8 = 10x + 14$

    61. $2w - 5 = 21$ **62.** $\dfrac{a-2}{3} = 6$

    63. $4m - 10 = 2m - 16$ **64.** $1 - 2x = 7$

    65. $-2 - \dfrac{t}{3} = 4$ **66.** $2(1 - 4y) = 10$

    67. $2b + 4(b - 1) = 32$ **68.** $5 - 2(x - 1) = 7$

3–8 Solve the equations for the indicated variables.

69. $6n - t = 1$, for n **70.** $A = \dfrac{1}{2}bh$, for b

3–9 For each problem, write these four things:

    a. What the variable represents (always a *number* of something).
    b. An equation that fits the problem.
    c. The solution to the equation.
    d. The answer to the question (with proper units).

71. The length of a rectangle is 8 cm more than its width. What is the width if the perimeter of the rectangle is 40 cm?

72. The combined weight of the Simpson twins is 235 pounds. If Wyomia Simpson weighs 25 pounds less than Ralph Simpson, how much does each weigh?

4 Polynomials

The prefix poly- from the Greek word "polys" means much or many. Several mathematical terms contain this prefix — polygon, polyhedron, and polynomial.

A famous sculpture created by Ron Resch, a computer scientist and artist, is the Vegreville Alberta Easter Egg. This polyhedral egg is 31 feet high, 18 feet wide, and weighs 5000 pounds. Its 3512 facets are made from 524 star-shaped and 2208 triangular pieces of aluminum.

4–1 Adding and Subtracting Polynomials

Preview

A child's nursery rhyme goes like this:

> As I was going to St. Ives,
> I met a man with seven wives;
> Each wife had seven sacks,
> Each sack had seven cats,
> Each cat had seven kits:
> Kits, cats, sacks, and wives,
> How many were going to St. Ives?

The answer is one.

A more difficult question is, "Man, kits, cats, sacks, and wives: How many were met by the person going to St. Ives?"

In this lesson, you will learn how to represent expressions such as the one for the nursery rhyme.

■ LESSON

A **monomial** is a number, a variable, or the product of numbers, and/or variables. In a monomial, only 0 and the positive integers can be used as exponents. The following expressions are examples of monomials:

$$y \qquad -x \qquad ab \qquad \frac{1}{3}z \qquad x^2 \qquad 8 \qquad xy^2 \qquad cd^2 \qquad (abcd^2)^3$$

No monomial has a variable as an exponent, nor does it have a variable in the denominator of a fraction. The following expressions are not monomials, because the numbers and variables are involved in operations other than multiplication and because the variables appear as powers and as denominators of fractions:

$$t + 2 \qquad bc - 8 \qquad x^2 + 1 \qquad a + b \qquad \frac{7}{y} \qquad 4^x$$

A **polynomial** is the sum (or difference) of monomials. Monomials are also polynomials. These expressions are examples of polynomials:

$$a + b \qquad 7 - x \qquad -2x^2 + yx - 3 \qquad a^2b^3 - \frac{1}{7}a \qquad r + 9$$

Monomials joined by addition (or subtraction) are called the **terms** of the resulting polynomial. The following polynomial has 6 terms:

$$\overset{\text{terms}}{5x^2 + 7x + 9 - 6x^2 + 2x - 5}$$

We have used the word "term" before when referring to like and unlike terms. Remember that like terms have identical variable parts.

Like terms	Unlike terms
$5x^2$ and $-6x^2$	$5x$ and $7x^2$
$7a$ and a	$3x$ and $3y$
9 and 4	$2x$ and -5

We added and subtracted polynomials in Chapter 2. Study these examples as a review.

Example 1 Add $5x + 7$ and $8 - 2x$.

Solution $\quad (5x + 7) + (8 - 2x)$

$= (5x + 7) + [8 + (-2)x]$ *Change subtraction to addition.*

$= [5x + (-2)x] + (7 + 8)$ *Commutative and associative properties of addition*

$= 3x + 15$ *Distributive property and computation*

Note that when two polynomials are added, the parentheses around the polynomials can be omitted.

$$(5x + 7) + (8 - 2x) = 5x + 7 + 8 - 2x$$
$$(3a - 5) + (-4 + 2a) = 3a - 5 + (-4) + 2a$$

Example 2 Subtract $3a + b$ from $7a + 5b$.

Solution $\quad (7a + 5b) - (3a + b)$

$= (7a + 5b) + [-(3a + b)]$ *Change subtraction to addition.*

$= (7a + 5b) + (-1)(3a + b)$ -1 *property of multiplication*

$= (7a + 5b) + [(-3)a + (-1)b]$ *Distributive property*

$= [7a + (-3)a] + [5b + (-1)b]$ *Commutative and associative properties of addition*

$= 4a + 4b$ *Distributive property and computation*

In Examples 3 and 4, note that we add and subtract polynomials in much the same way that we add and subtract numbers.

Example 3 Add $3x + 5y - 8$ and $4x - 2y - 5$.

Solution Arrange like terms in columns and add.

$$\begin{array}{r} 3x + 5y - 8 \\ 4x - 2y - 5 \\ \hline 7x + 3y - 13 \end{array}$$

Answer $7x + 3y - 13$

Example 4   Subtract $5c + 6d - 7$ from $2c - 3d + 1$.

Solution   Arrange like terms in columns.

$$2c - 3d + 1$$
$$5c + 6d - 7$$

Change the signs of the polynomial to be subtracted in order to get its opposite. Then *add* like terms in columns.

$$2c - 3d + 1$$
$$\underline{-5c - 6d + 7}$$
$$-3c - 9d + 8$$

Answer $-3c - 9d + 8$

■ CLASSROOM EXERCISES

Which of these expressions are monomials?

1. x^2y^2 **2.** $\dfrac{x^2}{y^2}$ **3.** $\dfrac{1}{7}$ **4.** $ax^2 + bx + c$ **5.** $\dfrac{1}{x} + y$

Which of these expressions are polynomials?

6. $x^2 + y^2$ **7.** x^3 **8.** $x^2 - \dfrac{1}{3}$ **9.** $ax^2 + bx + c$ **10.** $\dfrac{1}{x} + y$

Add.

11. $(x + y)$ and $(y + x)$ **12.** $(2x + 7)$ and $(3x - 4)$

Subtract.

13. $(x + y)$ from $(x - y)$ **14.** $(4c - 1)$ from $(2c - 8)$

■ WRITTEN EXERCISES

Ⓐ **1.** Which of these expressions are monomials?

$$2a \qquad 5b^2 \qquad \frac{3}{c} \qquad \frac{8}{9}d \qquad e + f$$

2. Which of these expressions are polynomials?

$$p \qquad \frac{q}{2} \qquad r^3 \qquad -5s \qquad t - u$$

How many terms does each polynomial contain?

3. $7xyz + 8x$ **4.** $5abc + 3ab$ **5.** $2x + 5y - 8z$

6. $7a + 5b + 2c$ **7.** $5x^3 + 4x^2 - 9x - 2$ **8.** $8a^3 + a^2 - 9a + 18$

Add these polynomials.

9. $5x^2 + 3x + 8$ **10.** $2b^2 + 3b + 4$ **11.** $2a + 3b - c$ **12.** $8x - y$
    $\underline{2x^2 - 7x + 3}$ $\underline{3b^2 + 7b - 6}$ $\underline{6a \qquad\quad - 4c}$ $\underline{x - 5y - z}$

Simplify each sum.

13. $(x^2 + 5x - 3) + (3x^2 + 2x + 4)$ **14.** $(7a^2 + 4a + 8) + (2a^2 + a - 9)$

15. $(2a + 3b) + (4b + c)$ **16.** $(4x + 2y) + (3x + 5w)$

Write and simplify an expression for the *perimeter* of each of the following figures.

17.
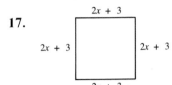

18.
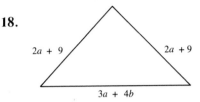

19.
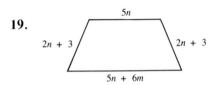

20.
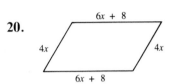

Subtract these polynomials.

21. $8a^2 + 7a + 6$ **22.** $10x^2 + 8x + 6$ **23.** $5x - 6y + 3z$ **24.** $7a \qquad - 3c$
     $\underline{6a^2 + 6a + 1}$ $\underline{9x^2 + 3x + 2}$ $\underline{3y - 3z}$ $\underline{2a - 2b + 2c}$

Simplify each difference.

25. $(5x^2 + 2x + 7) - (x^2 - 5x - 9)$ **26.** $(7a^2 + 3a + 5) - (a^2 - 4a + 9)$

27. $(3a + 4b) - (b + 2c)$ **28.** $(5x + 3y) - (y + 3z)$

Simplify.

B **29.** $(2x^2 - 3x + 5) + (5x^2 + 8x - 5) - (x^2 + 5x + 6)$

30. $(8a^3 + 6a - 5) - (8a^3 + 5a^2 + 6a + 5) + (6a^2 + 3)$

31. $(7a^2 - 6ab + 5b^2 - 4a + 3b - 2) - (7a^2 - 6ab + 5b^2) - (4a + 3b)$

32. $(5x^2 + 5xy + 5y^2 - 5x - 5y - 5) - (5x^2 - 5xy + 5y^2) - (5x + 5y)$

33. $(4t^3 + 2t^2 + 7t + 4) + (9t^3 + 6t^2 + 2)$

34. $(8t^4 + 5t^2 + 7) + (3t^3 + 3t^2 + 3t)$

35. $(5t^3 + 6t^2 + 7t + 8) - (5t^2 + 6t + 8)$

36. $(8t^4 + 7t^3 + 6t^2) - (7t^3 + 6t^2 + 1)$

37. $(8t^3 + 5t^2 + 9t + 5) + (7t^2 + 9) - (8t^2 + 8)$

38. $(3t^3 + 2t^2 + 7t + 8) - (5t^2 + 9t + 3) + (4t^2 + 8t)$

C **39.** If n is an integer, write and simplify an expression for the sum of n and the next 4 consecutive integers.

40. If n is an even integer, write and simplify an expression for the sum of n, the preceding smaller integer, and the next larger integer.

41. If n is an odd integer, write and simplify an expression for the sum of n and the next 5 consecutive odd integers.

Write and simplify an expression for the total *surface area* of each of the following solid figures.

42. Cube

43. Rectangular solid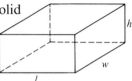

44. Triangular prism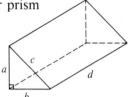

45. Cylinder

46. Write and simplify an expression for "everything" on the St. Ives road. [See the Preview on page 143.]

■ REVIEW EXERCISES

Simplify. [1–2, 1–3]

1. $8 + 6 \div 2$

2. $\dfrac{6 + 15}{3}$

3. $8 + \dfrac{24}{4} - 2$

4. $2 \cdot 5^2$

5. $(2 + 5)^2$

6. $(7 + 3)^2 - (7^2 + 3^2)$

a	b	c	d
2	5	10	20

Substitute and simplify.

7. $a + bc - d$

8. $(a + b)c + ad$ [1–2]

9. $(ab)^2$

10. $(c + d)^2 - (c^2 + d^2)$ [1–3]

List the order in which the operations are performed in evaluating this expression.

11. $40 - 3^2 \cdot 4 + 2$ [1–3]
  ↑ ↑↑ ↑

4-2 Types of Polynomials

Preview

Consider these polynomials. Find two different ways to group them into three distinct categories.

$7x^2 + 4x - 5$ $8x^2$ 6 $5x + 3$

$5x^2 + 3$ $-3 + 4x - x^2$ $6x^2 - 4x$ $7x$

In this lesson you will learn some ways to classify polynomials.

■ LESSON

Polynomials can have one or more terms. One way to classify polynomials is according to the number of terms they have.

Monomials *have one term.*	**Binomials** *have two terms.*	**Trinomials** *have three terms.*
6	$5x + 3$	$3x^2 + 5x - 6$
$7a$	$6y^2 - 2$	$3b^2 + 2a - 2ab$
$5x^2$	$a - b$	$-3m + m^3 - 2$
$-4m^3n^2$	$2x^2y + 3xy^2$	$4x^2 - 4xy + 3y^2$

Polynomials with more than three terms are simply called *polynomials.*

A polynomial can also be classified according to the powers of its variables. The **degree of a monomial** is the sum of the exponents of the variables. Monomials that are numbers are called **constants.** The degree of a constant is 0.

Monomial	*Degree*	
x^3	3	
x^3y^2	5	
$3x^3y^2$	5	
$3^2x^3y^2$	5	
9	0	
x	1	[Remember that $x = x^1$.]
$5xy$	2	

Strategy for Success   Finding information

At the back of the book is a glossary that gives definitions of the mathematical terms and symbols used in the book. For example, look up the term *variable* in the glossary.

The **degree of a polynomial** is the highest degree of any of its terms *after* the polynomial has been simplified.

Polynomial	Degree
$3x^2 + 5x + 7$	2

↑
highest-degree term

$3x^2 - 9xyz + y + z$	3

↑
highest degree term

$x + y + 7$	1

↑ ↑
highest degree terms

$2x^2 + 7x - 3 - 2x^2$	1

The polynomial $2x^2 + 7x - 3 - 2x^2$ has second-degree terms, but it can be simplified to $7x - 3$. The polynomial $7x - 3$ is a first-degree polynomial.

Terms of a polynomial in one variable are usually written in order of their degrees, as follows.

Ascending order (from lowest to highest degree):

$$3 + 4x - x^2 \qquad 4y - 4y^2 + 2y^3 \qquad 7 - 3a + 2a^2 - 5a^6$$

Descending order (from highest to lowest degree):

$$-x^2 + 4x + 3 \qquad 2y^3 - 4y^2 + 4y \qquad -5a^6 + 2a^2 - 3a + 7$$

If the terms of a polynomial have more than one variable, the terms are arranged in ascending or descending order for one of the variables.

$3x^2 + 3xy^2 - 2y \leftarrow$ descending order in x
$3xy^2 - 2y + 3x \leftarrow$ descending order in y
$2a^2 - 3ab + 2b^2 \leftarrow$ descending order in a and ascending order in b

■ CLASSROOM EXERCISES

1. Which of these expressions are monomials?

$$6a \qquad 3a + 1 \qquad x^2y \qquad \frac{3}{x}$$

2. Which of these expressions are binomials?

$$3x \qquad 5a + 2b \qquad 6x^2y \qquad 8x^2 + x + 1$$

3. Which of these expressions are trinomials?

$$5 + 3x + 2x^2 \qquad 5ab \qquad 7x - 3y - z \qquad a^3 + 3$$

State the degree of each monomial.

4. $5x^3$ **5.** 6 **6.** $3a$ **7.** $2mn$ **8.** $3a^2b^3$

State the degree of each polynomial.

9. $4x^2 + 3x - 8$ **10.** $3a^2b - 2a^2b^2 + 4b$ **11.** $3x^2 + 5 - 3x^2 + 2$

12. Arrange the terms in ascending order: $5x + 3 - 2x^2$.

13. Arrange the terms in descending order: $5a - 3a^2 + 2a^3 - 5$.

14. Arrange the terms in descending order in x: $4x^2y + 3x^3 - 2xy + y$.

Strategy for Success   Doing assignments

Set aside a time and a place to do your mathematics each day. Choose a time and a place free of distractions so that you can have the best conditions for concentrating.

■ WRITTEN EXERCISES

A   **1.** Which of these expressions are monomials?

$$5x \qquad 5 + x \qquad xy \qquad \frac{x}{y} \qquad -6 \qquad 6^2$$

2. Which of these expressions are binomials?

$$wx \qquad 2 + y \qquad z - x \qquad 2xy + 3w - z$$

3. Which of these expressions are trinomials?

$$3xy^3 \qquad 2x + 2y + 2z \qquad x^7 - 2x^6 - x^5$$

4. Which of these expressions are polynomials?

$$x \qquad 3a^2b + 1 \qquad 5 \qquad a + 2b - c \qquad \frac{1}{y}$$

State the degree of each term.

5. $3x^2$ **6.** $7x^3$ **7.** 8 **8.** 2 **9.** $2r^2s^3$

10. $5a^3b^2$ **11.** $\frac{1}{2}xyz$ **12.** $\frac{1}{3}abc$ **13.** $-2x^2y$ **14.** $-4xy^2$

State the degree of each polynomial.

15. $3x^2 + 4x + 5$ **16.** $6a^2 + 5a - 4$

17. $a + b + c + d$ **18.** $w + x - y - z$

19. $5 + 9$ **20.** $2 + 3$

21. $2a^2 + abc - 5d^2$ **22.** $8x + 7xy - 5y$

23. $-5xy^3 + 7$ **24.** $\frac{2}{3}a^2b - 4b^2$

Write each polynomial in ascending order.

25. $5x + 9x^2 + 4$

26. $3a^2 + 8 - 7a$

27. $9b^3 - 3b + 5$

28. $4v^4 + 3v^2 + 8 - v$

Write each polynomial in descending order.

29. $2x + 5 + 8x^2$

30. $5a^2 - 7 - 9a$

31. $-7y^2 + 5y^3 + 4y$

32. $3w^4 + 4w - 7 - 2w^2$

State the degree of each polynomial.

B **33.** $2x^2 - 3x + 5 - 2x^2$

34. $3b - 7 + 5 + 3b^3 - 3b + 2$

35. $2ab - 3b^2 + 2ab + 5$

36. $5mn + 3m^2 - 2m - 3m^2$

Write each polynomial in ascending order in x.

37. $3xy - 5y^2 + 7x^2$

38. $2x^3 - 3xy^2 + 5y^3 + 6x^2y$

39. $6x^2y^2 - 7xy + 8$

40. $9 + 5x^2y^2 - 4xy$

41.–44. Write each polynomial in Exercises 37–40 in descending order in y.

C **45.** **a.** What is the degree of $(3x^2 + 2x - 7)$?
   b. What is the degree of $(5x - x^2 + 2)$?
   c. Add the polynomials from parts (a) and (b). What is the degree of the sum?

46. **a.** What is the degree of $(2x - 3x^2 + 4)$?
   b. What is the degree of $(x^3 - 2x + 1)$?
   c. What is the degree of the sum of the polynomials from parts (a) and (b)?

47. **a.** What is the degree of $(4x^2 - 7x + 2)$?
   b. What is the degree of $(x - 4x^2 + 1)$?
   c. What is the degree of the sum of the polynomials from parts (a) and (b)?

48. How are the degrees of two polynomials related to the degree of their sum?

■ REVIEW EXERCISES

Indicate whether the expressions are equivalent. [1–4]

1. $3(x + 2)$ and $3x + 6$ **2.** $3x^2$ and $6x$ **3.** $4(a + 8)$ and $4a + 8$

4. $a^2 + b^2$ and $(a + b)^2$ **5.** $a^2 - b^2$ and $(a - b)^2$ **6.** $\dfrac{a+1}{b+1}$ and $\dfrac{a}{b}$

Indicate the property that is illustrated by each equation. [1–5]

7. $2a + a = 2a + 1a$

8. $2a + 1a = (2 + 1)a$

9. $10 + (3a + 7a) = (3a + 7a) + 10$

10. $a \cdot 0 = 0 \cdot a$

11. $6a + (3a + 5) = (6a + 3a) + 5$

12. $8a(3a \cdot 6) = (3a \cdot 6)8a$

If you need to evaluate the trinomial

$$4x^2 - 10xy + 3y^2$$

for a number of different values of x and y, you might use the following BASIC program. The statements could be numbered from 1–6, but it is customary to use multiples of ten so that additional statements can be inserted without re-numbering the entire program.

```
10  PRINT "WHAT VALUE FOR X";
20  INPUT X
30  PRINT "WHAT VALUE FOR Y";
40  INPUT Y
50  PRINT 4 * X ↑ 2 - 10 * X * Y + 3 * Y ↑ 2
60  END          [The last line in a BASIC program is always END.]
```

The quotation marks in lines 10 and 30 indicate to the computer to print the text between them exactly as it is typed. The computer on reading the command INPUT in lines 20 and 40 will type a ?, then wait for you to enter values for X and Y. The semicolons at the end of lines 10 and 30 indicate to the computer to type the ? from the INPUT statements on the same line as the PRINT statements.

1. Type the program and run it to evaluate the polynomial for the following values of x and y. (Type RUN each time you want to enter a pair of values.)

x	10	-10	10	1	1	1
y	10	10	-10	1	2	3
$4x^2 - 10xy + 3y^2$	?	?	?	?	?	?

2. If x has the value 1, what value of y (to the nearest tenth) results in the smallest value for the polynomial?

3. Change line 50 so that the program can be used to evaluate the polynomial $x^2 + 7xy - y^2$.

4. Change line 50 so that the program can be used to evaluate the polynomial $x^4 + 3x^2y - y^2$.

4-3 Multiplying Monomials

Complete these multiplication equations. Then use the table to write the equations using powers of 2. Look for a pattern.

$4 \cdot 8 = \underline{\quad ? \quad}$

$2^2 \cdot 2^3 = 2^?$

$16 \cdot 16 = \underline{\quad ? \quad}$

$2^4 \cdot 2^4 = 2^?$

$256 \cdot 8 = \underline{\quad ? \quad}$

$2^? \cdot 2^? = 2^?$

$8 \cdot 16 = \underline{\quad ? \quad}$

$2^? \cdot 2^? = 2^?$

Guess; then check your guess.

$3^2 \cdot 3^3 = 3^?$ $\qquad$ $10^3 \cdot 10^5 = 10^?$ $\qquad$ $10^7 \cdot 10^4 = 10^?$

Guess.

$a^3 \cdot a^5 = a^?$ $\qquad$ $b^7 \cdot b^2 = b^?$ $\qquad$ $x^3 \cdot x^9 = x^?$

$$
\begin{aligned}
2^1 &= 2 \\
2^2 &= 4 \\
2^3 &= 8 \\
2^4 &= 16 \\
2^5 &= 32 \\
2^6 &= 64 \\
2^7 &= 128 \\
2^8 &= 256 \\
2^9 &= 512 \\
2^{10} &= 1024 \\
2^{11} &= 2048 \\
2^{12} &= 4096
\end{aligned}
$$

■ LESSON

The exercises in the Preview illustrate the **addition property of exponents.**

The Addition Property of Exponents

For all real numbers a, and all positive numbers m and n,

$$a^m \cdot a^n = a^{m+n}.$$

For example,

$$
2^3 \cdot 2^4 = \overbrace{(2 \cdot 2 \cdot 2)}^{3 \text{ factors}} \overbrace{(2 \cdot 2 \cdot 2 \cdot 2)}^{4 \text{ factors}}
$$

$$
= \overbrace{2 \cdot 2 \cdot 2 \cdot 2 \cdot 2 \cdot 2 \cdot 2}^{3 + 4 \text{ factors}}
$$

$$
= 2^7
$$

In general, we have

$$
x^m \cdot x^n = \overbrace{(x \cdot x \cdot \ldots)}^{m \text{ factors}} \ \overbrace{(x \cdot x \cdot \ldots)}^{n \text{ factors}}
$$

$$
= \overbrace{x \cdot x \cdot x \cdot \ldots}^{m + n \text{ factors}}
$$

$$
= x^{m+n}
$$

The addition property of exponents and the commutative and associative properties of multiplication can be used to simplify the product of two monomials.

Example 1 Multiply. $3x^2$ and $-5x^4$

  Solution $(3x^2)(-5x^4) = 3(-5)(x^2 \cdot x^4)$ *Commutative and associative properties*
  $= -15\ (x^2 \cdot x^4)$ *Computation*
  $= -15x^6$ *Addition property of exponents*

To simplify the product of monomials, first multiply the numerical factors and then find the product of variable factors with like bases by using the addition property of exponents.

Example 2 Simplify. $(4ab)\left(\frac{1}{3}a^2b\right)$

  Solution $4ab\left(\frac{1}{3}a^2b\right) = \left(4 \cdot \frac{1}{3}\right)(a \cdot a^2)(b \cdot b)$

  $= \frac{4}{3}a^3b^2$

Example 3 Simplify. $(-2m^3n)(-8m^2n^2)(3mn)$

  Solution $(-2m^3n)(-8m^2n^2)(3mn) = (-2)(-8)(3)(m^3 \cdot m^2 \cdot m)(n \cdot n^2 \cdot n)$
  $= 48m^6n^4$

■ CLASSROOM EXERCISES

Complete.

1. $3^2 \cdot 3^4 = 3^?$

2. $3^{10} \cdot 3^2 = ?$

3. $4 \cdot 4^6 = ?$

4. $x^3 \cdot x^5 = x^?$

5. $y^? \cdot y^3 = y^{12}$

6. $r^a \cdot r^b = ?$

Simplify.

7. $y^7 \cdot y^4$

8. $(3x^2)(4x^3)$

9. $(5x^2y)(-2xy^2)$

■ WRITTEN EXERCISES

Complete.

A
1. $2^3 \cdot 2^4 = 2^?$

2. $2^5 \cdot 2^2 = 2^?$

3. $5^4 \cdot 5^4 = 5^?$

4. $5^3 \cdot 5^6 = 5^?$

5. $3 \cdot 3^4 = 3^?$

6. $3^5 \cdot 3 = 3^?$

7. $4^? \cdot 4^2 = 4^8$

8. $4^3 \cdot 4^? = 4^{12}$

9. $2 \cdot 2^3 \cdot 2^6 = 2^?$

10. $2^8 \cdot 2^4 \cdot 2 = 2^?$

11. $\left(\frac{2}{3}\right)^2 \cdot \left(\frac{2}{3}\right)^2 = \left(\frac{2}{3}\right)^?$

12. $\left(\frac{3}{4}\right)^4 \cdot \left(\frac{3}{4}\right)^3 = \left(\frac{3}{4}\right)^?$

Simplify.

13. $a^5 \cdot a^{15}$ **14.** $a^4 \cdot a^{12}$ **15.** $(2b^3)(3b^2)$

16. $(4c^4)(3c^3)$ **17.** $(-2q^{10})(-6q^{10})$ **18.** $(-8q^9)(-3q^9)$

19. $(5r)(20r^{100})$ **20.** $(25s)(4s^{50})$ **21.** $\left(\frac{1}{2}t^2\right)\left(\frac{1}{4}t^4\right)$

22. $\left(\frac{1}{3}t^3\right)\left(\frac{1}{6}t^6\right)$ **23.** $(5ab^2)(3a^2b^3)$ **24.** $(2xy^3)(6x^4y^6)$

Write the degrees of each product.

25. $(5a^2)(10a^8)$ **26.** $(7a^4)(3a^5)$ **27.** $(4y^3)(2y^{20})$

28. $(5y^{30})(6y^5)$ **29.** $(4xy)(2x^2y)$ **30.** $(5ab)(2ab^2c)$

31. $(-2x^2y)(-3xy^2)(2x)$ **32.** $(4mn)(-3m^2n)(-2mn)$

Use the table in the Preview to answer these questions.

33. In the quiz show "Double or Nothing," a contestant won \$2 for answering a question correctly, \$4 for answering 2 questions correctly, \$8 for answering 3 questions correctly, and so on, doubling the winnings for each correct answer. How many questions did the contestant answer correctly to win \$64?

34. Under the right conditions the mass of a bacteria culture doubles every hour. After 10 hours, what will be the mass of a culture that began with a mass of 1 gram?

B 35. An investment of \$100 doubles in value every decade. What is the value of an investment after 8 decades?

36. A rumor was spread by one person telling two others. An hour later each of the new people told two others. In another hour each of those four told two others; and each hour thereafter, the process continued. If one person started the rumor at noon, how many would have heard the rumor by 6 P.M.?

37. One bean is placed on the lower lefthand square of a checkerboard. On the square to its right are placed two beans, to their right four beans, and to their right eight beans. If the process of doubling continues, how would you represent the number of beans piled on the final (64th) square?

Simplify.

38. $(2a^2b)(3ab^2)(2a)$ **39.** $(-5xy^3)(-3x^2)(-24)$ **40.** $(3m^2n)(-2mn)(-2n^2)$

41. $a^3 + a^3$ **42.** $a^3 \cdot a^3$ **43.** $-2a^4 \cdot \dfrac{a^2}{2}$

44. $\left(-\frac{1}{2}a^6\right)(8a^3)$ **45.** $(x^2)^4$ **46.** $(x^3)^2$

Are these like terms? (Answer yes or no.)

C **47.** $(3xy^2)(x)$ and $(4x^2y)(y)$ **48.** $(5ab^2)(-3a^2b)$ and $(11ab)(ab^2)$

49. $(2x^2)^3$ and $(5x^3)^2$ **50.** $(3x^3)^4$ and $(4x^4)^3$

Write an expression for the area of each figure.

51.

3a

2b

52.

1.5x

2x

53.

3k

4k

Write an expression for the volume of each figure.

54.

a

2a

b

55.

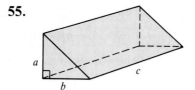

a

b

c

56.

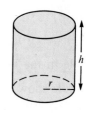

h

r

▇ REVIEW EXERCISES

Simplify.

1. $6x - x$ **2.** $4x - 5 + x$ [1–6]

3. $x + 7 + x$ **4.** $3(x + 4) + 4x$

Self-Quiz 1

4–1   Identify each of the following polynomials as a monomial, binomial, or trinomial.

     1. $3x^2 + 5$ **2.** $4xy^3$ **3.** $8x^2y + 5xy - 3xy^2$

4–2   Simplify.

     4. $(3x^2 + 5x - 8) + (8x^2 - 2x + 3)$ **5.** $(a^2 + 2ab + b^2) + (3a^2 - ab - b^2)$

     6. $(5x^2 + 6x + 7) - (2x^2 + 4x - 3)$ **7.** $(7x^2 - 3x - 5) - (2x^2 - 4x - 7)$

4–3   Simplify.

     8. $x^2 \cdot x^4 \cdot x^3$ **9.** $(-4y^2)\left(\frac{1}{2}y^3\right)$

     10. $(3a^2b)(5ab^3)$ **11.** $(-8x^2y^3)\left(\frac{3}{4}x^4y^2\right)$

4–4 Scientific Notation

Preview Application

The closest star to the earth is the sun, at a distance of about 149 million kilometers. The next closest star is Alpha Centauri, a distance of about 4.3 light years from earth. A *light year* is a unit for measuring astronomical distances. It represents the distance light travels in one year at a speed of about 300,000 kilometers per second (km/s).

To determine the distance in kilometers from the earth to Alpha Centauri, we must first compute the number of kilometers in a light year.

$$\begin{array}{ccccc} 300{,}000 \times & 60 & \times & 60 & \times & 24 & \times 365.25 \\ \text{km/s} & \text{s/min} & \text{min/h} & \text{h/day} & \text{day/yr} \end{array}$$

$$\approx 9{,}467{,}280{,}000{,}000 \text{ km}$$

If we used a computer or calculator to perform the above computation, the calculator or computer would print the result as follows:

9.46728 E+12   or   9.46728   12

Alpha Centauri, a double-star system, appear as a single star when viewed with the naked eye.

Both of the expressions are ways of writing the product:

$$9.46728 \times 10^{12}$$

This last expression is written in *scientific notation*, a simple way of writing very large or very small numbers. In this lesson you will learn to use scientific notation.

■ LESSON

Scientific notation is a way of writing a number as the product of two factors. One factor is a number greater than or equal to 1 and less than 10. The other factor is a power of 10 written in exponential form. For example,

The speed of light is about 300,000 km/s.

$$300{,}000 = 3 \times 10^5$$

$$\begin{array}{cc} \uparrow & \uparrow \\ \text{between} & \text{power} \\ \text{1 and 10} & \text{of 10} \end{array}$$

The earth is about 93,000,000 miles from the sun.

$$93{,}000{,}000 = 9.3 \times 10^7$$

$$\begin{array}{cc} \uparrow & \uparrow \\ \text{between} & \text{power} \\ \text{1 and 10} & \text{of 10} \end{array}$$

To change from standard decimal notation to scientific notation, factor the number so that one factor is a number between 1 and 10 and the other is a power of 10. Then write the power of 10 using exponents.

Example 1 Change to scientific notation. 5400

   Solution $5400 = 5.4 \times 1000 = 5.4 \times 10^3$

Example 2 Change to scientific notation. 765,000,000

   Solution $765{,}000{,}000 = 7.65 \times 100{,}000{,}000 = 7.65 \times 10^8$

To change from scientific notation to standard decimal notation, write the power of 10 in standard form. Then simplify the remaining product.

Example 3 Change to standard decimal notation. 4.53×10^5

   Solution $4.53 \times 10^5 = 4.53 \times 100{,}000 = 453{,}000$

Look for shortcuts. Note the relationship between the exponent and the number of places that the decimal point is moved.

$$4.53 \times 10^5 = \underset{\text{5 places}}{453{,}000}$$

The basic number properties as well as the addition property of exponents can be used to simplify the product of two numbers that are expressed in scientific notation.

Example 4 Write the product in scientific notation. $(3.4 \times 10^2)(2.8 \times 10^3)$

   Solution $(3.4 \times 10^2)(2.8 \times 10^3)$

   $= (3.4 \times 2.8)(10^2 \times 10^3)$ *Commutative and associative properties of multiplication*

   $= (3.4 \times 2.8)(10^5)$ *Addition property of exponents*

   $= 9.52 \times 10^5$ *Computation*

Note in Example 4 that the product of 3.4 and 2.8 was between 1 and 10. Therefore, the simplified product was already in scientific notation. Now consider $(5 \times 10^3)(4 \times 10^5)$. The product simplifies to 20×10^8, which is not in scientific notation. Therefore, two more steps are required to write the product in scientific notation.

Example 5 Write the product in scientific notation. $(5 \times 10^3)(4 \times 10^5)$

Solution

$$(5 \times 10^3)(4 \times 10^5)$$
$$= (5 \times 4)(10^3 \times 10^5)$$
$$= 20 \times 10^8$$
$$= (2 \times 10^1) \times 10^8$$
$$= 2 \times 10^9$$

■ CLASSROOM EXERCISES

State whether the number is expressed in scientific notation.

1. 2.3×10^7 **2.** 34×10^1 **3.** 0.55×10^4 **4.** 6007×10^1

Change to scientific notation.

5. 6000   **6.** 468,000   **7.** 567.5×10^5 **8.** 26,000,000

Change to standard decimal notation.

9. 3×10^4 **10.** 8.9×10^5 **11.** 7.07×10^1 **12.** 4.505×10^9

Write the product in scientific notation.

13. $(4 \times 10^3)(2 \times 10^5)$ **14.** $(3.3 \times 10^4)(8.1 \times 10^3)$

■ WRITTEN EXERCISES

State whether the number is expressed in scientific notation.

Ⓐ **1.** 56×10^5 **2.** 742×10^4 **3.** 8.6×10^8 **4.** 7.5×10^3

5. 0.368×10^3 **6.** 0.42×10^7 **7.** 6.2×2^8 **8.** 1.6×2^6

Write the number in scientific notation.

9. 530,000   **10.** 1,200,000,000,000

11. 328×10^6 **12.** 662×10^{12}

13. 0.25×10^8 **14.** 0.97×10^6

15. 217,000,000,000,000   **16.** 82,800,000,000,000,000

Write in standard decimal notation.

17. 5×10^6 **18.** 8×10^9 **19.** 6.503×10^2 **20.** 9.027×10^1

21. 8.123×10^1 **22.** 1×10^{12} **23.** 4.00×10^9 **24.** 7.3×10^6

Express these quantities using scientific notation.

Ⓑ **25.** Maximum distance of the earth from the sun: 152,000,000 km

26. Maximum distance of Pluto from the sun: 7,323,000,000 km

27. Distance of the star Alpha Centauri from the earth: 40,710,000,000,000 km

Express these quantities using scientific notation.

28. Distance of the star Capella from the earth: 425,000,000,000,000 km

29. Area of the Pacific Ocean: 166,000,000 km^2

30. Area of the Atlantic Ocean: 86,520,000 km^2

Write the product in scientific notation.

31. $(5 \times 10^4)(4 \times 10^5)$

32. $(8 \times 10^5)(5 \times 10^3)$

33. $(1.1 \times 10^{20})(1.1 \times 10^{30})$

34. $(2.3 \times 10^{15})(1.2 \times 10^{10})$

35. $(9 \times 10^3)(9 \times 10^6)(9 \times 10^9)$

36. $(3 \times 10^4)(7 \times 10^6)(8 \times 10^8)$

37. $(2 \times 10^4)^3$

38. $(3 \times 10^5)^4$

39. $400,000 \times 5,000,000$

40. $60,000 \times 45,000 \times 700,000$

41. $6,200,000 \times 3,100,000$

42. $5,000 \times 75,000 \times 80,000$

The volume of a sea can be estimated by multiplying its area (in square meters) by its average depth (in meters). Write the volume (in cubic meters) of these seas using scientific notation.

Sea	Area (m^2)	Average depth (m)
43. Mediterranean	2.5×10^{12}	1500
44. Red	4.5×10^{11}	580
45. Yellow	2.9×10^{11}	37
46. Bering	2.25×10^{12}	1500

■ REVIEW EXERCISES

Simplify these absolute value expressions.   [2–1]

1. $\left| -3\frac{1}{2} \right|$

2. $|-1.5| + |1.5|$

Simplify.   [2–2, 2–3]

3. $4 + (-7)$

4. $(-8 + (-5)) + 7$

5. $|-3 + 5| + |-3| + |5|$

6. $-6 - (-5)$

7. $-7.5 - 6.25$

8. $|-4 - 10|$

r	s	t
-3	4	-12

Substitute and simplify.

9. $|r| + |s|$ [2–1]

10. $(r + s) + t$ [2–2]

11. $s - t$ [2–3]

EXTENSION   Scientific notation on a calculator

The numbers displayed on most calculators are limited to 8 or 10 digits. However, you can compute with numbers that have more digits by using scientific notation on a calculator that has a key labeled $\boxed{\text{EXP}}$ or $\boxed{\text{EE}}$.

Example 1   Enter 4,764,000,000,000 on a calculator.

Solution   Express the number using scientific notation.

$$4{,}764{,}000{,}000{,}000 = 4.764 \cdot 10^{12}$$

Key Sequence	Display
4.764 $\boxed{\text{EXP}}$ 12	4.764 12

Example 2   Simplify.   1,725,900,000,000 × 3,940,000

Express the product using scientific notation.

Solution   *Key Sequence*   *Display*

1.7259 $\boxed{\text{EXP}}$ 12 × 3.94 $\boxed{\text{EXP}}$ 6 = 6.800046   18

Write the number using standard decimal notation.

1. 5.218   07
2. 3.946281   9
3. 4.6   05
4. 9.57   02
5. 4.87   01
6. 3.001   05

Simplify. Express the answer using scientific notation.

7. 482,000 × 3000

8. 485,000 × 36,170,000

9. 2,198,073 × 328,651

10. 685,000 ÷ 50,000

11. 729,800,000,000 ÷ 89,000,000

12. 23,278,000,000,000,000 ÷ 565,000,000,000

Use your knowledge of exponents to complete these equations.

$(a^3)^2 = (a^3)(a^3) = a^?$

$(b^4)^3 = (b^?)(b^?)(b^?) = b^?$

$(y^2)^4 = y^?$

$(a^2)^3 = (a^2)(a^2)(a^2) = a^?$

$(x^5)^2 = (x^?)(x^?) = x^?$

$(a^m)^2 = a^?$

The following expressions consist of two factors raised to a power. Write each expression without using parentheses.

$(2x)^3 = (2x)(2x)(2x) = ?$

$(ab^2)^3 = a^?b^?$

$(a^5b^m)^2 = a^{10}b^?$

$(3a^2)^2 = ?a^4$

$(x^2y^3)^4 = ?$

$(x^my^n)^2 = ?$

■ LESSON

The first group of problems in the Preview illustrates this property of exponents.

The Multiplication Property of Exponents

For all numbers a, and all positive integers m and n,

$$(a^m)^n = a^{mn}.$$

The multiplication property of exponents is a consequence of the definition of an exponent and the basic number properties.

$$(4^3)^2 = (4^3)(4^3) = 4^{3+3} = 4^6$$

$$(a^m)^n = \overbrace{a^m \cdot a^m \cdot \ \cdots \ \cdot a^m}^{n \text{ factors}} = a^{\overbrace{m+m+\ldots+m}^{n \text{ addends}}} = a^{mn}$$

If a number with an exponent is raised to a power, multiply the exponents.

> **Example 1** Simplify. **a.** $(x^2)^4$ **b.** $(y^3)^n$
>
>   *Solution*   **a.** $(x^2)^4 = x^{2\cdot4} = x^8$ **b.** $(y^3)^n = y^{3n}$

The second group of problems in the Preview illustrates this property of exponents.

> ## The Distributive Property of Exponents over Multiplication
>
> For all numbers a and b, and all positive integers n,
> $$(ab)^n = a^n b^n.$$

The distributive property of exponents over multiplication is also a consequence of the definition of an exponent and basic number properties.

$$(2x)^3 = (2x)(2x)(2x) = (2 \cdot 2 \cdot 2)(xxx) = 2^3 x^3$$

$$(a^m b^n)^p = \overbrace{(a^m b^n) \cdot \quad \cdots \quad \cdot (a^m b^n)}^{p \text{ factors of } a^m b^n}$$

$$= \overbrace{(a^m \cdot a^m \cdot \quad \cdots \quad \cdot a^m)}^{p \text{ factors of } a^m} \overbrace{(b^n \cdot b^n \cdot \quad \cdots \quad \cdot b^n)}^{p \text{ factors of } b^n}$$

$$= a^{mp} b^{np}$$

If the product of two factors is raised to a power, *each* factor is raised to that power.

Example 2   Simplify. $(3a^2 b^3)^4$

   Solution $(3a^2 b^3)^4 = 3^4 (a^2)^4 (b^3)^4$
   $$= 81 a^8 b^{12}$$

Some students confuse these two expressions:

$$a^2 \cdot a^3 \quad \text{and} \quad (a^2)^3$$

Remember! If you are not sure how to simplify an expression involving exponents, think about what the exponent means.

$$a^2 \cdot a^3 = (aa)(aaa) = a^5 \qquad (a^2)^3 = (aa)(aa)(aa) = a^6$$

■ CLASSROOM EXERCISES

Simplify.

1. $a^2 \cdot a^4$

2. $(a^2)^4$

3. $(m^3)^3$

4. $m^3 \cdot m^3$

5. $(x^3)^5$

6. $(3x)^2$

7. $(2x)^3$

8. $(2x^2 y)^4$

■ WRITTEN EXERCISES

Simplify.

A 1. $(x^4)^5$ 2. $(y^3)^4$ 3. $(a^3)^7$ 4. $(b^5)^6$

5. $(2c)^4$ 6. $(3d)^3$ 7. $(4y^2)^3$ 8. $(2x^3)^5$

9. $\left(\frac{1}{2}b^3\right)^3$ 10. $\left(\frac{1}{3}c\right)^3$ 11. $(0.1r^4)^2$ 12. $(0.01s^3)^2$

13. $(-2x^3)^2$ 14. $(-3y^2)^4$ 15. $(-w^5)^3$ 16. $(-t^4)^5$

17. $(x^2y)^4$ 18. $(a^3b^2)^3$ 19. $(2c^3d)^5$ 20. $(3rs^4)^4$

Which number is larger? Why?

21. $(5 \cdot 7)^3$ or $5 \cdot 7^3$ 22. $5^4 \cdot 2^5$ or $(5 \cdot 2)^5$

23. $(4^5 \cdot 4^{10})$ or 4^{50} 24. $(4^{10} \cdot 4^{20})$ or $(4^{10})^3$

Simplify.

B 25. $2x^3 \cdot (3x)^2$ 26. $3y^2 \cdot (2y)^3$

27. $(-ab)(a^2b)^2$ 28. $(-rs)(rs^3)^2$

29. $(-2xy)^3(-x^2)$ 30. $(-3cd)^3(-d^2)$

31. $(4a^2)^3\left(\frac{1}{2}a^3\right)^2$ 32. $(8b^3)^2\left(\frac{1}{4}b^2\right)^2$

33. $(-x)^5(-x)^2(-x)^3$ 34. $(-y)^4(-y)^3(-y)^2$

35. $(2t)^3(-t^2)$ 36. $(-w^3)(3w^2)^2$

37. $(abc^2)^3(a^2b)^2$ 38. $(r^2st^3)^2(s^4t)^3$

39. $(-3xy^2)^3(-2x^2y)^2$ 40. $(-5c^2d)^2(-2cd^2)^3$

Rewrite as a single number with an exponent.

Sample $8 \cdot 6^3 = 2^3 \cdot 6^3 = (2 \cdot 6)^3 = 12^3$

C 41. $16 \cdot 5^4$ 42. $64 \cdot 27$ 43. $25 \cdot 2^5 \cdot 5^3$ 44. $3^2 \cdot 18 \cdot 2^3$

■ REVIEW EXERCISES

Simplify.   [2–4, 2–5, 2–6]

1. $(-5)(-6)$ 2. $(-3)(-6) + (-2)(-8)$ 3. $(3)(-4) + (-3)(-4)$

4. $\frac{-24}{-8}$ 5. $-\frac{12}{3} + -\frac{12}{4}$ 6. $-\frac{2}{3} \div \frac{3}{4}$

7. $-5 - (-3x) - (-x)$ 8. $3x - (-7) - (-x)$

r	s	t
-3	4	-12

Substitute and simplify.   [2–4, 2–5, 2–6]

9. $rs + rt$ 10. $\frac{t}{r} + \frac{t}{s}$ 11. $r - s - t$

4–6 Multiplying a Polynomial by a Monomial

Preview

The multiplication algorithm is based on the distributive property.

$$\begin{array}{r} 231 \\ \times\ 3 \\ \hline 693 \end{array}$$

By writing the numbers in expanded form, we can see how the distributive property is applied.

$$\begin{aligned}
3(231) &= 3(200 + 30 + 1) \\
&= 3 \cdot 200 + 3 \cdot 30 + 3 \cdot 1 \\
&= 600 + 90 + 3 \\
&= 693
\end{aligned}$$

Note that the equation shows the product of a monomial and a polynomial. In this lesson, you will use the same skills with polynomials that have variables in them.

■ LESSON

To find the product of a monomial and a polynomial, use the distributive property to multiply each term of the polynomial by the monomial. Then simplify.

Example 1   Simplify. $7(3x - 2y)$

Solution $7(3x - 2y) = (7 \cdot 3x) - (7 \cdot 2y)$ *Distributive property*

$\qquad\qquad\qquad = 21x - 14y$ *Computation*

Example 2   Simplify. $4a(2a + 8)$

Solution $4a(2a + 8) = (4a \cdot 2a) + (4a \cdot 8)$

$\qquad\qquad\qquad\ = 8a^2 + 32a$

When using the distributive properties to expand expressions, be careful when multiplying negative factors.

Example 3   Simplify. $-2x(-3x + 5y)$

Solution $-2x(-3x + 5y) = (-2x)(-3x) + (-2x)(5y)$

$\qquad\qquad\qquad\qquad = (6)x^2 + (-10)xy$

$\qquad\qquad\qquad\qquad = 6x^2 - 10xy$

Example 4 Simplify. $2x^2(3x^2 - x + 7)$

Solution $2x^2(3x^2 - x + 7) = (2x^2 \cdot 3x^2) - (2x^2 \cdot x) + (2x^2 \cdot 7)$
$$= 6x^4 - 2x^3 + 14x^2$$

Example 5 Simplify. $-3a(5a - b + 2)$

Solution $-3a(5a - b + 2) = -3a[5a + (-b) + 2]$
$$= (-3a) \cdot (5a) + (-3a) \cdot (-b) + (-3a) \cdot 2$$
$$= -15a^2 + 3ab + (-6a)$$
$$= -15a^2 + 3ab - 6a$$

By substituting simple values for variables, we can check to determine whether the polynomial product is equivalent to the original expression.

Check Let $a = 1$ and $b = 2$. Then
$$-3a(5a - b + 2) = -3(5 - 2 + 2) = -15$$
$$-15a^2 + 3ab - 6a = -15 + 6 - 6 = -15 \quad \text{It checks.}$$

■ CLASSROOM EXERCISES

Simplify.

1. $3(4x + 2y)$ **2.** $2(3a - 5b)$ **3.** $-5(3x - 4)$ **4.** $x(3x + 4)$

5. $3a(2a + 1)$ **6.** $3c^2(-9c - 4c^2)$ **7.** $-3ab(a^2 + b^2)$ **8.** $-y(7 - 3y)$

9. $4a^2(2a - 3)$ **10.** $7(6a + 2b - c)$ **11.** $-2a(-5b + 3c + 2d)$

■ WRITTEN EXERCISES

Simplify.

A **1.** $4(a + 3)$ **2.** $3(b + 5)$ **3.** $5(2d + 4)$ **4.** $5(2d - 4)$

5. $-5(n - 2m)$ **6.** $-4(a - 2b)$ **7.** $-5d(3d - 2)$ **8.** $-2x(8x - 10)$

Multiply the monomial and the polynomial. Check.

9. $3(100 + 30 + 2)$ *Check:* $3 \cdot 132$ **10.** $2(400 + 20 + 3)$ *Check:* $2 \cdot 423$

11. $5(300 - 1)$ *Check:* $5 \cdot 299$ **12.** $6(200 - 1)$ *Check:* $6 \cdot 199$

13. $\frac{1}{2}(200 + 80 + 6)$ *Check:* $\frac{1}{2} \cdot 286$ **14.** $\frac{1}{3}(600 + 90 + 3)$ *Check:* $\frac{1}{3} \cdot 693$

15. $0.1(5 + 0.2)$ *Check:* $0.1 \cdot 5.2$ **16.** $0.6(4 + 0.5)$ *Check:* $0.6 \cdot 4.5$

Simplify.

B **17.** $-5r(3r^2 - 2r - 8)$ **18.** $-4s(2s^2 - 3s - 4)$ **19.** $10t^2(3t^2 + 6t + 5)$

20. $3t^2(10t^2 + 20t + 30)$ **21.** $a^2(a + b)$ **22.** $y^2(x + y)$

23. $2y^3(2x + y)$ **24.** $3a^3(2a + b)$ **25.** $-ab(a + b)$

26. $-xy(x + y)$ **27.** $-st(2s - t)$ **28.** $-mn(m - 2n)$

29. $x^2y(x + y)$ **30.** $ab^2(a + b)$ **31.** $x^2y(x + xy + y)$

32. $-ab^2(a - ab + b)$ **33.** $-ab^2(a - 2ab + b)$ **34.** $3x^2y(2x + xy - y)$

For each figure below:

a. Express the area of the shaded region as a product of a monomial and a polynomial.

b. Simplify the product.

Sample

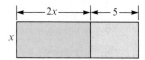

Solution **a.** $x(2x + 5)$
 b. $2x^2 + 5x$

C **35.**

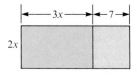

36.

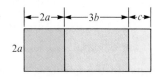

37.

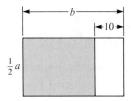

For each solid pictured below:

a. Express the volume of the solid as a product.

b. Simplify the product.

38. Triangular prism

39. Cylinder

[*Hint:* The volume equals the area of the base times the height.]

Strategy for Success Using review exercises

The review exercises after each lesson help you review mathematics from previous lessons and previous courses. Lesson numbers are given along with the review exercises so that if you have difficulty with an exercise, you can review the related lesson itself.

■ REVIEW EXERCISES

Simplify.

1. $5x - x$ **2.** $7x - x - 7$ [2–7]

3. $3x + 6y - 2x - 6$ **4.** $4x + 3y - 3x - 2y$

5. $7(x - 2)$ **6.** $3(x + 5) - (x - 6)$ [2–8]

7. $2(3a - 5) - 3(2a - 6)$ **8.** $4(a - b) - (a + 4b)$

9. $3(5a - 1) - 2(a - 3)$ **10.** $2(a - b) - (a - b)$

11. $2(a + b) - 3(b - 1)$ **12.** $3(a + b) + (a + b)$

Solve this problem by making a table of data.   [2–9]

13. Michelle deposits $1 in the bank on the first day, $2 on the second day, $4 on the third day. On which day will she have a total of $1023 in the bank?

Self-Quiz 2

4–4   **1.** Express 2,890,000 in scientific notation.

       2. Express 5.7×10^4 in standard decimal notation.

       3. Find the product $2400 \times 190,000$. Express the answer in scientific notation.

4–5   Simplify.

       4. $(y^4)^3$ **5.** $(2a^2)^4$ **6.** $(a^2b^3)^5$ **7.** $(-3b^3)^3$

4–6   Simplify.

       8. $4(2x + 5y)$ **9.** $3(8a - 6b)$

      10. $-2r(3r - 8)$ **11.** $3c^2(2a - 3b + c)$

Preview

The distance between town A and town B is 160 kilometers. Mrs. Carter and Mrs. Brown left town A in separate cars at 8:00 A.M. Mrs. Carter drove at a very steady rate and arrived in town B at 10:00 A.M. Mrs. Brown did not drive steadily. She sometimes drove 100 kilometers per hour (km/h) and sometimes 30 kilometers per hour. She even stopped once for 5 minutes to watch a hot-air balloon. However, Mrs. Brown also arrived in town B at 10:00 A.M.

- How many kilometers did each person drive?
- How much time did each trip require?
- What was each driver's average rate in kilometers per hour?
- Where was each driver at 9:00 A.M.?
- Did Mrs. Brown ever pass Mrs. Carter?

In this lesson you will learn to solve problems involving uniform motion.

■ LESSON

In the Preview, Mrs. Carter's motion is an example of **uniform motion.** That is, her speed was constant. When motion is uniform, the relationship among distance traveled (d), rate of speed (r), and time traveled (t) is given by this formula:

$$d = rt$$

The units used for the time and the "time element" of the rate must be the same. For example, if the rate is given in miles per hour and the time is given in minutes, then the time units are not the same. At least one of the units *must* be changed.

> Example 1   Mrs. Green drove at a rate of 80 km/h for 30 min. How far did she drive?
>
> *Solution 1*   Change 30 min to 0.5 h.
> $$d = rt$$
> $$d = 80(0.5)$$
> $$d = 40$$
>
> *Answer*   Mrs. Green drove 40 km.
>
> *Solution 2*   Change 80 km/h to km/min.
> $$80 \text{ km/h} = \frac{80}{60} \text{ km/min} = \frac{4}{3} \text{ km/min}$$
> $$d = rt$$
> $$d = \frac{4}{3}(30)$$
> $$d = 40$$
> *Answer*   Mrs. Green drove 40 km.

As motion problems become more complex, a drawing and/or table can make the problems easier to solve.

Example 2   Radar picked up an unidentified flying object (UFO) 5000 miles away approaching at a rate of 800 miles per hour (mph). In 15 minutes an intercepter plane (INT) was on its way to meet the UFO at a rate of 1200 mph. How long did it take the INT to reach the UFO? How far away did they meet?

Solution   First draw a figure.

UFO

$\longleftarrow$ _____

INT $\text{rate}_{UFO} = 800$ mph

_____ $\longrightarrow$

     $\text{rate}_{INT} = 1200$ mph

_____

5000 miles

The figure shows that the sum of the distances flown by the UFO and the INT is 5000 miles.

Let x = the number of hours flown by the INT.
Then $x + 0.25$ = the number of hours flown by the UFO after radar sighting. [*Remember:* 15 min = 0.25 h.]

	Rate (mph)	Time (h)	Distance (mi)
UFO	800	$x + 0.25$	$800(x + 0.25)$
INT	1200	x	$1200x$

Total: 5000

$$1200x + 800(x + 0.25) = 5000$$
$$1200x + 800x + 200 = 5000$$
$$2000x + 200 = 5000$$
$$2000x = 4800$$
$$x = 2.4$$

Answer   It took the INT 2.4 h to reach the UFO.

Now find the distance traveled by the INT.

$$d = rt$$
$$d = 1200 \cdot 2.4$$
$$d = 2880$$

Answer   The INT traveled 2880 mi before it reached the UFO.

Example 3 Mr. and Mrs. Kamm were moving from Houston to Toledo. Mr. Kamm left Houston driving one car, pulling a trailer filled with their belongings. He drove at a steady 50 mph. One-half hour later, Mrs. Kamm left, driving the other car. They both agreed that she could drive 55 mph because she was not pulling the trailer. After three hours Mr. Kamm became worried because his wife had not yet caught up with him. Should he have been worried?

Solution One way to solve this problem is to find out how long it should take Mrs. Kamm (MRS) to catch Mr. Kamm (MR).

$$\text{MR 50 mph} \longrightarrow$$

$$\text{MRS 55 mph} \longrightarrow$$

When MRS catches up with MR, they will have driven the same distance.

Let $x =$ the number of hours driven by MRS.

Then $x + 0.5 =$ the number of hours driven by MR.

	Rate (mph)	Time (h)	Distance (mi)	
MR	50	$x + 0.5$	$50(x + 0.5)$	$\}$ Distances
MRS	55	x	$55x$	are equal

$$55x = 50(x + 0.5)$$
$$55x = 50x + 25$$
$$5x = 25$$
$$x = 5$$

Answer · It will take Mrs. Kamm 5 hours to catch up with Mr. Kamm. Mr. Kamm should not have worried after only 3 hours.

Mathematics and Your Future

Some students are tempted to take easier courses in high school in order to earn a higher grade-point average. Using that reasoning to avoid mathematics courses is counterproductive. People responsible for college admissions and scholarships are interested in the courses you have taken as well as your average grade. They also look at scores on college entrance tests, all of which have major mathematics components. If you are interested in obtaining a scholarship or entering a selective college, you should take plenty of mathematics.

■ CLASSROOM EXERCISES

1. State a formula that relates distance, rate, and time.

Copy and complete the table.

	Distance	Rate	Time
2.	?	60 km/h	$\frac{1}{2}h$
3.	?	x mph	3.5 h
4.	100 km	?	2 h
5.	y km	?	5 min
6.	5 mi	40 mph	?
7.	?	55 mph	$(x + 0.5)h$

Solve.

8. Two airplanes left the same airport at the same time. One flew east at 150 mph. The other flew west at 250 mph. They are able to maintain radio contact until they are 1000 miles apart. For how many hours can they maintain radio contact?

■ WRITTEN EXERCISES

Complete these sentences.

A **1.** 300 km/h = ___?___ km/min

2. 240 km/h = ___?___ km/min

3. 300 m/min = ___?___ m/s

4. 240 m/min = ___?___ m/h

Copy and complete the tables.

	Distance	Rate	Time
5.	?	50 mph	3 h
6.	?	55 mph	2 h
7.	?	30 m/s	1.5 s
8.	?	20 m/s	2.5 s
9.	?	600 mph	x h
10.	?	350 mph	t h

	Distance	Rate	Time
11.	280 km	70 km/h	?
12.	400 km	80 km/h	?
13.	100 m	?	2 s
14.	200 m	?	10 s
15.	?	x mph	10 h
16.	?	$(x + 10)$ mph	8 h

172   Chapter 4   Polynomials

The distances for A and B are equal. Write an equation for each problem and solve for x.

17. A: Rate is x mph and time is 10 h
    B: Rate is $(x + 10)$ mph and time is 8 h

18. A: Rate is x km/h and time is 4 h
    B: Rate is $(x - 20)$ km/h and time is 5 h

19. A: Rate is 60 km/h and time is x h
    B: Rate is 80 km/h and time is $(x - 1)$ h

20. A: Rate is 80 mph and time is x h
    B: Rate is 50 mph and time is $(x + 3)$ h

The sum of the distances for A and B is 1000 km. Write an equation for each problem and solve for x.

21. A: Rate is 40 km/h and time is x h
    B: Rate is 60 km/h and time is x h

22. A: Rate is x km/h and time is 12 h
    B: Rate is x km/h and time is 13 h

23. A: Rate is 40 km/h and time is x hours
    B: Rate is 60 km/h and time is $(x + 5)$ hours

24. A: Rate is x km/h and time is 2 hours
    B: Rate is $(x + 60)$ km/h and time is 5 hours

Solve.

25. Two planes left O'Hare International Airport in Chicago at 2:00 P.M. One was flying east at 300 mph and the other was flying west at 350 mph. How many miles apart are they at 4 P.M.?

26. Two planes left Will Rogers World Airport in Oklahoma City at 3:00 P.M. One was flying south at 150 mph and the other was flying south at 250 mph. How many miles apart are they at 5 P.M.?

B **27.** A car left the Mile High Stadium in Denver driving north at 70 km/h. One hour later a second car left the Mile High Stadium driving south at 80 km/h. How long had the second car been driving when the two cars were 520 km apart?

28. A bus left the Astrodome in Houston driving west at 60 km/h. Two hours later another bus left the Astrodome driving east at 80 km/h. How long had the second bus been driving when the two buses were 820 km apart?

29. A zebra leaves a water hole running at 60 km/h. After the zebra has been running for 1.5 min, a wild dog leaves the water hole and chases after the zebra at 70 km/h. If both animals maintain their speed, how long would it take the hunting dog to catch the zebra?

30. A zebra who is running at 60 km/h passes a resting cheetah. After the zebra has been running for 1 minute, the cheetah chases after the zebra at 100 km/h. If both animals maintain their speed, in how many minutes will the cheetah catch the zebra?

31. A garden snail, crawling at 60 cm/min, goes through a chicken coop. After the snail has been crawling for 299 min, a chicken leaves the coop and chases after the snail at 18,000 cm/min. If both animals maintain their speed, in how many minutes will the chicken catch the snail?

32. A tortoise, moving at a rate of 4 m/min, passes a rabbit. The rabbit plans to catch the tortoise after 100 m by running 800 m/min. How many minutes head start can the rabbit give the tortoise?

33. Mr. Douglas and Mrs. Douglas left their home in New Orleans at the same time. Mrs. Douglas drove her car at 95 km/h and Mr. Douglas drove his car at 90 km/h. Mrs. Douglas stopped in Jackson. Mr. Douglas drove 4 hours longer and stopped in Memphis. They drove 915 km in all.
  a. How many hours did each drive?
  b. How many kilometers is Jackson from New Orleans?
  c. How many kilometers is Memphis from New Orleans?

■ REVIEW EXERCISES

State which of the numbers $\{-2, -1, 0, 1, 2\}$ are solutions to the following equations.   [3–1]

  1. $|x| + 1 = 2$ **2.** $x^2 = x$ **3.** $x^2 + x = 2$

What equations do you get by following these directions in order?   [3–2]

  4. $4x + 5 = 25$ **a.** Subtract 5 from both sides.
  b. Divide both sides by 4.

  5. $\frac{1}{2}x - 5 = 20$ **a.** Add 5 to both sides.
  b. Multiply both sides by 2.

  6. $\frac{1}{3}(x - 12) = 6$ **a.** Use the distributive property.
  b. Add 4 to both sides.
  c. Multiply both sides by 3.

Solve.   [3–3]

  7. $5x = 4x + 7$ **8.** $6x + 7 = 7x + 6$ **9.** $7x + 5.5 = 6x - 3$

174   Chapter 4   Polynomials

4–8 Dividing Monomials

Preview Discovery

In Section 4–3, we multiplied numbers with like bases by adding exponents. Determine a rule for *dividing* numbers with like bases. You should discover the rule after solving the following equations.

$\dfrac{64}{16} = \underline{\quad ? \quad}$

$\dfrac{128}{4} = \underline{\quad ? \quad}$

$\dfrac{2^6}{2^4} = 2^?$

$\dfrac{2^7}{2^2} = 2^?$

$\dfrac{1024}{128} = \underline{\quad ? \quad}$

$\dfrac{64}{64} = \underline{\quad ? \quad}$

$\dfrac{2^{10}}{2^7} = 2^?$

$\dfrac{2^6}{2^6} = ?$

$2^1 = 2$
$2^2 = 4$
$2^3 = 8$
$2^4 = 16$
$2^5 = 32$
$2^6 = 64$
$2^7 = 128$
$2^8 = 256$
$2^9 = 512$
$2^{10} = 1024$

Does your rule seem to work?

$\dfrac{4}{8} = \underline{\quad ? \quad}$ $\dfrac{32}{128} = \underline{\quad ? \quad}$ $\dfrac{64}{1024} = \underline{\quad ? \quad}$

$\dfrac{2^2}{2^3} = \dfrac{1}{2^?}$ $\dfrac{2^5}{2^7} = \dfrac{1}{2^?}$ $\dfrac{2^6}{2^{10}} = \dfrac{1}{2^?}$

In this lesson you will learn how to divide numbers with like bases.

■ LESSON

The exercises in the Preview illustrate the subtraction property of exponents.

$$\dfrac{2^7}{2^4} = \dfrac{\overbrace{2 \cdot 2 \cdot 2 \cdot 2 \cdot 2 \cdot 2 \cdot 2}^{7 \text{ factors}}}{\underbrace{2 \cdot 2 \cdot 2 \cdot 2}_{4 \text{ factors}}} = \dfrac{2 \cdot 2 \cdot 2 \cdot 2}{2 \cdot 2 \cdot 2 \cdot 2} \cdot 2 \cdot 2 \cdot 2 = 1 \cdot 2^3 = 2^3 \text{ or } 2^{7-4}$$

$$\dfrac{2^5}{2^9} = \dfrac{2 \cdot 2 \cdot 2 \cdot 2 \cdot 2}{2 \cdot 2 \cdot 2 \cdot 2 \cdot 2 \cdot 2 \cdot 2 \cdot 2 \cdot 2} = \dfrac{2 \cdot 2 \cdot 2 \cdot 2 \cdot 2}{2 \cdot 2 \cdot 2 \cdot 2 \cdot 2} = \dfrac{1}{2 \cdot 2 \cdot 2 \cdot 2} = 1 \cdot \dfrac{1}{2^4} = \dfrac{1}{2^4} \text{ or } \dfrac{1}{2^{9-5}}$$

The Subtraction Property of Exponents

For all numbers a (except 0) and for all positive integers m and n,

$$\text{if } m > n, \text{ then } \dfrac{a^m}{a^n} = a^{m-n}$$

$$\text{if } m < n, \text{ then } \dfrac{a^m}{a^n} = \dfrac{1}{a^{n-m}}.$$

The subtraction property of exponents helps us to simplify the quotient $\dfrac{a^m}{a^n}$ ($a \neq 0$)) when m and n are *not* equal. We already know that when m and n *are* equal, the quotient is 1 (that is, any number (except 0) divided by itself is equal to 1). Note what happens if we subtract equal exponents.

$$\text{If } m = n, \frac{a^m}{a^n} = \frac{a^m}{a^m} = a^{m-m} = a^0.$$

The example given above suggests this definition.

Definition: Zero Exponents

For all numbers a (except 0),

$$a^0 = 1.$$

For example,

$$\frac{2^5}{2^5} = 2^{5-5} = 2^0 = 1$$

$$\frac{x^3}{x^3} = x^{3-3} = x^0 = 1 \quad (x \neq 0)$$

Study these examples to see how you can use the subtraction property of exponents to simplify the quotient of two monomials.

Example 1 Simplify. $\dfrac{x^4}{x^3}$ $(x \neq 0)$

Solution $\qquad \dfrac{x^4}{x^3} = x^{4-3} = x^1 = x$

Example 2 Simplify. $\dfrac{-12m^3n^2}{-9mn^4}$ $(m \neq 0,\ n \neq 0)$

Solution $\dfrac{-12m^3n^2}{-9mn^4} = \left(\dfrac{-12}{-9}\right)\left(\dfrac{m^3}{m^1}\right)\left(\dfrac{n^2}{n^4}\right) = \dfrac{4}{3}(m^2)\left(\dfrac{1}{n^2}\right) = \dfrac{4m^2}{3n^2}$

Example 3 Simplify. $\dfrac{-x^2y^3}{xy}$ $(x \neq 0,\ y \neq 0)$

Solution $\qquad \dfrac{-x^2y^3}{xy} = (-1)\left(\dfrac{x^2}{x}\right)\left(\dfrac{y^3}{y}\right) = (-1)(x)(y^2) = -xy^2$

Example 4 Simplify. $\dfrac{2x^7}{5x^7}$ $(x \neq 0)$

Solution $\qquad \dfrac{2x^7}{5x^7} = \left(\dfrac{2}{5}\right)\left(\dfrac{x^7}{x^7}\right) = \dfrac{2}{5}x^0 = \dfrac{2}{5} \cdot 1 = \dfrac{2}{5}$

Example 5 Write the quotient using scientific notation.

$$\frac{3.7 \times 10^8}{5.4 \times 10^5}$$

Solution

$$\frac{3.7 \times 10^8}{5.4 \times 10^5} = \left(\frac{3.7}{5.4}\right)\left(\frac{10^8}{10^5}\right) \approx (0.69)(10^3)$$

$$= (0.69)(10)(10^2)$$

$$= 6.9 \times 10^2$$

■ CLASSROOM EXERCISES

Simplify. Assume that no variable is equal to 0.

1. $\dfrac{x^5}{x^3}$

2. $\dfrac{x^3}{x^5}$

3. $\dfrac{2y^4}{8y^2}$

4. $\dfrac{3^2y^5}{3y}$

5. $\dfrac{8a^2}{12a^4}$

6. $\dfrac{6a^3b}{9ab^2}$

7. $\dfrac{-15x^3y^3}{6x^4y}$

8. $\dfrac{-8m^3n^5}{-6m^2n^7}$

9. x^0

10. $\dfrac{x^7}{x^7}$

Write the quotient using scientific notation.

11. $\dfrac{6 \times 10^5}{3 \times 10^2}$

12. $\dfrac{3.2 \times 10^4}{8 \times 10^1}$

■ WRITTEN EXERCISES

Simplify. Assume that no variable is equal to 0.

A 1. $\dfrac{c^8}{c^2}$

2. $\dfrac{a^{12}}{a^4}$

3. $\dfrac{b^3}{b^{12}}$

4. $\dfrac{x^2}{x^{10}}$

5. $\dfrac{15a^8}{10a^6}$

6. $\dfrac{18b^{10}}{12b^6}$

7. $\dfrac{24c^3}{18c^9}$

8. $\dfrac{24d^4}{15d^8}$

9. $\dfrac{8j^5}{8j^4}$

10. $\dfrac{6k^7}{6k^6}$

11. $\dfrac{10p^5}{2p^5}$

12. $\dfrac{9q^6}{3q^6}$

13. $\dfrac{-15r^2s^3}{10rs}$

14. $\dfrac{-12x^4y^6}{8xy}$

15. $\dfrac{x^3y^6}{x^3y^7}$

16. $\dfrac{a^2y^4}{a^3y^4}$

17. $\dfrac{-bc}{b^2c}$

18. $\dfrac{-rs}{rs^2}$

19. $\dfrac{5x^4}{5x^4}$

20. $\dfrac{3y^3}{3y^3}$

Solve.

21. $\dfrac{2^{12}}{2^4} = 2^x$

22. $\dfrac{2^{10}}{2^2} = 2^x$

23. $\dfrac{3^4}{3^3} = 3^x$

24. $\dfrac{3^5}{3^4} = 3^x$

25. $\dfrac{4^5}{4^6} = x$

26. $\dfrac{5^3}{5^4} = x$

27. $\dfrac{2^{10}}{2^7} = x$

28. $\dfrac{2^8}{2^6} = x$

Solve.

B 29. $\dfrac{3 \cdot 2^5}{6 \cdot 2^3} = x$ **30.** $\dfrac{5 \cdot 2^7}{3 \cdot 2^9} = x$ **31.** $\dfrac{9 \cdot 2^8}{3 \cdot 2^9} = x$ **32.** $\dfrac{15 \cdot 2^8}{5 \cdot 2^{10}} = x$

33. $\dfrac{3^8 \cdot 2^{100}}{3^9 \cdot 2^{100}} = x$ **34.** $\dfrac{3^x \cdot 2^5}{3^7 \cdot 2^4} = 6$ **35.** $\dfrac{5^4 \cdot 2^6}{5^5 \cdot 2^x} = 0.1$ **36.** $\dfrac{2^5 x}{2^3} = -12$

Write the quotient using scientific notation.

37. $\dfrac{7.5 \times 10^7}{2.5 \times 10^2}$ **38.** $\dfrac{5.3 \times 10^8}{4.6 \times 10^5}$ **39.** $\dfrac{7.1 \times 10^5}{8.8 \times 10^2}$ **40.** $\dfrac{8.7 \times 10^9}{9.2 \times 10^7}$

Each number in a box is the product of the numbers in the two boxes directly below it. Copy the figures and use the pattern to fill in the blank boxes.

41.

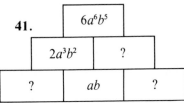

42.

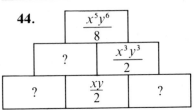

43.

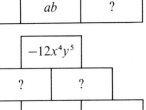

44.

C The average depth of these bodies of water may be estimated by dividing the volume by the area. Write the average depth (in meters) using scientific notation.

	Body of water	Volume (m^3)	Area (m^2)
45.	Persian Gulf	2.3×10^{13}	2.3×10^{11}
46.	Pacific Ocean	6.8×10^{17}	1.7×10^{14}
47.	Gulf of California	10.5×10^{14}	1.5×10^{11}
48.	Caribbean Sea	6.5×10^{15}	2.5×10^{12}

■ REVIEW EXERCISES

Solve each equation.

1. $4x = -7$ **2.** $\dfrac{1}{3}x = 15$ **3.** $\dfrac{3}{4}x = 24$ [3–4]

4. $5x + 7 = 7x$ **5.** $\dfrac{x + 4}{5} = 6$ **6.** $\dfrac{x - 5}{2} = 10$ [3–5]

7. $3x + 4 = 5(4 - x) + 8$ **8.** $5x - 10 = 3(x + 5)$ **9.** $6(x - 4) = 4x$ [3–6]

4-9 Problem Solving — Money Problems

Preview

Many practical problems encountered every day involve money. Some puzzles involve money, too. Try solving this one:

> A student had 14 coins — some dimes and some nickels. Their value was 90 cents. How many of the coins were nickels?

The puzzle can be solved by systematic trials. But in this lesson you will learn another method. Using equations is a powerful problem-solving approach that has many applications.

■ LESSON

In most money problems, you are given two kinds of quantities: *numbers* of items and *values* of items.

Items	Number of items	Value in cents	Value in dollars
Dimes	5	10×5	0.1×5
Nickels	n	$5n$	$0.05n$
\$35 auto tires	m	$3500m$	$35m$
25¢ pens	$x + 2$	$25(x + 2)$	$0.25(x + 2)$

A table can help organize information in money problems. It can also make it easier to recognize expressions that must be equivalent. These equivalent expressions can then be used to write an equation for the problem.

Example 1 Mrs. Cooper bought four tires for her car. The cost, including \$11.04 in taxes, was \$194.96. What was the price for one tire before taxes?

Solution Let x = the cost of one tire in dollars.

Cost of 1 tire in dollars	Cost of 4 tires in dollars	Total cost in dollars
x	$4x$	$4x + 11.04$ or 194.96

Example 1 (continued)

Since the cost of the tires plus tax equals $194.96, we have:

$$4x + 11.04 = 194.96$$
$$4x = 183.92$$
$$x = 45.98$$

Answer The cost of 1 tire before taxes is $45.98.

Check Cost of 4 tires = 4 × $45.98 = $183.92
(before taxes)
Tax = $11.04
Total cost = $183.92 + $11.04 = $194.96   It checks.

Recall these two important points for solving written problems.

1. When solving a written problem using an equation, you are not finished when you have solved the equation. The solution of an equation is a number. You must interpret the number in terms of the question asked. The problem in Example 1 did not ask for a number but an amount of money.

2. When checking a written problem, go back to the problem itself instead of just substituting the answer into the equation. You may have solved the equation correctly, but it could have been the wrong equation.

Example 2 Twenty-five coins (dimes and nickels) are worth $1.80. How many coins of each kind are there?

Solution 1 Let d = the number of dimes.

	Number	Value in cents
Dimes	d	$10d$
Nickels	$25 - d$	$5(25 - d)$

Total value: 180

The sum of the values of the dimes and nickels must equal 180 cents. The table gives the following equation:

$$10d + 5(25 - d) = 180$$
$$10d + 125 - 5d = 180$$
$$5d = 55$$
$$d = 11 \qquad 25 - 11 = 14$$

Answer There are 11 dimes and 14 nickels.

Check Value of dimes = 11 × $0.10 = $1.10

Value of nickels = 14 × $0.05 = $0.70

Total value = $1.10 + $0.70 = $1.80   It checks.

Example 2 (continued)

Solution 2   Let n = the number of nickels.

	Number	Value in cents
Nickels	n	$5n$
Dimes	$25 - n$	$10(25 - n)$

Total value: 180

The table gives the following equation:

$$5n + 10(25 - n) = 180$$
$$5n + 250 - 10n = 180$$
$$-5n = -70$$
$$n = 14$$
$$25 - n = 11 \qquad 25 - 14 = 11$$

Answer   There are 14 nickels and 11 dimes.

■ CLASSROOM EXERCISES

1. What is the value in cents of x quarters?

2. What is the value in dollars of y dimes?

3. Soup costs 39 cents per can. What is the cost in cents of g cans of soup?

4. Soup costs c cents per can. What is the cost in cents of 5 cans of soup?

5. If 4 apples cost 64¢, what does 1 apple cost?

6. If b apples cost 68¢, what does 1 apple cost?

7. If c apples cost d¢, what does 1 apple cost?

8. There are twice as many dimes as quarters. If there are q quarters, then:
   a. How many dimes are there?
   b. What is the value of the dimes in cents?
   c. What is the value of the quarters in cents?

Define the variables, write the equations, and then solve.

9. There are 2 times as many quarters as dimes in a stack worth $4.20. How many coins of each type are there?

10. There are 4 more nickels than dimes in a stack worth $1.55. How many coins of each type are there?

A 1. What is the value in cents of h half dollars?

2. What is the value in cents of d dimes?

3. What is the value in dollars of d dimes?

4. What is the value in dollars of q quarters?

5. Stamps cost 30¢ each. What is the cost in cents of s stamps?

6. Frozen vegetables cost 79¢ per package. What is the cost in cents of p packages?

7. If 6 cans of juice cost j dollars, what does 1 can of juice cost?

8. If 12 rolls cost r cents, what does 1 roll cost?

9. There are 4 more nickels than dimes. If there are d dimes, then:
   a. How many nickels are there?
   b. What is the value of the nickels in cents?
   c. What is the value of the dimes in cents?

10. There is a total of 25 dimes and quarters. If there are d dimes, then:
    a. How many quarters are there?
    b. What is the value of the dimes in cents?
    c. What is the value of the quarters in cents?

Write an expression for each situation. Let x represent the cost in dollars for 1 tire.

11. The cost of 2 tires

12. The cost of 4 tires

13. Half the cost of a tire

14. $\frac{3}{4}$ the cost of a tire

15. The cost of a tire and $4.20 tax

16. The cost of a tire and $5.00 for mounting and balancing

Solve.

17. The cost of 3 bottles of milk is $6.15. What is the cost of 1 bottle of milk?

18. The cost of 4 notebooks is $9.00. What is the cost of 1 notebook?

19. The cost of 5 cans of oil, including 26¢ tax, is $4.51. What is the cost of 1 can of oil before taxes?

20. The cost of 2 boxes of laundry detergent, including $0.63 tax, is $11.13. What is the cost of 1 box of detergent before taxes?

21. An electrician charges $25, plus $20 per hour, for making a service call. How many hours were charged if the total bill was $75?

22. An auto mechanic charges $20, plus $25 per hour. How many hours were charged if the total bill was $57.50?

B 23. There are 2 times as many dimes as nickels in a stack of coins worth $1.50. How many nickels are in the stack?

24. There are 2 times as many nickels as dimes in a stack of coins worth $1.60. How many dimes are in the stack?

25. A stack of coins is worth $3.25. There are 5 more nickels than quarters. How many quarters are in the stack?

26. A stack of coins is worth $2.75. There are 5 more quarters than nickels. How may nickels are in the stack?

27. A stack of 30 coins consists of dimes and quarters. How many of the coins are dimes if the stack is worth $4.65?

28. A stack of 30 coins consists of dimes and quarters. How many of the coins are dimes if the stack is worth $5.10?

C 29. A shopper bought some cans of soup at 39¢ per can and some packages of frozen vegetables at 59¢ per package. Twice as many packages of vegetables were purchased as cans of soup. How many packages of vegetables were purchased if the total bill was $9.42?

30. A club sold student tickets for $3.25 and full-price tickets for $5.25, and 16 more student tickets were sold than full-price tickets. How many student tickets were sold if the total sales were $2559.50?

31. Solve this problem in two ways: First, let x represent the number of dimes; then let x represent the number of quarters. Explain which of the five steps in the two solutions are different and which are alike.

A stack of dimes and quarters is worth $7.60. It has 6 more dimes than quarters. How many coins of each type are in the stack?

■ REVIEW EXERCISES

Solve.

1. $5(x - 2) = 40$

2. $6a + 2(a + 3) = 10$ [3–7]

3. $5(y - 6) = 30$

4. $3(x - 5) = 2(5 - x)$

Solve for the indicated variable.   [3–8]

5. $A = \frac{1}{2}bh$, for h

6. $V = lwh$, for w

7. $A = \frac{1}{2}h(b_1 + b_2)$, for h

8. $A = \frac{1}{2}h(b_1 + b_2)$, for b_1

Self-Quiz 3

4–7   Copy and complete the chart.

	r (rate)	t (time)	d (distance)
1.	42 mph	1.5 h	?
2.	1.6 m/s	?	14.4 m
3.	?	2h 15 min $\left(2\frac{1}{4} \text{ h}\right)$	189 km

Solve.

4. A bicyclist riding at 30 mph leaves Metro City at 10 A.M. One hour later, a second cyclist riding at 25 mph travels in the opposite direction. At what time will the cyclists be 250 miles apart?

4–8   Simplify.

5. $\dfrac{x^3}{x^4}$

6. $\dfrac{2z^{10}}{6z^2}$

7. $\dfrac{14p^4q^7}{7p^4q^5}$

8. $\dfrac{-ab^2}{a^2b}$

4–9   Complete.

4–1   9. Write an expression for the value in cents of $(n + 1)$ quarters.

10. A stack of dimes and quarters is worth $2.85. If there are 4 more dimes than quarters, how many of each kind of coin are there?

EXTENSION   Fuel economy

The fuel economy of an automobile depends on many factors, including air resistance, the gear being used, and the rate of acceleration. Air resistance depends on the speed and shape of the automobile. Air resistance increases by a factor that is the square of the factor by which the speed is increased. For example, if the speed is tripled, the air resistance is multiplied by 9. If the speed is increased by $\frac{5}{4}$, the air resistance is increased by a factor of $\frac{25}{16}$.

By what factor is the air resistance increased when the speed is increased from 50 mph to 55 mph?

The best fuel economy is achieved when the automobile is driven at the lowest constant speed possible in the highest gear possible.

4-10 Problem Solving — Integer Problems

Preview

In cross-country running, the number of points a runner earns for the team is equal to the number of the runner's finishing position. For example, the 3rd-place finisher earns 3 points and the 27th-place finisher earns 27 points. The team score is the sum of the scores of its top five finishers and the team with the lowest score wins. In one meet, five runners for one team finished one right after another. Their team's score was 55. In what places did the runners finish?

In this lesson you will learn to solve problems like this one by writing equations.

■ LESSON

Tuesday, Wednesday, and Thursday are *consecutive* days. October, November, and December are *consecutive* months. Integers that "follow" one after another are called **consecutive integers.** In a sequence of consecutive integers, each is one greater than the integer that precedes it.

$$7, 8, 9, 10 \qquad \text{are consecutive integers}$$
$$-4, -3, -2, -1, 0 \qquad \text{are consecutive integers}$$

If n is an integer, then:

$$n, n + 1, n + 2, n + 3 \qquad \text{are consecutive integers}$$
$$n - 3, n - 2, n - 1, n, n + 1 \qquad \text{are consecutive integers}$$

The difference between two consecutive *even* integers *or* two consecutive *odd* integers is *always* two.

$$-4, -2, 0, 2, 4, 6, 8, \ldots, \qquad \text{are } \textit{consecutive even} \text{ integers, and}$$
$$-3, -1, 1, 3, 5, 7, 9, \ldots, \qquad \text{are } \textit{consecutive odd} \text{ integers}$$

If n is an *even* integer, then $n - 4, n - 2, n, n + 2, n + 4, n + 6, \ldots,$ are consecutive *even* integers.

However, if n is an *odd* integer, then $n - 4, n - 2, n, n + 2, n + 4, n + 6, \ldots,$ are consecutive *odd* integers.

Example 1 The sum of three consecutive integers is 132. What are the integers?

Here are three solutions based on three different ways to define variables. Study each of these solutions.

Solution 1 Let $n =$ the smallest integer. Then $n + 1 =$ the next larger integer, and $n + 2 =$ the largest integer.

$$n + (n + 1) + (n + 2) = 132$$
$$3n + 3 = 132$$
$$3n = 129$$
$$n = 43 \leftarrow \text{the smallest integer}$$

Answer The integers are 43, 44, 45.

Solution 2 Let $n =$ the middle integer. Then $n - 1 =$ the smallest integer, and $n + 1 =$ the largest integer.

$$n + (n - 1) + (n + 1) = 132$$
$$3n = 132$$
$$n = 44 \leftarrow \text{the middle integer}$$

Answer The integers are 43, 44, 45.

Solution 3 Let $n =$ the largest integer. Then $n - 1 =$ the middle integer, and $n - 2 =$ the smallest integer.

$$n + (n - 1) + (n - 2) = 132$$
$$3n - 3 = 132$$
$$3n = 135$$
$$n = 45 \leftarrow \text{the largest integer}$$

Answer The integers are 43, 44, 45.

Example 2 Find two consecutive odd integers whose sum is $- 36$.

Solution Let $n =$ the smaller integer. Then $n + 2 =$ the larger integer.

$$n + (n + 2) = - 36$$
$$2n + 2 = - 36$$
$$2n = - 38$$
$$n = - 19 \leftarrow \text{the smaller integer}$$
$$n + 2 = - 17 \leftarrow \text{the larger integer}$$

Answer The integers are $- 19$ and $- 17$.

Always be aware that the solution of the equation may be one number, but the answer to the problem may be more than one number. In Example 1, the solution to the problem is 43, 44, and 45, not just 43.

7. If x and y are positive integers and $a \neq 0$, then $\dfrac{(a^x)^y}{a^x} = \underline{\ ?\ }$.

A. $a^{x(y-1)}$ **B.** a^y **C.** $a^{xy} - a^x$

D. $a^{x+y/x}$ **E.** $a^{x(y-x)}$

8. If $(2^5)(3^4) = 2(6^n)$, then $n = \underline{\ ?\ }$.

    A. 2   **B.** 3   **C.** 4   **D.** 5   **E.** 8

9. $xy = x + \underline{\ ?\ }$

    A. $x(y-1)$ **B.** $x(1-y)$ **C.** $(x-1)y$ **D.** $(1-x)y$ **E.** $y-x$

10. For all numbers m and all numbers $n \neq 0$, $\begin{Bmatrix} m \\ n \end{Bmatrix}$ is defined as $\dfrac{m^2}{n}$. Then $\begin{Bmatrix} 2 \\ 3 \end{Bmatrix} \cdot \begin{Bmatrix} 3 \\ 4 \end{Bmatrix} = \underline{\ ?\ }$.

A. $\dfrac{1}{4}$ **B.** $\dfrac{4}{3}$ **C.** $\dfrac{3}{2}$ **D.** 3   **E.** $\dfrac{9}{4}$

11. The first number of a sequence is $1^2 + 1$. The second number is $2^2 + 1$, the third number is $3^2 + 1$, and so on. Which of the following numbers in the sequence will be odd?

    I. The 99th number   II. The 100th number   III. The 101st number

A. I only   **B.** II only   **C.** III only

D. I and III only   **E.** I, II, and III

12. How many different four-digit numbers can be formed using the digits 1, 2, 3, and 4 one time each? [*Example:* 1234, 1243.]

    A. 4   **B.** 6   **C.** 8   **D.** 16   **E.** 24

13. $\dfrac{\dfrac{2}{3} + \dfrac{2}{3} + \dfrac{2}{3}}{\dfrac{3}{2} + \dfrac{3}{2} + \dfrac{3}{2} + \dfrac{3}{2}} = ?$

A. $\dfrac{1}{3}$ **B.** $\dfrac{4}{9}$ **C.** $\dfrac{2}{3}$ **D.** 1   **E.** $\dfrac{9}{4}$

Strategy for Success   Beginning a chapter

> Think about the "major plot" that is being developed in the chapter. Turn to the table of contents, read the chapter title, and read each section title. Think about how one section leads to the next, and how this chapter relates to the other chapters.

5 Graphing Linear Equations and Functions

The slope is the ratio of the change in height (rise or decline) for a given distance (run). It is used in many descriptions of steepness.

$$\text{slope} = \frac{\text{rise (or decline)}}{\text{run}}$$

The rate of change in the height of a riverbed is used to divide the course of a river into three parts—upper, middle, and lower. In the upper course of a river, the slope of the channel, called the gradient, is steep (usually more than 50 feet per mile) and the current is swift.

Preview

What do these three statements have in common?

- My address is 1234 W. 25th Street.
- Our ship is located at 48° N latitude and 30° W longitude.
- A theater ticket is for row 10 seat 15.

Each statement is an example of locating a position by giving a *pair* of quantities: 1234 and W. 25th St., 48° N latitude and 30° W longitude, and 10th row, 15th seat.

In this lesson you will study a system for locating points by using pairs of numbers.

■ LESSON

A number line is constructed by drawing a line, picking a point for 0, and then picking a point for 1.

Once that is done, the unit distance between 0 and 1 determines the location of other numbers. Each point on the line has a number associated with it, and every number has a point associated with it. The number is the **coordinate** of the point and the point is the **graph** of the number. The coordinate is the "address" of the point.

If we wish to assign "addresses" to points in a plane, we must use pairs of numbers. To assign a pair of numbers to each point, select two perpendicular number lines, called **axes.** The horizontal axis is usually called the **x-axis,** and the vertical axis is usually called the **y-axis.** The point of intersection of the two axes is called the **origin.** The axes divide the plane into four regions called **quadrants.** The axes are not a part of any quadrant; they are boundaries of the quadrants.

Point A is 3 units to the left of the y-axis and 4 units above the x-axis. The "address" of point A is $(-3, 4)$. Note the order of the two numbers in the number pair. The first component of the pair gives the direction and distance to move from the y-axis, and the second component gives the direction and distance to move from the x-axis. The number pair $(-3, 4)$ is called an **ordered pair** since the order of the numbers is important.

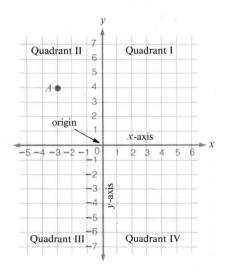

Point *B* is the graph of $(4, -1)$ and the numbers 4 and -1 are the coordinates of point *B*. The first coordinate, 4, is called the **x-coordinate.** The second coordinate, -1, is called the **y-coordinate.** (The *x*-coordinate is also called the **abscissa** of the point, and the *y*-coordinate the **ordinate** of the point.)

The coordinates of point *C* are $(2, 5)$ and of point *D* are $(5, 2)$. Note that points *C* and *D* are different points and that $(2, 5)$ and $(5, 2)$ are different ordered pairs. Ordered pairs are equal only if they have both the same **first components** and the same **second components.**

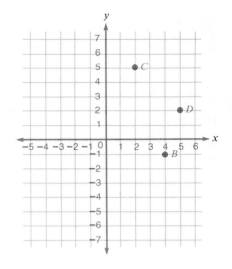

Definition: Equal Ordered Pairs

For all real numbers *a, b, c,* and *d,*

$(a, b) = (c, d)$ if and only if $a = c$ and $b = d$.

Example 1 Graph $(-2, -3)$.

Solution

Draw a pair of axes. Start at the origin $(0, 0)$. The *x*-coordinate indicates the direction and distance to move horizontally (move *left* 2 units). The *y*-coordinate indicates the direction and distance to move vertically (move *down* 3 units).

Make a dot and label it $(-2, -3)$.

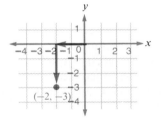

Example 2 Graph $\left(3, 1\frac{1}{2}\right)$.

Solution

Start at the origin. Move 3 units to the right, then $1\frac{1}{2}$ units up.

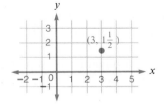

■ CLASSROOM EXERCISES

State the coordinates of each point.

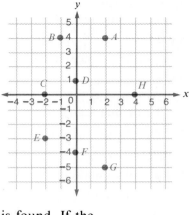

1. A 2. B

3. C 4. D

5. E 6. F

7. G 8. H

9. Name the points that have the same x-coordinate.

10. Name the points that have the same y-coordinate.

11. What are the coordinates of the origin?

Identify the quadrant in which the graph of each ordered pair is found. If the graph is on an axis, state which one.

12. $(5, 3)$ 13. $(5, -3)$ 14. $\left(-2, 1\frac{1}{2}\right)$ 15. $(0, -2)$ 16. $(-3, -2)$

■ WRITTEN EXERCISES

State the coordinates of each point.

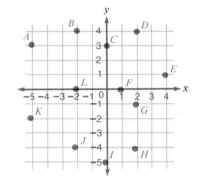

 1. A 2. B 3. C 4. D

5. E 6. F 7. G 8. H

9. I 10. J 11. K 12. L

Identify the quadrant each point is in or the axis it is on.

13. A 14. B 15. C 16. D

17. E 18. F 19. G 20. H

21. I 22. J 23. K 24. L

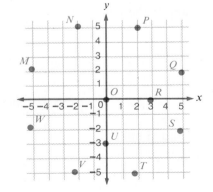

Name the point having these coordinates.

25. $(5, 2)$ 26. $(2, 5)$

27. $(-5, 2)$ 28. $(-2, 5)$

29. $(3, 0)$ 30. $(0, -3)$

31. $(-5, -2)$ 32. $(-2, -5)$

33. Graph these points: $A(-3, 4)$, $B(2, 4)$, $C(-5, -2)$, and $D(4, -2)$.
Draw line segments from A to B, B to C, C to D, and D to A.

34. Graph these points: $E(-2, 4)$, $F(0, 8)$, $G(4, 8)$, and $H(0, 0)$.
Draw line segments from E to F, F to G, G to H, and H to E.

35. Graph these points: $J(-5, 0)$, $K(-3, 3)$, $L(3, 3)$, $M(3, -2)$, and $N(-1, -3)$.
Draw line segments from J to K, K to L, L to M, M to N, and N to J.

36. Graph these points: $P(3, -4)$, $Q(-2, -4)$, $R(-4, 2)$, and $S(5, 2)$.
Draw line segments from P to Q, Q to R, R to S, and S to P.

Identify the quadrant in which each point is located.

B 37. The x-coordinate is positive and the y-coordinate is negative.

38. The x-coordinate is negative and the y-coordinate is positive.

39. The x-coordinate is positive and the y-coordinate is greater than the x-coordinate.

40. The x-coordinate is negative and the y-coordinate is less than its x-coordinate.

41. The sum of the coordinates is positive and the product of the coordinates is positive.

42. The x-coordinate is negative and the y-coordinate is -5 times the x-coordinate.

Coordinates are given for three vertices of a square. Give the coordinates of the fourth vertex.

43. $(0, 4)$, $(0, 0)$, $(4, 0)$

44. $(2, 3)$, $(-3, 3)$, $(-3, -2)$

C 45. $(2, 3)$, $(0, 0)$, $(3, -2)$

46. $(2, -3)$, $(-3, -2)$, $(-2, 3)$

Graph and label five points that meet the given condition(s).

47. The x-coordinate and the y-coordinate are equal.

48. The x-coordinate and the y-coordinate are opposites.

49. The sum of the coordinates is 5.

■ REVIEW EXERCISES

a	b	x	y
3	-4	12	-6

Substitute and simplify.

1. $ax + by$

2. $x - 3y$

3. $a^2 + b^2$

4. $(a + b)^2$ [2–4]

5. $\dfrac{x}{b} + \dfrac{y}{a}$

6. $\dfrac{(-ax + b)}{y}$

7. $\dfrac{x}{a} + \dfrac{y}{b}$

8. $\dfrac{(x + y)}{(a + b)}$ [2–5]

5-2 Graphing Relations

Preview History

The principles of coordinate geometry based on ordered pairs of numbers were developed by René Descartes in 1637 while he was still in his teens. A popular, but possibly fictional, version of his discovery is told as follows. Because of poor health, he had to spend much of his time in bed. One day while lying on his back he watched a fly crawling around on the ceiling. He discovered that the fly's position on the ceiling could be determined by an ordered pair of numbers, the first number being the distance of the fly from one wall and the second number being the fly's distance from a wall perpendicular to the first wall. Extending this idea, he saw that all points in a plane could be represented as ordered pairs of real numbers. This invention by Descartes was called by one English writer "the greatest single step ever made in the progress of exact science."

■ LESSON

A **relation** is a set of ordered pairs. Three common ways of describing a relation are as follows.

1. Listing the pairs.

$$\left\{ (2,\ 3),\ (-1,\ 4),\ (0,\ 1),\ \left(5,\ \frac{1}{4}\right) \right\}$$

2. Making a table.

x	y
2	3
-1	4
0	1
5	$\frac{1}{4}$

3. Making a graph.

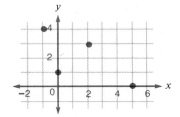

For the relation described above, the first components of the ordered pairs are 2, -1, 0, and 5. The set of first components of a relation is called the **domain of the relation.** The second components of the above relation are 3, 4, 1, and $\frac{1}{4}$. The set of second components of a relation is called the **range of the relation.**

Example 1 Given the following table, list the ordered pairs in the relation. State the domain and the range. Then graph the relation.

x	-2	0	1	3	$\frac{1}{2}$
y	-1	-3	3	0	$1\frac{1}{2}$

Example 1 (continued)

Solution The relation is $\left\{(-2, -1), (0, -3), (1, 3), (3, 0), \left(\frac{1}{2}, 1\frac{1}{2}\right)\right\}$.

The domain is $\left\{-2, 0, 1, 3, \frac{1}{2}\right\}$.

The range is $\left\{-1, -3, 3, 0, 1\frac{1}{2}\right\}$.

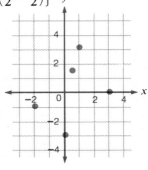

Example 2 List the ordered pairs in the relation shown on the graph. State the domain and the range.

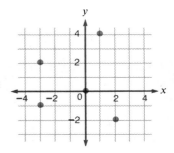

Solution The relation is $\{(-3, -1), (-3, 2), (0, 0), (2, -2), (1, 4)\}$.
The domain is $\{-3, 0, 2, 1\}$.
The range is $\{-1, 2, 0, -2, 4\}$.

Relations are sometimes described in words.

Example 3 List and graph five ordered pairs in the relation.

Relation: The set of all ordered pairs with first and second components equal.

Solution

x	y
1	1
-3	-3
2	2
$\dfrac{1}{2}$	$\dfrac{1}{2}$
0	0

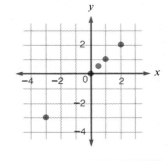

In Example 3 only five ordered pairs were listed and graphed. The graph of the whole relation is a straight line since the domain and the range is the set of real numbers.

■ CLASSROOM EXERCISES

1. Define "relation."

2. Is the set $\{-2, 0, 2, 4, 6\}$ a relation? Why or why not?

3. Is the set $\{(2, 1), (3, 2)\}$ a relation? Why or why not?

4. State the domain of the relation $\{(3, 1)\}$.

List the ordered pairs in each relation. State the domain and the range.

5.

x	y
-1	4
-2	2
-2	3
2	4

6.

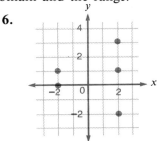

7. Graph the relation $\{(1, -1), (3, 2), (5, -1), (-3, 0)\}$.

8. Relation T is the set of all ordered pairs in which the first and second components are opposites.

    a. List 6 ordered pairs in Relation T.
    b. Graph the ordered pairs that you listed.
    c. Guess what the graph of the whole relation would be.
    d. Test your guess by listing more ordered pairs in the relation and graphing them.
    e. What are the domain and range of Relation T?

■ WRITTEN EXERCISES

State whether the set is a relation.

A   **1.** $\{(1, 2), (1, 3), (1, 4)\}$ **2.** $\{(1, 4), (2, 3), (3, 2), (4, 1)\}$

3. $\left\{0, \frac{1}{2}, 1, 1\frac{1}{2}\right\}$ **4.** $\{-3, -2, -1\}$

5. $\{(1, 2, 3), (4, 5, 6)\}$ **6.** $\{(1, 3, 5), (2, 4, 6)\}$

List the ordered pairs in the relation shown by the table.

7.

x	y
2	-2
-3	3
0.5	-0.5
-100	100

8.

x	y
-1	1
-10	10
5	-5
-0.8	0.8

9.

x	y
6	6
-6	6
8	8
-8	8

10.

x	y
-3	3
-1	1
1	1
3	3

List the domain and range of each relation.

11. {(1, 2), (3, 4), (5, 6)}

12. {(10, 9), (8, 7)}

13. $\left\{\left(\dfrac{1}{2}, \dfrac{1}{3}\right)\right\}$

14. {(0, 100)}

15.

x	5	-1	-4
y	6	0	-3

16.

x	3	-4	0
y	-1	6	2

17. Relation P is the set of ordered pairs in which both components are positive integers having a sum of 4.

18. Relation Q is the set of ordered number pairs in which the first component is a positive integer less than 5 and the second component is twice as large as the first component.

19. The graph of Relation A is shown.
  a. List the ordered pairs in Relation A.

  b. List the domain.

  c. List the range.

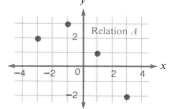

20. The graph of Relation B is shown.
  a. List the ordered pairs in Relation B.

  b. List the domain.

  c. List the range.

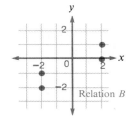

Graph the following relations.

21. {(−3, 3), (−1, 1), (0, −2), (0, 4)}

22. {(−3, −3), (−1, −1), (0, 2), (2, 4)}

23.

x	4	3	2	1
y	0	−1	−2	−3

24.

x	−4	−2	0	2
y	0	2	4	6

The following exercises describe a relation.

a. List the ordered pairs in the relation.
b. Graph the relation.

c. List the domain.
d. List the range.

B **25.** The components are integers having a product of 10.

26. The components are integers having a product of 6.

27. The first component is a negative integer greater than − 6 and the second component is the absolute value of the first component.

28. The first component is a positive integer less than 6 and the second component is the opposite of the first component.

29. The first component is a negative integer greater than -5 and the second component is the square of the first component.

30. The first component is a negative integer greater than -5 and the second component is the cube of the first component.

C **31.** The first and second components of a relation are positive integers less than 6 and the second component is less than the first component.

32. The first and second components of a relation are positive integers less than 6 and the second component is equal to the first component.

33. The first component x and the second component y of a relation are integers such that $|x|$ and $|y| = 3$.

34. The first component x and the second component y of a relation are integers such that $x^2 + y^2 = 5$.

35. The first component x and the second component y of a relation are integers such that $x^2 + y^2 = 25$.

■ REVIEW EXERCISES

w	x	y	z
-2	8	-6	-12

Substitute and simplify.

1. $3x - 2y$

2. $x - y - 2z$ [2–4]

3. $2w - 3x + 4y$

4. $\dfrac{1}{2}x - \dfrac{2}{3}y$

5. $\dfrac{w}{x} + \dfrac{y}{z}$

6. $\dfrac{w + x}{y - z}$ [2–5]

7. $\dfrac{x - y}{w - z}$

8. $\dfrac{1}{x} + \dfrac{1}{y}$

Solve by making and using a table of data.

9. Andrea earns $1 the first day, $3 the second day, $5 the third day, and so on, earning $2 more each day. How many days will it take her to earn a total of $100?   [2–9]

10. Jane starts with $60 and earns $5 a day. Sue starts with $25 and earns $10 a day. In how many days will Sue have as much money as Jane?   [2–9]

5–3 Solving Equations in Two Variables

This question was on a history exam.

> The European explorer __?__ came to North
> (name)
> America in the ship __?__.
> (name)

Which of these answers could be correct?
a. Christopher Columbus, Santa Maria
b. Santa Maria, Christopher Columbus
c. Henry Hudson, Half Moon
d. Half Moon, Henry Hudson

■ LESSON

The solutions of equations containing two variables, x and y, are ordered pairs of numbers. We can determine whether an ordered pair is a solution by substituting the first component for x and the second component for y. If the resulting equation is true, then the ordered pair is a solution. If the resulting equation is false, then the ordered pair is not a solution.

Given the equation $x + 2y = 5$:

The ordered pair $(1, 2)$ is a solution since $1 + 2 \cdot 2 = 5$ is true.
The ordered pair $(2, 1)$ is not a solution since $2 + 2 \cdot 1 = 5$ is false.
The ordered pair $(5, 0)$ is a solution since $5 + 2 \cdot 0 = 5$ is true.
The ordered pair $(7, -1)$ is a solution since $7 + 2 \cdot (-1) = 5$ is true.

To find solutions of equations in two variables, pick a number and substitute it for one of the variables. Then solve the equation for the other variable.

Example 1 Make a table of some solutions of this equation.

$$3x - 2y = 6$$

Solution **1.** Pick a value, say 4, and substitute for x.

$$3(4) - 2y = 6$$

Now solve for y.
$$12 - 2y = 6$$
$$-2y = -6$$
$$y = 3$$

This gives one solution of the original equation:

x	y
4	3

Example 1 (continued)

    2. Pick another value for x, say -2, and substitute.

$$3(-2) - 2y = 6$$

Solve for y.
$$-6 - 2y = 6$$
$$-2y = 12$$
$$y = -6$$

This gives a second solution.

x	y
4	3
-2	-6

    3. Pick a value for y, say $\frac{1}{2}$, and substitute.

$$3x - 2\left(\frac{1}{2}\right) = 6$$

Solve for x.
$$3x - 1 = 6$$
$$3x = 7$$
$$x = \frac{7}{3}$$

Answer

x	y
4	3
-2	-6
$\frac{7}{3}$	$\frac{1}{2}$

Example 2   Make a table of some solutions of this equation.

$$y = |x|$$

Solution   **1.** Pick 5 for x.
$$y = |5|$$
$$y = 5$$

    One solution is $(5, 5)$.

    2. Pick -3 for x.
$$y = |-3|$$
$$y = 3$$

    Another solution is $(-3, 3)$.

    3. Pick 0 for y.
$$0 = |x|$$
$$0 = x$$

    Another solution is $(0, 0)$.

Answer

x	y
5	5
-3	3
0	0

If the variables are not x and y, then the problem will usually indicate which variable is associated with the first component of the ordered pairs, and which variable is associated with the second component. In the following example, (m, n) indicates that m is the variable for the first component and n is the variable for the second component. In word problems, you may have to make your own decision about the order of the variables.

Example 3   Make a table of solutions (m, n) of the equation $5m - 7n = 70$ for the domain $\{-14, -7, 0, 7, 14\}$.

Solution   Remember that the numbers in the domain are first-component values to be substituted for m.

m	-14	-7	0	7	14
n					

Answer   The table of solutions is:

m	-14	-7	0	7	14
n	-20	-15	-10	-5	0

■ CLASSROOM EXERCISES

State whether the ordered pairs are solutions of the equation.

1. $x + y = 7$
  a. $(4, 3)$
  b. $(3, 4)$
  c. $(-2, 9)$

2. $x = |y|$
  a. $(-3, 3)$
  b. $(3, -3)$
  c. $\left(\frac{1}{2}, \frac{1}{2}\right)$

3. $y = x^2$
  a. $(2, 4)$
  b. $(3, 6)$
  c. $(16, 4)$

4. How many numbers are solutions of the equation $2x + 3 = 11$?

5. How many ordered pairs are solutions of the equation $2x + 3y = 11$?

Find the missing number of each solution pair.

6. $2x - 2y = 12$

x	4	$?$	-3	$?$
y	$?$	4	$?$	-3

7. $q - p = -3$

p	7	$?$	-2	$?$
q	$?$	7	$?$	-2

■ WRITTEN EXERCISES

State whether the ordered pairs are solutions of the equation.

A   **1.** $3x + y = 6$
  a. $(0, 3)$
  b. $(2, 0)$
  c. $(4, -6)$

2. $x + 3y = 6$
  a. $(0, 2)$
  b. $(3, 0)$
  c. $(9, -1)$

3. $x - y = 4$
  a. $(4, 0)$
  b. $(6, -2)$
  c. $(8, 4)$

4. $x - y = 6$
  a. $(10, 4)$
  b. $(0, 6)$
  c. $(6, 0)$

Write the set of solutions (x, y) of each equation for the domain $\{-2, 0, 2\}$.

5. $x = y$ **6.** $x = 2y$ **7.** $y = 2x$

Write the set of solutions (x, y) of each equation for the domain $\{-4, 0, 4\}$.

8. $x + y = 4$ **9.** $y = 4x$ **10.** $y = \frac{1}{4}x$

Find the missing number for each solution.

11. $x - 2y = 6$

x	y
12	?
6	?
?	1
?	5

12. $x - 3y = 6$

x	y
12	?
6	?
?	1
?	5

13. $2x - y = 8$

x	y
6	?
4	?
?	2
?	10

14. $3a - b = 8$

a	b
6	?
4	?
?	1
?	7

15. $2a + 3b = 12$

a	b
0	?
?	0
1	?
?	1

16. $3a + 2b = 12$

a	b
0	?
?	0
1	?
?	1

17. $2a - 3b = 12$

a	b
0	?
?	0
1	?
?	1

18. $3a - 2b = 12$

a	b
0	?
?	0
1	?
?	1

Copy and complete each table.

19. $5x + 2y = 10$

x	0	2	4	-2
y	?	?	?	?

20. $3x + 2y = 6$

x	0	1	3	-1
y	?	?	?	?

21. $3y = x + 6$

x	0	3	-3	1
y	?	?	?	?

22. $3y = x + 9$

x	0	3	-3	1
y	?	?	?	?

B **23.** $y = 3x^2$

x	0	-4	4	$\frac{1}{3}$
y	?	?	?	?

24. $y = (3x)^2$

x	0	-4	4	$\frac{1}{3}$
y	?	?	?	?

25. $\frac{x}{y} = 6$

x	12	6	-6	3
y	?	?	?	?

26. $\frac{1}{2}x - \frac{1}{3}y = 2$

x	4	6	-4	0
y	?	?	?	?

Equivalent equations have the same solutions. Find the solution set of each equation using the given domain. State whether the two equations are equivalent.

27. Domain: {0, 1, 2}
$y = x + 3$
$x = y + 3$

28. Domain: {−2, 0, 2}
$y = x^2$
$y = -x^2$

29. Domain: {−3, 0, 3}
$y = |x|$
$y = |-x|$

C 30. Domain: {6, 0, −6}
$y = 3 - x$
$x = 3 - y$

31. Domain: {0, 3, 5}
$y = (2x)^2$
$y = 4x^2$

32. Domain: {−3, 0, 3}
$y = -|x^2|$
$y = -x^2$

Find three solutions of each equation. State whether you think the two equations are equivalent.

33. $y = \frac{1}{2}x + 6$

$y = \frac{x + 3}{2}$

34. $y = (x + 3)^2$
$y = x^2 + 9$

35. $y = (x - 2)(x + 2)$
$y = x^2 - 4$

■ **REVIEW EXERCISES**

Simplify.

1. $a - (-5)$

2. $5x - (-8) + (-y)$ [2–6]

3. $6a - 3 - 5a + 3b$

4. $3x - 4y - 4x - 3y$ [2–7]

5. $4a + 4b - 3a - b$

6. $-(a - b) + 4(2a + b)$ [2–8]

Self-Quiz 1

5–1 **1.** Graph these points: $A(-2, 3)$, $B(0, -4)$, $C(3, 1)$, $D(-6, 0)$. Draw line segments from A to C, C to B, B to D, and D to A.

2. Identify the quadrant each point is in or the axis it is on.
  a. $P(-4, -2)$ **b.** $Q(0, -5)$ **c.** $R(-9, 6)$

5–2 The following exercises describe a relation.

a. Graph the relation.   **b.** List the domain.   **c.** List the range.

3. {(1, 5), (−2, 1), (0, 4)}

4. The first component is a negative integer greater than −4 and the second component is the absolute value of the first component.

5–3 Copy and complete the table of solutions.

5. $2x - y = 1$

x	−2	?	0.5	3
y	?	1	?	?

6. $4a + 3b = 12$

a	0	?	?	−3
b	?	2	0	?

5–4 Graphing Equations

Two students went for a bike ride. This table shows how far they had traveled from home at the end of each 10-minute interval during the first hour.

Time from home, in minutes	0	10	20	30	40	50	60
Distance from home, in miles	0	2	4	6	6	8	10

Here is the same information plotted on a graph. From this information we know exactly where they were only at the start and the end of each 10-minute interval. However, we can use our imagination to fill in the spaces between the dots. If we assume that they traveled at a steady rate for the first half hour, rested for 10 minutes, and then traveled at a steady rate for the last 20 minutes, we can connect the dots and use this graph to represent their trip.

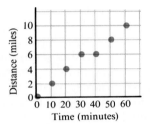

In this lesson you will learn to graph equations by plotting points and then connecting the points.

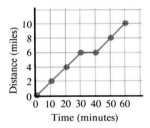

◼ LESSON

The equations in one variable that we have solved usually have had just one solution. For example,

the equation $2x + 3 = 8$ has only one solution, $\dfrac{5}{2}$.

However, equations in two variables usually have infinitely many solutions. For example,

the equation $2x + y = 8$ has infinitely many solutions

including $(-3, 14)$, $(-2, 12)$, $(0, 8)$, $\left(\dfrac{1}{2}, 7\right)$,

$(10, -12)$, and $(4, 0)$.

Since the solutions of the equations in two variables are ordered pairs, they can be graphed in the coordinate plane. The graph can show the patterns of the solution set and make the equations easier to understand. The graph of the set of all solutions of an equation is called the **graph of the equation.**

Example 1 Graph this equation. $x + y = 4$

Solution

Here is a table of some solutions and a graph of those solutions.

x	y
1	3
3	1
-2	6
6	-2
0	4
4	0

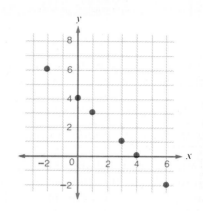

The graph of the six number pairs shows six points. Since the equation has infinitely many solutions, its graph has infinitely many points. We make an educated guess about the complete graph. The graph looks like a straight line. We test that guess by finding more solutions and checking whether their graphs are on the same straight line.

x	y
2	2
$3\frac{1}{2}$	$\frac{1}{2}$
-3	7

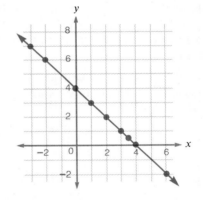

The graphs of these three points are on the line, so the guess seems correct.

Example 2 Graph. $y = |x|$

Solution

Here are some solutions and their graph.

x	1	0	3	4	5
y	1	0	3	4	5

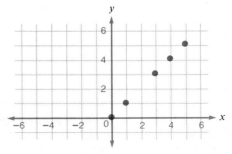

It looks as if the graph of all the solutions may again be a straight line. This guess can be tested by finding more solutions. It is always wise to try some negative numbers. Since none have been used yet, they should be used to test the guess.

Example 2 (continued)

x	-1	-3	$-\frac{1}{2}$
y	1	3	$\frac{1}{2}$

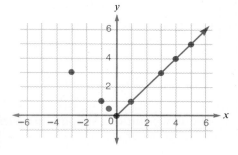

The guess that the graph was a straight line was not correct. Now it seems that the graph is shaped like a "v" with the tip of the v at (0, 0). This guess should be tested.

x	-6	$-3\frac{1}{2}$	-5
y	6	$3\frac{1}{2}$	5

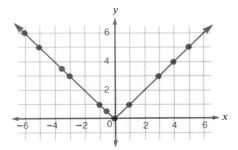

The solutions support the second guess. The graph is two rays that have (0, 0) as a common endpoint.

Example 3 Graph. $y = \dfrac{x^2}{4}$

Solution

Here are some solutions and their graph.

x	0	2	4
y	0	1	4

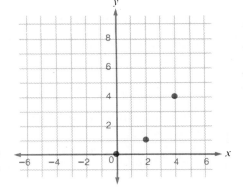

These three points are certainly not on a straight line, and it is a little difficult to visualize the whole graph.

Here are five more solutions and their graph.

x	-2	-4	1	-1	6
y	1	4	$\frac{1}{4}$	$\frac{1}{4}$	9

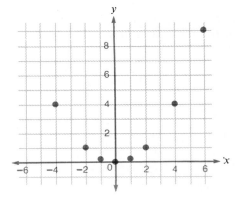

These extra points make the graph clearer, and a guess about the whole graph can be made. The graph appears to be a curve with the origin at its bottom.

Example 3 (continued)

Check the guess with these three solutions.

x	$-\dfrac{1}{2}$	-6	3
y	$\dfrac{1}{16}$	9	$\dfrac{9}{4}$

The new solutions support the guess. The graph is a curve known as a **parabola**.

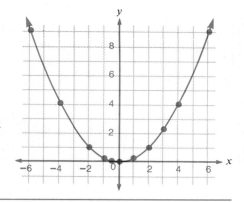

Summary: Follow these steps to graph an equation in two variables.

1. Find at least three solutions of the equation.

2. Graph these solutions.

3. Guess the pattern for the graph of all the solutions.

4. Sketch a graph of all solutions.

5. Check by finding and plotting at least two additional solutions.

In Steps 1 and 5 be sure to use fractions, zero, and negative numbers.

■ CLASSROOM EXERCISES

Graph these equations.

1. $y = 3x$

2. $y = 3x + 1$

3. $x + 2y = 5$

4. $y = x$

5. $y = x^2$

6. $y = \dfrac{1}{x}$

■ WRITTEN EXERCISES

For each equation, copy and complete the table of solutions, graph the solutions in the table, and graph the equation.

A **1.** $y = 2x + 3$

x	-2	0	2
y	?	?	?

2. $y = 2x + 5$

x	-3	0	1
y	?	?	?

3. $y = \dfrac{1}{4}x + 3$

x	-8	0	8
y	?	?	?

4. $y = \dfrac{1}{3}x + 5$

x	-6	0	3
y	?	?	?

5. $2x + 3y = 6$

x	0	3	6
y	?	?	?

6. $3x + 2y = 12$

x	0	2	4
y	?	?	?

Graph each equation.

7. $y = 3x - 3$ **8.** $y = 3x - 5$ **9.** $y = \frac{1}{2}x^2$ **10.** $y = 2x^2$

11. $y = -2x^2$ **12.** $y = -\frac{1}{2}x^2$ **13.** $xy = 6$ **14.** $xy = 12$

B **15.** $3y - 2x = -6$ **16.** $4y - 2x = -8$ **17.** $y = \frac{1}{2}x^3$ **18.** $y = \frac{1}{2}x^4$

19. $y = (x + 1)^2$ **20.** $y = 2(x - 1)$ **21.** $y = x^2 + 1$ **22.** $y = 2x - 1$

Graph the formula or equation that describes each situation.

23. Tom charges $1 plus $1.50 per hour when he babysits for a child. He uses the formula $C = 1 + 1.5h$ to find the total charge C, where h represents the number of hours he babysits. Graph the pairs (h, C).

24. Tom charges $2 plus $3.00 per hour when he babysits for three children at once. He uses the formula $C = 2 + 3h$ to find the total charge C, where h represents the number of hours he babysits. Graph the pairs (h, C).

25. The formula $A = 6e^2$ shows the relationship between the length of an edge of a cube (e cm) and the area of the surface of the cube (A cm^2). Graph the pairs (e, A).

26. The formula $V = e^3$ shows the relationship between the length of an edge of a cube (e cm) and the volume of the cube (V cm^3). Graph the pairs (e, V).

27. The formula $2l + 2w = 12$ shows the relationship between the length (l cm) and the width (w cm) of a rectangle that has a perimeter of 12 cm. Graph the pairs (l, w).

28. The formula $lw = 12$ shows the relationship between the length (l cm) and the width (w cm) of a rectangle that has an area of 12 cm^2. Graph the pairs (l, w).

C **29.** Barbara and Ed were 10 miles from home and started walking home at the rate of 3 miles per hour. The formula $d = -3t + 10$ gives their distance d from home at any time t. Draw a graph with t as the first component. Start at $t = 0$ and end the graph when they reach home.

30. Peggy and Sally were 2 miles from home when they began hiking toward a nature preserve at 4 miles per hour. The nature preserve is 12 miles from home. The formula $d = 4t + 2$ indicates their distance d from home at any time t. Draw the graph with t as the first component. Start at $t = 0$ and end the graph when they reach the nature preserve.

■ REVIEW EXERCISES

Solve.

1. $4x + 3.75 = x - 6$ **2.** $7x + 4\frac{1}{4} = 6x - 3\frac{1}{2}$ [3–3]

3. $\frac{2}{3}x = -18$ **4.** $0.06x = 7.20$ [3–4]

5-5 Slopes of Lines

Preview

Whether a climber climbs a gentle slope or a steep slope, the motion has two components, a vertical component and a horizontal component. The climber moves both up and forward at the same time.

Explain in terms of the upward and forward movement how a steep slope differs from a gentle slope.

Explain in terms of a downward component and a forward component how these two ski slopes differ.

■ LESSON

In moving from one point to another on a line, there is a vertical component and a horizontal component to the motion. For example, in going from point A to point B, the vertical change is $5 - 1$, or 4, while the horizontal change is $4 - 2$, or 2. The ratio of the vertical change to the horizontal change is called the **slope** of the line.

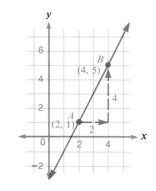

$$\text{slope of } \overleftrightarrow{AB} \text{ (line } AB) = \frac{5-1}{4-2} = 2$$

The slope indicates that in moving from point to point on the line (in either direction), the vertical change is 2 times as great as the horizontal change.

Definition: Slope of a Line

The slope of a line is the ratio of the change in y to the corresponding change in x between two points on the line.

$$\text{slope of } \overleftrightarrow{MN} = \frac{y_2 - y_1}{x_2 - x_1}$$

or

$$\text{slope of } \overleftrightarrow{MN} = \frac{y_1 - y_2}{x_1 - x_2}$$

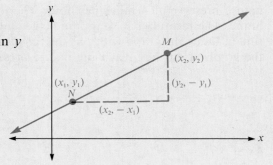

The slope of a line is the difference in two *y*-values divided by the difference in the two *corresponding* *x*-values.

[**Remember:** In y_1 and y_2 the numbers 1 and 2 are *subscripts* that show that y_1 and y_2 are different components. Be sure not to confuse subscripts with exponents.]

$$\frac{y_2 - y_1}{x_2 - x_1}$$

This represents ↑ ↑ This represents
one ordered pair.—┘ └—another ordered pair.

Lines that *rise* as you move from left to right along the *x*-axis have positive slopes.

Example 1 Find the slope of $\overleftrightarrow{CD}$.

Solution Substitute the coordinates of points *C* and *D* in the formula.

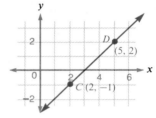

$$\text{slope} = \frac{y_2 - y_1}{x_2 - x_1}$$

$$= \frac{-1 - 2}{2 - 5} = \frac{-3}{-3} = 1$$

or

$$\text{slope} = \frac{2 - (-1)}{5 - 2} = \frac{3}{3} = 1$$

Horizontal lines, lines parallel to the *x*-axis, have 0 slope.

Example 2 Find the slope of $\overleftrightarrow{GH}$.

Solution $\text{slope} = \dfrac{1 - 1}{3 - (-2)} = 0$

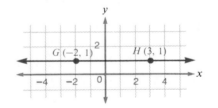

Vertical lines, lines parallel to the *y*-axis, have no slope.

Example 3 Find the slope of $\overleftrightarrow{MN}$.

Solution $\text{slope} = \dfrac{2 - (-3)}{2 - 2} = \dfrac{5}{0} \leftarrow$ not defined

Since division by 0 is not defined, $\overleftrightarrow{MN}$ has no slope.

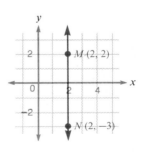

Lines that descend as you move from left to right along the *x*-axis have negative slopes.

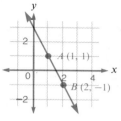

Example 4 Find the slope of $\overleftrightarrow{AB}$.

Solution slope $= \dfrac{-1-1}{2-1} = \dfrac{-2}{1} = -2$

Example 5 Draw a line through point $A(1, -2)$ with slope 3.

Solution Since the slope of the line is 3, the change in *y* is 3 times the change in *x* between any two points on the line.

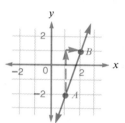

1. Start at point *A*.

2. Go up 3 units.

3. Move 1 unit to the right to get a second point, *B*.

4. Draw a line through the two points.

■ CLASSROOM EXERCISES

1. Explain the difference between "no slope" and "0 slope."

2. In moving from left to right on a line, state whether the line rises, falls, or is horizontal in these situations.
   a. Negative slope   **b.** 0 slope   **c.** Positive slope

Use the labeled points to find the slopes of these lines.

3.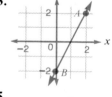

4.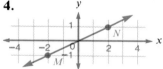

5.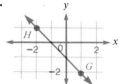

6.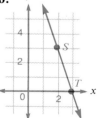

7. Draw a line through point $(1, -8)$ with slope $\dfrac{1}{2}$.

8. Draw a line through point $(-1, 2)$ with slope -1.

218   Chapter 5   Graphing Linear Equations and Functions

■ WRITTEN EXERCISES

In moving from one point on a line to another, follow the given directions.
State whether the slope is positive, negative, or zero.

A

1. Go up 5 and go right 3.
2. Go up 3 and go right 5.
3. Go down 5 and go right 3.
4. Go down 3 and go right 5.
5. Go up 5 and go left 3.
6. Go up 3 and go left 5.

Find the slopes of these lines.

7. $\overleftrightarrow{AB}$
8. $\overleftrightarrow{CD}$
9. $\overleftrightarrow{EF}$
10. $\overleftrightarrow{GH}$
11. x-axis

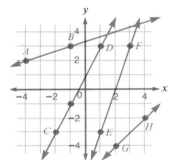

Find the slopes of these lines.

12. $\overleftrightarrow{PQ}$
13. $\overleftrightarrow{RS}$
14. $\overleftrightarrow{TU}$
15. $\overleftrightarrow{VW}$
16. y-axis

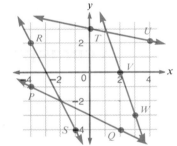

17. Find the slope of each side of square *JKLM*.
    a. $\overline{JK}$
    b. $\overline{KL}$
    c. $\overline{LM}$
    d. $\overline{MJ}$

18. Find the slope of each side of rectangle *ABCD*.
    a. $\overline{AB}$
    b. $\overline{BC}$
    c. $\overline{CD}$
    d. $\overline{DA}$

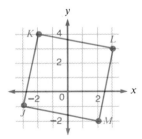

19. Find the slope of each side of right triangle *EFG*.

a. $\overline{EF}$
b. $\overline{FG}$
c. $\overline{GE}$

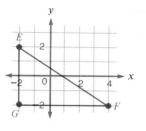

20. Find the slope of each side of right triangle *HIJ*.

a. $\overline{HI}$
b. $\overline{IJ}$
c. $\overline{JH}$

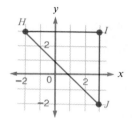

Graph the point, and then draw lines with the stated slopes through the point. Label the lines *a, b,* and *c.*

21. Point (2, 1) a. slope 2 b. slope −1 c. slope 0

22. Point (1, 3) a. slope 3 b. slope $-\frac{1}{2}$ c. no slope

23. Point (−4, 0) a. slope $-\frac{1}{2}$ b. slope $\frac{1}{4}$ c. no slope

24. Point (0, −3) a. slope −3 b. slope $\frac{1}{2}$ c. slope 0

B 25. Point (0, 2) a. slope $\frac{2}{5}$ b. slope $-\frac{5}{2}$ c. slope $2\frac{1}{2}$

26. Point (4, 0) a. slope $\frac{2}{7}$ b. slope $-\frac{2}{7}$ c. slope $3\frac{1}{2}$

Find the slope of the line that goes through the two points.

27. *A*(0, 0) **28.** *C*(0, 0) **29.** *E*(1, 2) **30.** *G*(2, 1)
   B(3, 2) *D*(2, 8) *F*(2, 5) *H*(3, 10)

31. *K*(−2, 2) **32.** *P*(3, −2) **33.** *R*(5, 7) **34.** *T*(−5, 7)
   L(−4, 1) *Q*(−1, 3) *S*(5, −7) *U*(−5, −7)

35. Draw a line with slope 2 that passes through the given point.
   a. (−1, 2) b. (1, 2) c. (3, 2)

C 36. a. What is the relationship between the slopes of the opposite sides of square *JKLM* in Exercise 17?
   b. What is the relationship between the slopes of the *opposite* sides of rectangle *ABCD* in Exercise 18?
   c. What is the relationship between the slopes of *parallel* lines?

37. a. For each pair of *adjacent* sides of square *JKLM* in Exercise 17, find the product of their slopes (that is, find the product of the slopes of $\overline{JK}$ and $\overline{KL}$, etc.).
  b. For each pair of adjacent sides of rectangle *ABCD* in Exercise 18, find the product of their slopes.
  c. What is the relationship between the slopes of *perpendicular* lines that are neither vertical nor horizontal?

38. On a set of axes like this, draw graphs for the following:
  a. A person crawling at 1 mph
  b. A person walking at 4 mph
  c. A person running at 8 mph
  d. Describe the relationship between the slopes of the lines and the rates.

■ **REVIEW EXERCISES**

Solve.

1. $8 - 3x = 20$

2. $6 + \dfrac{x}{2} = 10$

3. $\dfrac{a}{3} + 4 = 8$ [3–6]

4. $12 - 3(2 - x) = x$

5. $6x - (5 - x) = 5(x + 3)$

6. $\dfrac{1}{2}(x - 8) = 20$ [3–7]

Self-Quiz 2

5–4 Copy and complete the table of solutions. Then graph the equation.

1. $x + 2y = 6$

x	-1	0	2
y	?	?	?

2. $y = x^3$

x	-2	-1	0	1	2
y	?	?	?	?	?

3. A babysitter charges $3 plus $2 per hour. The formula $C = 3 + 2h$ is used to find the total charge C, where h represents the number of hours spent babysitting. Graph the equation.

5–5 Determine the slopes of the lines.

4. $\overleftrightarrow{AB}$

5. $\overleftrightarrow{CD}$

6. Draw lines through the point $(3, 2)$ with the stated slopes. Label the lines a, b, and c.

  a. $m = \dfrac{1}{3}$

  b. $m = -2$

  c. $m = 0$

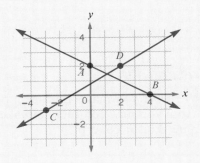

EXTENSION   Finding slopes on a calculator

The expression for slope

$$m = \frac{y_2 - y_1}{x_2 - x_1}$$

can be evaluated easily on a calculator that has memory keys.

- The keys labeled $\boxed{\text{STO}}$ or $\boxed{\text{M+}}$ or $\boxed{\text{MIN}}$ store a number in memory.
- The keys labeled $\boxed{\text{RCL}}$ or $\boxed{\text{MR}}$ or $\boxed{\text{RM}}$ recall a number from memory.

Example 1   Find the slope of the line passing through (2, 8) and (6, 20).

Solution   To evaluate the slope, first compute the difference of the *x*-values and store the result in memory.

ENTER: 6   −   2 $\boxed{\text{STO}}$

The difference 4 is stored in memory.

Next, compute the difference in the *y*-values (in the same order) and divide by the number recalled from memory.
The complete calculation is:

Key Sequence	Display
6 − 2 $\boxed{\text{STO}}$	4
20 − 8 =	12
÷ $\boxed{\text{RCL}}$ =	3

The slope of the line is 3.

Use a calculator to find the slope of the line passing through the two points.

1. (5, 1) and (9, 3)

2. (4, 0) and (6, −8)

3. (1, 2) and (4, 17)

4. (−6, 2) and (−2, 4)

5. (14, −9) and (12, 6)

6. (89, −11) and (84, −13)

7. (−21, 7) and (21, −7)

8. (14.5, 3.5) and (2, 9)

Strategy for Success   Learning from mistakes

Whenever you make a mistake, be sure to find out why you made it. Otherwise you may continue to make mistakes on other problems that use the same skill or concept.

5-6 Slope-Intercept Form

Preview

Even though the axes below do not show a scale, can you match these lines with their slopes?

Line	Slope
a	− 1
b	− 4
c	3
d	$\frac{1}{4}$

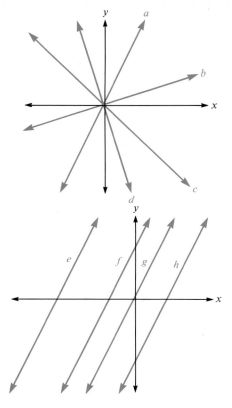

These lines all have the same slope. Can you describe how the lines differ from one another?

In this lesson you will learn to describe how lines *e*, *f*, *g*, and *h* differ.

■ LESSON

This figure shows the graph of the equation $y = 2x + 3$. The line crosses the *y*-axis at the point (0, 3). The *y*-value of the point at which a graph crosses the *y*-axis is called the **y-intercept.** The *y*-intercept of this graph is 3. The slope of the line can be computed using the coordinates of points *A* and *B*:

$$\frac{7-3}{2-0} = 2$$

Note that for the equation $y = 2x + 3$, the value of the slope (2) is the coefficient of the *x*-term and that the value of the *y*-intercept (3) is the constant term in the equation. In general, when the equation of a line is written in the form $y = mx + b$, where *m* is the slope of the line and *b* is the *y*-intercept, the equation is said to be written in **slope–intercept form.**

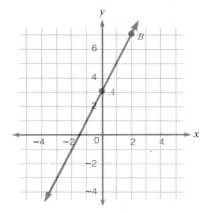

y-intercept: 3   slope: 2

Definition: Slope–Intercept Form of a Linear Equation

The equation $y = mx + b$ is called the slope–intercept form of a linear equation. The graph of the equation is a straight line with slope m and y-intercept b.

If an equation is written in slope–intercept form, the form clearly displays the slope and y-intercept. If the equation is not already in slope–intercept form, it can be transformed into an equivalent equation in slope–intercept form in order to find the slope and y-intercept.

Example 1   Find the slope and y-intercept of the line with equation $2y + x = 6$.

Solution   Transform $2y + x = 6$ to an equivalent equation in the form $y = mx + b$.

$$2y + x = 6$$
$$2y = -x + 6$$
$$y = -\frac{1}{2}x + 3$$

Answer   The slope is $-\frac{1}{2}$ and the y-intercept is 3.

Example 2   Write the equation of the line with slope -3 and y-intercept 1.

Solution   Substitute -3 for m and 1 for b in $y = mx + b$.

Answer $y = -3x + 1$

This figure shows the graphs of three lines. The lines appear to be parallel. Note that their slopes are equal, but that their y-intercepts are different.

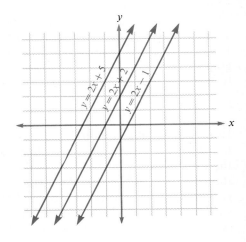

This figure shows the graphs of four vertical lines. These lines have undefined slopes.

The graphs on these two pages illustrate an important relationship between lines and their slopes:

Parallel lines have the same slope, or they are vertical lines and have undefined slopes.

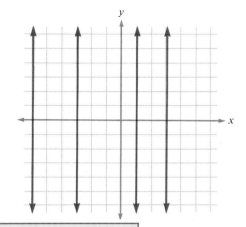

Property: Slopes of Parallel Lines

Two lines are parallel if and only if they have the same slope or they have no slope.

Example 3   Show that $y = 2x - 1$ and $4x = 2y - 1$ are equations of parallel lines.

Solution   The lines are parallel if they both have the same slope.

$$y = 2x - 1 \qquad\qquad 4x = 2y - 1$$
$$\text{slope: } 2 \qquad\qquad 4x + 1 = 2y$$
$$2x + \frac{1}{2} = y$$
$$y = 2x + \frac{1}{2}$$
$$\text{slope: } 2$$

Answer   Since their slopes are equal, the lines are parallel.

Example 4   Write the equation of the line that is parallel to the graph of $y - 5 + 3x = 0$ and that has a y-intercept of -5.

Solution   Transform the equation $y - 5 + 3x = 0$ to slope–intercept form.

$$y - 5 + 3x = 0$$
$$y = -3x + 5$$

The slope of the given line is -3.
The desired line has slope -3 and y-intercept -5.

Answer   The equation of the line is $y = -3x - 5$.

Consider the graphs of these two equations.

$$y = \frac{3}{4}x - 1 \quad \text{and} \quad y = -\frac{4}{3}x + 2$$

The lines appear to be perpendicular. The slopes of the given lines are $\frac{3}{4}$ and $-\frac{4}{3}$. The product of the slopes is -1. The graphs illustrate another important relationship between lines and their slopes: Two lines are perpendicular if the product of their slopes is -1.

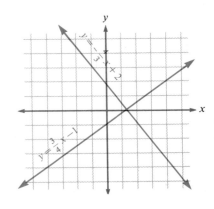

Horizontal lines, which have 0 slope, are perpendicular to vertical lines, which have undefined slopes.

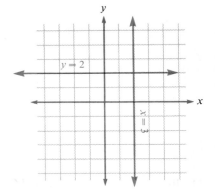

Property: Slopes of Perpendicular Lines

Two lines are perpendicular if and only if the product of their slopes is -1, or if one slope is zero and the other is undefined.

Example 5 Show that $y = 3x - 2$ and $3y + x = 4$ are equations of perpendicular lines.

Solution Write both equations in slope–intercept form and find the slopes.

$$y = 3x - 2 \qquad 3y + x = 4$$
$$\text{slope: } 3 \qquad \quad\; 3y = -x + 4$$

$$y = -\frac{1}{3}x + \frac{4}{3}$$

$$\text{slope: } -\frac{1}{3}$$

The product of the slopes is $3\left(-\frac{1}{3}\right) = -1$.

Answer The lines are perpendicular since the product of their slopes is -1.

■ CLASSROOM EXERCISES

1. Write the equation $6 - 2y = 5x$ in slope–intercept form.

State the slope and y-intercept of the line with the given equation.

2. $y = -0.5x + 7$ **3.** $3y - 5x = 2$

4. Write an equation of the line with slope -5 and y-intercept 0.

5. Are the graphs of the equations $y = -2x$ and $y = -2x + 1$ parallel?

6. Are the graphs of the equations $y = 3x$ and $y = \dfrac{1}{3}x + 1$ perpendicular?

7. Are the graphs of the equations $y = 3x$ and $y = -3x + 1$ perpendicular?

■ WRITTEN EXERCISES

State the slope and y-intercept of the line with the given equation.

1. $y = 5x + 3$ **2.** $y = -3x + 4$ **3.** $y = -\dfrac{1}{2}x - 6$

4. $y = \dfrac{1}{3}x - 5$ **5.** $y = \dfrac{1}{2}x$ **6.** $y = -8x$

State the slope and y-intercept of the line.

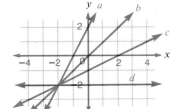
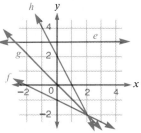

7. line a

9. line b

11. line c

13. line d

8. line e

10. line f

12. line g

14. line h

Use the given slope and y-intercept to write an equation of the line in slope–intercept form.

15. Slope 3; y-intercept 10   **16.** Slope 2; y-intercept 7

17. Slope -6; y-intercept 0.75   **18.** Slope -5; y-intercept 2.25

19. Slope 0; y-intercept -3 **20.** Slope 0; y-intercept -10

21. Slope $\dfrac{2}{3}$; y-intercept -2 **22.** Slope $\dfrac{3}{4}$; y-intercept -3

State whether the graphs of these pairs of equations are parallel, perpendicular, or neither.

23. $y = 2x + 3$
$\quad y = -2x - 3$

24. $y = 2x + 3$
$\quad y = -2x + 3$

25. $y = -2x + 3$
$\quad y = 2x - 3$

26. $y = \dfrac{1}{2}x + 5$

$\quad y = -2x - 5$

27. $y = -\dfrac{1}{2}x + 5$

$\quad y = -2x + 5$

28. $y = 3x + 5$

$\quad y = 3x - 5$

Write an equivalent equation in slope–intercept form. State the slope and y-intercept of the line.

29. $2x + 3y = 6$

30. $3x + 2y = 6$

31. $4y - 6x = 12$

32. $6y - 4x = 12$

33. $2y + 3 = 4x$

34. $3y + 4 = 15x$

State whether the graphs of these pairs of equations are parallel, perpendicular, or neither.

35. $2x + 3y = 6$
   $-3x + 2y = 6$

36. $2x + 3y = 6$
   $-2x + 3y = 6$

37. $2x - 3y = 6$
   $-2x + 3y = 6$

38. $2x - 3y = 6$
   $3x + 2y = 6$

39. $y = \frac{1}{2}x + 3$
   $2y = -2x + 4$

40. $y = \frac{2}{5}x + 3$
   $5y - 2x = 10$

Write an equivalent equation in slope–intercept form. State the slope and y-intercept of the line.

B **41.** $5x - 3y = 9$

42. $7x - 2y = 8$

43. $2(x + 3y) = 7$

44. $4(2x + y) = 6$

45. $3x + \frac{1}{2}y = 4$

46. $7x + \frac{1}{3}y = 6$

47. $3(x + 2y) = 8$

48. $2x + 3y = 5y + x + 1$

49. $2(y + 3) = 3(2y + x)$

Match the graphs with the given descriptions.

50. $m = \frac{2}{3};\ b = 2$

51. $m = -\frac{2}{3};\ b = 2$

52. $m = -\frac{3}{2};\ b = -2$

53. $m = \frac{2}{3};\ b = -2$

54. Slope undefined

55. $m = 0;\ b = 0$

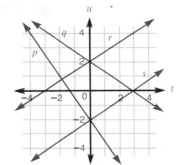

Write an equation in slope–intercept form for each line.

56. A line parallel to the graph of $y = 3x + 4$ and with y-intercept 5.

57. A line parallel to the graph of $y = -2x + 8$ and passing through the origin.

58. A line perpendicular to the graph of $y = 3x + 4$ and with y-intercept 5.

59. A line perpendicular to the graph of $y = -2x + 8$ and passing through the origin.

60. A line perpendicular to the graph of $x = 5$ and with y-intercept 3.

The **x-intercept** of a line is the *x*-coordinate of the point at which the line crosses the *x*-axis. State the *x*-intercept of the line with the given equation.

C **61.** $2x + 3y = 6$ **62.** $y = \frac{1}{2}x - 5$ **63.** $7x - 5y = 14$

64. If an equation is written in the form $x = ny + c$, where n and c are real numbers, what does the value c represent?

65. Find the slope of the line with *y*-intercept 2 that passes through the point (5, 5). Write the equation of the line in slope–intercept form.

66. Find the *y*-intercept of the line with slope $-\frac{1}{2}$ that passes through the point (4, −1). Write the equation of the line in slope–intercept form.

■ REVIEW EXERCISES

Simplify.

1. $\frac{3}{4} + \frac{5}{6}$ **2.** $\frac{3}{4} - \frac{5}{6}$ **3.** $\frac{3}{4} \cdot \frac{5}{6}$ **4.** $\frac{3}{4} \div \frac{5}{6}$

Draw a figure and write an equation to help you solve this problem.   [3–10]

5. The two congruent sides of an isosceles triangle are each 10 cm longer than the base. What are the lengths of the sides and the base if the perimeter of the triangle is 125 cm?

EXTENSION   Unit pricing

Consumers are often confronted by questions like the following:

The same brand of mustard comes in two sizes, 9 oz for 47 cents and 6.5 oz for 35 cents. Which is the better buy?

The question can be answered by computing a "unit price" for each size. In this case we compute the cost per ounce.

$$\frac{47}{9} \approx 5.2 \qquad \frac{35}{6.5} \approx 5.4$$

The larger size costs about 5.2 cents per ounce. The smaller size costs about 5.4 cents per ounce. The larger size is the better buy.

1. Graph the two ordered pairs (9, 47) and (6.5, 35). Draw a line from each point through the origin.

2. Find the slope of each line.

3. How can graphs be used to determine the better buy?

5-7 Writing Equations for Lines

Preview

The line shown at the right is clearly identi-
fied by the name "line AB" because there is
only one line that contains both point A and
point B. What other ways can be used to
identify a particular line?

In this lesson you will learn ways of writing
equations to identify lines.

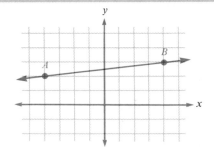

■ LESSON

There are two basic ways to identify a line: (1) by stating the slope of the line
and the coordinates of a point on the line, or (2) by stating the coordinates of
two points on the line. In both cases, an equation can be written for the line by
determining the slope and y-intercept.

Example 1   Write an equation of the line that has slope 2 and that passes through point
(3, 4).

Solution   Since the slope of the line is 2, the equation of the line is in the form

$$y = 2x + b$$

Next, determine the value of b. Since the line contains the point (3, 4), the
values 3 for x and 4 for y satisfy the equation. Substitute those values into
the equation $y = 2x + b$ and solve for b.

$$4 = 2(3) + b$$
$$4 = 6 + b$$
$$b = -2$$

The y-intercept is -2.

Answer   An equation of the line is $y = 2x - 2$.

Example 2   Write an equation of the line that has x-intercept 6 and that is parallel to the
graph of $y = -\frac{1}{2}x + 3$.

Solution   Determine the slope of the line. The line is parallel to the graph of
$y = -\frac{1}{2}x + 3$, which has slope $-\frac{1}{2}$. Therefore, the slope of the desired line is
$-\frac{1}{2}$. The equation of the line has the form $y = -\frac{1}{2}x + b$.

Example 2 (continued)

Next determine the y-intercept. Since the x-intercept is 6, the coordinates $(6, 0)$ satisfy the equation. Substitute those values in the equation $y = -\frac{1}{2}x + b$ and solve for b.

$$0 = -\frac{1}{2}(6) + b$$

$$0 = -3 + b$$
$$b = 3$$

Answer   An equation of the line is $y = -\frac{1}{2}x + 3$.

Note the steps in determining an equation of a line.

1. Determine the slope of the line and write the slope–intercept form of the equation, using the value for m.

2. Substitute the coordinates of a point on the line into the equation and solve for b, the y-intercept.

3. Write the slope–intercept form of the equation using the slope for m and the y-intercept for b.

Example 3   Write an equation of the line that passes through the points $(-2, 3)$ and $(4, 1)$.

Solution   Determine the slope by substituting in the formula.

$$m = \frac{y_2 - y_1}{x_2 - x_1}$$

$$m = \frac{3 - 1}{-2 - 4}$$

$$= \frac{2}{-6}$$

$$= -\frac{1}{3}$$

The equation is of the form $y = -\frac{1}{3}x + b$.

Substitute -2 for x and 3 for y, and solve for b.

$$3 = -\frac{1}{3}(-2) + b$$

$$b = \frac{7}{3}$$

Substitute $-\frac{1}{3}$ for m and $\frac{7}{3}$ for b.

Answer $y = -\frac{1}{3}x + \frac{7}{3}$

■ CLASSROOM EXERCISES

Write an equation for the line with the given characteristics.

1. Has slope 4 and passes through (3, 1).

2. Passes through $(-1, 2)$ and $(1, -2)$.

■ WRITTEN EXERCISES

Write an equation in slope–intercept form for the line that has the given slope and that passes through the point (4, 2).

1. $m = 1$ **2.** $m = 2$ **3.** $m = \dfrac{1}{4}$ **4.** $m = \dfrac{1}{2}$

5. $m = -2$ **6.** $m = -1$ **7.** $m = -\dfrac{1}{2}$ **8.** $m = -\dfrac{1}{4}$

Write an equation in slope–intercept form for the line that has slope 2 and that passes through the given point.

9. (1, 4)   **10.** (1, 6)   **11.** (3, 10)   **12.** (3, 8)

13. $(-3, 2)$ **14.** $(-5, 0)$ **15.** $(-1, -2)$ **16.** $(-3, -10)$

Write an equation in the form $y = mx + b$ for the line with the given characteristics.

17. Has slope 3 and passes through (4, 6).   **18.** Has slope 0 and passes through (2, 4).

19. Has slope $\dfrac{1}{2}$ and passes through (4, 6).   **20.** Has no slope and passes through (2, 4).

21. Has slope $-\dfrac{2}{3}$ and passes through $(-3, 6)$.   **22.** Has slope $-\dfrac{3}{2}$ and passes through $(-6, 6)$.

23. Is parallel to the graph of $y = -2x + 5$ and passes through $(-3, 10)$.   **24.** Is parallel to the graph of $y = -3x - 10$ and passes through $(-4, 0)$.

25. Is perpendicular to the graph of $y = \dfrac{1}{3}x + 2$ and passes through (5, 6).   **26.** Is perpendicular to the graph of $y = -\dfrac{1}{4}x + 8$ and passes through (2, 3).

Find the slope of the line that passes through the two points.

27. (5, 6), (6, 9)   **28.** (3, 1), (4, 8)

29. (8, 2), (11, 8)   **30.** (2, 6), (5, 15)

31. $(4, -2)$, $(6, -6)$ **32.** $(-4, 3)$, $(-2, 5)$

33. $(-3, 2)$, $(-5, 3)$ **34.** $(1, -5)$, $(-1, -4)$

Write an equation in the form $y = mx + b$ for the line that passes through the two points.

35. (2, 3), (3, 1)   **36.** (3, 5), (4, 2)

37. (5, 0), (8, 6)   **38.** (2, 0), (4, 6)

39. (0, 5), (8, 6)   **40.** (0, 2), (4, 6)

41. (−3, 5), (−1, 8)   **42.** (−2, 1), (1, 5)

Write an equation in the form $y = mx + b$ for each situation.

B **43.** The U-Smash-M repair service charges $95 for a job that takes 3 hours. It charges $145 for a 5-hour job. Write an equation that shows how the charge (y) is related to the number of hours (x).

44. The I-Fix-M garage charges $95 for a 3-hour job and $135 for a 5-hour job. Write an equation that shows how the amount charged (y) is related to the number of hours worked (x).

45. When the temperature is 32° Fahrenheit, it is 0° Celsius. When it is 212° Fahrenheit, it is 100° Celsius. Write an equation that shows how degrees Fahrenheit (x) are related to degrees Celsius (y).

46. When the temprature is 37° Celsius, it is 98.6° Fahrenheit. When it is −40° Celsius, it is −40° Fahrenheit. Write an equation that shows how degrees Celsius (x) are related to degrees Fahrenheit (y).

Point A has coordinates (x_1, y_1) and point B has coordinates (x_2, y_2).

C **47.** What is the slope of the line that passes through points A and B?

48. What is the y-intercept of the line that passes through points A and B?

49. What is the y-intercept of the line with slope m that passes through point A?

■ REVIEW EXERCISES

1. State the degree of the polynomial $4x - 3x^2 - 7$.   [4–2]

2. Which of these expressions are binomials?   [4–2]

$$2xy \qquad 3\frac{a}{b} \qquad x^2 + 3x + 2 \qquad x - 3 \qquad a^2 + b^2$$

3. Write the polynomial $4 - 3x^2 + 5x$ in descending order.   [4–2]

4. Solve: $2 \cdot 2^3 \cdot 2^x = 2^{10}$ [4–3]

Simplify.

5. $a^2 \cdot a^4$ **6.** $(3ab^2) \cdot (-4ab)$ [4–3]

Complete.

7. 3 km/min = _?_ km/h   [4–7]

Solve.

8. Two planes left Denver at the same time, one flying east at 560 mph and one flying west at 500 mph. How many miles apart will they be after an hour and a half?   [4–7]

Preview

The Classic Gift Company charges its customers a postage and handling fee based on the amount of the order.

Amount of order	Postage and handling fee
Up to $10.00	$2.75
10.01 to 25.00	4.25
25.01 to 50.00	6.95
50.01 to 75.00	8.95
75.01 to 100.00	10.95
100.01 to 150.00	14.95
Over $150.00	16.95

It would be poor public relations for a company to charge two customers different amounts for the same purchases. In this lesson you will learn that the relationship shown in the table above is an example of a *function*.

What is the total bill for each amount of order?

- $9.00
- $10.05
- $101.00
- $10.00
- $100.00
- $110.00

True or false?

- Two orders for the *same* amount could have *different* total bills.
- Two orders for *different* amounts could have the *same* total bills.

■ LESSON

Any set of ordered pairs is a relation, but the relations $\{(0, 0), (1, 3), (4, 6)\}$ and $\left\{(1, 1), \left(2, \frac{1}{2}\right), \left(3, \frac{1}{3}\right)\right\}$ have an important property that the relation $\left\{(1, 3), \left(1, \frac{1}{3}\right), (2, 4), \left(2, \frac{1}{2}\right)\right\}$ does not have. No two ordered pairs have the same first component. The relations $\{(0, 0), (1, 3), (4, 6)\}$ and $\left\{(1, 1), \left(2, \frac{1}{2}\right), \left(3, \frac{1}{3}\right)\right\}$ are examples of a kind of relation called a **function.**

Definition: Function

A relation is a function if and only if each first component in the relation is paired with exactly one second component.

Functions and relations can be described by graphs. The relation shown in this figure is not a function, because the ordered pairs (4, 2) and (4, − 2) are both members of the relation. Note that some vertical lines of the grid cross the graph twice since there are first components that are matched with two second components.

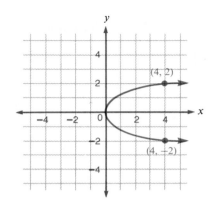

The relation shown in this figure is a function. No vertical line crosses the graph of the relation more than once. A vertical line can never cross the graph of a function more than once since there are no two ordered pairs with the same first component. This method of using vertical lines to determine whether a relation is a function is called the **vertical-line test.**

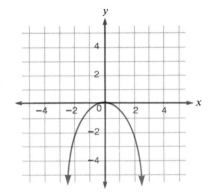

Functions and relations can also be described by equations. The relation defined by the equation $y = 2x$ is a function since each real number x is matched with its double, $2x$, and no real number has more than one double.

The relation defined by the equation $x = |y|$ is not a function since two ordered pairs (4, 4) and (4, − 4) with the same first component are in the relation. The graph of the relation fails the vertical-line test.

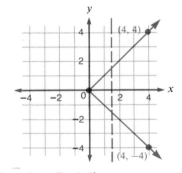

A function whose graph is a straight line is called a **linear function.** Each linear function can be defined by an equation in the slope–intercept form:

$$y = mx + b$$

Equations that define linear functions are called **linear equations in two variables.** In a linear equation the degree of both variables is 1.

$$y = 5x + 13$$

first degree

Example 1   State whether $2y - 3x = 4$ is the equation of a linear function.

Solution   The equation $2y - 3x = 4$ can be transformed to the equivalent equation in slope–intercept form $y = \frac{3}{2}x + 2$. Therefore, it is an equation of a linear function.

Example 2   State whether $y = x^2$ is the equation of a linear function.

Solution   The equation $y = x^2$ is the equation of a function since each number has exactly one square. However, the function is not a linear function since the degree of the x-term is 2, not 1.

Example 3   State whether $x = 0y + 2$ is the equation of a linear function.

Solution   These are some solutions of the equation $x = 0y + 2$:

$$\{(2, 8), (2, 3), (2, -5)\}$$

Therefore, the relation defined by $x = 0y + 2$ is not a function. Its graph is a vertical line.

■ CLASSROOM EXERCISES

1. State the definition of a function.

State whether these relations are functions. Give a reason for your answer.

2. $\{(2, 3), (4, 5), (-2, 5), (0, 4)\}$

3. $\{(1, 5), (-1, 4), (1, 3), (5, -2)\}$

4. $\{(2, 6), (3, 6), (4, 6), (5, 6)\}$

5. $\{(3, 0), (0, 3), (4, 8), (8, 4)\}$

6.
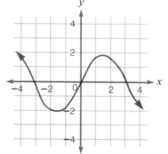

7.
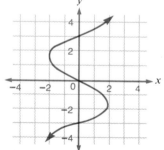

State whether the equation represents a function. Give a reason for your answer.

8. $y = -2x$ **9.** $y = x^2$ **10.** $x = y^2$ **11.** $y = |x|$

State whether the equation represents a linear function.

12. $y = -3x$ **13.** $y = 2x - 6$ **14.** $y + x = -2$ **15.** $y = |x|$

■ WRITTEN EXERCISES

State whether the relation is a function.

A **1.** {(1, 2), (2, 3), (3, 4)} **2.** {(4, 3), (3, 2), (2, 1)}

3. {(3, 5), (− 3, 5)} **4.** {(− 7, 8), (7, 8)}

5. {(16, 4), (16, − 4)} **6.** {(25, − 5), (25, 5)}

State whether the graph represents a function.

7.  **8.** 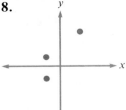 **9.**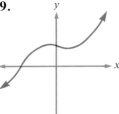

10. 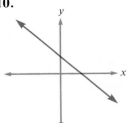 **11.** 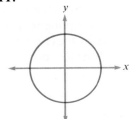 **12.**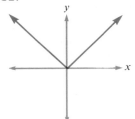

State whether the relation defined by the table is a function.

13.

x	− 4	− 1	2
y	3	5	3

14.

x	− 3	1	5
y	− 2	− 2	3

15.

x	0.25	$\frac{1}{4}$	1
y	0.5	$\frac{1}{2}$	1

16.

x	$\frac{4}{6}$	$\frac{2}{3}$	$-\frac{1}{2}$
y	4	4	4

State whether the equation defines a function.

B **17.** $y = 10x + 9$ **18.** $y = 7x + 2$ **19.** $y = |x + 2|$ **20.** $y = |x - 3|$

21. $y = (x + 1)^2$ **22.** $y = (5 - x)^2$ **23.** $x^2 = y^2$ **24.** $y^3 = x^3$

State whether the equation defines a *linear* function.

25. $2x - 3y = 5$ **26.** $3y - 5x = 8$ **27.** $y = \frac{1}{2}x + 4$

28. $y = \frac{1}{3}x + 2$ **29.** $y = x^4$ **30.** $y = x^2 + 2$

Graph each relation and state whether it is a function.

31. {(5, 3), (4, 2), (4, 1)}

32. {(3, 0), (3, 2), (5, 5)}

33.

x	2	-2	3	-3
y	4	6	6	4

34.

x	5	6	7	8
y	3	3	1	1

Graph these equations and state whether the relation is a function. If it is a function, state whether it is a linear function.

35. $y = x^2$ **36.** $y = x^3$ **37.** $x + y = 4$ **38.** $x + y = 8$

39. $y = x + 2$ **40.** $y = x - 4$ **41.** $2x + 4y = 8$ **42.** $3x + 6y = 18$

State whether the equation defines a function. If it does, state whether it is a linear function.

43. $x + y = 4$ **44.** $x - y = 4$ **45.** $xy = 4$

46. $\dfrac{y}{x} = 4$ **47.** $\dfrac{1}{x} + \dfrac{1}{y} = 4$ **48.** $\dfrac{1}{x + y} = 4$

49. The table in the Preview describes a function whose first component is the amount of the order and whose second component is the postage and handling fee. This function is a special kind of function called a step function. Graph the number pairs (amount of the order, postage and handling fee).

A phone call costs 50¢ for the first minute and 25¢ for each additional minute or part of a minute.

50. Calculate the cost for a phone call that lasts:

    a. $2\frac{1}{2}$ min   **b.** 3 min 1 s   **c.** 6 min 59 s

51. Draw a graph of the phone-call costs for calls lasting up to 10 min.

■ REVIEW EXERCISES

1. Write 30,000,000,000 in scientific notation.   [4–4]

2. Write $3.45 \cdot 10^5$ in standard decimal notation.   [4–4]

3. Simplify. Write the product in scientific notation.   [4–4]

$$(4 \cdot 10^4) \cdot (6 \cdot 10^3)$$

Simplify.   [4–5]

4. $(a^2)^4$ **5.** $(2x^3)^2$ **6.** $(10a^2b^3)^4$

7. What is the value in cents of q quarters?   [4–9]

Solve.

8. There are three times as many dimes as quarters in a pile worth $3.30. How many dimes and how many quarters are in the pile?   [4–9]

Self-Quiz 3

5–6 1. State the slope and y-intercept of the line whose equation is $3y - 2x = 12$.

2. Write the equation of the line that is parallel to the graph of $y = 2x - 5$ and passes through $(0, 5)$.

3. Show that the lines whose equations are $5x - 2y = 10$ and $y = -\frac{2}{5}x + 1$ are perpendicular by finding their slopes.

5–7 Find the equation of the line meeting these conditions. Write the equation in slope–intercept form.

4. Has slope $\frac{2}{3}$ and passes through $(3, 6)$.

5. Is perpendicular to $y = -\frac{1}{2}x + 5$ and has y-intercept -3.

6. Passes through $(1, 2)$ and $(4, -4)$.

5–8 State whether each of the following represents a function.

7.

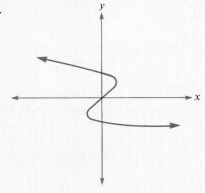

8.

x	y
-3	5
-1	3
0	2
-1	1

9. State whether the equation $2(x - y) = 5$ defines a linear function.

Strategy for Success   Taking tests

Write your solutions in a neat, orderly fashion so your teacher can help you diagnose the kinds of errors you are making, and also perhaps award partial credit.

EXTENSION Rounding numbers on a computer

There is no BASIC command to round a number, but a number can be rounded by using the greatest integer function. The function is written in BASIC as INT(X).

Examples

PRINT INT (3.67)
3

PRINT INT (4.38)
4

PRINT INT (5)
5

For positive numbers, INT deletes the digits to the right of the decimal point. To use INT to round a number to the nearest integer, add 0.5 to the number and find the greatest integer of the sum.

Examples

PRINT INT (3.67 + 0.5)
4

PRINT INT (4.38 + 0.5)
4

PRINT INT (5 + 0.5)
5

The following program rounds numbers to the nearest integer.

```
10  PRINT "WHAT IS THE NUMBER";
20  INPUT X
30  Y = INT (X + 0.5)
40  PRINT X; " ROUNDED TO THE NEAREST INTEGER IS "; Y
50  END
```

To round a number to the nearest tenth, lines 30 and 40 can be changed to

```
30  Y = INT (10 * X + 0.5)/10
40  PRINT X; " ROUNDED TO THE NEAREST TENTH IS "; Y
```

Run each program to round these numbers to the nearest integer and nearest tenth.

1. 7.283　　　　　　　　　**2.** 4.609　　　　　　　　　**3.** 0.514

4. Rewrite lines 30 and 40 so that the program will round a number to the nearest hundredth. Check your program.

Strategy for Success Believing in yourself

The chances are great that if you believe that you can learn algebra and you work at it, then you will learn algebra. Everyone has difficulty in mathematics at one time or another. However, those who believe they can learn mathematics and persevere usually do succeed.

5-9 Problem Solving with Linear Functions

Preview   Charles's Law

Gases expand when heated and contract when cooled. This fact can be tested by blowing up a balloon, putting it in a very warm place (in bright sunlight) or a very cold place (a freezer), and noting how the diameter of the balloon changes. The law that relates the volume of a gas to its temperature was first stated by scientist Jacques Charles in 1787.

Charles measured the volumes of a quantity of gas at various temperatures. These data can be graphed.

Temperature (°C)	15	30	45	60	75
Volume (cm³)	480	505	530	555	580

Charles noted that the data appeared to be linear and wrote a linear formula relating the temperature and volume of the gas. After this lesson you will be able to write Charles's formula.

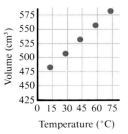

■ LESSON

A naturalist noted that a certain species of crickets chirps faster as the temperature rises. Here are some data about the number of chirps.

		20	20	20	
Chirps per minute	40	60	80	100
Temperature (°F)	47	52	57	62
		5	5	5

For equal increases in the number of chirps per minute, there are equal increases in the temperature. This means that the formula relating the two quantities is linear. The ratio of the increase in temperature to the increase in chirps is the slope of the graph of the formula.

$$\text{slope} = \frac{\text{change in temperature}}{\text{change in chirps}} = \frac{5}{20} = \frac{1}{4}$$

If T represents the number of degrees Fahrenheit and c represents the number of chirps per minute, the set of number pairs (c, T) is a linear function. The equation of the function is of the form

$$T = \frac{1}{4} c + b$$

To determine the value of b, that is, the T-intercept, substitute one of the (c, T) number pairs from the table in the formula and then solve for b.

$$T = \frac{1}{4} c + b$$

Substitute (40, 47). $47 = \frac{1}{4} \cdot 40 + b$

Solve for b. $47 = 10 + b$
$$37 = b$$

Here is a graph of the formula $T = \frac{1}{4}c + 37$.

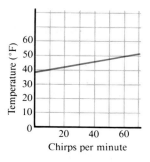

Example 1 Determine whether the (x, y) number pairs are a linear function.

x	-2	-1	0	1	2	3
y	1	$-\dfrac{1}{2}$	-1	$-\dfrac{1}{2}$	1	$3\dfrac{1}{2}$

Solution First find the successive differences of the x-values and of the y-values.

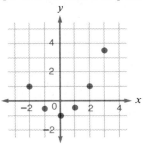

The successive differences in x-values are constant, but the successive differences in y-values are not constant. The ordered pairs are not part of a linear function. The graph shows that the points do not lie on a straight line.

Example 2 Show that the pairs in the table belong to a linear function.

x	y
-1	-8
0	-5
2	1
5	10
20	55

Solution Find the slopes of the lines joining successive pairs of points.

$$m = \frac{y_2 - y_1}{x_2 - x_1}$$

x	y	
-1	-8	
		$m_1 = \dfrac{3}{1} = 3$
0	-5	
		$m_2 = \dfrac{6}{2} = 3$
2	1	
		$m_3 = \dfrac{9}{3} = 3$
5	10	

Since the slopes are equal, the points are on the same line. Therefore, the function is linear.

Example 3 Write a linear equation in the form $y = mx + b$ that relates x and y.

x	2	0	-1	3	4	1
y	1	3	4	0	-1	2

Solution Arrange the data so that the x-values are in order. Then find successive differences.

$$1 \quad 1 \quad 1 \quad 1 \quad 1$$

x	-1	0	1	2	3	4
y	4	3	2	1	0	-1

$$-1 \quad -1 \quad -1 \quad -1 \quad -1$$

The ordered pairs belong to a linear function. Selecting the fourth and fifth number pairs we have:

slope: $\quad m = \dfrac{y_2 - y_1}{x_2 - x_1} = \dfrac{1 - 0}{2 - 3} = \dfrac{1}{-1} = -1$

y-intercept: $\quad b = 3$

Answer $y = -1x + 3$

■ CLASSROOM EXERCISES

State whether the ordered pairs belong to a *linear* function.

1.

x	y
−2	−3
0	1
4	9
6	13

2.

x	y
−2	−10
−1	−3
0	−2
1	−1
2	6

3.

x	y
2	−3
5	−1.5
−2	−5
4	−2
0	−4

4. In Exercises 1–3, if the function is linear, determine its slope, *y*-intercept, and equation in slope–intercept form.

■ WRITTEN EXERCISES

State whether the ordered pairs belong to a *linear* function.

A

1.

x	−2	−1	0	1	2
y	−8	−6	−4	−2	0

2.

x	−2	−1	0	1	2
y	−8	−5	−2	1	4

3.

x	−2	−1	0	1	2
y	5	2	−1	−4	−7

4.

x	−2	−1	0	1	2
y	10	8	6	4	2

5.

x	−2	−1	0	1	2
y	4	1	0	1	4

6.

x	−2	−1	0	1	2
y	6	8	10	10	12

7.

x	3	4	1	2	5
y	10	13	4	7	16

8.

x	3	4	2	5	1
y	11	15	7	19	3

For these linear functions, determine the slope, *y*-intercept, and equation in slope–intercept form.

9.

x	4	2	0
y	11	7	3

10.

x	6	3	0
y	19	10	1

11.

x	3	2	1
y	13	8	3

12.

x	6	4	2
y	21	13	5

13.

x	5	3	1
y	14	8	2

14.

x	5	3	1
y	9	5	1

15.

x	−6	−4	1
y	15	11	1

16.

x	−5	−1	2
y	23	11	2

The tables show the amount of weight attached to a spring and the length of the extended spring. Show whether the number pairs belong to a linear function.

17.

weight (in grams)	0	100	200	300
length (in centimeters)	50	55	60	65

18.

weight (in grams)	0	1000	2000
length (in centimeters)	40	80	100

The tables show the number of books printed and the total cost of printing the books. Show whether the number pairs belong to a linear function.

19.

books printed	1000	2000	3000
total cost	2075	4075	6075

20.

books printed	1000	2000	3000
total cost	1600	3100	4600

The tables show the number of miles driven in a month and the cost of owning a car for the month. Show whether the number pairs belong to a linear function.

21.

miles	100	200	300	400
cost (in dollars)	125	145	162	170

22.

miles	100	200	300	400
cost (in dollars)	150	175	200	225

Show whether the number pairs belong to a linear function.

B **23.**

x	-2	0	5	10
y	-8	-2	13	28

24.

x	-5	-1	0	4
y	-5	3	5	12

25.

x	-3	1	6	-2
y	10	-6	-26	6

26.

x	-2	-1	1	2
y	4	2	2	4

27. In the Preview, the number pairs (15, 480) and (30, 505) were found to be members of a linear function. Write an equation in slope–intercept form for the volume V in terms of the temperature T.

28. Two temperature pairs that show how the Celsius and Kelvin temperature scales are related are shown in the table.

	Oxygen boils	Aspirin melts
Degrees Celsius, C	− 183	135
Degrees Kelvin, K	90	408

Write an equation in slope–intercept form for the Kelvin temperature K in terms of the Celsius temperature C.

29.–32. Write equations of the form $y = mx + b$ for the data tables in Exercises 17, 19, 20, and 22.

These results were obtained in an experiment investigating the effect of temperature on the resistance of a copper wire.

Temperature (°C)	15	30	45	100
Resistance (ohms)	23.5	25	26.5	32

33. Do the number pairs appear to belong to a *linear* function? Explain.
34. Estimate the resistance of the wire at 0°C.
35. Estimate the resistance of the wire at − 10°C.

C Here is a table of first, second, and third successive differences for the function $y = x^2$.

x	$y = x^2$	First differences	Second differences	Third differences
0	0			
1	1	1		
2	4	3	2	
3	9	5	2	0
4	16	7	2	0
5	25	9	2	0

36. Complete this table of first and second successive differences for the function $y = 2x + 3$.

x	$y = 2x + 3$	First differences	Second differences
0	?		
		?	
1	?		?
		?	
2	?		?
		?	
3	?		?
		?	
4	?		?
		?	
5	?		

37. How can successive differences be used to determine whether a table of values represents a linear function? Pick a linear function and check your guess.

38. Complete this table of first, second, and third successive differences for the function $y = x^2 + x - 4$.

x	$y = x^2 + x - 4$	First differences	Second differences	Third differences
0	?			
		?		
1	?		?	
		?		?
2	?		?	
		?		?
3	?		?	
		?		?
4	?		?	
		?		
5	?			

39. How can successive differences be used to determine whether a table of values represents a second-degree function? Pick a second-degree function and check your guess.

■ REVIEW EXERCISES

Simplify. Assume that no denominator equals zero. [4–6]

1. $6x(x - 3)$

2. $(x^2 - 4x + 3) \cdot 2x$

3. $3a(2a - b + c)$

4. $\dfrac{8a^6}{2a^2}$

5. $\dfrac{-16xy^3}{4x^2y^2}$ [4–8]

6. Write the quotient in scientific notation. $\dfrac{6.4 \cdot 10^6}{8 \cdot 10^2}$ [4–8]

7. The sum of five consecutive integers is 100. What are the integers? [4–10]

8. The sum of three consecutive even integers is -72. What are the integers? [4–10]

Write an equation that fits the problem. Do not solve the equation. [3–9]

9. John worked two weeks during winter holidays, earning a total of $100. If he earned $10 more the second week than he did the first week, how much did he earn each week?

Write an equation that fits the problem and solve it. [3–9]

10. If a rectangle is 3 times as long as it is wide, and the perimeter is 40 cm, what are the length and width?

EXTENSION   A computer program for finding the _____
equation of a line

If the coordinates of two points are known, an equation of the line that passes through them can be determined. When the coordinates of the two points are entered into the program below,

first, the slope M is determined;
next, the slope is used to determine the *y*-intercept B;
finally, the equation of the line is determined.

```
10   PRINT "WHAT ARE THE X- AND Y-COORDINATES OF THE FIRST
     POINT";
20   INPUT X1, Y1
30   PRINT "WHAT ARE THE X- AND Y-COORDINATES OF THE SECOND
     POINT";
40   INPUT X2, Y2
50   M = (Y1 - Y2)/(X1 - X2)
60   PRINT "M =   "; M
70   B = Y1 - M * X1
80   PRINT "B =   "; B
90   PRINT "THE EQUATION OF THE LINE IS Y =   "; M; "X +   ";
     B
100  END
```

Run the program for each pair of points.

1. $(2, 6), (7, -4)$ **2.** $(0, -1), (-4, 3)$

3. $(-2, 1), (-1, 5)$ **4.** $(4, -3), (6, 0)$

5. What happens when you enter two points, such as $(-1, 2)$ and $(4, 2)$, that lie on a horizontal line?

6. What happens when you enter two points, such as $(3, 7)$ and $(3, 1)$, that lie on a vertical line?

Strategy for Success   Preparing for a test _____

Review with a friend can be helpful. You can quiz each other on the meanings of mathematical symbols and words. You can give each other the kinds of questions you expect on a test.

■ CHAPTER SUMMARY

- To graph the equation of a line, graph three ordered pairs that satisfy the equation and then draw the line through them. (Two ordered pairs are enough. The third pair is used as a check.)   [5–4]

- To graph an equation in two variables, find as many ordered pairs as necessary to determine the pattern of the graph. Use positive integers, zero, negative integers, and possibly fractions as first components. Graph them and connect the points according to an observed pattern. (More points may be needed in places where the graph turns sharply.)   [5–4]

- The *slope m* of the line through points $A(x_1, y_1)$ and $B(x_2, y_2)$ is equal to   [5–5]

$$\frac{y_2 - y_1}{x_2 - x_1} \quad \text{or} \quad \frac{y_1 - y_2}{x_1 - x_2}.$$

- An equation in the form $y = mx + b$ is in *slope–intercept form.* The value of m is the slope of the line and the value of b is the y-intercept.   [5–6]

- A relation of ordered pairs is a *function* if and only if each first component in the relation is paired with exactly one second component.   [5–8]

- A function can be described by:
  Listing the ordered pairs.
  Using a table of solutions.
  A graph.
  An equation.
  A worded description.

■ CHAPTER REVIEW

5–1 **Objective:** To graph ordered pairs of real numbers.

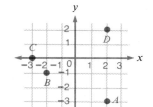

1. Identify the coordinates of each point.
   a. A
   b. B
   c. C
   d. D

2. Identify the quadrant the point is in or the axis it is on.
   a. A **b.** B **c.** C **d.** D

5–2 **Objective:** To graph relations.

The following exercises describe a relation.
   a. List the domain.   **b.** List the range.   **c.** Graph the relation.

3. $\{(1, 2), (2, 2), (3, 6)\}$

4.

x	-2	-1	0	1	2
y	4	1	0	1	4

5–3 **Objective:** To identify solutions to equations in two variables.

5. State whether the ordered pair is a solution of the equation $2x + y = 8$.
   a. $(1, 6)$ **b.** $(0, 4)$ **c.** $(-2, 12)$

6. Copy and complete the solution table for the equation $3a - b = 9$.

a	0	?	1	?
b	?	0	?	3

5–4 **Objective:** To graph equations.

7. Select three values of x, copy and complete the table, plot the points, and then graph the equation.

$y = 2x - 3$

x	?	?	?
y	?	?	?

8. Copy and complete the table. Then use the ordered pairs to graph the equation.

$y = 2 - x^2$

x	-3	-2	-1	$-\frac{1}{2}$	0	$\frac{1}{2}$	1	2	3
y	?	?	?	?	?	?	?	?	?

5–5 **Objective:** To find the slope of a line.

9. State the slope of each line.

    a. $\overleftrightarrow{AB}$ **b.** $\overleftrightarrow{BC}$

    c. $\overleftrightarrow{EF}$ **d.** $\overleftrightarrow{GE}$

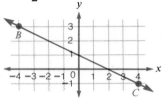

10. Draw lines with the given slopes through the point $(0, 2)$. Label the lines a, b, and c.

    a. Slope $\dfrac{1}{2}$ **b.** Slope -2 **c.** Slope 0

5–6 **Objective:** To determine the slope and y-intercept of a line.

11. State the slope and y-intercept of the line with equation $y = \frac{1}{2}x - 3$.

12. State the slope and the y-intercept of line BC.

13. Graph the line with slope $\frac{1}{3}$ and y-intercept 2.

14. Write the equation $2x + 3y = 12$ in slope–intercept form.

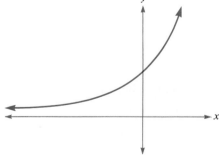

Objective: To use slopes to determine whether two lines are parallel, perpendicular, or neither.

15. State whether the lines with these equations are parallel.
$y = 3x - 7,$ $2y - 6x = 4$

16. State the slope of a line perpendicular to the graph of $y = 4x - 2$.

5–7 **Objective:** To write the equation, in slope–intercept form, of a line given two characteristics of the line.

17. Write an equation of the line that passes through $(2, 3)$ and $(-2, -7)$.

18. Write an equation of the line through $(-1, 2)$ that is parallel to the graph of $y = -x - 5$.

5–8 **Objective:** To identify functions. To identify linear functions.

19. State whether the relation $\{(2, 3), (1, 0), (2, 4)\}$ is a function. Explain.

20. How can you show that the graph represents a function?

21. State whether the equation $x^2 + y^2 = 4$ represents a function.

22. State whether the equation $4x - 3y = 12$ represents a linear function.

5–9 **Objective:** To determine whether a set of ordered pairs belongs to a linear function.

23. State whether these ordered pairs belong to a linear function.

x	-2	0	2	4
y	9	5	9	13

24. State the slope, y-intercept, and equation in slope–intercept form of this linear function.

x	2	4	-1
y	1	-3	7

■ CHAPTER 5 SELF-TEST

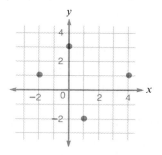

5–1 **1.** State the coordinates of point A.

5–5 **2.** State the slope of $\overleftrightarrow{AB}$.

5–1 **3.** In what quadrant are the points whose x-coordinates are negative and y-coordinates are -3 times their x-coordinates?

5–2 Exercises 4 and 5 refer to the graph of the relation at the right.

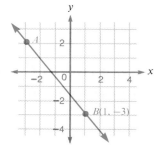

4. List the ordered pairs in the relation.

5. State the domain and range of the relation.

6. Write the ordered pairs of the relation whose components are positive integers with a product of 12.

5–3 **7.** Find the missing numbers in the table of solutions for the equation $2x + y = 4$.

x	6	-3	?
y	?	?	0

State whether the ordered pair is a solution of the equation.

8. $y = 3x + 2$,
   $(-2, 3)$

9. $y = (3x)^2$,
   $(-1, 9)$

5–4 Graph each equation.

10. $3y = 2(x - 1)$ **11.** $y = |2x|$

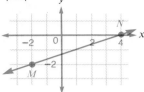

5–5 **12.** Find the slope of $\overleftrightarrow{MN}$.

13. Draw a line with slope 2 through the point $(-1, 4)$.

5–6 **14.** What is the slope of any line perpendicular to $2x + 3y = 5$?

Write the slope–intercept form of the equation of the line determined by these conditions.

15. Has slope $-\dfrac{1}{4}$ and y-intercept 3.

16. Is parallel to the graph of $y = 3x$ and crosses the y-axis at $(0, -1)$.

5–7 **17.** Passes through $(1, 4)$ and $(2, 0)$.

18. Is perpendicular to the graph of $y = 2x + 4$ and passes through $(-2, 5)$.

5–8 **19.** Which of these graphs represent functions?

a. 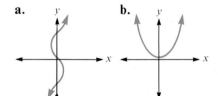   **b.**   **c.** 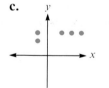   **d.** 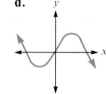   **e.**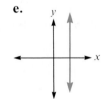

20. Do the ordered pairs determine a linear function?

x	-2	-1	0	1
y	-10	-7	-3	-1

5–9 **21.** Two temperature pairs that show how the Celsius and Fahrenheit scales are related are shown in the table. Write an equation in slope–intercept form for the Celsius temperature in terms of the Fahrenheit temperature.

	Only temperature where $°C = °F$	Water boils
Celsius, $°C$	-40	100
Fahrenheit, $°F$	-40	212

1. In the figure, the coordinates of point A are $(3, -2)$. If $\overline{AB}$ is the diameter of the circle with center O, what are the coordinates of point B?

   A. $(-2, 3)$ **B.** $(2, -3)$ **C.** $(-3, 2)$

   D. $(-3, -2)$ **E.** $(-2, -3)$

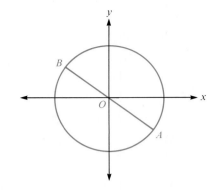

2. The coordinates of points A and B are shown in the figure. If $\overline{AB}$ and $\overline{BC}$ are the same length and $\overline{AB}$ is perpendicular to $\overline{BC}$, what are the coordinates of point C?

   A. $(2, -2)$ **B.** $(3, -2)$ **C.** $(2, -3)$

   D. $(3, -3)$ **E.** $(-3, 3)$

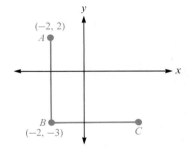

3. If the area of triangle OPQ is 24 square units, what is the value of x? (The figure is not drawn to scale.)

   A. 3 **B.** 4 **C.** 6

   D. 8 **E.** 12

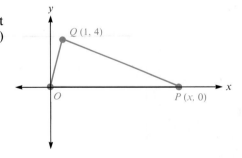

4. What is the length of $\overline{PQ}$?

   A. 0.08 unit **B.** 0.5 unit **C.** 0.8 unit

   D. 7 units **E.** 8 units

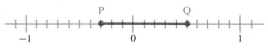

5. Which quadrants contain pairs (x, y) that satisfy the condition $y = -x$?

   A. III only **B.** II and III only **C.** III and IV only

   D. II and IV only **E.** I and III only

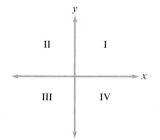

6. The coordinates of points B and C are shown. What is the value of x?

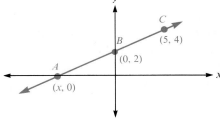

A. -2 **B.** $-2\frac{1}{2}$ **C.** -4

D. $-4\frac{1}{2}$ **E.** -5

7. A line containing the points $(0, 0)$ and $(2, 3)$ will also contain which of the following points?

A. $(3, 2)$ **B.** $(6, 4)$ **C.** $(10, 15)$ **D.** $(12, 16)$ **E.** $(15, 20)$

8. Let the symbol $\overline{X}$ represent one more than the number of digits in an integer X. If $X = 1{,}234{,}567{,}890$, then $(\overline{\overline{X}}) = \underline{\ ?\ }$.

A. 0 **B.** 1 **C.** 2 **D.** 3 **E.** 4

9. If Heidi is x years old and Anita is 4 years younger, what will Anita's age be 6 years from now?

A. $x + 2$ **B.** $x - 2$ **C.** $x - 4$ **D.** $x + 6$ **E.** $x + 10$

10. The perimeter of a 5-sided figure is 11 units. If the length of each side of the figure is increased by 3 units, what is the perimeter of the new figure?

A. 16 units **B.** 19 units **C.** 26 units **D.** 33 units **E.** 70 units

11.

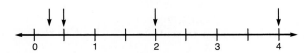

Which of the following values is the closest approximation to the product of the four numbers indicated by the arrows?

A. 1 **B.** $\frac{1}{2}$ **C.** 2 **D.** $\frac{1}{4}$ **E.** 4

12. If the odometer of an automobile reads 7777.7, what is the least number of miles the automobile must travel before all the digits will again be the same?

A. 1000.0 **B.** 1111.1 **C.** 8888.8 **D.** 9999.9 **E.** 111.1

Strategy for Success **Taking tests** _____

On multiple-choice tests, find out whether points are deducted for wrong answers. If not, answer every question even though you are not sure which answer is right. You can pick up some extra points that way.

6 Systems of Linear Equations

The Golden Gate Bridge is considered to be one of the most beautiful suspension bridges in the world. The main cable is in the shape of a parabola. The supporting towers are 1280 meters apart and rise 166 meters above the roadway. The lowest point of the main cable is 2 meters above the roadway. Suspension cables hang from the main cable and support the roadway.

The mathematical problems faced in the design of the bridge involved the stresses and forces that would act on the towers. The solution of the problem required a system of 33 linear equations in as many variables.

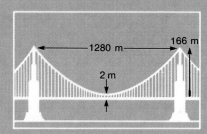

6-1 Solving Systems by Graphing

Preview

The Federal Communications Commission used a mobile direction finder to locate an illegal radio transmitter. The direction finder was positioned near the point where the FCC thought the transmitter was located. The antenna was then turned until it pointed in the direction from which the illegal signal was strongest. This position and direction were marked on a map. The FCC now knew that the transmitter was located somewhere along the line drawn on the map, but they could not tell at which point.

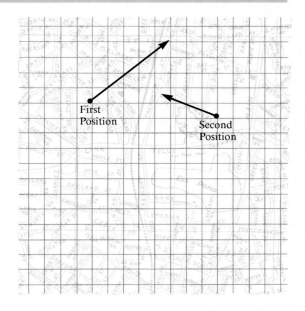

The direction finder was moved to another position and the process was repeated. The FCC knew that the illegal transmitter was located along the second direction line, too. Where do you think that the transmitter was located? Why do you think that?

■ LESSON

In Chapter 3 you solved problems by writing an equation in one variable. Many problems can be solved more easily by using two equations in two variables. For example, let's look at the following problem.

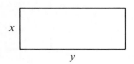

A rectangle is 2 times as long as it is wide. Its perimeter is 30 cm. What are the dimensions of the rectangle?

Let x = the width in centimeters.
Let y = the length in centimeters.

The first condition can be written: $y = 2x$ ("... is 2 times as long as it is wide").
The second condition can be written: $2x + 2y = 30$ ("Its perimeter is 30 cm").

There are infinitely many ordered pairs that are solutions of the first equation. These solutions lie on the graph of $y = 2x$.

There are also infinitely many ordered pairs that are solutions of the second equation, and they lie on the graph of the second equation, $2x + 2y = 30$.

The graphs intersect at point (5, 10). The point where the two graphs intersect is a solution of both equations. Therefore, the rectangle is 5 cm wide and 10 cm long.

The pair of equations

$$y = 2x$$
$$2x + 2y = 30$$

is called a **system of equations**. The system above is a system with two equations in two variables. The **solution of the system** is an ordered pair that is a solution of each of the two equations. We have illustrated that one way of finding the solution of a system with two equations is to graph the two equations and find the points of intersection.

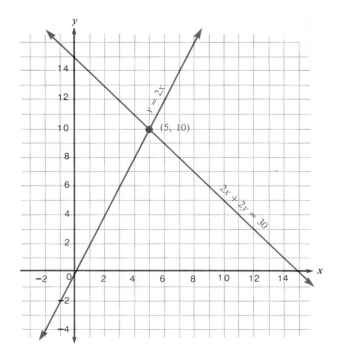

Example 1   Solve this system by graphing. $x + y = 7$
$$y = 2x - 5$$

Solution   Graph the equations using the same pair of axes.

The point of intersection appears to be (4, 3).

Answer   {(4, 3)}

Check   Substitute 4 for x and 3 for y in both equations.

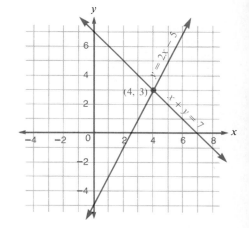

$$x + y = 7 \qquad\qquad y = 2x - 5$$
$$4 + 3 \stackrel{?}{=} 7 \qquad\qquad 3 \stackrel{?}{=} 2(4) - 5$$
$$7 = 7 \qquad\qquad 3 = 3$$

The answer checks in both equations.

Example 2 Solve this system by graphing. $y = 3x$
$$y = 3x + 2$$

Solution Graph the equations using the same axes.

The two lines are parallel. They have no points in common. This indicates that there are no solutions of the system.

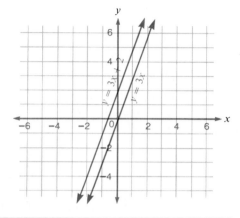

Answer ∅

■ CLASSROOM EXERCISES

State whether the number pair is a solution of the system of equations.

1. (4, 2)
$y = x - 2$
$y = \frac{1}{2}x$

2. $(-4, -8)$
$x + y = -12$
$y = 3x$

Use the graphs to solve each of these systems.

3. $y = \frac{1}{3}x$
$x + y = 4$

4. $x + y = 4$
$y = x$

5. $y = \frac{1}{3}x$
$y = x$

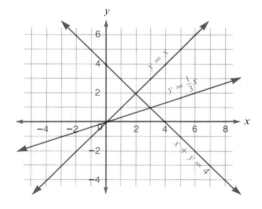

Solve these systems by graphing.

6. $y = -2x$
$x - y = 9$

7. $y = x^2$
$y = x$

■ WRITTEN EXERCISES

Use the graphs at the right to solve these systems of equations.

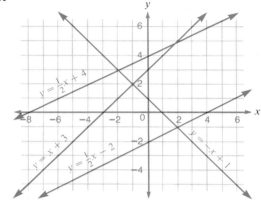

A   **1.** $y = -x + 1$

$y = \frac{1}{2}x + 4$

2. $y = -x + 1$

$y = x + 3$

3. $y = \frac{1}{2}x + 4$

$y = x + 3$

4. $y = -x + 1$

$y = \frac{1}{2}x - 2$

Use the graphs at the right to solve these systems of equations.

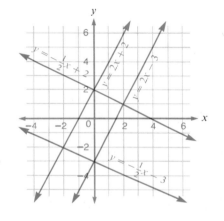

5. $y = -\frac{1}{2}x - 3$

$y = 2x + 2$

6. $y = -\frac{1}{2}x + 2$

$y = 2x - 3$

7. $y = -\frac{1}{2}x + 2$

$y = 2x + 2$

8. $y = -\frac{1}{2}x - 3$

$y = 2x - 3$

9. $y = 2x + 2$

$y = 2x - 3$

10. $y = -\frac{1}{2}x + 2$

$y = -\frac{1}{2}x - 3$

State whether the number pair is a solution of the system of equations.

11. (3, 11)
$y = 3x + 2$
$y = 2x + 5$

12. (2, 9)
$y = 4x + 1$
$y = 3x + 3$

13. (5, 3)
$y = 2x - 7$
$y = x - 2$

14. (12, 2)
$y = x - 10$
$y = 3x - 34$

15. (2, 6)
$y = 5x - 4$
$y = 2x - 2$

16. (4, 8)
$y = 3x - 4$
$y = x - 4$

17. (4, 2)
$y + x = 6$
$y + x = 0$

18. (1, 2)
$x + 2y = 5$
$x - 2y = 3$

Solve these systems of equations by graphing.

19. $y = 2x - 3$
$y = x$

20. $y = x + 2$
$y = 2x$

21. $y = \frac{1}{2}x$
$y = x - 3$

22. $y = \frac{1}{3}x$
$y = x - 2$

23. $x + y = 5$
$x - y = 3$

24. $x + y = 6$
$x - y = 4$

25. $2x + y = 7$
$x + y = 3$

26. $x + 2y = 7$
$x + y = 1$

A system of equations is given for each problem. Solve the system by graphing and then solve the problem.

27. The length of a rectangle is 3 times the width. Find the dimensions of the rectangle if its perimeter is 16 cm.

Let x = the number of centimeters in the width.
Let y = the number of centimeters in the length.

$$y = 3x$$
$$2x + 2y = 16$$

28. Crystal is two years older than Sabrina. The sum of their ages is 10. Find their ages.

Let x = the number of years in Sabrina's age.
Let y = the number of years in Crystal's age.

$$y = x + 2$$
$$x + y = 10$$

29. Brian has a total of 8 nickels and dimes. Their total value is 70 cents. How many coins of each type does he have?

Let x = the number of nickels.   Let y = the number of dimes.

$$x + y = 8$$
$$5x + 10y = 70$$

30. The sum of two consecutive integers is 5. What are the numbers?

Let x = the first integer.   Let y = the next integer.

$$y = x + 1$$
$$x + y = 5$$

One equation is given for each problem. Write a second equation that describes the situation. Solve the system by graphing and then answer the question.

B 31. Vince charges $1 plus $3 per hour for babysitting with the Young Mother's Club. Shelley charges $3 plus $2 per hour for the same job. For how many hours worked would Vince and Shelly receive the same pay?

Let x = the number of hours each worked.
Let y = the number of dollars paid.

Vince: $y = 3x + 1$
Shelley: ?

32. The length of a rectangle is twice the width. Find the dimensions of the rectangle if its perimeter is 15 cm.

Let x = the number of centimeters in the width.
Let y = the number of centimeters in the length.

Length: $y = 2x$
Perimeter: ?

Solve these systems of equations by graphing. There may be more than one solution.

33. $y = |x|$

$y = -\dfrac{1}{2}x + 3$

34. $y = x^2$

$y = x + 2$

35. $y = x^2$
$y = 2x + 3$

36. $y = (x + 2)^2$
$y = x + 8$

The following systems of equations each have one solution but the solution is difficult to find by graphing. Try to solve each system by graphing and explain what makes finding the solution difficult.

37. $2x + 3y = 8$

$6x - 3y = 4$

38. $y = \dfrac{1}{5}x - 12$

$y = \dfrac{1}{10}x - 2$

39. $y = \dfrac{1}{9}x - \dfrac{1}{3}$

$y = \dfrac{1}{8}x - \dfrac{1}{2}$

■ REVIEW EXERCISES

a	b	c	d
−2	3	−6	12

Substitute and simplify.

1. $(a + b) + (c + d)$

2. $a - b - c$ [2–2]

3. $ad \div bc + ac$

4. $\dfrac{c}{a} + \dfrac{d}{c}$ [2–5]

Solve.   [3–6]

5. $3x + 2(2x - 4) = 7$

6. $2(x + 3) + 3x = 26$

Solve each equation for the indicated variable.   [3–8]

7. $3x + 4y = 24$, for x

8. $3x - 4y = 12$, for y

Mathematics and Your Future

Teaching is an example of a field in which there are more employment opportunities for people with a background in mathematics. Teachers tend to enter the field because they like to work with people and because they wish to make a significant contribution to society. Many school systems have difficulty finding and hiring mathematics teachers; those same systems may have surpluses of teachers in other disciplines. You will find that as you increase your knowledge of mathematics, you will increase your opportunities.

6–2 Solving Systems by Substitution

Preview

The graph of the system

$$y = 2x + 1$$
$$3y = 3x - 2$$

is shown at the right.

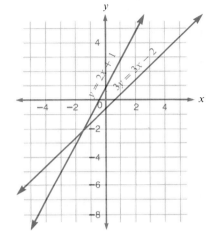

- What point appears to be the solution of the system?
- Does that apparent solution satisfy the system?

In this lesson, you will learn a method for solving systems that does not depend on graphing.

■ LESSON

If the solutions of a system are not pairs of integers, it is often difficult to find exact answers by graphing. However, we can find exact answers by working with the equations directly.

$$y = 3x$$
$$2y = x - 5$$

The first equation indicates that the values of the expressions y and $3x$ are equal. This means that y and $3x$ can be used interchangeably. Therefore, we can substitute $3x$ for y in the second equation and get an equation in one variable that we can solve.

$$2(3x) = x - 5$$
$$6x = x - 5$$
$$5x = -5$$
$$x = -1$$

The final equation, $x = -1$, indicates that -1 is the x-coordinate of the point of intersection of the graphs of the original equations. To find the y-coordinate, substitute -1 for x in either equation of the system. For example, substituting -1 for x in the first equation gives

$$y = 3x$$
$$y = 3(-1)$$
$$y = -3$$

Therefore the solution of the system is $\{(-1, -3)\}$.

We already know that this solution checks in the first equation of the system. Now we must check it in the second equation.

$$2y = x - 5$$
$$2(-3) \overset{?}{=} -1 - 5$$
$$-6 = -6 \qquad \text{It checks!}$$

This method of finding the solutions of systems of equations is called the **substitution method.**

Example 1   Solve the system by substitution.

$$2x + 3y = 7$$
$$y + 1 = x$$

Solution   The second equation indicates that x is equal to $y + 1$. Therefore, we substitute $y + 1$ for x in the first equation.

$$2x + 3y = 7$$
$$2(y + 1) + 3y = 7$$
$$2y + 2 + 3y = 7$$
$$5y + 2 = 7$$
$$5y = 5$$
$$y = 1$$

Now substitute 1 for y in the second equation to find the value for x.

$$y + 1 = x$$
$$1 + 1 = x$$
$$2 = x$$
$$x = 2$$

Answer $\{(2, 1)\}$

Check   We use the first equation to check the solution.

$$2x + 3y = 7$$
$$2(2) + 3(1) \overset{?}{=} 7$$
$$4 + 3 = 7 \qquad \text{It checks!}$$

Mathematics and Your Future

Mathematics has been described as the "critical filter" that determines one's opportunities for the future. A sound knowledge of mathematics is necessary to succeed in the information society of the future. Those with a limited background in mathematics are more likely to have limited opportunities. The choice you make today about how much mathematics you will take in high school is very important for your future.

Example 2   Solve the system by substitution.

$$2x - y = 4$$
$$3x + 2y = 7$$

Solution   Solve the first equation for y.

$$2x = y + 4$$
$$2x - 4 = y$$

Substitute $2x - 4$ for y in the second equation and solve for x.

$$3x + 2y = 7$$
$$3x + 2(2x - 4) = 7$$
$$3x + 4x - 8 = 7$$
$$7x = 15$$
$$x = \frac{15}{7}$$

Substitute $\frac{15}{7}$ for x in one of the equations (the first) and solve for y.

$$2x - y = 4$$
$$2\left(\frac{15}{7}\right) - y = 4$$
$$\frac{30}{7} - y = 4$$
$$-y = 4 - \frac{30}{7}$$
$$-y = \frac{28}{7} - \frac{30}{7}$$
$$-y = -\frac{2}{7}$$
$$y = \frac{2}{7}$$

Answer $\left\{\left(\frac{15}{7}, \frac{2}{7}\right)\right\}$

Check   Use one of the equations (the second) to check.

$$3x + 2y = 7$$
$$3\left(\frac{15}{7}\right) + 2\left(\frac{2}{7}\right) \stackrel{?}{=} 7$$
$$\frac{45}{7} + \frac{4}{7} \stackrel{?}{=} 7$$
$$\frac{49}{7} = 7 \qquad \text{It checks!}$$

Example 3   Solve the system by substitution.

$$2x + 3y = 7$$
$$2x - 2y = 2$$

Solution   There are several ways of doing substitution. You should look ahead to see if you can save yourself some work. For example, study the system. Note that $2x$ occurs in both equations. An easy way to proceed is to solve the first (or the second) equation for $2x$ and then substitute for $2x$ in the other equation.

First equation: $2x + 3y = 7$
Solve for $2x$. $2x = 7 - 3y$

Substitute $(7 - 3y)$ for $2x$ in the second equation.

Second equation: $2x - 2y = 2$
$$(7 - 3y) - 2y = 2$$
$$7 - 5y = 2$$
$$-5y = -5$$
$$y = 1$$

Substitute 1 for y in the first equation and solve for x.

First equation: $2x + 3y = 7$
$$2x + 3(1) = 7$$
$$2x = 4$$
$$x = 2$$

Answer $\{(2, 1)\}$

Check   The check is left to the student.

■ CLASSROOM EXERCISES

Use the first equation to make a substitution in the second equation.

1. $y = 3x + 1$
$2x + 3y = 4$

2. $y = \frac{1}{2}x - 2$
$y - 3 = x + 2$

3. $y + 6 = x$
$y + 7 = 2x$

4. $y = \frac{x}{2}$
$2y - x = 0$

Solve for y.

5. $x + y = 5$

6. $x - y = -3$

7. $3x - 2y = 6$

8. $3y + x = 1$

Solve for x.

9. $3x + 5y = 9$

10. $2x - y = 6$

11. $2y - 3x = 6$

12. $y + 2x = 1$

Solve by substitution.

13. $y = 2x$
$2x + y = 7$

14. $x = y + 2$
$3x - 2y = 8$

15. $3x + y = -3$
$6x - 2y = 2$

16. $2x - y = -4$
$x + 3y = 12$

266   Chapter 6   Systems of Linear Equations

WRITTEN EXERCISES

Use the first equation to make a substitution in the second equation. Solve the system of equations.

A 1. $y = 3x$
$y + 4x = 28$

2. $y = 4x$
$y + 5x = 18$

3. $y = 2x + 3$
$y + 3x = 23$

4. $y = 3x - 2$
$y + x = 18$

5. $x = 2y - 4$
$x + 3y = 11$

6. $x = 4y + 2$
$x + 2y = 14$

7. $3x = y - 5$
$3x + 2y = 7$

8. $2x = y - 2$
$2x + 3y = 10$

Solve for y.

9. $2x + 4y = 8$

10. $2x + 3y = 18$

11. $6x - 3y = 15$

12. $7x - 2y = 28$

Solve for x.

13. $2x + 4y = 8$

14. $6x + 3y = 18$

15. $3y - 6x = 15$

16. $2y - 7x = 28$

Solve by substitution.

17. $2x + 4y = 8$
$3x + 5y = 14$

18. $6x + 3y = 18$
$5x + 2y = 16$

19. $6x - 3y = 15$
$4x + 5y = 3$

20. $7x - 2y = 28$
$3x + 2y = 2$

21. $y = 3x$
$2y + 5x = 33$

22. $y = 5x$
$3y + x = 32$

23. $y = x + 3$
$2y + 3x = 26$

24. $y = x + 4$
$3y + 2x = 32$

B 25. $x + y = 8$
$2x + 3y = 15$

26. $y - 2x = 12$
$2y + 3x = 10$

27. $2x + 3y = 10$
$3x + 3y = 13$

28. $2x + 3y = 10$
$2x + 5y = 16$

29. $y = 2x + 1$
$3y = 3x - 2$

30. $y = 3x - 36$
$y = 2x + 64$

31. $x = 50 - 4y$
$5y - 2x = 1200$

32. $50x + 10y = 0$
$50x - 10y = 1$

33. $y = 2 - x$
$2y + x = 5$

34. $x = 5 - 2y$
$3x + 10y = 11$

35. $y = \frac{1}{2}x$
$4y + 3x = 15$

36. $2y = 3x$
$6y - 5x = 2$

Write two equations in two variables that fit the problem. Solve the system by substitution and then solve the problem.

37. The length of a rectangle is 3 times the width. Find the dimensions if the perimeter is 20 cm.

Let $w =$ the number of centimeters in the width.
Let $l =$ the number of centimeters in the length.

38. Hal is 10 years younger than Colleen. In 4 years he will be half her age. Find their ages now.

Let $h =$ the number of years in Hal's age now.
Let $c =$ the number of years in Colleen's age now.

Write two equations in two variables that fit the problem. Solve the system by substitution and then solve the problem.

39. Greg has 4 more nickels than dimes. The total value of his nickels and dimes is $4.10. How many coins of each type does he have?

Let n = the number of nickels.   Let d = the number dimes.

40. Tami said, "I am thinking of two numbers. The second number is 4 times the first number. If I multiply the first number by 4 and subtract that product from the product of 4 and the second number, I get 6." What two numbers was Tami thinking of?

Let x = the first number.   Let y = the second number.

41. The sum of two numbers is 80. The first number is 10 more than the second. What are the numbers?

42. Bob has a total of 30 coins, all dimes and quarters. How may coins of each type does he have if their total value is $6.00?

The solutions of these systems of equations are number triples, that is, a value of x, a value of y, and a value of z. Solve each system by substitution.

43. $y = z$
$y = x + z$
$x + y = 12$

44. $x = y + 2$
$x + y + z = 8$
$z = 4y$

45. $x = 6$
$y = x + z$
$2y = 4z + 9$

Write two equations in two variables for each of the following problems. Solve the system of equations by substitution, and then answer the question.

46. Geraldine is 3 times as old as Kate. In 10 years she will be twice as old as Kate. What are their ages today?

47. An automobile starts out on a trip at an average speed of 50 mph. A half hour later a second automobile follows the first at 55 mph. How long will it take the second automobile to catch up to the first?

■ REVIEW EXERCISES

1. Write 37% as a decimal.

2. Write $\frac{7}{20}$ as a percent.

3. Write 0.4 as a percent.

4. What is 40% of 80?

5. State which of these numbers $\{-2, -1, 0, 1, 2\}$ are solutions of the equation $x^2 = x$.

[3–1]

Solve.

6. $3x - 7 = 14$

7. $\frac{x + 4}{5} = 20$

[3–5]

8. $3(x - 4) = 2(4 - x)$

9. $3(2a + 1) = \frac{3}{2}(a - 2)$

[3–6]

6–3 Systems with No Solutions or Infinitely Many Solutions

Preview

In solving first-degree equations in one variable, we usually obtained exactly one solution. Sometimes, however, we encountered an equation that had no solution, or an equation that had many solutions.

Which of these equations have no solutions, which have one solution, and which have many solutions?

- $3x + 5 = 2x + 10$
- $4x + 3 = 4x + 3$
- $x + 6 = x$
- $2(x + 3) = 2x + 6$

In this lesson we will study a similar situation for systems of first-degree equations in two variables.

■ LESSON

If we attempt to solve this system by substitution, something unusual happens.

$$y = 2x$$
$$y = 2x + 3$$

First, substitute from the first equation into the second equation. $2x = 2x + 3$

Next, subtract $2x$ from both sides. $0 = 3$

The variables drop out and the resulting sentence is false. The fact that the substitution method resulted in the false sentence $0 = 3$ suggests that there is no solution to the system.

Here are the graphs of the equations.

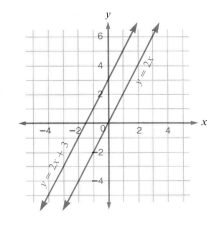

Note that the lines have the same slope, 2. Therefore, the lines are parallel. Parallel lines have no points in common, so there are no solutions to the system of equations.

Note what happens when we try to solve this system by substitution.

$$y = 2x + 3$$
$$2y - 4x = 6$$

Substitute from the first equation into the second equation.

$$2(2x + 3) - 4x = 6$$

Simplify. $\qquad 4x + 6 - 4x = 6$
$$6 = 6$$

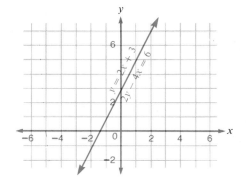

The variables again dropped out. However, this time the resulting sentence is true. The graphs of the equations show why this happened.

The graphs of the two equations are the same line. The equations are equivalent and have the same set of solutions. A system of the two equivalent equations has infinitely many solutions because the ordered pair for any point on the line is a solution for both equations.

Example 1 How many solutions does this system have?

$$y - x = 3$$
$$6 + 2x = 2y$$

Solution *Solve the first equation for y.* $\qquad y - x = 3$
$$y = 3 + x$$

Substitute in the second equation. $\quad 6 + 2x = 2(3 + x)$
Solve. $\qquad\qquad\qquad\qquad\quad 6 + 2x = 6 + 2x$
$$6 = 6$$

The variables drop out and the resulting sentence is true.

Answer The equations are equivalent and the system has infinitely many solutions.

Example 2 How many solutions does this system have?

$$y = 3x + 5$$
$$4y - 12x = 5$$

Solution *Substitute in the second equation.* $\quad 4(3x + 5) - 12x = 5$
Solve. $\qquad\qquad\qquad\qquad\qquad\quad 12x + 20 - 12x = 5$
$$20 = 5$$

The variables drop out and the resulting sentence is false.

Answer The system has no solution.

Example 3   How many solutions does this system have?

$$y = 2x - 5$$
$$3x - y = 9$$

Solution   *Substitute in the second equation.*

$$3x - (2x - 5) = 9$$
$$3x - 2x + 5 = 9$$
$$x + 5 = 9$$
$$x = 4$$

Substitute 4 for x in the first equation.

$$y = 2(4) - 5$$
$$y = 3$$

The solution is $\{(4, 3)\}$.

Answer   The system has one solution.

■ CLASSROOM EXERCISES

A system of equations and the graphs of the equations are given. How many solutions does each system have?

1. $x + y = 6$
   $y = 2x + 1$

2. $2(x - 1) = y$
   $2x - 2 = y$

3. $y = 2x + 1$
   $y = 2x - 1$

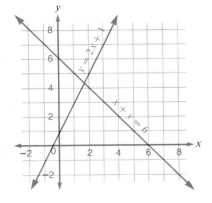
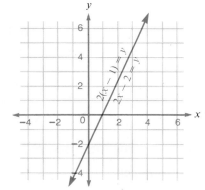
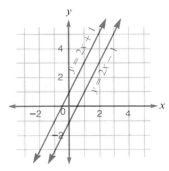

How many solutions does each system of equations have?

4. $y + x = 3$
   $y = 3 - x$

5. $x + y = 5$
   $x - y = 5$

6. $y + x = 5$
   $y + x = 3$

7. $y - x = 3$
   $4y = 12 + 4x$

8. $2x + y = 10$
   $x + y = 8$

9. $y = 2x + 3$
   $y = 2x - 3$

The graphs of equations *a*, *b*, *c*, *d*, and *e* are shown at the right. How many solutions does each system of equations have?

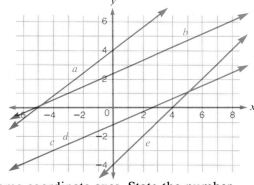

A
1. Equations *a* and *b*
2. Equations *d* and *e*
3. Equations *a* and *e*
4. Equations *b* and *c*
5. Equations *c* and *d*
6. Equations *d* and *b*
7. Equations *b* and *e*
8. Equations *a* and *c*

Graph each pair of equations using the same coordinate axes. State the number of solutions each system has.

9. $y = 2x + 3$
$y = 2x + 1$

10. $y = x + 1$
$y = x - 1$

11. $x + y = 4$
$x - y = -4$

12. $x - y = 5$
$x + y = -5$

13. $y = 2x - 3$
$y + 3 = 3x$

14. $y = x - 4$
$y + 4 = x$

15. $x + y = 6$
$2x + 2y = 12$

16. $y = 3x - 3$
$3y = 9x - 9$

If possible, solve the system by substituting from the first equation into the second equation. State the number of solutions each system has.

17. $y = 2x$
$2y = 4x$

18. $y = 3x$
$8y = 24x$

19. $y = 2x - 3$
$3y + 9 = 6x$

20. $y = 2x + 1$
$2y - 4x + 5 = 6$

21. $y = 3x - 2$
$3y + 2 = 9x$

22. $y = 2x - 3$
$2y + 6 = 4x$

23. $2y = 5x$
$2y + 3x = 16$

24. $3y = 12x$
$3y - 2x = 20$

B **25.** $y = \frac{1}{2}x + 1$
$2y - 3x = 1$

26. $y = \frac{1}{4}x + 3$
$8y - 3x = 12$

27. $y = 3x - 5$
$6x - 2y = 10$

28. $y = 5x - 3$
$15x - 3y = 9$

Four babysitters gave these rules for setting their fees, *F*.

Kari charges 3 dollars plus 2 dollars per hour. $F = 3 + 2h$

Justin charges 2 dollars plus 2 dollars per hour. $F = 2 + 2h$

Lana charges 2 dollars for every hour worked and for
1 hour of traveling. $F = 2(h + 1)$

Mike charges 2 dollars plus 3 dollars per hour. $F = 2 + 3h$

State the number of solutions each system of rules has.

29. Kari's and Justin's rules

30. Kari's and Lana's rules

31. Kari's and Mike's rules

32. Justin's and Lana's rules

33. Justin's and Mike's rules

34. Lana's and Mike's rules

C For each system described below, list all the possibilities for the number of solutions. [*Hint:* Make a sketch.]

Sample Both equations have the same *y*-intercept.
Answer One solution or infinitely many solutions.

35. Both equations have the same slope.

36. The equations have different slopes.

37. The *y*-intercepts are different.

38. The *x*-intercepts are the same.

39. The *x*-intercepts are different.

40. The slopes and *y*-intercepts are the same.

41. The slopes are different and the *y*-intercepts are the same.

42. The slopes are the same and the *y*-intercepts are different.

■ **REVIEW EXERCISES**

Simplify.

1. $3 + 4 \cdot 5 - 2$

2. $3^2 + 4^2 + (3 + 4)^2$ \hfill [1–2, 1–3]

3. $6x - 5x + 7y - y$

4. $2(a - 3) - (3a + 4)$ \hfill [2–7, 2–8]

Solve the problem by making and using a table of data.

5. What number can be added to the numerator and denominator of $\frac{1}{5}$ so that the resulting fraction will be equivalent to $\frac{1}{3}$? \hfill [2–9]

Self-Quiz 1

6–1 Solve by graphing.

    1. $2x - y = 5$
          $3x + 2y = 18$

    2. $x - 2y = 5$
          $14x + 4y = 6$

6–2 Solve by substitution.

    3. $x = 3y$
          $2x - y = 5$

    4. $y = 2x - 5$
          $3x + 2y = 11$

6–3 State whether the graphs of these systems of equations have 0, 1, or infinitely many points of intersection.

    5. $3x - 2y = -1$
          $2x + y = 11$

    6. $4x + 2y = 18$
          $-12x - 9y = 54$

How many solutions do these systems of equations have?

    7. $8x - 4y = -32$
          $-2x + y = 8$

    8. $3x + 2y = -5$
          $-9x - 6y = -15$

Preview

Mr. Williams said, "I am thinking of a two-digit positive integer." Which one of these clues is sufficient to allow you to figure out Mr. Williams's number?

Clue 1: The tens digit is 3 more than the units digit.
Clue 2: The tens digit is 4 times the units digit.
Clue 3: If the tens digit and the units digit are reversed, the new number is 27 less than the original number.
Clue 4: The sum of the digits is 5.

In this lesson we will learn how to use systems of equations to solve problems, including puzzles about digits.

■ LESSON

To solve a word problem, we must note the conditions stated in the problem. If we express these conditions as an equation containing only *one* variable, that equation can be solved and the problem answered. But if we use *two* variables, then we must write a system of two equations. Each of the equations must express a different condition stated in the problem.

Example 1   Solve the following problem.

Kent and Carol started at the same place and rode their bicycles in opposite directions. Carol rode 5 km/h faster than Kent. After 2 h they were 94 km apart. How fast did each ride?

Solution

One equation	*Two equations*
Let x = Kent's rate (km/h).	Let x = Kent's rate (km/h).
Then $x + 5$ = Carol's rate (km/h).	Let y = Carol's rate (km/h).
Diagram	**Diagram**

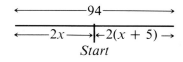

Start

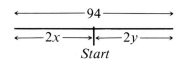

Start

Example 1 (continued)

Table

	Rate (km/h)	Time (h)	Distance (km)
Kent	x	2	$2x$
Carol	$x + 5$	2	$2(x + 5)$

Table

	Rate (km/h)	Time (h)	Distance (km)
Kent	x	2	$2x$
Carol	y	2	$2y$

Equation

$$2x + 2(x + 5) = 94$$
$$2x + 2x + 10 = 94$$
$$4x + 10 = 94$$
$$4x = 84$$
$$x = 21$$

$$x + 5 = 26$$

Equations

$$y = x + 5$$
$$2x + 2y = 94$$

Solve for x by substitution.

$$2x + 2(x + 5) = 94$$
$$2x + 2x + 10 = 94$$
$$4x + 10 = 94$$
$$4x = 84$$
$$x = 21$$

$$y = x + 5$$
$$y = 21 + 5$$
$$y = 26$$

The solution of the system is $\{(21, 26)\}$.

Answer   Kent rode 21 km/h and Carol rode 26 km/h.

There are some problems for which it is easier to use two equations than one. In this lesson we will practice using systems of equations to solve problems.

Example 2   Solve this "puzzle" problem.

The digit in the ones place of a two-digit number is 2 times the digit in the tens place. If the two digits are reversed, the new number is 27 more than the original number. What is the original number?

Solution

Let t = the tens digit and u = the ones digit. (For example, if the number were 64, t would equal 6 and u would equal 4.)

	Tens digit	Ones digit	Value
First number	t	u	$10t + u$
Second number	u	t	$10u + t$

Example 2 (continued)

The two conditions given in the statement of the problem can be expressed as this system of equations.

$$u = 2t$$
$$(10u + t) = (10t + u) + 27$$

Substitute $2t$ for u in the second equation and solve for t.

$$[10(2t) + t] = (10t + 2t) + 27$$
$$21t = 12t + 27$$
$$9t = 27$$
$$t = 3$$

Solve for u.
$$u = 2t$$
$$u = 2(3)$$
$$u = 6$$

The solution of the system is $\{(3, 6)\}$.

Answer The number is 36.

Check The ones digit is 2 times the tens digit.
$63 - 36 = 27$ It checks!

■ CLASSROOM EXERCISES

Let t = the tens digit and u = the ones digit of a two-digit number.

1. Write an expression for the value of the number.

2. Write an expression for the value of the number formed by reversing the digits.

Write an equation for each statement.

3. The sum of the digits of the two-digit number is 7.

4. The tens digit is 3 greater than the ones digit.

5. The sum of the number and the number formed by reversing the digits is 85.

Write two equations. Solve the system and answer the questions.

6. One number is 15 greater than another. The sum of the numbers is 123. What are the numbers?

7. There are 8 more nickels than dimes in a stack of coins worth $4.00. How many nickels and how many dimes are there?

■ WRITTEN EXERCISES

Write an equation for each statement. Let t represent the tens digit and let u represent the ones digit of a two-digit number.

A **1.** The tens digit is 3 times the ones digit.

2. The ones digit is 3 times the tens digit.

3. The ones digit is 4 more than the tens digit.

4. The tens digit is 4 more than the ones digit.

5. The two-digit number is 58.

6. The two-digit number is 97.

7. If the digits are reversed, the new number is 31.

8. If the digits are reversed, the new number is 19.

9. If the digits are reversed, the new number is 9 more than the original number.

10. If the digits are reversed, the new number is 72 more than the original number.

For each problem, let t represent the tens digit and u represent the ones digit of a two-digit number. Write two equations. Solve the system and answer the question.

11. The tens digit is 3 more than the ones digit. The sum of the digits is 5. What is the number?

12. The ones digit is 4 less than the tens digit. The sum of the digits is 10. What is the number?

13. The tens digit is three times the ones digit. The sum of the digits is 12. What is the number?

14. The tens digit is three times the ones digit. The sum of the digits is 8. What is the number?

15. The tens digit is 2 more than the ones digit. If the digits are reversed and the new number is doubled, the result is 17 more than the original number. What is the original number?

16. The ones digit is 3 less than the tens digit. If the digits are reversed and the new number is doubled, the result is 20 more than the original number. What is the original number?

Solve.

17. One number is 3 times as large as another. If their difference is 42, what are the numbers? [*Hint:* Let x and y represent the numbers.]

18. One number is 4 times as large as another. If their difference is 105, what are the numbers? [*Hint:* Let x and y represent the numbers.]

19. The length of a rectangle is 25 cm greater than its width. If the perimeter of the rectangle is 250 cm, what are the length and width?

20. The length of a rectangle is 30 cm greater than its width. If the perimeter of the rectangle is 300 cm, what are the length and width?

B 21. In a two-digit number, the tens digit is 2 more than twice the ones digit. If the digits are reversed and the new number is doubled, the result is 7 less than the original number. What was the original number?

22. In a two-digit number, the ones digit is 3 more than twice the tens digit. If the digits are reversed, the result is 9 less than the original number tripled. What was the original number?

23. In a three-digit number, the hundreds digit is 1 more than the tens digit and the ones digit is 5. If the digits are reversed, the new number is 198 greater than the original number. What was the original number?

24. In a three-digit number, the hundreds digit is 3 more than the tens digit and the ones digit is 8. If the digits are reversed, the new number is 99 greater than the original number. What was the original number?

For each problem, let b represent Barbara's rate in kilometers per hour and let c represent Carl's rate in kilometers per hour. Write two equations. Solve the system and answer the question.

25. Carl jogs 4 km/h faster than Barbara walks. At noon, Carl jogged straight west and Barbara walked straight east from the same point. Two hours later they were 32 km apart. What were Carl's and Barbara's rates?

26. Barbara runs 3 times as fast as Carl walks. At 10 A.M. Carl walked straight north and Barbara ran straight south from the same point. Two hours later they were 32 km apart. What were Barbara's and Carl's rates?

C 27. At noon Carl drove east at a constant rate. One hour later Barbara left from the same point following the same route as Carl and driving at a constant rate 11 km/h faster than Carl. She drove for 7 hours before she caught up to him. What were Carl's and Barbara's rates?

28. Carl finished an auto race in 4 h. Barbara finished 10 min later with a 5 km/h slower rate. What were Carl's and Barbara's rates and what distance did each of them drive?

Solve.

29. Pat earns 25¢ per hour more than Carol. One weekend Carol worked 14 hours and earned $10 more than Pat who worked 11 hours. What were Pat's and Carol's hourly pay rates?

■ REVIEW EXERCISES

1. Write 125% as a decimal.

2. Write $\dfrac{3}{100}$ as a percent.

3. Write 0.08 as a percent.

4. What is 4% of 520?

Simplify.

5. $x^2 \cdot x^3$

6. $(2xy^2)(-3x^3)$

7. $(x^2)^4$

[4–3, 4–5]

6-5 Solving Systems by Addition

Preview

Graph this system of linear equations.

$$y = 2x$$
$$y = x + 3$$

Now form a new equation by adding the left sides and the right sides of these equations.

$$\begin{array}{l} y = 2x \\ \underline{y = x + 3} \\ 2y = 3x + 3 \end{array}$$

Graph the new equation on the same axes that you used to graph the system. What happened?

■ LESSON

The Preview indicates that the solution of a system of equations is also a solution of the equation formed by adding the left sides and the right sides of the equations of the system. This fact is a consequence of the following property of equality.

The Addition Property of Equality

For all numbers a, b, c, and d,

$$\text{if } a = b \text{ and } c = d, \text{ then } a + c = b + d.$$

The addition property of equality provides a method of solving some systems of equations. For example, consider this system:

$$2x + y = -4$$
$$-2x + 3y = 12$$

Note that the coefficients of the x-terms are opposites. When the left sides and the right sides are added, the result is an equation in one variable.

$$\begin{array}{l} 2x + y = -4 \\ \underline{-2x + 3y = 12} \\ 4y = 8 \end{array}$$

Solving the equation $4y = 8$ gives $y = 2$.

The second component of the ordered-pair solution of the system is 2. Substituting 2 for y in the first equation and solving for x gives the first component of that solution.

$$2x + y = -4$$
$$2x + 2 = -4$$
$$2x = -6$$
$$x = -3$$

The solution of the system is $\{(-3, 2)\}$. We can check the solution by substituting in the second equation.

$$-2x + 3y = 12$$
$$-2(-3) + 3(2) \overset{?}{=} 12$$
$$6 + 6 = 12 \qquad \text{Check.}$$

Whenever the coefficients of one of the variables in the equations of a system are opposites, the system can be solved easily by addition.

Example 1 Solve. $2x + 3y = 7$
$x - 3y = 8$

Solution Since the coefficients of y are opposites, add the equations and solve for x.

$$2x + 3y = 7$$
$$x - 3y = 8$$
$$\overline{3x = 15}$$
$$x = 5$$

Now substitute 5 for x in the first equation and solve for y.

$$2x + 3y = 7$$
$$2(5) + 3y = 7$$
$$3y = -3$$
$$y = -1$$

Answer $\{(5, -1)\}$

Example 2 Solve. $3a = 2b + 7$
$2b = 9 - a$

Solution *Rearrange terms and add.* $\qquad 3a - 2b = 7$
$$a + 2b = 9$$
$$\overline{4a = 16}$$

Solve for a. $\qquad\qquad\qquad\qquad a = 4$
Solve for b. $\qquad\qquad\qquad\quad 3a = 2b + 7$
$$3(4) = 2b + 7$$
$$12 = 2b + 7$$
$$5 = 2b$$

$$\frac{5}{2} = b$$

Answer $\left\{\left(4, \dfrac{5}{2}\right)\right\}$

CLASSROOM EXERCISES

Solve by addition.

1. $c + 2d = 6$
$-c + 3d = -1$

2. $a + b = 7$
$a - b = 9$

3. $5x + y = 17$
$-y = x - 1$

WRITTEN EXERCISES

Solve by addition.

A **1.** $-x + 2y = 1$
$x + 3y = 14$

2. $x + 3y = 19$
$-x + 4y = 16$

3. $x + y = 8$
$-x + y = 4$

4. $-x + y = 5$
$x + y = 11$

5. $x + y = 8$
$x - y = 6$

6. $x - y = 3$
$x + y = 13$

7. $2x - 3y = 4$
$8x + 3y = 46$

8. $3x - 2y = 7$
$5x + 2y = 33$

9. $2x + y = 5$
$3x - y = 15$

10. $3x + y = 5$
$2x - y = 10$

11. $4x + y = -6$
$3x - y = -1$

12. $12x + y = 6$
$9x - y = 1$

13. $x - 2y = 7$
$-x - 3y = 18$

14. $-x + 3y = 3$
$x - 8y = 2$

15. $4x - y = 5$
$y = 3x - 3$

16. $5x - y = 5$
$y = 2x + 1$

17. $3x - 2y = 1$
$2y = x + 5$

18. $4x - 3y = 6$
$3y = x + 3$

19. $2y + x = 17$
$3y = x + 8$

20. $2y + x = 4$
$5y = x + 14$

21. $3y + 2x = 5$
$5x = 3y - 19$

22. $3y + x = 7$
$2x = 3y - 31$

23. $5x + 3y = 15$
$7x - 3y = -15$

24. $3x - 4y = 9$
$2x + 4y = 6$

For each puzzle, let x represent the first number and let y represent the second number. Write two equations and solve the puzzle.

25. I am thinking of two numbers whose sum is 11. If I subtract the second number from the first, I get 7. What are the numbers?

26. I am thinking of two numbers whose sum is 11. If I double the first number and then subtract the second number, I get 10. What are the numbers?

B **27.** I am thinking of two numbers. If I double the first number and add the second number, I get 1. If I add 7 to the first number, I get the second number. What are the numbers?

28. I am thinking of two numbers. If I double the second number and subtract the product from the first number, I get 10. Doubling the second number gives the same result as subtracting the first number from 2. What are the numbers?

29. I am thinking of two numbers. If I multiply the first number by 3 and subtract twice the second number, I get 1. If I multiply the first number by 3 and add twice the second number, I get 3. What are the numbers?

30. I am thinking of two numbers. If I multiply the first number by 2 and subtract 3 times the second number, I get 2. If I multiply the first number by 4 and add 3 times the second number, I get 1. What are the numbers?

For each problem, write two equations, solve the equations by addition, and use the result to solve the problem.

31. John is 5 years older than Mary, and the sum of their ages is 45. How old is each?

32. Bill weighs 20 pounds more than Suzanne, and the sum of their weights is 230 pounds. How much does each weigh?

33. Sally earns $20 more per week than John, and her earnings equal $100 minus John's earnings. How much does each earn?

34. The sum of the temperatures at noon for yesterday and today in Frozen Falls, Minnesota, is $-18°F$. If it was $8°$ warmer yesterday than today, what were the noon temperatures on the two days?

Graph the two given equations. Then graph the equation formed by adding the left sides and right sides of the given equations. Graph all three equations on the same coordinate system.

C **35.** $y = 4x + 6$
$y = 2x + 4$

36. $y = x + 3$
$y = -x + 7$

37. a. Add the equations $y = m_1 x + b_1$ and $y = m_2 x + b_2$ and then rewrite the new equation in slope–intercept form.
  b. If equations with slopes 1 and 2 are added, what is the slope of the new equation?
  c. If equations with y-intercepts 2 and -4 are added, what is the y-intercept of the new equation?

■ REVIEW EXERCISES

1. Write 256,000 in scientific notation. [4–4]

2. Write $2.5 \cdot 10^6$ in standard decimal notation. [4–4]

Simplify.

3. $(3 \cdot 10^2) \cdot (4 \cdot 10^4)$ (Express the answer in scientific notation.) [4–4]

4. $3a(a^2 - 2ab + b^2)$ [4–6]

5. $-2x(x^2 - 4x + 5)$ [4–6]

6. $\dfrac{6x^2}{8x^3}$ $(x \neq 0)$ [4–8]

7. $\dfrac{a^2 b^4}{a^5 b^2}$ $(a \neq 0, b \neq 0)$ [4–8]

8. A long-distance phone call costs $.25 for the first minute plus $.15 for each additional minute. How long can you talk for $2.50? [4–9]

6-6 Solving Systems by Multiplication and Addition

Preview

This system of equations was in a homework assignment. It was supposed to be solved by addition.

$$4x - y = 13$$
$$x + 2y = 1$$

At first, one student thought that the textbook was wrong. When the equations were added, the new equation still had two variables.

$$4x - y = 13$$
$$\underline{x + 2y = 1}$$
$$5x + y = 14$$

Suddenly the student had an idea. What would happen if the first equation is added again?

$$5x + y = 14$$
$$\underline{4x - y = 13}$$
$$9x \quad = 27$$

Complete the work and then check the answer in the original system. Did the idea work?

■ LESSON

Suppose that we wish to solve this system by addition.

$$x + y = 9$$
$$3x - 2y = 32$$

When we add, we get this equation, which still has two variables.

$$4x - y = 41$$

None of the variables drops out because neither the coefficients of the x-terms nor the coefficients of the y-terms are opposites. However, we can use the multiplication property of equations to transform one of the equations so that some pair of coefficients are opposites. For example, if we multiply both sides of the first equation by 2, the coefficients of y in the equations become opposites.

$$2x + 2y = 18$$
$$\underline{3x - 2y = 32}$$

Adding results in one equation in one variable.
$$5x \quad = 50$$
$$x = 10$$

Substituting 10 for x in the first equation and solving for y gives:

$$x + y = 9$$
$$10 + y = 9$$
$$y = -1$$

The solution of the system is $\{(10, -1)\}$.

Example 1   Solve. $5x - 2y = 45$
                            $3x + y = 5$

Solution   *Multiply both sides of the second equation by 2.*

           Add.
           Solve for x.
           Substitute 5 for x in the second equation and solve for y.

$$5x - 2y = 45$$
$$6x + 2y = 10$$
$$\overline{11x \quad\quad = 55}$$
$$x = 5$$
$$3x + y = 5$$
$$3(5) + y = 5$$
$$y = -10$$

Answer $\{(5, -10)\}$

Check   The check is left to the student.

Example 2   Solve. $x + y = 8$
                            $x + 2y = 13$

Solution   *Multiply both sides of the first equation by −1.*

           Add.
           Substitute 5 for y in the first equation and solve for x.

$$-x - y = -8$$
$$x + 2y = 13$$
$$\overline{y = 5}$$
$$x + y = 8$$
$$x + 5 = 8$$
$$x = 3$$

Answer $\{(3, 5)\}$

Check   The check is left to the student.

Note, in Example 2, that multiplying both sides of the first equation by -1 and then adding the sides of the first equation to the sides of the second equation is the same as subtracting the sides of the first equation from the sides of the second equation. Just as we can solve systems by adding, we can also solve them by subtracting. Sometimes it is necessary to use the multiplication property of equations on both equations in order to make pairs of coefficients into opposites.

Example 3   Solve. $2x + 3y = -1$
                            $3x + 5y = -1$

Solution 1   *Multiply both sides of the first equation by −3.*
                Multiply both sides of the second equation by 2.
                Add.
                Substitute 1 for y and solve for x.

$$-6x + (-9y) = 3$$
$$6x + 10y = -2$$
$$\overline{y = 1}$$
$$2x + 3y = -1$$
$$2x + 3(1) = -1$$
$$2x = -4$$
$$x = -2$$

Answer $\{(-2, 1)\}$

Check   The check is left to the student.

There are several ways of solving a system of equations by multiplication and addition. Here are two other solutions of the system in Example 3.

$$2x + 3y = -1$$
$$3x + 5y = -1$$

Solution 2

Multiply both sides of the first equation by 3 and the second equation by -2.

$$
\begin{array}{r}
6x + 9y = -3 \\
-6x + (-10y) = 2 \\
\hline
-y = -1 \\
y = 1
\end{array}
$$

Solve for x.

$$
\begin{array}{r}
3x + 5y = -1 \\
3x + 5(1) = -1 \\
3x = -6 \\
x = -2
\end{array}
$$

Answer $\{(-2, 1)\}$

Solution 3

Multiply both sides of the first equation by 5 and the second equation by -3.

$$
\begin{array}{r}
10x + 15y = -5 \\
-9x + (-15y) = 3 \\
\hline
x = -2
\end{array}
$$

Solve for y.

$$
\begin{array}{r}
3x + 5y = -1 \\
3(-2) + 5y = -1 \\
5y = 5 \\
y = 1
\end{array}
$$

Answer $\{(-2, 1)\}$

Here is a summary of the steps for solving a system of equations by multiplication and addition:

1. Apply the multiplication property of equations to one or both equations of the system to obtain a new system in which the coefficients of one variable are opposites.

2. Add the sides of one equation to the sides of the other equation to obtain one equation in one variable.

3. Solve for the variable.

4. Substitute the value found in step 3 for the appropriate variable in one of the original equations.

5. Solve for the other variable.

6. Check your answer in the original system.

■ CLASSROOM EXERCISES

Explain how to eliminate the x-terms by multiplication and addition.

Sample
$$3x + y = 5$$
$$6x - 4y = 7$$

Answer Multiply the first equation by -2 and add.

1. $x + y = 11$
$ x + 2y = 13$

2. $3x - y = -5$
$ x + 2y = -5$

3. $2x + y = 7$
$ 3x - 2y = 7$

Explain how to eliminate the *y*-terms by multiplication and addition.

4. $3x + 2y = 4$
$2x - y = 3$

5. $3x - 2y = 7$
$2x - 2y = 4$

6. $3x + 5y = 4$
$3x + 3y = 12$

Solve by using multiplication and addition.

7. $c - 5d = 0$
$2c - 3d = 7$

8. $3x + 5y = -1$
$2x - 6y = -10$

■ WRITTEN EXERCISES

Explain how to eliminate the *x*-terms by multiplication and addition.

Ⓐ **1.** $2x - y = 8$
$6x - 5y = 12$

2. $x + 2y = 18$
$3x - 4y = 4$

3. $-4x + 3y = 7$
$x + 2y = 12$

4. $-8x + 11y = 1$
$4x - 3y = 7$

Explain how to eliminate the *y*-terms by multiplication and addition.

5. $x + 2y = 18$
$3x - 4y = 4$

6. $3x - 4y = 1$
$5x + 8y = 31$

7. $3x - 5y = 15$
$2x - 3y = 9$

8. $3x - 2y = 17$
$4x - 3y = 26$

Solve these systems by multiplication and addition.

9. $3x + 2y = 13$
$5x + 2y = 19$

10. $3x + 4y = 19$
$3x + 2y = 11$

11. $3x + 3y = 3$
$4x + 2y = 18$

12. $3x + 2y = 4$
$2x + 4y = 16$

13. $3x - 2y = 6$
$5x - 6y = 26$

14. $x - 3y = 2$
$3x - 8y = 3$

15. $2x - y = 3$
$6x + y = 1$

16. $x + 10y = 1$
$-x + 2y = 5$

17. $6x + 4y = 7$
$3x + 8y = 5$

18. $3x - 4y = 1$
$6x + 16y = 11$

19. $2x + 3y = 28$
$3x + 2y = 27$

20. $4x + 3y = 29$
$3x + 4y = 34$

21. $3x + 8y = 1$
$2x + 7y = 4$

22. $5x + 11y = 3$
$3x + 7y = 3$

23. $7x - 2y = 4$
$9x - 3y = 3$

24. $8x - 5y = 4$
$5x + 3y = 27$

25. $2x - 7y = 21$
$3x + 5y = 16$

26. $3x + 2y = 6$
$4x - 3y = 25$

Write each equation in the form $ax + by = c$. Then solve the system of equations.

Ⓑ **27.** $x + 2(x - y) = 2$
$5x = 3x + y + 3$

28. $8x - y = 6x - 2$
$4x + y = 2x + 4$

29. $2x - 6y = 4y + 5$
$4(x - 2y) = 3x + 1$

30. $3(x - 2y) = x + y$
$3(x - 5) = 2(y + 1)$

Solve these systems of equations. If the system does not have exactly one solution, state whether there are no solutions or infinitely many solutions.

31. $2x + 3y = 6$
$4x + 5y = 3$

32. $5x - y = 8$
$10x - 2y = 16$

33. $-x + y = 7$
$x - y = 7$

34. $3x + 7y = 50$
$7x + 3y = 50$

35. $y = 3$
$y = -3$

36. $-2x - 3y = 12$
$2x + 3y = -12$

Solve.

C **37.** Twice a first number plus 3 times a second number equals 87. Three times the first number minus twice the second number equals -6. What are the numbers?

38. Five years ago John was 3 times as old as Carla. Five years from now John will be twice as old as Carla. How old is each now?

39. For a school play student tickets cost $2.00 and adult tickets cost $5.00. If ticket sales amounted to $1480 and 500 tickets were sold, how many were adult tickets and how many were student tickets?

40. When Judy worked 8 hours and Ben worked 10 hours, their combined pay was $80. When Judy worked 9 hours and Ben worked 6, their combined pay was $69. What was the hourly rate of pay for each?

41. Solve for x and y in terms of a_1, b_1, c_1, a_2, b_2, c_2.

$$a_1x + b_1y = c_1$$
$$a_2x + b_2y = c_2$$

42. Use the results from Exercise 41 to show the computations needed to solve this system. Do not carry out the calculations.

$$0.26x + 5.7y = 4.68$$
$$2.95x - 4.3y = 1.97$$

■ REVIEW EXERCISES

1. Identify the quadrant each point is in or the axis it is on.   [5–1]

a. A
b. B
c. C

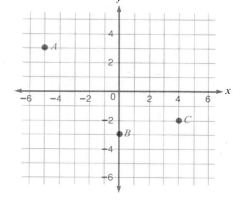

2. a. List the domain and range of this relation. [5–2]

x	-2	-1	0	1	2
y	0	1	2	1	0

b. Graph the relation.

3. Find the missing number for each solution pair. [5–3]

$4c - d = 10$

c	0	?	2	?
d	?	0	?	2

Copy and complete the table. Then use the ordered pairs to graph the equation. [5–4]

4. $y = |x - 1|$

x	-3	-2	-1	0	1	2	3
y	?	?	?	?	?	?	?

5. $y = x^3$

x	-2	-1	$-\frac{1}{2}$	0	$\frac{1}{2}$	1	2
y	?	?	?	?	?	?	?

Self-Quiz 2

6–4 Write an equation for each statement, letting t represent the tens digit and letting u represent the ones digit of a two-digit number.

1. The ones digit of a two-digit number is 6 more than the tens digit.

2. A two-digit number is equal to 47.

Solve.

3. The ones digit of a two-digit number is 3 times the tens digit. If the digits are reversed and the new number is added to the original number, the sum is 132. Find the original number.

6–5 Solve by addition.

4. $x - 3y = 5$
$2x + 3y = 1$

5. $-5x + 4y = 12$
$5x - 8y = -14$

6–6 Solve by multiplication and addition.

6. $-3x + 8y = 34$
$5x - 2y = 0$

7. $4x + 6y = 12$
$6x - 4y = -21$

6-7 Problem Solving — Mixture Problems

Preview

Select the most reasonable choice in each mixture situation.

- One pound of $1.34 candy is mixed with one pound of $2.50 candy. What is a reasonable price per pound for the mixture?

    $3.84   or $1.92

- A quiz contestant is asked to select 30 coins totaling $6.00 in value from a pile of dimes and a pile of quarters. Should more dimes or quarters be selected?

    dimes   or   quarters

- Two acid solutions are to be mixed to produce a 35% acid mixture. Which choice of acid solutions could be used to make the mixture?

    90% solution and   or   10% solution and   or   20% solution and
    50% solution   25% solution   40% solution

In this lesson, you will use algebra to investigate mixture problems.

■ LESSON

Mixture problems can be solved by using two equations in two variables. However, we must be careful to distinguish among quantities, price per unit, and values of quantities. Example 1 shows how to set up a table of values that will be used to write the equations.

Example 1   Solve using a system of equations.

Dee paid $6.00 for 5 pounds of candy. She bought two kinds of candy, one that cost $1.10 per pound and one that cost $1.50 per pound. How many pounds of each kind did she buy?

Solution   Let x = the number of pounds of $1.10 candy.
Let y = the number of pounds of $1.50 candy.

Table

	Number of pounds	Cost per pound	Value in dollars
$1.10 candy	x	1.10	$1.1x$
$1.50 candy	y	1.50	$1.5y$
Mixture	5	1.20	6.00

Example 1 (continued)

System of equations
$$x + y = 5$$
$$1.1x + 1.5y = 6$$

Solve the system.

Multiply the first equation by 1.5.	$1.5x + 1.5y = 7.50$
Multiply the second equation by −1.	$-1.1x - 1.5y = -6.00$
Add.	$0.4x = 1.50$
Solve for x.	$x = 3.75$
Substitute in the first equation.	$3.75 + y = 5$
Solve for y.	$y = 1.25$

The solution of the system is $\{(3.75, 1.25)\}$.

Answer   Dee bought 1.25 pounds of the $1.50 per pound candy and 3.75 pounds of the $1.10 per pound candy.

Check $1.25 + 3.75 = 5$

1.25 pounds of the $1.50 per pound candy is worth $1.875 and 3.75 pounds of the $1.10 per pound candy is worth $4.125. Together they are worth $1.875 + $4.125, or $6.00. The answer checks.

Example 2   Solve using a system of equations.

Tom wanted to mix antifreeze that was 60% alcohol with antifreeze that was 40% alcohol to make a solution that was 52% alcohol. How much of the alcohol solution was needed to make 5 liters of the new solution?

Solution   Let $x =$ the number of liters of the 60% solution.
Let $y =$ the number of liters of the 40% solution.

Table

	Number of liters of solution	*Number of liters of alcohol*
60% solution	x	$0.6x$
40% solution	y	$0.4y$
52% solution	5	$0.52(5)$

System of equations
$$x + y = 5$$
$$0.6x + 0.4y = 0.52(5)$$

Solve the system.

Multiply the first equation by −0.4.	$-0.4x + 0.4y = -2$
Add the second equation.	$0.6x + 0.4y = 2.6$
	$0.2x = 0.6$
Solve for x.	$x = 3$

Example 2 (continued)

Substitute in the first equation.
Solve for y.

$$x + y = 5$$
$$3 + y = 5$$
$$y = 2$$

The solution of the system is $\{(3, 2)\}$.

Answer   Tom must mix 2 liters of the 40% solution with 3 liters of the 60% solution.

Check   The final solution will have $2 + 3 = 5$ liters. There are $0.4(2) = 0.8$ liters of alcohol in the 2 liters of the 40% solution, and $0.6(3) = 1.8$ liters of alcohol in 3 liters of the 60% solution. There are $0.8 + 1.8 = 2.6$ liters of alcohol in the final solution and $2.6 \div 5 = 0.52 = 52\%$. It checks.

■ CLASSROOM EXERCISES

Suppose that solution A is 55% alcohol in water and solution B is 70% alcohol in water. Answer these questions.

1. a. How many milliliters of alcohol are in 100 milliliters of solution A?
   b. How many milliliters of alcohol are in 40 milliliters of solution B?

2. a. If 40 milliliters of solution A are mixed with 60 milliliters of solution B, how many milliliters of alcohol are in the new solution?
   b. What percent of the new solution is alcohol?

3. How many liters of alcohol are in x liters of solution A?

4. How many liters of alcohol are in y liters of solution B?

5. a. How many liters of alcohol are in a mixture of x liters of solution A and y liters of solution B?
   b. How many liters of the new solution are there?

6. Write a system of equations for this problem: How many liters of solutions A and B are needed to make 10 liters of a solution that is 60% alcohol?

■ WRITTEN EXERCISES

A   **1.** A birdseed mixture contains two kinds of seeds. Type A costs 20¢ per pound and type B costs 40¢ per pound.
   a. How much would x pounds of type A seed cost?
   b. How much would x pounds of type A seed and y pounds of type B seed cost?
   c. How many pounds of seed are in a mixture of x pounds of type A seed and y pounds of type B seed?
   d. Write an equation for this condition: A 10-pound mixture contains x pounds of type A seed and y pounds of type B seed.
   e. Write an equation for this condition: The cost of a mixture of x pounds of type A seed and y pounds of type B seed is $3.50.

2. A pet food contains two ingredients, cereal and meat. The cereal costs 25¢ per pound and the meat costs 75¢ per pound.

   a. How much would x pounds of cereal cost?

   b. What is the total cost of x pounds of cereal and y pounds of meat?

   c. How many pounds of ingredients are in a mixture of x pounds of cereal and y pounds of meat?

   d. Write an equation for this condition: A 20-pound bag of pet food contains x pounds of cereal and y pounds of meat.

   e. Write an equation for this condition: The cost of a mixture of x pounds of cereal and y pounds of meat is $7.00.

Write two equations. Solve the system and answer the question.

3. Some $4.00-per-pound candy and some $1.00-per-pound candy are to be mixed to produce 5 pounds of a mixture that sells for $2.20 per pound. How many pounds of each type of candy should be used?

	Number of pounds	Value in dollars
$4.00/lb candy	x	$4x$
$1.00/lb candy	y	$1y$
$2.20/lb candy	5	(2.2)5

4. Some $4.00-per-pound candy and some $1.00-per-pound candy are to be mixed to produce 5 pounds of a mixture that sells for $2.80 per pound. How many pounds of each type of candy should be used?

	Number of pounds	Value in dollars
$4.00/lb candy	x	$4x$
$1.00/lb candy	y	$1y$
$2.80/lb candy	5	(2.8)5

5. A shopowner plans to mix $2.00-per-pound nuts and $3.00-per-pound nuts to produce 10 pounds of a mixture that sells for $2.30 per pound. How many pounds of each type of nuts are required?

6. A social club mixed some $3.00-per-pound tea and some $4.00-per-pound tea to produce a 2-pound package of a mixture valued at $6.25. How many pounds of each type of tea were used?

7. Some $1.75-per-pound coffee and some $2.25-per-pound coffee are mixed to produce 100 pounds of a coffee mixture valued at $1.98 per pound. How many pounds of each type of coffee were used?

8. Cheeses costing $2.75 per pound and $3.50 per pound are mixed to produce a 50-pound cheese tray that sells for $2.96 per pound. How many pounds of each type of cheese are used?

9. Sandwich meats costing $3.25 per pound and $2.25 per pound are mixed to produce 100 pounds of party platters that sell for $2.69 per pound. How many pounds of each type of sandwich meat are used?

10. Ground beef costing $2.40 per pound is mixed with ground pork costing $1.80 per pound to produce 60 pounds of a meatloaf mixture that sells for $1.99 per pound. How many pounds of each type of meat are used?

11. Grass seed valued at $3.00-per-pound is mixed with $2.20-per-pound grass seed to produce 25 pounds of a mixture that sells for $2.40 per pound. How many pounds of each type of grass seed are used?

12. Milk that is 9% butterfat is mixed with milk that is 1.5% butterfat to produce 100 liters of milk that is 3% butterfat. How much milk of each type is used?

	Liters of milk	Liters of butterfat
Milk with 9% butterfat	x	$0.09x$
Milk with 1.5% butterfat	y	$0.015y$
Milk with 3% butterfat	100	$(0.03)100$

13. An alloy that is 40% copper is mixed with an alloy that is 60% copper to produce 80 kilograms of an alloy that is 56% copper. How many kilograms of each type of alloy are used?

B 14. Liquid that is 80% antifreeze is mixed with a liquid that is 40% antifreeze to produce 10 L of a liquid that is 76% antifreeze. How many liters of each liquid are used?

15. Liquid that is 80% antifreeze is mixed with a liquid that is 40% antifreeze to produce 10 L of a liquid that is 50% antifreeze. How many liters of each liquid are used?

16. An investor put a total of $10,000 in stocks and bonds for one year. The stocks paid $9\frac{1}{2}$% and the bonds paid 6%. The earnings for the year were $824. How much was invested in stocks and how much in bonds?

17. A chemist mixed a solution that was 25% acid with a solution that was 40% acid and got 50 mL of a solution that was 37% acid. How many milliliters of each solution were used?

18. A 10-L container is filled with an alcohol–water solution that is 40% alcohol. How much of the solution should be poured out and replaced with pure water in order to have a container filled with a solution that is 35% alcohol?

19. A 100-L container is filled with a solution that is 20% acid. How much of the solution should be poured out and replaced with pure acid in order for the container to have a solution that is 25% acid?

C **20.** A 12-L gasoline–alcohol solution is 10% alcohol. How many liters of gasoline must be added to the solution to make it 8% alcohol?

21. A butcher had 100 lb of ground beef that was 85% lean (15% fat). How many pounds of fat must be added to the ground beef so that the mixture can be sold as 70% lean (30% fat)? (Round your answer to the nearest pound.)

22. How many pounds of nuts that cost $2.25/lb must be added to 100 lb of nuts that cost $4.75/lb to make a mixture that costs $3.25/lb?

23. How many liters of a 90%-alcohol solution must be added to 3 L of a 60%-alcohol solution to obtain a 72%-alcohol solution?

■ **REVIEW EXERCISES**

1. Give the slopes of the lines. Write "none" if the line has no slope.   [5–6]

   a. $\overleftrightarrow{AB}$

   b. $\overleftrightarrow{BC}$

   c. $\overleftrightarrow{AC}$

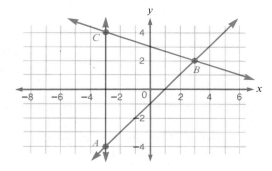

2. Graph the line with slope 2 and y-intercept -4.   [5–6]

3. State the slope and y-intercept of the line for the equation $y = \frac{2}{3}x - 6$.   [5–6]

4. Write the equation in slope–intercept form of the line with slope 2 that passes through (2, 3).   [5–7]

5. State whether the relation $\{(2, 3), (3, 4), (4, 4)\}$ is a function.   [5–8]

6. State whether the ordered pairs belong to a linear function.   [5–8]

x	-3	-1	0	1
y	2	1	0	1

7. Cities A and B are 380 mi apart. At noon, a train leaves city A heading for city B at 50 mph. An hour later, another train leaves city B heading for city A at 60 mph. When will the trains meet each other? How far from city A will they be?   [4–7]

The program below illustrates how a computer can be used to solve a problem using systematic trials in BASIC. It uses the greatest integer function to round the cost to the nearest cent.

A 10-lb mixture of nuts is to consist of nuts that sell for $4.59 per lb and peanuts that sell for $1.25 per lb. If the mixture is to be sold for $3.99 per lb, how many pounds of nuts and how many pounds of peanuts should be in the mixture?

```
10   PRINT "LB AT", "LB AT", "MIX"
20   PRINT "$4.59", "$1.25", "COST"
30   FOR X = 0 TO 10
40   Y = 10 - X
50   Z = (4.59 * X + 1.25 * Y)/10
60   Z = (INT(100 * Z + 0.5))/100
70   PRINT X, Y, Z
80   NEXT X
90   END
```

[Commas in PRINT statements separate the output into columns, so line 10 prints the table headings.]

[Lines 30 and 80 set up a ''FOR-NEXT'' loop in which X is successively given the values 0, 1, 2, . . . , 10. The computations in lines 40 through 70 are made on these successive values of X.]

[Line 50 computes the price per pound of the mixture.]

[Line 60 rounds the price per pound to the nearest cent.]

[Line 70 prints the table.]

Run the program.

Notice that the cost of the mixture is below $3.99 when there are 8 lb of $4.59 nuts and above $3.99 when there are 9 lb of $4.59 nuts. Therefore, the correct answer is between 8 and 9 lb of $4.59 nuts.

To get an approximation to the answer between successive tenths of a pound, change line 30 to the following and run the program again.

```
30 FOR X = 8 TO 9 STEP 0.1
```

What do you get for an answer?

■ CHAPTER SUMMARY

- To solve a system of linear equations in two variables by graphing, graph each of the equations. If the graphs intersect, the point of intersection is the solution of the system. [6–1]

- To solve a system of equations by substitution: [6–2]

  1. Using whichever equation is more convenient, express one variable in terms of the other.

  2. Substitute this expression for the variable in the other equation to obtain a simpler equation in one variable that can be solved.

  3. Substitute this value in either equation to solve for the other variable.

  4. Check the solution pair in both equations of the original system.

- A system of two linear equations can have 0, 1, or infinitely many solutions. [6–3]

  1. Solve the system by using substitution or graphing. If the system has one solution, it can be determined.

  2. If, after substitution, the variables drop out and the resulting sentence is false, the system has no solutions. The graph of the equations is parallel lines with the same slope and different y-intercepts.

  3. If, after substitution, the variables drop out and the resulting sentence is true, the system has an infinite number of solutions. The equations are equivalent and have the same slope and the same y-intercept. The graphs of the equations are the same line.

- The Addition Property of Equations [6–5]

  For all numbers a, b, c, and d,
  if $a = b$ and $c = d$, then $a + c = b + d$.

- Wherever the coefficients of one of the variables in the equations of a system are opposites, the system can be solved by using the addition property of equality and substitution. [6–5]

- To solve a system of equations by multiplication and addition: [6–6]

  1. Apply the multiplication property of equations to one or both equations of the system to obtain a new system in which the coefficients of one variable are opposites.

  2. Add the sides of one equation to the sides of the other equation to obtain one equation in one variable.

  3. Solve for the variable.

  4. Substitute the value found in step 3 for the appropriate variable in one of the original equations.

  5. Solve for the other variable.

  6. Check your answer in the original system.

■ CHAPTER REVIEW

6–1 **Objective:** To solve systems of equations by graphing.

Use the graphs to solve each of these systems.

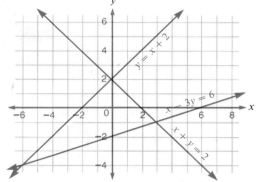

1. $y = x + 2$
   $x + y = 2$

2. $x + y = 2$
   $x - 3y = 6$

3. Solve this system by graphing.
   $y = -x + 2$

   $y = \frac{1}{3}x - 2$

6–2 **Objective:** To solve systems of equations by substitution.

Solve by substitution.

4. $y = 2x - 5$
   $x + 3y = 20$

5. $3x = y$
   $3x - 4y = 9$

6–3 **Objective:** To determine the number of solutions for a system of equations.

State whether each of the following systems has 0, 1, or infinitely many solutions.

6. $x + y = 4$
   $x + y = 5$

7. $y = 3x + 2$
   $2y = 6x + 4$

8. $y = 2x + 4$
   $y = 3x + 4$

6–4 **Objective:** To use systems of equations to solve problems.

For each problem below, write two equations, solve the system, and solve the problem.

9. The tens digit of a two-digit number is 3 less than twice the ones digit. The difference between the number and the number formed by reversing the digits is 18. What is the number?

10. One number is 4 times another. If the sum of the two numbers is 1000, what are the numbers?

11. The width of a rectangle is 20 cm less than its length. The perimeter is 68 cm. Find the length and width.

6–5 **Objective:** To solve systems of equations by addition.

Solve by addition.

12. $x + y = 14$
$x - y = 6$

13. $2y + 3x = 14$
$2y = 3x + 10$

14. I am thinking of two numbers. If I double the second number and subtract the product from the first, the difference is 5. If I double the second number and add the product to the first, the sum is 65. What are the two numbers?

6–6 **Objective:** To solve systems of equations by multiplication and addition.

15. $3a + b = 5$
$2a + 3b = 1$

16. $3x + 4y = -8$
$5x - 3y = 35$

6–7 **Objective:** To solve mixture problems.

17. Some $3.75-per-pound candy is mixed with $5.00-per-pound candy to obtain 10 pounds of $4.25-per-pound candy. How many pounds of each kind of candy are used?

■ CHAPTER 6 SELF-TEST

6–1 **1.** State whether the ordered pair $\left(2, \frac{1}{2}\right)$ is a solution of the system.

$$3x + 4y = 8$$
$$x - 6y = 5$$

Refer to the graphs to solve each system of equations.

2. $5x + 2y = 25$
$3x - 3y = -6$

3. $5x + 2y = 25$
$y = x - 5$

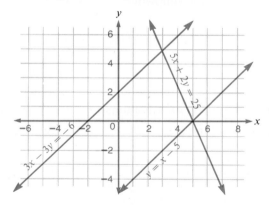

6–2 **4.** Solve the equation $2y + 4x = 12$ for y.

5. Solve by substitution.

$$y = x - 5$$
$$2x + 3y = 0$$

6–3 State whether each system has 0, 1, or infinitely many solutions.

6. $x + y = 5$
   $x + y = 10$

7. $2x - 3y = 6$
   $-4x + 6y = 3$

6–5 Solve by addition.

8. $2x + 3y = 0$
   $3x - 3y = -6$

6–6 Solve by multiplication and addition.

9. $7x + 3y = 0$
   $5x + 2y = 25$

6–4,
6–7 Solve.

10. A solution is 45% alcohol in water. How much alcohol is in x milliliters of the solution?

11. Two numbers have a sum of 84. One number is 8 less than 3 times the other. Find the numbers.

12. For a given two-digit number, the ones digit is 4 less than the tens digit. If the digits are reversed and the new number is doubled, the result is 1 more than the original number. Find the original number.

13. Ms. Macaw, who works for Sunset Birdseed Corporation, must mix first-quality seed worth 60¢ per pound with filler seed worth 35¢ per pound. How much of each type must she mix to produce 100 pounds of a mixture worth 50¢ per pound?

■ PRACTICE FOR COLLEGE ENTRANCE TESTS

Choose the best answer for each question.

1. If the sum of two numbers is 21 and the difference between the two numbers is 11, then the average of the two numbers is ___?___.

   A. 9   **B.** $10\frac{1}{2}$ **C.** 12   **D.** 21   **E.** $40\frac{1}{2}$

2. The cost C in dollars of producing n units of a product is given by the formula $C = kn + 1000$. If it costs \$3000 to produce 10 units, what does it cost to produce 20 units?

   A. \$4000   **B.** \$4500   **C.** \$5000   **D.** \$5500   **E.** \$6000

3. If $2x - 1 = 7$ and $x + y = 12$, then $y = $ ___?___.

   A. 3   **B.** 4   **C.** 6   **D.** 8   **E.** 9

4. Carla had 3 times as much money as Emily. After Carla gave Emily $10, Carla still had $4 more than Emily. How much money did Emily have originally?

    A. $6   **B.** $8   **C.** $9   **D.** $10   **E.** $12

5. If $x + y + z = 14$ and $x + y - z = 10$, then $x + y = \underline{}$.

    A. 4   **B.** 10   **C.** 12   **D.** 14   **E.** 24

6. If $xy = 24$ and $y = 3$, then $x - y = \underline{}$.

    A. 5   **B.** 8   **C.** 11   **D.** 18   **E.** 21

7. The sum of two numbers is 14. The difference of the two numbers is 6. The product of the two numbers is $\underline{}$.

    A. 21   **B.** 28   **C.** 30   **D.** 32   **E.** 40

8. If $a = 12b$ and $b = 3c$, then what is a in terms of c?

    A. $\frac{1}{4}c$ **B.** $4c$ **C.** $9c$ **D.** $15c$ **E.** $36c$

9. Each term of a sequence is obtained by doubling the preceding term and adding 1. If the third term is 15, what is the first term?

    A. 2   **B.** 3   **C.** 4   **D.** 5   **E.** 7

■ CUMULATIVE REVIEW (*Chapters* 1–6)

a	B	f	H	m
14	3	9	7	4

1–1,
1–3   Substitute and simplify.

    1. $H + a$ **2.** $a - \dfrac{f}{B}$ **3.** $m \cdot B^2$

1–2   Simplify.

    4. $19 - (14 - 2)$ **5.** $8 + 4 \cdot 2$ **6.** $16 \div 4 \cdot 2$

1–5,
1–6   **7.** Which of these expressions is equivalent to $7a + a$?

    a. $6a$ **b.** 7 **c.** $7a^2$ **d.** $8a$ **e.** $14a$

1–7   Write an expression.

    8. Luana has 43 cents less than Trini. If Trini has T cents, how many cents does Luana have?

2–1   True or false?

    9. $24 > -25$ **10.** $|-4| + |-8| < 0$

2–2,
2–3,
2–4, Substitute and simplify.
2–5

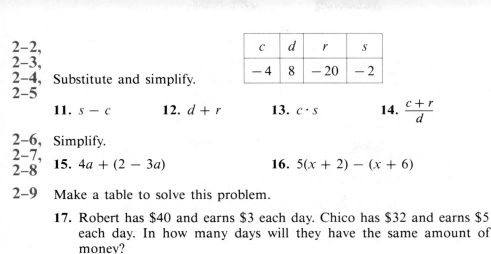

	c	d	r	s
	-4	8	-20	-2

11. $s - c$ **12.** $d + r$ **13.** $c \cdot s$ **14.** $\dfrac{c + r}{d}$

2–6, Simplify.
2–7,
2–8 **15.** $4a + (2 - 3a)$ **16.** $5(x + 2) - (x + 6)$

2–9 Make a table to solve this problem.

17. Robert has $40 and earns $3 each day. Chico has $32 and earns $5 each day. In how many days will they have the same amount of money?

3–3, Solve.
3–4,
3–5, **18.** $8y = 7y - 4$ **19.** $\dfrac{2w}{3} = -4$
3–6,
3–7 **20.** $5(x + 3) = 15$ **21.** $3(2x + 1) = 8x - 7$

3–9 **22.** Gloria made twice as many gizmos for a Junior Achievement project as Rob. Together, they made 48 gizmos. How many did Gloria make?

4–1 **23.** Simplify. $(9x^2 + 3x - 6) + (4x^2 - x + 1)$ **24.** Subtract. $\begin{aligned} 5b^2 - 6b + 7 \\ \underline{2b^2 + 3b + 1} \end{aligned}$

4–2 **25.** Which of these expressions is a binomial?

    a. 7 **b.** $4x^2$ **c.** $1 - 6y^2 + 8y$ **d.** $11s^2 - 25$ **e.** $\dfrac{3}{x^2}$

26. What is the degree of $a^2b - 7b^2 + 4a$?

27. Arrange the terms in descending order. $12x^2 + 6 + 9x^3 - x$

4–3 Simplify.

    28. $x^5 \cdot x^4$ **29.** $(-4y^2)(3y)$ **30.** $(m^2n)(m^3n^2)$

4–4 **31.** Write in scientific notation. $573{,}000$

32. Write the product in scientific notation. $(3 \cdot 10^4)(5 \cdot 10^3)$

4–5, Simplify.
4–6,
4–8 **33.** $(a^2)^5$ **34.** $(5x^3)^2$ **35.** $(xy^4)^3$

    36. $2r(r + 6)$ **37.** $b^2(1 - 5b^2)$ **38.** $xy(x^2 + y)$

    39. $\dfrac{x^{10}}{x^5}$ **40.** $\dfrac{32y^2}{4y^5}$ **41.** $-\dfrac{ab}{ab^2}$ **42.** $\dfrac{8t^3}{2t^3}$

4–7 Solve.

43. Mr. Elmo drove at x miles per hour for 5 hours. Mrs. Elmo drove at $(x - 9)$ miles per hour for 6 hours. If they drove the same distance, what was Mr. Elmo's rate of driving?

4–10 **44.** The sum of three consecutive integers is 105. What is the smallest integer?

5–1 **45.** Graph the points $A(1, 2)$, $B(-2, -4)$, $C(-4, 0)$, $D(3, -3)$, and $E(0, 5)$.

46. Which point in Exercise 45 is on the y-axis?

47. Which point in Exercise 45 is in quadrant IV?

5–2 **48.** List the domain and range of the relation shown here.

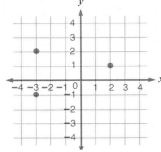

5–3 **49.** Is the ordered pair $(2, 3)$ a solution of the equation $y = 3x - 7$?

50. Find the missing number for each solution.

$$4x - y = 7$$

x	?	2
y	5	?

5–4 Graph the equation.

51. $y = 2x + 1$ **52.** $y + x = 4$

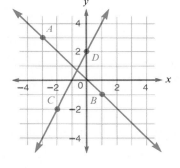

5–5 **53.** What is the slope of $\overleftrightarrow{AB}$?

54. What is the slope of $\overleftrightarrow{CD}$?

55. Draw a line through the origin with slope $\frac{1}{2}$.

5–6 **56.** What are the slope and y-intercept of the line with the equation

$$y = -\frac{2}{3}x + 4?$$

57. Write the equation $4x - 2y = 5$ in slope-intercept form.

5–8 **58.** Which graph represents a function?

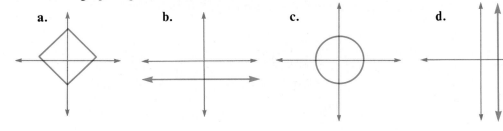

a.   b.   c.   d.

59. Does $5x - 4y = 6$ represent a linear function?

60. Ordered pairs of a linear function are shown. Find the missing number.

x	3	2	1	-1
y	1	3	5	?

61. Write the equation of the linear function satisfied by these ordered pairs.

x	1	0	-1
y	3	0	-3

6–1 **62.** State whether the number pair (1, 5) is a solution of the system.

$$y = 2x + 3$$
$$y = -3x + 2$$

63. Solve the system by graphing.

$$y = 2x - 1$$
$$y = -x + 5$$

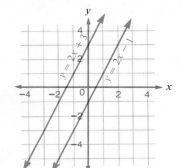

6–2 Solve by substitution.

64. $y = -x$
$\quad\;\, y = 3x - 4$

65. $x + y = 8$
$\quad\;\;\;\, y = x + 2$

6–3 **66.** How many solutions does the graphed system have?

67. How many solutions does this system have?

$$y = 3x + 4$$
$$2y = 6x + 8$$

6–5 Solve the system by addition.

68. $x + y = 3$
$\quad\;\, 3x - y = 5$

69. $\quad 4x - \;\; y = 2$
$\quad -4x + 3y = 2$

6–6 **70.** Tell how to eliminate the x-terms by multiplication and addition.

$$x + 2y = 4$$
$$3x - y = -9$$

71. Solve this system by multiplication and addition.

$$4x - y = 6$$
$$2x + 3y = 10$$

Solve.

6–1 **72.** Marcia is 3 years older than Donna. The sum of their ages is 21. Find Marcia's age.

6–4, **73.** The ones digit of a two-digit number
6–7 $\quad\;\;\,$ is 3 times the tens digit. The difference of the digits is 6. Find the number.

74. Some $5.00-per-lb candy is mixed with $3.00-per-lb candy to make a mixture that sells for $3.40 per lb. How many pounds of each type of candy are used?

7 Multiplying and Factoring Polynomials

When the water in a fountain is projected upward with an initial velocity v, it rises more and more slowly until it reaches a maximum height of h meters. Then it falls back to the ground, picking up speed as it falls. The formula $h = vt - 5t^2$ (where t is the time in seconds) and techniques learned in this chapter can be used to help determine when the water will hit the ground.

7–1 Prime Factorization

Preview

Imagine a row of wall lockers numbered from 1 to 1000 in a school with 1000 students. The locker doors are all closed. Student number 1 walks by the row of lockers, opening each one of them. Student number 2 walks by the lockers and closes every second locker beginning with locker 2. Student number 3 changes the state of the doors (opening closed doors and closing open doors) of every third locker beginning with locker 3. Student number 4 changes the state of the doors of every fourth locker beginning with locker 4. In general, student number n changes the state of every nth door beginning with locker n.

- Which lockers will be touched exactly twice?
- After the 1000th student has passed by the row of lockers, which locker doors will be closed?

The numbers that answer the first question will be used extensively in this lesson.

■ LESSON

You have probably used the word "factor" to refer to "numbers that are multiplied."

4 is a **factor** of 12 since there is a positive integer n such that $4n = 12$.

5 is not a factor of 12 since there is no positive integer n such that $5n = 12$.

The positive factors of 12 are 1, 2, 3, 4, 6, and 12.

Here is a table of the first ten positive integers and their positive factors.

Positive Integer	Factors
1	1
2	1, 2
3	1, 3
4	1, 2, 4
5	1, 5
6	1, 2, 3, 6
7	1, 7
8	1, 2, 4, 8
9	1, 3, 9
10	1, 2, 5, 10

Some numbers have exactly two factors. They are called **prime numbers.** The prime numbers listed in the table above are 2, 3, 5, and 7.

Positive numbers with more than two factors are called **composite numbers.** Composite numbers listed in the table above are 4, 6, 8, 9, and 10.

The number 1 is special. It is the only positive integer with just one factor and it is neither prime nor composite.

The word "factor" is also used as a verb. To **factor a number** is to write the number and the product of two or more of its factors.

> **Example 1** Factor 6.
>
> *Solution* To factor 6, we can write:
>
> $$1 \cdot 6 \qquad 6 \cdot 1 \qquad 2 \cdot 3 \quad \text{or} \quad 3 \cdot 2$$

> **Example 2** Factor 12.
>
> *Solution* Some answers are:
>
> $$3 \cdot 4 \qquad 12 \cdot 1 \qquad 2 \cdot 6 \qquad 2 \cdot 2 \cdot 3 \qquad 3 \cdot 2 \cdot 2 \quad \text{and} \quad 2 \cdot 3 \cdot 2$$

For the factorization $2 \cdot 2 \cdot 3$, each of the factors is prime. Such a factorization is called a **prime factorization.**

Positive integer	Prime factorization
6	$2 \cdot 3$
9	$3 \cdot 3$, or 3^2
24	$2 \cdot 2 \cdot 2 \cdot 3$, or $2^3 \cdot 3$
36	$2 \cdot 2 \cdot 3 \cdot 3$, or $2^2 \cdot 3^2$

Factoring is often assumed to involve only positive integers. This is called factoring **over the set of positive integers.** We can also factor over the set of *all* integers.

> **Example 3** Factor 6 over the set of integers.
>
> *Solution* Some answers are:
>
> $$1 \cdot 6 \qquad -1 \cdot -6 \qquad 2 \cdot 3 \quad \text{and} \quad -2 \cdot -3$$

> **Example 4** Factor -8 over the set of integers.
>
> *Solution* Some answers are:
>
> $$-1 \cdot 2 \cdot 2 \cdot 2 \qquad -2 \cdot -2 \cdot -2 \quad \text{and} \quad -4 \cdot 2$$

■ CLASSROOM EXERCISES

True or false? Explain your answer.

1. 2 is a factor of 18.

2. 2 is a factor of 17.

3. 38 is a factor of 38.

4. 1 is a factor of 209.

5. 5 is a prime number.

6. 7 is a composite number.

7. 1 is a prime number.

8. 1 is a composite number.

State the prime factorization of each number.

9. 12   **10.** 15   **11.** 2   **12.** 27

State three factorizations over the set of integers.

13. 4   **14.** -8 **15.** -6 **16.** 10

■ WRITTEN EXERCISES

A List the numbers from the given set that are factors of the given number.

1. 45; {1, 2, 3, 4, 5}   **2.** 28; {1, 2, 3, 4, 5}

3. 52; {4, 13, 14, 21}   **4.** 33; {3, 11, 13, 33}

5. 49; {2, 7, 14}   **6.** 34; {1, 17, 19}

List all positive integer factors of the given number.

7. 18   **8.** 20   **9.** 49   **10.** 25

11. 35   **12.** 39   **13.** 42   **14.** 70

15. 84   **16.** 31   **17.** 47   **18.** 28

State whether the number is prime or composite.

19. 99   **20.** 50   **21.** 97   **22.** 53

23. 57   **24.** 93   **25.** 43   **26.** 59

27. 17   **28.** 29   **29.** 34   **30.** 43

31. 101   **32.** 57   **33.** 61   **34.** 39

Write each number as the product of two factors in three different ways.

35. 44   **36.** 68   **37.** 63   **38.** 12

39. 75   **40.** 98   **41.** 76   **42.** 242

Write the prime factorization of each number. List the factors in increasing order.

43. 30   **44.** 66   **45.** 24   **46.** 40

47. 54   **48.** 88   **49.** 91   **50.** 86

51. 50   **52.** 100   **53.** 72   **54.** 28

55. 56   **56.** 49   **57.** 81   **58.** 30

Write three factorizations over the set of integers.

59. -15 **60.** -14 **61.** 10   **62.** 6

63. -77 **64.** -74 **65.** 69   **66.** 65

A "factor tree" can be used to determine the prime factorization of a composite number. First, write the number as the product of two factors. Next, write each composite factor as the product of two factors. Continue the process until only prime factors are used. In the example, the prime factors are circled.

Sample Find the prime factorization of 48.

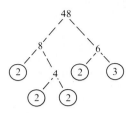

Solution The prime factorization of 48 is $2^4 \cdot 3$.

Use a factor tree to determine the prime factorization of each number.

B **67.** 60   **68.** 80   **69.** 84   **70.** 72

71. 81   **72.** 32   **73.** 90   **74.** 92

C **75.** How many different factors does each number have?

  a. 2^1 **b.** 2^2 **c.** 2^3 **d.** 2^4 **e.** 2^5

76. Complete this rule for the number of different factors of a power of 2.

  For all positive integers n, 2^n has ___?___ different factors.

77. How many factors does each number have?

  a. 5^1 **b.** 5^2 **c.** 5^3 **d.** 5^4 **e.** 5^5

78. Complete this rule for the number of different factors of a power of 5.

  For all positive integers n, 5^n has ___?___ different factors.

79. Complete this rule.

  For each prime number p and each positive integer n, p^n has ___?___ different factors.

80. How many factors does each number have?

  a. $2^1 \cdot 5^1$ **b.** $2^2 \cdot 5^1$ **c.** $2^3 \cdot 5^1$ **d.** $2^2 \cdot 5^2$

81. Complete this rule for the number of different factors for a product of powers of 2 and powers of 5.

  For all positive integers m and n, $2^m \cdot 5^n$ has ___?___ different factors.

82. Complete this rule.

  For all prime numbers p_1 and p_2 ($p_1 \neq p_2$) and for all positive integers m and n, $p_1{}^m \cdot p_2{}^n$ has ___?___ different factors.

■ REVIEW EXERCISES

State the property illustrated in each equation.

1. $3x + (4x + 2) = (3x + 4x) + 2$ **2.** $4a + 5a = (4 + 5)a$ [1–5]

Which of the numbers $-2, -1, 0, 1,$ and 2 are solutions of the following equations?

3. $x^3 = x$ **4.** $\dfrac{1}{x} = \dfrac{x}{4}$ [3–1]

Simplify.

5. $2a^2(-2ab)$ **6.** $(3xy)(3xy)$ [4–3]

7. $\dfrac{12x^2y^3}{4xy}$ $(x \neq 0, \ y \neq 0)$ **8.** $\dfrac{-18a^2b}{6ab}$ $(a \neq 0, \ b \neq 0)$ [4–8]

EXTENSION   Determining prime numbers on a computer

Recall that the greatest integer function INT(X) finds the greatest integer that is less than or equal to X. We can use this function to determine whether a whole number A is divisible by a whole number B. If B does not divide A evenly, then A/B and INT(A/B) are not equal. If A/B = INT(A/B), then B is a factor of A. This fact is used in the following program to determine whether a whole number greater than 1 is prime.

```
10   PRINT "WHAT IS THE WHOLE NUMBER";
20   INPUT X
30   FOR Y = 2 TO X - 1
40   IF X/Y = INT(X/Y) THEN 80
50   NEXT Y
60   PRINT X; "   IS A PRIME NUMBER."
70   GOTO 90
80   PRINT X; "   IS NOT A PRIME NUMBER."
90   END
```

[If the statement in line 40 is true, the computer skips lines 50, 60, and 70 in the program. If the statement is false, the computer reads lines 50, 60, and 70.]

Use the program to determine whether the whole number is prime.

1. 103   **2.** 203   **3.** 403   **4.** 503

Although the program tells you when a number is not prime, it does not help you find the prime factors of that number. You can have the program print the smallest prime factor of the number just by adding this line:

```
85   PRINT "A PRIME FACTOR IS   "; Y
```

If the number is not prime, find its prime factors.

5. 127   **6.** 129   **7.** 131   **8.** 133

7-2 Greatest Common Factor

Preview Euclidean algorithm

In the branch of mathematics known as number theory, a step-by-step process has been developed to find the largest number that divides two given numbers exactly. Suppose we wish to find the largest number that exactly divides 495 and 572.

Follow this process.

1. Divide the larger number by the smaller.

$$
\begin{array}{r}
1 \\
495\overline{)572} \\
495 \\
\hline
77
\end{array}
$$

2. Divide the previous divisor by the remainder.

$$
\begin{array}{r}
6 \\
77\overline{)495} \\
462 \\
\hline
33
\end{array}
$$

3. Repeat step 2 until the remainder is 0.

$$
\begin{array}{r}
2 \\
33\overline{)77} \\
66 \\
\hline
11
\end{array}
$$

4. The divisor that produces the 0 is the answer.

$$
\begin{array}{r}
3 \\
11\overline{)33} \\
33 \\
\hline
0
\end{array}
$$

The largest number that exactly divides 495 and 572 is 11. This step-by-step process is called the **Euclidean algorithm.** This method was included in the writings of Euclid (about 300 B.C.), but most likely was invented even earlier than that. Use the Euclidean algorithm to find the largest number that exactly divides 330 and 1260.

In this lesson you will study a similar process in algebra.

■ LESSON

These are the factors of 12: 1, 2, 3, 4, 6, 12.
These are the factors of 16: 1, 2, 4, 8, 16.
 Some numbers are factors of both 12 and 16. They are called **common factors** of 12 and 16.

Common factors of 12 and 16: 1, 2, 4.

 The greatest of the common factors is called the **greatest common factor (GCF).**

The greatest common factor of 12 and 16 is 4.

Definition: The Greatest Common Factor (of Integers)

The greatest common factor of two or more integers is the greatest integer that is a factor of each given integer.

Example 1   Find the greatest common factor of 36 and 48.

Solution   Find the prime factorization of each number and identify common prime factors.

Prime factorization of 36: $2 \cdot 2 \cdot \qquad 3 \cdot 3$
Prime factorization of 48: $2 \cdot 2 \cdot 2 \cdot 2 \cdot 3$

Answer   The greatest common factor of 36 and 48 is $2 \cdot 2 \cdot 3$, or 12.

We can use a similar method to find the greatest common factor of expressions that contain variables.

Definition: The Greatest Common Factor (of Monomials)

The greatest common factor of two or more monomials is a monomial whose coefficient is the greatest common factor of the coefficients of the given monomials and whose variables have the greatest common degree of each variable in the given monomials.

Example 2   Find the GCF of $9x^2y$ and $12xy^2$.

Solution   *Prime factorization of $9x^2y$:* $\qquad 3 \cdot 3 \cdot x \cdot x \cdot y$
Prime factorization of $12xy^2$: $\qquad 2 \cdot 2 \cdot 3 \cdot x \cdot \qquad y \cdot y$

Answer   The GCF of $9x^2y$ and $12xy^2$ is $3xy$.

Example 3   Find the GCF of $15a^3b^2$, $30a^2b^2c$, and $10a^2b^3$.

Solution
$15a^3b^2$: $\qquad 3 \cdot 5 \cdot a \cdot a \cdot a \cdot b \cdot b$
$30a^2b^2c$: $\quad 2 \cdot 3 \cdot 5 \cdot a \cdot a \cdot \quad b \cdot b \cdot c$
$10a^2b^3$: $\quad 2 \cdot \quad 5 \cdot a \cdot a \cdot \quad b \cdot b \cdot b$

Answer   The GCF of $15a^3b^2$, $30a^2b^2c$, and $10a^2b^3$ is $5 \cdot a \cdot a \cdot b \cdot b$, or $5a^2b^2$.

■ CLASSROOM EXERCISES

List the factors of each number (over the set of positive integers).

1. 14

2. 17

List the common factors of each pair of numbers.

3. 12, 18

4. 45, 46

List the greatest common factor of each pair of numbers.

5. 15, 21

6. 24, 56

List the six factors of each monomial.

7. $25x$

8. $7x^2$

List the common factors of each pair of monomials.

9. $x^2y,\ xy^3$

10. $4xy,\ 6x^2$

List the greatest common factor of each pair of monomials.

11. $3a^2b,\ 6a^2b$

12. $12m^2n^3,\ 16mn^4$

■ WRITTEN EXERCISES

List the factors of each number (over the set of positive integers).

A **1.** 58

2. 55

3. 31

4. 41

List the common factors of each pair of numbers.

5. 48, 60

6. 30, 45

7. 39, 65

8. 66, 55

9. 26, 33

10. 51, 38

11. 20, 30

12. 30, 40

List the greatest common factor of each pair of numbers.

13. 8, 12

14. 9, 12

15. 65, 68

16. 33, 38

17. 50, 150

18. 20, 120

19. 46, 62

20. 38, 26

List the six different factors of each monomial.

21. 12

22. 18

23. $2x^2$

24. $3y^2$

25. $9a$

26. $4b$

27. ab^2

28. a^2b

List the common factors of each pair of monomials.

29. $3x^2,\ 2xy$

30. $4ab,\ 5ab^2$

31. $6a^2b^2,\ 3ab$

32. $2xy^2,\ 4x^2y$

List the greatest common factor of each pair of monomials.

33. $2x^3y,\ 3x^3$

34. $5xy^3,\ 7y^3$

35. $6x^2y,\ 8xy^2$

36. $6xy^2,\ 9x^2y$

37. $10abc,\ 15ad$

38. $25abc,\ 20cd$

39. $18a^2bc^2,\ 45abc^2$

40. $35ab^2c^2,\ 21ab^2c$

List the greatest common factor of the three monomials.

B **41.** 10, 16, 20   **42.** 8, 12, 14

43. 24, 18, 30   **44.** 16, 20, 40

45. $2x^2y$, $3xy$, $4xy^2$ **46.** $3a^2b^2$, $5ab^2$, $6a^2b$

47. $6x^2y^2$, $8xy^2$, $12x^3y^2$ **48.** $4a^3b$, $8a^2b$, $12a^2$

49. $4xy$, $8x^2y$, $12xy^2$ **50.** $5a^2b^3$, $2ab^2$, $8a^3b^3$

True or false? If the answer is false, provide a specific example to illustrate your answer.

51. Two is a factor of every even number.

52. If two numbers are even, their greatest common factor is 2.

53. If two numbers are prime, their greatest common factor is 1.

54. If the greatest common factor of two numbers is 1, the two numbers are prime.

55. One is a factor of every number.

56. Zero is a factor of every positive number.

List the greatest common factor of each pair of monomials.

C **57.** $x^{50}y^{50}$ and $x^{37}y^{87}$ **58.** $a^{10}b^{20}c^{30}$ and $a^{20}b^{10}c$

59. 132,132 and 543,543   **60.** 132,132 and 234,234

The symbol $n!$, "n factorial," is defined to be

$$n \cdot (n - 1) \cdot (n - 2) \cdot \cdots \cdot 1$$

where n is any positive integer. For example,

$$6! \text{ is } 6 \cdot 5 \cdot 4 \cdot 3 \cdot 2 \cdot 1, \text{ or } 720$$

Find the greatest common factor of each pair of factorials.

61. 6! and 7!   **62.** 10! and 5!   **63.** 4! and 8!   **64.** 100! and 6!

■ REVIEW EXERCISES

Solve.

1. $4x + 12 = 0$ **2.** $2x - 5 = 0$ [3–5]

3. $6(x - 4) = 30$ **4.** $\dfrac{x + 4}{3} = 12$ [3–6]

5. Solve for I: $E = IR$.   **6.** Solve for y: $ax + by = c$.   [3–8]

7-3 The Zero-Product Property

Preview

Which of these number pairs (a, b) are solutions of the equation $ab = 0$?

$(5, 0)$	$(5, -5)$	$(0, -5)$	$(0, 0)$
$(0.5, 0)$	$(0, 0.5)$	$(-0.5, 0.5)$	$(0, 5000)$

Can you list additional solutions?

Although there are infinitely many solutions of the equation $ab = 0$ and you could not possibly list them all, can you describe what all solutions have in common?

In this lesson you will solve equations that are similar to $ab = 0$.

■ LESSON

For the product of two numbers to be zero, one or both numbers must be zero. This property is called the **zero-product property**.

The Zero-Product Property

For all numbers a and b,

$$\text{if } ab = 0, \text{ then } a = 0 \text{ or } b = 0.$$

Mathematicians use the word "or" to mean "and/or." Therefore, in the zero-product property, if $ab = 0$, then both a and b can be 0.

The zero-product property permits us to solve some equations.

Example 1   Solve. $x(x + 1) = 0$

Solution   The product of the two factors x and $x + 1$ is 0. Therefore, one or both factors is equal to 0. This means that the following sentence is equivalent to $x(x + 1) = 0$:

$$x = 0 \quad \text{or} \quad x + 1 = 0.$$

Therefore 0 is a solution of the first equation, and -1 is a solution of the second equation.

Answer $\{0, -1\}$

Check
$$x(x + 1) = 0 \qquad\qquad x(x + 1) = 0$$
$$0(0 + 1) \stackrel{?}{=} 0 \qquad\qquad -1(-1 + 1) \stackrel{?}{=} 0$$
$$0(1) = 0 \quad \text{It checks.} \qquad\qquad -1(0) = 0 \quad \text{It checks.}$$

Example 2 Solve. $(2a - 1)(a + 3) = 0$

    Solution

        Use the zero-product property.
        Solve both equations.

$$(2a - 1)(a + 3) = 0$$
$$2a - 1 = 0 \quad \text{or} \quad a + 3 = 0$$
$$2a - 1 = 0 \qquad a + 3 = 0$$
$$2a = 1 \qquad a = -3$$
$$a = \frac{1}{2}$$

    Answer $\left\{ \frac{1}{2}, -3 \right\}$

    Check $\qquad (2a - 1)(a + 3) = 0$

$$\left(2\left(\frac{1}{2}\right) - 1 \right) \left(\frac{1}{2} + 3 \right) \stackrel{?}{=} 0$$

$$(1 - 1) \left(\frac{1}{2} + 3 \right) \stackrel{?}{=} 0$$

$$0\left(3\frac{1}{2} \right) = 0 \qquad \text{It checks.}$$

$$(2a - 1)(a + 3) = 0$$
$$(2(-3) - 1)(-3 + 3) \stackrel{?}{=} 0$$
$$(-6 - 1)(0) = 0$$
$$\text{It checks.}$$

■ CLASSROOM EXERCISES

Solve.

1. $x(x - 2) = 0$ **2.** $a(a + 3) = 0$

3. $y(2y - 1) = 0$ **4.** $b(3b + 1) = 0$

5. $(x - 1)(x - 2) = 0$ **6.** $(a - 5)(a - 2) = 0$

7. $(2a + 1)(a - 3) = 0$ **8.** $(3x + 2)(2x + 1) = 0$

■ WRITTEN EXERCISES

Solve.

A   **1.** $(x - 3)x = 0$ **2.** $(x - 5)x = 0$

     3. $a(a + 4) = 0$ **4.** $a(a + 7) = 0$

     5. $(y - 3)(y - 8) = 0$ **6.** $(y - 6)(y - 2) = 0$

     7. $(x + 5)(x + 5) = 0$ **8.** $(x - 6)(x - 6) = 0$

     9. $(x - 5)(x + 5) = 0$ **10.** $(x + 9)(x - 9) = 0$

    11. $(3n + 1)(2n - 1) = 0$ **12.** $(2n + 1)(3n - 1) = 0$

    13. $(3x + 5)(4x - 3) = 0$ **14.** $(5x - 3)(3x + 4) = 0$

    15. $5(2x - 3)(3x + 2) = 0$ **16.** $6(5x - 8)(x + 7) = 0$

    17. $x(2x - 8)(3x - 9) = 0$ **18.** $6x(x - 4)(x - 3) = 0$

For each of the following problems:

   a. Write an equation that fits the problem, letting x represent the missing number.
   b. Solve the equation and answer the question.
   c. Check your answers in the original problem.

B **19.** The product of a number decreased by 5 and the number increased by 7 is zero. What is the number?

20. The product of a number increased by 10 and the number decreased by 10 is zero. What is the number?

21. The product of a number decreased by 7 and the number increased by 3 is zero. What is the number?

22. The product of a number increased by 6 and the number decreased by 2 is zero. What is the number?

23. The product of twice a number and the number increased by 2 is zero. What is the number?

24. The product of a number and twice the number is zero. What is the number?

25. The product of 3 less than a number and 5 less than twice the number is zero. What is the number?

26. The square of 5 more than a number is zero. What is the number?

C **27.** In the coming 162-game baseball season, Sarah is willing to bet that the *sum* of the runs scored per game by the St. Louis Cardinals for the season will be greater than the *product* of the runs scored per game by the Chicago Cubs for the season. Explain why Sarah is willing to make this bet.

How many different solutions does each of the following equations have?

28. $x(x - 1)(x - 2)(x - 3) = 0$

29. $(x - 1)(x - 1) = 0$

30. $(x + 2)(x - 2) = 0$

31. $x \cdot x \cdot x \cdot x = 0$

32. $(2x - 1)(2x + 1) = 0$

33. $(x + 3)^2(x - 4)^5 = 0$

34. $(x - 1)(x - 2)(x - 1)(x - 2) = 0$

35. $(x + 3)^2(x - 4)^5 = 0$

36. $5(x + 5)(x - 5)(5x)\left(\dfrac{x}{5}\right) = 0$

37. $\dfrac{(x - 5)(x - 4)(x - 3)}{(x - 2)(x - 1)} = 0$

38. $\dfrac{(x - 5)(x - 4)}{(x - 3)(x - 2)(x - 1)} = 0$

39. $\dfrac{x - 5}{(x - 4)(x - 3)(x - 2)(x - 1)} = 0$

■ REVIEW EXERCISES

Simplify.

1. $4 + 6 \cdot 2 - 2$ 2. $4 \cdot 5 - 3 \cdot 2$ [1–2]

3. $5 \cdot 2^3$ 4. $(3 \cdot 2)^2$ [1–3]

Simpify.

5. $2x(x + 2)$ 6. $a(3bc - 2 + 3b)$ [4–6]

7. $6mn(2m - 3n)$ 8. $10(2x - 1)$ [4–6]

The formula for the area of a circle is $A = \pi r^2$.

9. Write an equation for the area of a circle with radius 10 units.   [3–9]

10. Write an equation for the area of a semicircle with radius x units.   [3–9]

Self-Quiz 1

7–1   List all positive integer factors of the given number.

     1. 45   2. 52

     Write the prime factorization of each number.

     3. 120   4. 126

7–2   Find the greatest common factor.

     5. 45, 75   6. 70, 105, 350   7. $8x^2y,\ 18xy^3$

7–3   Use the zero-product property to solve each equation.

     8. $5a(a - 4) = 0$ 9. $(n - 7)(n + 2) = 0$ 10. $(2x - 3)(5x + 2) = 0$

Mathematics and Your Future

In recent years the fields that require more preparation in mathematics have also been those paying higher salaries. For example, surveys of recent college graduates indicate that engineering and science majors receive higher salaries than do business and economics majors, who in turn receive higher salaries than do social science and humanities majors. This trend is likely to continue as the information society generates greater needs for people with strong backgrounds in mathematics. You can keep your options open for higher-paying careers by continuing to take mathematics in high school.

7-4 Factoring Monomials from Polynomials

Preview **Application: Projectiles**

Imagine an object projected upward from the ground. As it rises its speed gets less and less, until it stops rising at its peak. Then it falls back toward the ground, with its speed increasing until it strikes the ground.

If the object is projected at an initial velocity of 40 meters per second, its height h, in meters, after t seconds is given by the formula

$$h = t(40 - 5t)$$

- Complete this table.

t	0	1	2	3	4	5	6	7	8
h	0	35	?	?	?	?	?	?	?

- Graph the ordered pairs (t, h).
- Solve the equation: $t(40 - 5t) = 0$.
- What do the table, the graph, and the equation indicate about how long it will take the object to fall back to Earth?

In this lesson you will learn to write polynomials in factored form and solve some second-degree equations like the one above.

■ LESSON

In earlier lessons we used the distributive properties to change products from factored form to expanded form. The process of changing from factored form to polynomial form is called **expanding**.

Factored form	*Polynomial form*
$2(3x + 7)$	$6x + 14$
$5a(2a - 4 + b)$	$10a^2 - 20a + 5ab$
$2x^2y(3x + 2y)$	$6x^3y + 4x^2y^2$

We can also use the distributive properties to accomplish the opposite result—changing from polynomial form to factored form. This process is called **factoring**. The following examples show how to factor a monomial term from a polynomial.

Example 1 Factor. $2x^2 + 4x$

Solution Find the greatest common factor of the terms of the polynomial.

$2x^2{:}\quad 2 \cdot x \cdot x$
$4x{:}\quad 2 \cdot 2 \cdot x$
GCF: $2x$

Write each term as the product of the GCF and another factor.
Use the distributive property to factor out the GCF.

$2x^2 + 4x = 2x(x) + 2x(2)$
$= 2x(x + 2)$

Answer The factored form is $2x(x + 2)$.

Check Check by expanding.

Example 2 Factor. $3abc - 2a + 3ab$

Solution Find the GCF of the terms.

$3abc$: $3 \cdot a \cdot b \cdot c$
$2a$: $2 \cdot a$
$3ab$: $3 \cdot a \cdot b$
GCF: a

Write each term as the product of the GCF and another factor.

$3abc - 2a + 3ab = a(3bc) - a(2) + a(3b)$

Use the distributive property to factor out the GCF.

$= a(3bc - 2 + 3b)$

Answer $a(3bc - 2 + 3b)$

Check Check by expanding.

Example 3 Factor. $20x - 10$

Solution Find the GCF of the terms.

$20x$: $2 \cdot 2 \cdot 5 \cdot x$
10: $2 \cdot 5$
GCF: $2 \cdot 5 = 10$

Write each term as the product of the GCF and another factor.
[*Note:* Use 1 as the other factor if the term is equal to the GCF.]
Factor out the GCF.

$20x - 10 = 10(2x) - 10(1)$

$= 10(2x - 1)$

Answer $10(2x - 1)$

Check Check by expanding.

Factoring can be used with the zero-product property to solve some polynomial equations. If the polynomial can be expressed as two or more factors whose product is zero, the zero-product property indicates that one or more of these factors must be zero.

Example 4 Solve. $x^2 - 3x = 0$

Solution Factor out the greatest common monomial factor.

$x^2 - 3x = 0$
$x(x - 3) = 0$

Use the zero-product property.

$x = 0$ or $x - 3 = 0$

Solve each equation.

$x = 0$ or $x = 3$

Answer $\{0, 3\}$

Check

$x^2 - 3x = 0$
$0^2 - 3(0) \overset{?}{=} 0$
$0 - 0 = 0$ It checks.

$x^2 - 3x = 0$
$3^2 - 3(3) \overset{?}{=} 0$
$9 - (9) \overset{?}{=} 0$
$0 = 0$ It checks.

■ CLASSROOM EXERCISES

Expand.

1. $2x(3x + 4)$ **2.** $3a(a - 2ab + b)$

Factor.

3. $3x + 3b$ **4.** $6x + 3y$ **5.** $4x^2 + x$

6. $7a^2 - 3a$ **7.** $21b^2c - 14bc$ **8.** $6a^3b + 4a^2bc - 8abc$

Solve.

9. $x^2 + 4x = 0$ **10.** $2y - y^2 = 0$ **11.** $5a^2 = 10a$

■ WRITTEN EXERCISES

Find the greatest common factor of the terms of the polynomial.

A   **1.** $5x + 15y$ **2.** $3x + 21y$ **3.** $14x^2 - 3x$ **4.** $10x^3 - 7x^2$

     5. $12x^2 + 6x$ **6.** $15x^2 + 5x$ **7.** $6a^2b + 10ab$ **8.** $21ab^2 + 18ab$

Factor. If the expression cannot be factored, write "not factorable."

9. $2x + 10b$ **10.** $5a + 25b$ **11.** $21a^2 - 7a$

12. $15a^2 - 3a$ **13.** $3x + 4y$ **14.** $5x + 6y$

15. $ab + a$ **16.** $ab + b$ **17.** $2xy + 4x^2y + 6xy^2$

18. $3x^2y + 6xy^2 + 9xy$ **19.** $4x^2 + 9$ **20.** $9x^2 + 16$

Solve.

21. $a^2 - 3a = 0$ **22.** $x^2 - 5x = 0$ **23.** $x^2 + 4x = 0$ **24.** $a^2 + 9a = 0$

25. $y^2 = 6y$ **26.** $p^2 = 10p$ **27.** $2x^2 - 3x = 0$ **28.** $3x^2 - 5x = 0$

Write an equation that fits the problem, letting x represent the number. Solve the equation and answer the question.

29. The square of a number is 10 times the number. What is the number?

30. The square of a number is 20 times the number. What is the number?

31. A number squared plus twice the number is zero. What is the number?

32. A number squared plus three times the number is zero. What is the number?

The following computations can be done mentally by first factoring the expressions. Write only the simplified answer.

B   **33.** $7 \cdot 49 + 7 \cdot 51$ **34.** $19 \cdot 89 + 19 \cdot 11$ **35.** $8 \cdot 6.53 + 8 \cdot 3.47$

     36. $43 \cdot 124 - 43 \cdot 24$ **37.** $43 \cdot 6.7 - 33 \cdot 6.7$ **38.** $113 \cdot 417 - 13 \cdot 417$

For each shaded region:

    a. Write an expression for the area in polynomial form.
    b. Write an expression for the area in factored form.

[*Remember:* The formula for the area of a circle is $A = \pi r^2$.]

39.

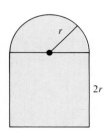

40.

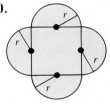

41.

42.

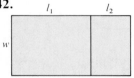

43.

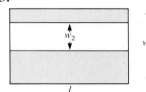

44.

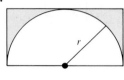

45.

46.

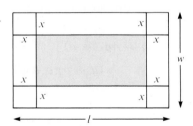

47.

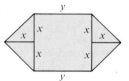

48.

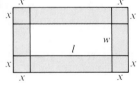

49.

■ REVIEW EXERCISES

Simplify.

1. $5x - x$

2. $4a - 3a$ [2–7]

3. $4(a - b) - (a - b)$

4. $x^2 + (-3)x + 4x + (-3)4$

5. $x^2 + 5x - 5x - 25$

6. $(x^2 + 3x - 2) + (2x^2 + 5)$ [2–8, 4–1]

7–5 Expanding Products of Binomials

Preview **A model for multiplying binomials** ▮▮▮▮▮▮▮▮▮▮▮▮▮▮▮▮

A rectangular pizza is to be shared by four people. They will divide it into four sections by making two cuts, one parallel to the long side and one parallel to the short side. To determine how much is to be paid for each section, it is necessary to find the area of the four sections. Find the areas of the four pieces for each pair of cuts.

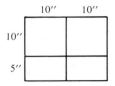
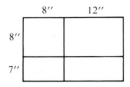
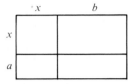

In this lesson we will use areas of regions of rectangles as a model for multiplying binomials.

■ LESSON

We have used the distributive properties of multiplication over addition and over subtraction to expand the product of a monomial and a binomial.

$$x(a + b) = xa + xb \quad \text{and} \quad x(a - b) = xa - xb$$

We can also use the distributive properties to expand the product of two binomials.

$$(a + b)(c + d)$$

Distribute (c + d) over the other sum.	$(a + b)(c + d) = a(c + d) + b(c + d)$
Distribute a over the sum (c + d).	$= ac + ad + b(c + d)$
Distribute b over the sum (c + d).	$= ac + ad + bc + bd$

Therefore, $(a + b)(c + d) = ac + ad + bc + bd$.

We can think of the binomials $(a + b)$ and $(c + d)$ as the sides of a rectangle. The product $(a + b)(c + d)$ represents the area of the rectangle.
The sum $ac + bc + ad + bd$ is the sum of the areas of the smaller regions that make up the large figure.

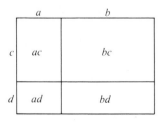

Area of a large rectangle = Sum of areas of small rectangles
$$(a + b)(c + d) \quad = \quad ac + ad + bc + bd$$

Here is a way to help you remember how to expand the product of binomials. This procedure is sometimes called the **FOIL method.**

The FOIL method

F *Multiply FIRST terms of the binomials.* $(a + b)(c + d)$ ac

O *Multiply OUTER terms of the binomials.* $(a + b)(c + d)$ $+ ad$

I *Multiply INNER terms of the binomials.* $(a + b)(c + d)$ $+ bc$

L *Multiply LAST terms of the binomials.* $(a + b)(c + d)$ $+ bd$

The sum of these products gives us the expanded form of $(a + b)(c + d)$.

$$(a + b)(c + d) = ac + ad + bc + bd$$

Example 1 Expand. $(2x + 3)(y + 5)$

Solution

	$(2x + 3)(y + 5)$	
Multiply first terms.	$(2x + 3)(y + 5)$	$2xy$
Multiply outer terms.	$(2x + 3)(y + 5)$	$10x$
Multiply inner terms.	$(2x + 3)(y + 5)$	$3y$
Multiply last terms.	$(2x + 3)(y + 5)$	15
Add products.	$2xy + 10x + 3y + 15$	

Answer $(2x + 3)(y + 5) = 2xy + 10x + 3y + 15$

Example 2 Expand and simplify. $(x + 3)^2$

Solution $(x + 3)^2 = (x + 3)(x + 3)$
$$= x^2 + 3x + 3x + 9$$
$$= x^2 + 6x + 9$$

Answer $(x + 3)^2 = x^2 + 6x + 9$

Binomials involving subtraction can be rewritten as sums before expanding.

Example 3 Expand. $(x + 4)(x - 3)$

Solution *Change subtraction to addition.* $(x + 4)(x - 3) = (x + 4)(x + (-3))$
Expand. $\quad = x^2 + (-3)x + 4x + 4(-3)$
Simplify. $\quad = x^2 + x - 12$

Answer $(x + 4)(x - 3) = x^2 + x - 12$

It is not necessary to change subtractions to additions in the written work. That change can be made mentally. Then the expansion is written with addition and subtraction signs.

> **Example 4** Expand and simplify. $(x - 2)(x - 3)$
>
> *Solution* *Expand.* $\quad (x - 2)(x - 3) = x^2 - 3x - 2x + 6$
> *Simplify.* $\qquad\qquad\qquad = x^2 - 5x + 6$
>
> *Answer* $(x - 2)(x - 3) = x^2 - 5x + 6$

■ CLASSROOM EXERCISES

Expand.

1. $(m + n)(p + q)$ **2.** $(m + 3)(m + 2)$ **3.** $(x - y)(x - y)$

4. $(a - 2)(a + 1)$ **5.** $(s - 3)(s + 3)$ **6.** $(2t + 1)(t - 3)$

■ WRITTEN EXERCISES

For each pair of binomials, list the product of (a) the first terms, (b) the outer terms, (c) the inner terms, and (d) the last terms.

A   **1.** $(2x + 3y)(3x + 2y)$ **2.** $(2x + 5y)(5x + 2y)$ **3.** $(3a + 4b)(a - 2b)$

4. $(a + 3b)(2a - 5b)$ **5.** $(p - q)(p - 6q)$ **6.** $(p - q)(5p - q)$

Expand the product.

7. $(a + b)(c + d)$ **8.** $(a + c)(b + d)$ **9.** $(a - c)(b + d)$

10. $(a - b)(c + d)$ **11.** $(a - d)(b - c)$ **12.** $(a - d)(c - b)$

Expand and simplify.

13. $(a + 3)(a + 7)$ **14.** $(a + 5)(a + 4)$ **15.** $(b - 2)(b - 3)$

16. $(b - 4)(b - 5)$ **17.** $(c - 8)(c + 8)$ **18.** $(c - 7)(c + 7)$

19. $(d + 3)(d - 3)$ **20.** $(d + 6)(d - 6)$ **21.** $(x + 3)^2$

22. $(x + 5)^2$ **23.** $(y - 5)^2$ **24.** $(y - 3)^2$

To express the area of each large rectangle:

    a. Write an expression for the area of each of the smaller rectangles.
    b. Add the areas of the smaller rectangles and simplify.

25. 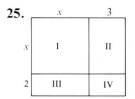   **26.** 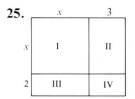   **27.** 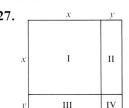   **28.**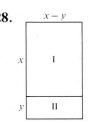

Write the letter of the area that matches each region.

Region	Area

B 29. I   **a.** $x^2 - xz - yx + yz$

30. II   **b.** $x^2 - xy$

31. III   **c.** $xz - yz$

32. IV   **d.** $x^2 - xz$

33. II and IV combined   **e.** $xy - zy$

34. I and II combined   **f.** yz

Expand and simplify.

35. $(2n + 5)(3n + 2)$ 36. $(5n + 2)(2n + 3)$ 37. $(3n - 7)(n + 3)$

38. $(2a - 3b)(2a + 3b)$ 39. $(5a - b)(a + 5b)$ 40. $(x + y)(2x + y)$

41. $(2x - 3)^2$ 42. $(x + 0.5)^2$ 43. $(x - 0.5)(x + 0.5)$

These two-digit numbers can be multiplied mentally by thinking of them as binomials. Find the products.

C 44. $88 \cdot 92 = (90 - 2)(90 + 2)$ 45. $42 \cdot 42 = (40 + 2)(40 + 2)$

46. $71 \cdot 71 = (70 + 1)(70 + 1)$ 47. $8.1 \cdot 7.9 = (8 + 0.1)(8 - 0.1)$

Solve. [*Hint:* Expand each side and collect like terms on one side of the equation.]

48. $(x + 2)(x + 6) = (x + 3)^2$ 49. $(x + 3)(x + 4) = (x + 1)(x + 2)$

50. $(x - 3)(x + 3) = (x - 1)(x - 4)$ 51. $(x + 5)^2 = (x + 1)(x + 7)$

■ REVIEW EXERCISES

Solve Exercises 1 and 2 by making and using a table.   [2–9]

1. What is the number n for which the sum of the whole numbers from 1 to n equals 105?

2. What is the number n for which 2^n equals 1024?

3. Write $3.14 \cdot 10^6$ in standard decimal notation.   [4–4]

4. Write 463,000 in scientific notation.   [4–4]

Solve.   [4–7]

5. How long will it take a cyclist traveling at 20 km/h to travel 25 km?

6. An automobile begins a trip traveling at 40 mph. A half hour later a second automobile leaves the same point traveling at 50 mph. How long will it take the second automobile to overtake the first?

7-6 The Difference of Two Squares

Preview

The shaded region in the figure below represents the area of the square with side x *minus* the area of the square with side y. The area of the shaded region is $x^2 - y^2$.

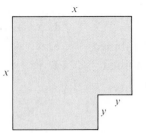

Area $= x^2 - y^2$

The area of the shaded region can be computed in another way. First divide the region into two rectangles.

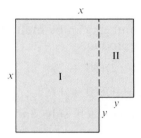

Then rearrange the two rectangles to form one rectangle.

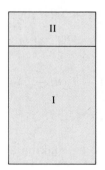

- How long is the vertical side of the rectangle?
- How long is the horizontal side of the rectangle?
- Express the area of the rectangle as the product of its length and width.

In this lesson you will use the relationship that you discovered to factor expressions.

■ LESSON

The product of the sum of two terms and the difference of those terms is a simple binomial.

$$(x + 3)(x - 3) = x^2 - 3x + 3x - 3^2$$
$$= x^2 - 3^2$$

The middle terms "drop out" (add to zero). The product is the *difference* of the squares of those two numbers.

The Difference of Squares Property

For all numbers a and b,

$$(a + b)(a - b) = a^2 - b^2$$

The left side of this equation is the factored form and the right is the expanded form.

$$\begin{array}{cc} \textbf{\textit{Factored form}} & \textbf{\textit{Expanded form}} \\ (x - 4)(x + 4) = & x^2 - 16 \end{array}$$

It is important to recognize both forms and to be able to change from either form to the other.

Example 1 Expand. $(2x + 3)(2x - 3)$

Solution $(2x + 3)(2x - 3) = (2x)^2 - 3^2$
$\qquad\qquad\qquad\qquad\quad = 4x^2 - 9$

Answer $4x^2 - 9$

Example 2 Factor. $m^2 - 49$

Solution Write as the difference of squares. $m^2 - 49 = m^2 - 7^2$
Factor. $\qquad\qquad\qquad\qquad\qquad\quad = (m + 7)(m - 7)$

Answer $(m + 7)(m - 7)$

Example 3 Factor. $9p^2 - 1$

Solution $9p^2 - 1 = (3p)^2 - (1)^2$
$\qquad\qquad\quad = (3p + 1)(3p - 1)$

Answer $(3p + 1)(3p - 1)$

Example 4 Factor. $2x^2 - 50$

Solution Factor out the monomial factor 2. $2x^2 - 50 = 2(x^2 - 25)$
Factor the difference of the two squares. $\qquad\quad = 2(x + 5)(x - 5)$

Answer $2(x + 5)(x - 5)$

The difference of squares sometimes can be used more than once to factor an expression.

Example 5   Factor. $x^4 - 81$

   Solution $x^4 - 81 = (x^2)^2 - 9^2$
                                  $= (x^2 + 9)(x^2 - 9)$
                                  $= (x^2 + 9)(x + 3)(x - 3)$

   Answer $(x^2 + 9)(x + 3)(x - 3)$

The sum of two perfect squares cannot be factored as the product of two binomials. For example, the binomials $x^2 + 9$, $a^2 + b^2$, and $p^2 + 5^2$ cannot be factored.

Factoring and the zero-product property can be used to solve some equations.

Example 6   Solve. $3x^2 = 300$

   Solution   *Divide both sides by 3.* $x^2 = 100$
                     Subtract 100 from both sides. $x^2 - 100 = 0$
                     Factor. $(x - 10)(x + 10) = 0$
                     Use the zero-product property. $x - 10 = 0$ or $x + 10 = 0$
                     Solve. $x = 10$ or $x = -10$

   Answer $\{10, -10\}$

   Check $3x^2 = 300$ $3x^2 = 300$
                     $3 \cdot 10^2 \stackrel{?}{=} 300$ $3(-10)^2 \stackrel{?}{=} 300$
                     $3 \cdot 100 = 300$ It checks. $3 \cdot 100 = 300$ It checks.

■ CLASSROOM EXERCISES

Write as squares.

1. $9x^2$

2. $16a^2b^2$

3. $\frac{4}{9}x^4$

4. $16a^2b^2$

5. $25m^4n^2$

6. $\frac{1}{9}a^2$

Expand.

7. $(x - 4)(x + 4)$

8. $(2a + 5)(2a - 5)$

9. $(2x + 1)(2x - 1)$

10. $(2a + 6)(2a - 6)$

11. $(4a - 3)(4a + 3)$

12. $(a - 1)(a + 1)$

Factor.

13. $9a^2 - 25$

14. $16x^2 - 1$

15. $4y^2 - z^2$

Solve.

16. $x^2 - 9 = 0$

17. $a^2 - 16 = 0$

18. $4x^2 = 100$

■ WRITTEN EXERCISES

Write as squares.

A **1.** $25a^2$ **2.** $36b^2$ **3.** $9c^2d^2$ **4.** $4r^2s^2$

5. p^4 **6.** q^6 **7.** $\dfrac{9}{25}s^{12}t^8$ **8.** $\dfrac{16}{49}g^{10}h^6$

Expand and simplify.

9. $(x + 5)(x - 5)$ **10.** $(x + 3)(x - 3)$ **11.** $(x - 6)(x + 6)$

12. $(x - 10)(x + 10)$ **13.** $(6 + x)(6 - x)$ **14.** $(10 + x)(10 - x)$

15. $(2x + 5)(2x - 5)$ **16.** $(3x + 4)(3x - 4)$ **17.** $(x - a)(x + a)$

18. $(x - r)(x + r)$ **19.** $(x + 5y)(x - 5y)$ **20.** $(5x + y)(5x - y)$

Factor.

21. $x^2 - 4$ **22.** $x^2 - 1$ **23.** $9x^2 - 1$ **24.** $25x^2 - 1$

25. $x^2 - 100$ **26.** $x^2 - 64$ **27.** $9x^2 - 16$ **28.** $16x^2 - 9$

29. $2x^2 - 18$ **30.** $3x^2 - 12$ **31.** $c^2d^2 - 49$ **32.** $r^2s^2 - 36$

Solve.

33. $x^2 - 49 = 0$ **34.** $x^2 - 25 = 0$ **35.** $x^2 = 64$

36. $x^2 = 36$ **37.** $9x^2 - 16 = 0$ **38.** $4x^2 - 25 = 0$

39. $16x^2 = 25$ **40.** $100x^2 = 1$ **41.** $2x^2 = 50$

42. $3x^2 = 48$ **43.** $2x^2 - 18 = 0$ **44.** $3x^2 - 12 = 0$

Write an equation that fits the problem. Solve the equation.

45. Five times the square of a number is 80. What is the number?

46. Seven times the square of a number is 63. What is the number?

Write the area of the shaded region in factored form.

B **47.** **48.** **49.**

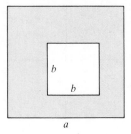

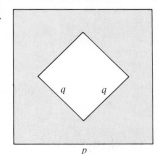

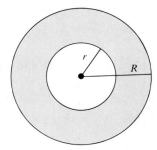

Factor.

50. $x^2 - \dfrac{1}{16}$ **51.** $x^2 - 0.25$ **52.** $x^2 - 1.44$ **53.** $x^2 - \dfrac{4}{9}$

Write as the product of three binomial factors.

54. $x^4 - 16$ **55.** $x^4 - 1$ **56.** $81x^4 - 1$ **57.** $16x^4 - 81$

The formula $d = 5t^2$ shows the distance d in meters that an object will fall in t seconds. Find the number of seconds it takes for an object to fall these distances.

58. 80 meters   **59.** 500 meters   **60.** 18,000 meters

Expand, simplify, and then factor these expressions.

[C] **61.** $(x + 2)(x - 5) + (3x + 1)$ **62.** $(x + 7)(x - 2) - (5x + 2)$

63. $(x + 5)^2 - 10(x + 5)$ **64.** $(x - 5)^2 + 2(5x - 17)$

■ REVIEW EXERCISES

Simplify.

1. $(a^3)^2$ **2.** $\left[\left(\frac{2}{3}\right)x^2y\right]^2$ **3.** $(2x^2)^3$ **4.** $(3xy^2)^3$ [4–5]

5. $(x + 3)^2$ **6.** $(y - 5)^2$ **7.** $(a + b)^2$ **8.** $(a - b)^2$ [7–5]

9. Draw and label a figure to illustrate this problem. Then solve the problem.   [3–10]

The length of a rectangle is 4 in. more than the width, and the perimeter is 48 in. What are the length and width?

Self-Quiz 2

7–4 Factor. If the expression cannot be factored, write "not factorable."

1. $25x^2 + 5x$ **2.** $21b - 14ab$ **3.** $9y^2 + 4$

4. Write an expression for the area in factored form.

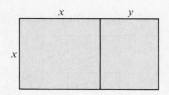

7–5 Expand and simplify.

5. $(c + 7)(c + 3)$ **6.** $(2r + 5)(r - 8)$

7–6 Factor.

7. $x^2 - 49$ **8.** $4y^2 - 81$ **9.** $2b^2 - 50$

**7–4,
7–6** Solve.

10. $x^2 + 7x = 0$ **11.** $4a^2 = 64$ **12.** $9x^2 - 100 = 0$

Preview

Show how these four rectangles can be arranged to form a square. What are the dimensions of the square? What is the area of the square?

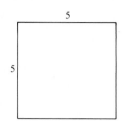

Let x and y be any positive numbers. Show how these four rectangles can be arranged to form a square. What are the dimensions of the square?

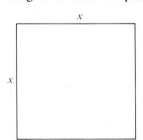
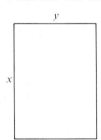
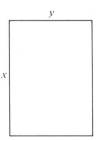
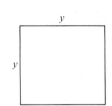

In this lesson you will study patterns that always produce squares.

■ LESSON

To find the square of a binomial, the FOIL method of expansion can be simplified.

F: *Multiply first terms.* $(3b + 5)^2 = (3b + 5)(3b + 5)$ $\quad 9b^2$

(The product of the first terms is the *square* of the binomial's first term.)

O: *Multiply outer terms.* $(3b + 5)^2 = (3b + 5)(3b + 5)$ $\quad 15b$

I: *Multiply inner terms.* $(3b + 5)^2 = (3b + 5)(3b + 5)$ $\quad 15b$

(The products of the inner and outer terms are the same. Each product is the product of the first and second terms of the binomial.)

L: *Multiply last terms.* $(3b + 5)^2 = (3b + 5)(3b + 5)$ 25

(The product of the last terms is the square of the binomial's last term.)

Add the products. $(3b + 5)^2 = 9b^2 + 30b + 25$

To square a binomial such as $(3b + 5)^2$

1. Square its first term. $(3b)^2 = 9b^2$

2. Double the product of its terms. $2(3b)(5) = 30b$

3. Square its last term. $5^2 = 25$

4. Add the products. $(3b + 5)^2 = 9b^2 + 30b + 25$

This property can be stated for any binomial.

The Perfect Square Property

For all numbers a and b,

$$(a + b)^2 = a^2 + 2ab + b^2$$
$$\text{and}$$
$$(a - b)^2 = a^2 - 2ab + b^2$$

Knowing this property, you can expand or factor some expressions quickly and easily. Study these examples.

Example 1   Expand. $(3x + 2)^2$

Solution $(3x + 2)^2 = (3x)^2 + 2(3x)(2) + 2^2$
$= 9x^2 + 12x + 4$

Answer $9x^2 + 12x + 4$

Mathematics and Your Future

As you learn algebra today, you are preparing to live and work in what has been called "the information society of tomorrow." To succeed in that society, you will need skills in the areas of mathematics and communication (speaking, listening, reading, and writing). Take enough mathematics in high school to prepare yourself for the future.

Example 2   Is the following trinomial a perfect square? That is, can we write it as the square of a binomial?

$$4x^2 + 12x + 9$$

Solution   *Factor the first and last terms of the trinomial into squares.* $4x^2 = (2x)^2$
$$9 = 3^2$$

If the trinomial is a perfect square then the factored form must be

$$(2x + 3)^2$$

[The binomial is a *sum* since all coefficients of the trinomial are positive.]

Expand this expression to see whether the $(2x + 3)^2 = 4x^2 + 12x + 9$
"middle" term is 12x.

Answer   The trinomial *is* a perfect square.

Example 3   Factor. $9x^2 - 6x + 1$

Solution   *Factor the first and last terms of the trinomial into squares.* $9x^2 = (3x)^2$
$$1 = 1^2$$

The coefficient of the middle term, -6, is negative, so the coefficient of the last term of the binomial must be negative. The factored form is

$$(3x - 1)^2$$

Expand to check the middle term. $(3x - 1)^2 = 9x^2 - 6x + 1$
It checks.

Answer $(3x - 1)^2$

Inspect the trinomial for common monomial factors before attempting other factoring.

Example 4   Factor. $2x^2 + 8x + 8$

Solution   *First factor out the GCF of the terms.* $2x^2 + 8x + 8 = 2(x^2 + 4x + 4)$
Factor the trinomial. $= 2(x + 2)^2$
Expand to check, squaring first. $2(x + 2)^2 = 2(x^2 + 4x + 4)$
$$= 2x^2 + 8x + 8$$
It checks.

Answer $2x^2 + 8x + 8 = 2(x + 2)^2$

Factoring and the zero-product property can be used to solve some equations.

Example 5 Solve. $x^2 - 10x + 25 = 0$

Solution *Factor the trinomial.* $(x - 5)(x - 5) = 0$
Use the zero-product property. $x - 5 = 0$
Solve. $x = 5$

Answer $\{5\}$

■ CLASSROOM EXERCISES

Write as squares.

1. $4x^2$
2. $9a^2b^2$
3. x^4
4. $\dfrac{9}{16}a^2b^4$

Expand and simplify.

5. $(x - 3)^2$
6. $2(x + 5)^2$

Is the trinomial a perfect square?

7. $-9x^2 - 6xy + y^2$
8. $9a^2 + 18ab + 9b^2$
9. $4x^2 - 4xy + y^2$

Factor.

10. $x^2 + 2xy + y^2$
11. $m^2 - 2mn + n^2$
12. $9a^2 + 12ab + 4b^2$
13. $x^4 - 4x^2y + 4y^2$
14. $9a^2 - 4b^2$
15. $3x^2 + 6x + 3$

Solve.

16. $x^2 + 14x + 49 = 0$
17. $x^2 - 81 = 0$

■ WRITTEN EXERCISES

Write as squares.

Ⓐ **1.** $9a^2$
2. $49b^2$
3. $16c^2d^2$
4. $64r^2s^2$

5. p^6
6. q^4
7. $\dfrac{16}{81}s^8t^{10}$
8. $\dfrac{9}{100}x^4y^6$

Is the trinomial a perfect square?

9. $x^2 + 4x - 4$
10. $y^2 - 6x - 9$
11. $4x^2 - 8xy + y^2$

12. $x^2 - 12xy + 36y^2$
13. $a^2 - a + \dfrac{1}{4}$
14. $a^2 + \dfrac{2}{3}a + \dfrac{1}{9}$

Expand and simplify.

15. $(x + 5)^2$
16. $(y + 4)^2$
17. $(a - 6)^2$
18. $(b - 3)^2$
19. $(3y + 2)^2$
20. $(4x + 3)^2$

21. $(3b - 4)^2$　　　　**22.**　$(2a - 3)^2$　　　　**23.** $(5c - d)^2$

24. $(8p - q)^2$　　　　**25.** $2(x + 10)^2$　　　　**26.** $3(x + 1)^2$

Factor.

27. $x^2 + 6x + 9$　　　　**28.** $x^2 + 12x + 36$　　　　**29.** $x^2 - 8xz + 16z^2$

30. $x^2 - 10xy + 25y^2$　　　　**31.** $2x^2 + 8xy + 8y^2$　　　　**32.** $3x^2 + 6xz + 3z^2$

33. $x^2 - 16$　　　　**34.** $x^2 - 25$　　　　**35.** $5x^2 - 45$

36. $3x^2 - 12$　　　　**37.** $4x^2 - 12x + 9$　　　　**38.** $9x^2 - 24x + 16$

Solve.

39. $x^2 - 10x + 25 = 0$　　　　**40.** $x^2 - 8x + 16 = 0$　　　　**41.** $x^2 + 12x + 36 = 0$

42. $x^2 + 6x + 9 = 0$　　　　**43.** $(x - 0.5)^2 = 0$　　　　**44.** $(x - 2.5)^2 = 0$

45. $2x^2 - 4x + 2 = 0$　　　　**46.** $3x^2 - 12x + 12 = 0$　　　　**47.** $(x + 1.5)^2 = 0$

48. $(x + 3.5)^2 = 0$　　　　**49.** $x^2 - 100 = 0$　　　　**50.** $x^2 - 64 = 0$

Write the area of each square as a binomial squared. Then expand and simplify.
Copy and subdivide each square into rectangles and write the area of each rectangle on your diagram.

 51.

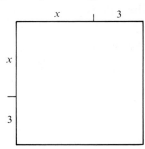

52.

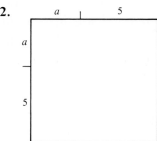

53.

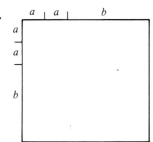

54.

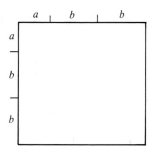

Write as the product of three factors.

55. $x^3 - 6x^2 + 9x$　　　　**56.** $x^3 + 22x^2 + 121x$　　　　**57.** $x^3 - 9x$

Write as the product of four factors.

58. $x^4 - 2x^2 + 1$　　　　**59.** $x^4 - 8x^2 + 16$　　　　**60.** $x^4 - 18x^2 + 81$

61. General Jones said, "$(x + y)^2 = x^2 + y^2$." Show that he is wrong.

62. Admiral Thompson said, "$(x - y)^2 = x^2 - y^2$." Show that she is wrong.

C **63.** Copy and complete the table.

n	15	25	35	45	55	65
n^2	225	625	?	?	?	?

If t represents the tens digit and u represents the ones digit, a two-digit number can be written as $10t + u$.

64. Expand and simplify $(10t + u)^2$.

65. Expand and simplify $(10t + 5)^2$.

66. Expand and simplify $100t(t + 1) + 25$.

67. Use the results of Exercise 65 to explain why the square of a whole number ending in 5 ends in 25.

68. Use the results of Exercises 65 and 66 to explain why the square of a whole number ending in 5 has digits to the left of the tens place that are the product of a number and its successor.

69. Expand and simplify $(10t + 2)^2$. Use the results to explain why the squares of two-digit numbers ending in 2 are divisible by 4.

70. Expand and simplify $(10t + 6)^2$. Use the results to explain why the squares of two-digit numbers ending in 6 are divisible by 4.

■ REVIEW EXERCISES

Expand and simplify.

1. $(x + 2)(x + 6)$ **2.** $(x - 3)(x - 5)$ [7–5]

3. State the quadrant each point is on or the axis it is on.   [5–1]

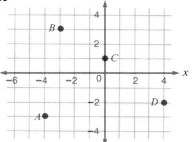

    a. A

    b. B

    c. C

    d. D

4. List the domain and range of the relation $\{(1, 4), (2, 4), (3, 5), (4, 5)\}$.   [5–2]

5. Graph the relation $\{(1, 2), (0, 2), (1, 1)\}$.   [5–2]

6. State whether the ordered pair is a solution of the equation $3x + y = 12$.   [5–3]

    a. $(2, 6)$ **b.** $(-4, 0)$ **c.** $(6, -2)$

Preview

Here is a list of all pairs of integers having the product 12.

1 and 12	2 and 6	3 and 4
-1 and -12	-2 and -6	-3 and -4

- Which of these pairs has the sum 7?
- Which of these pairs has the sum 13?
- Find a pair of integers that has the product 6 and the sum 5.

- List all pairs of integers having the product 15.
  a. Which pair has the sum 16?
  b. Which pair has the sum -8?

In this lesson you will use the skill of finding a pair of numbers that has a given product and a given sum to factor trinomials.

■ LESSON

In factoring the difference of two squares (such as $x^2 - 9$) or in factoring perfect squares (such as $x^2 + 6x + 9$), there is a step-by-step procedure to find the factors. For other polynomials, however, you will have to check several choices to find the pair that works. For example, consider factoring $x^2 + 8x + 12$.

We know that the first term of each binomial is x: $(x\quad)(x\quad)$.

We also know that the product of the last terms of the binomials is 12. Therefore, the second terms of the binomials are 3 and 4, 2 and 6, 1 and 12, -3 and -4, -2 and -6, or -1 and -12.

These are the possible factors.

$$(x + 3)(x + 4)$$
$$(x + 2)(x + 6)$$
$$(x + 1)(x + 12)$$
$$(x - 3)(x - 4)$$
$$(x - 2)(x - 6)$$
$$(x - 1)(x - 12)$$

Now we mentally expand each choice to determine whether any of them produces the middle term, $8x$.

$$(x + 3)(x + 4) = x^2 + 7x + 12$$
$$(x + 2)(x + 6) = x^2 + 8x + 12 \qquad \leftarrow$$
$$(x + 1)(x + 12) = x^2 + 13x + 12$$
$$(x - 3)(x - 4) = x^2 - 7x + 12$$
$$(x - 2)(x - 6) = x^2 - 8x + 12$$
$$(x - 1)(x - 12) = x^2 - 13x + 12$$

Only $(x + 2)(x + 6)$ correctly factors the polynomial. Note that $2 \cdot 6 = 12$ (last term) and $2 + 6 = 8$ (middle-term coefficient.)

We can usually eliminate some of the possibilities by noting the signs of the coefficients.

Example 1   Factor. $a^2 - 5a + 6$

Solution   The first terms of the binomials are a.

$$(a \qquad)(a \qquad)$$

The last term of the trinomial (6) is positive, so the last terms of the binomials are both positive or both negative. Write the pairs of numbers whose product is 6.

$$\begin{array}{cc} 1 \text{ and } 6 & 2 \text{ and } 3 \\ -1 \text{ and } -6 & -2 \text{ and } -3 \end{array}$$

The coefficient of the middle term of the trinomial (-5) indicates that both signs are negative, so the possible factors are

$$(a - 1)(a - 6)$$
$$(a - 2)(a - 3)$$

Mentally expand each choice to determine the one that gives the correct middle term.

$$(a - 1)(a - 6) = a^2 - 7a + 6$$
$$(a - 2)(a - 3) = a^2 - 5a + 6 \qquad \leftarrow$$

Note that $-2 \cdot -3 = 6$ (last term) and $-2 + (-3) = -5$ (middle-term coefficient).

Answer $(a - 2)(a - 3)$

At first, this process of "guess and check" can seem very tedious. With practice, most of the work can be done quickly "in your head."

Example 2   Factor. $y^2 + 8yz + 16z^2$

Solution   The first terms of the binomials are both y. The last terms of the binomials must contain the variable z.

$$(y \qquad z)(y \qquad z)$$

The coefficient of the middle term (8) and the last term (16) of the trinomial are both positive, so the coefficients of both terms of the binomials are positive. Write the pairs of positive numbers whose product is 16.

$$1 \text{ and } 16 \qquad 2 \text{ and } 8 \qquad 4 \text{ and } 4$$

The coefficient of the middle term of the trinomial indicates that the sum of the coefficients of the last terms of the binomials is 8. So choose the pair whose sum is 8.

$$4 \text{ and } 4$$

Example 2 (continued)

Write the binomial factors.

$$(y + 4z)(y + 4z)$$

Answer $(y + 4z)(y + 4z)$

Check $(y + 4z)(y + 4z) = y^2 + 8yz + 16z^2$ It checks.

Always remember to remove a common monomial factor before attempting other factoring.

Example 3   Factor. $5x^2 - 35x + 60$

Solution   Factor out the GCF of the terms. $5(x^2 - 7x + 12)$
Factor the trinomial. $5(x - 3)(x - 4)$

Answer $5(x - 3)(x - 4)$

Factoring and the zero-product property can be used to solve some equations.

Example 4   Solve. $y^2 + 15 = 8y$

Solution $y^2 + 15 = 8y$

Collect all terms on one side of the
equation and arrange the terms in de-
scending order. $y^2 - 8y + 15 = 0$
Factor. $(y - 3)(y - 5) = 0$
Apply the zero-product property. $y = 3$ or $y = 5$

Answer $\{3, 5\}$

Check

$y^2 + 15 = 8y$	$y^2 + 15 = 8y$
$3^2 + 15 \stackrel{?}{=} 8 \cdot 3$	$5^2 + 15 \stackrel{?}{=} 8 \cdot 5$
$9 + 15 \stackrel{?}{=} 24$	$25 + 15 \stackrel{?}{=} 40$
$24 = 24$ It checks.	$40 = 40$ It checks.

■ CLASSROOM EXERCISES

Factor.

1. $x^2 + 5x + 6$ **2.** $x^2 - 5x + 6$ **3.** $a^2 + 7a + 6$

4. $b^2 - 7b + 6$ **5.** $c^2 + 4c + 4$ **6.** $a^2 - b^2$

7. $x^2 + 6xy + 9y^2$ **8.** $3a^2 - 18a + 15$ **9.** $y^2 - 8yz + 7z^2$

Solve.

10. $x^2 + 7x + 12 = 0$ **11.** $x^2 + 5 = 6x$ **12.** $t^2 - 5t + 4 = 0$

■ WRITTEN EXERCISES

Factor.

A 1. $a^2 + 4a + 3$ | 2. $p^2 + 6p + 5$ | 3. $x^2 + 6x + 8$

4. $x^2 + 7x + 10$ | 5. $a^2 - 8a + 12$ | 6. $a^2 - 7a + 12$

7. $x^2 + 9x + 20$ | 8. $x^2 + 12x + 20$ | 9. $a^2 - 9a + 18$

10. $a^2 - 11a + 18$ | 11. $c^2 + 19c + 18$ | 12. $b^2 + 9b + 18$

13. $b^2 - 10b + 21$ | 14. $c^2 - 8c + 15$ | 15. $a^2 - 8a + 16$

16. $a^2 - 10a + 25$ | 17. $a^2 + 26a + 25$ | 18. $a^2 + 17a + 16$

19. $x^2 + 5x + 4$ | 20. $x^2 + 8x + 7$ | 21. $x^2 + 5xy + 6y^2$

22. $x^2 + 7xy + 6y^2$ | 23. $x^2 - 9xy + 8y^2$ | 24. $x^2 - 6xy + 8y^2$

Solve.

25. $(2t - 3)(t - 4) = 0$ | 26. $(3x - 5)(x - 1) = 0$ | 27. $x^2 - 4x + 3 = 0$

28. $y^2 - 6y + 5 = 0$ | 29. $x^2 + 7x + 6 = 0$ | 30. $x^2 + 5x + 6 = 0$

31. $x^2 + 6x + 9 = 0$ | 32. $x^2 + 4x + 4 = 0$ | 33. $v^2 - 9v + 14 = 0$

34. $w^2 - 10w + 16 = 0$ | 35. $x^2 - 11x + 24 = 0$ | 36. $x^2 - 10x + 24 = 0$

B 37. $y^2 + 8 = 6y$ | 38. $2y^2 + 14 = 16y$ | 39. $2x^2 + 12 = 10x$

40. $5x^2 + 15x + 10 = 0$ | 41. $3x^2 + 15x + 12 = 0$ | 42. $2x^2 + 10 = 12x$

43. $35 = 40t - 5t^2$ | 44. $40 = 30t - 5t^2$ | 45. $36 = 30x - 6x^2$

List all integer values for k for which these polynomials can be factored.

46. $x^2 + kx + 7$ | 47. $x^2 + kx + 3$ | 48. $x^2 + kx + 4$

49. $x^2 + kx + 9$ | 50. $x^2 + kx + 12$ | 51. $x^2 + kx + 18$

List all positive integer values of k for which these polynomials can be factored.

52. $x^2 + 3x + k$ | 53. $x^2 + 4x + k$ | 54. $x^2 - 4x + k$

55. $x^2 - 5x + k$ | 56. $x^2 - 7xy + ky^2$ | 57. $x^2 + 7xy + ky^2$

Suppose the equation $x^2 + bx + c = 0$ has two solutions, r and s.

58. What does rs equal? | 59. What does $r + s$ equal?

Solve.

60. If 9 more than the square of a number equals 10 times the number, what values could the number have?

61. If 12 more than the square of a number equals 7 times the number, what values could the number have?

62. The area of a rectangular dog pen is 600 ft² and the pen is enclosed with 100 ft of fencing. What are the length and width of the pen?

63. The area of a rectangle is 180 m² and the perimeter is 56 m. What are the length and width of the rectangle?

Expand, simplify, and then factor the expressions.

C **64.** $(x + 3)^2 + (x + 1)$

65. $(x + 3)(x + 4) + x$

66. $(x - 2)(x - 3) - 2x$

67. $(x - 2)^2 - 1$

68. $(2x + 3)(x + 4) - (x^2 + 3x)$

69. $(2x - 3)(3x - 2) - (5x^2 - 6x)$

Solve.

70. $(x - 4)^2 - 2x = 0$

71. $(x - 4)(x - 1) + x = 0$

72. $(x + 2)(x + 3) - 2 = 0$

73. $(x - 5)(x - 2) + 2 = 0$

■ REVIEW EXERCISES

Expand and simplify.

1. $(x - 3)(x + 5)$

2. $(x + y)(x - 4y)$ [7–5]

3. Select three values of x. Then copy and complete the table, plot the points, and graph the equation $y = \frac{1}{2}x - 2$. [5–4]

x	?	?	?
y	?	?	?

4. Copy and complete the table and then use the ordered pairs to graph the equation $y = x^2 - 2$. [5–4]

x	-3	-2	-1	$-\frac{1}{2}$	0	$\frac{1}{2}$	1	2	3
y	?	?	?	?	?	?	?	?	?

5. State the slopes of the lines. [5–5]

   a. $\overleftrightarrow{AC}$

   b. $\overleftrightarrow{BD}$

   c. $\overleftrightarrow{AB}$

   d. $\overleftrightarrow{BC}$

7–9 Factoring $x^2 + bxy + cy^2$ with c Negative

Preview

A large department store wants a display window installed. The Louden Glass Company installs a rectangular window with 96 ft² of viewing area and a length 4 ft more than the width. What are the length and width of the window?

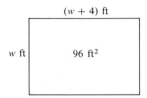

(w + 4) ft

w ft   96 ft²

To solve this problem, we could write this equation

$$w(w + 4) = 96$$

Expanding and collecting terms on one side, we have

$$w^2 + 4w - 96 = 0$$

We can solve this equation if we can factor the polynomial:

$$w^2 + 4w - 96$$

In this lesson you will learn to factor trinomials in which the last term is negative.

■ LESSON

Consider this equation:

Expanded form   **Factored form**
$$x^2 + (p + q)x + pq = (x + p)(x + q)$$

In Section 7–8, it was stated that when the *product pq* is positive, the values of p and q have the same sign. Thus with pq positive:

If the value of $(p + q)$ is positive, then p and q are both positive.

$$x^2 + 18x + 77 = (x + 7)(x + 11)$$

  ↑ ↑ ↑ ↑
positive positive both positive

If the value of $(p + q)$ is negative, then p and q are both negative.

$$y^2 - 10y + 9 = y^2 + (-10)y + 9 = (y + (-1))(y + (-9))$$

   ↑ ↑   ↑   ↑
negative positive both negative

Now consider the case in which the product pq is negative. This means that p and q differ in sign.

$$x^2 + 2x - 15 = (x - 3)(x + 5) \qquad y^2 - 2y - 24 = (y + 4)(y - 6)$$

↑	↑	↑	↑	↑	↑
negative	*different signs*		*negative*	*different signs*	

The above information helps us factor trinomials such as

$$a^2 + 2a - 15$$

We know that the first term of each binomial is a.

$$(a \quad)(a \quad)$$

The last term of the trinomial indicates that the *product* of the last terms of the binomials is -15. The factors of -15 have opposite signs. The possibilities are

$$-1 \text{ and } 15 \qquad 1 \text{ and } -15$$
$$3 \text{ and } -5 \qquad -3 \text{ and } 5$$

The coefficient of the middle term of the trinomial indicates that the *sum* of the last terms of the binomials must be 2. Therefore, the last terms are -3 and 5, and the correct factorization is

$$(a - 3)(a + 5)$$

Check by expanding: $\quad (a - 3)(a + 5) = a^2 + 5a - 3a - 15$
$$= a^2 + 2a - 15$$

Example 1   Factor. $\quad x^2 + 4x - 5$

Solution   The first term of each binomial is x. $(x \quad)(x \quad)$
The product of the last terms is -5.
Write pairs of numbers whose product is -5. 1 and -5, -1 and 5

The middle term of the trinomial indicates that the sum of the last terms of the binomials is 4. So, choose the pair whose sum is 4.

$$-1 \text{ and } 5$$

Write the binomial factors. $(x - 1)(x + 5)$

Answer $(x - 1)(x + 5)$

Check $(x - 1)(x + 5) = x^2 + 5x - x - 5$
$$= x^2 + 4x - 5 \quad \text{It checks.}$$

Example 2 Factor. $a^2 + a - 6$

Solution $a^2 + a - 6$

The first term of each binomial is a. $(a \quad)(a \quad)$
The product of the last terms of the bi-   1 and -6, 2 and -3,
nomials is -6. Write the pairs of numbers whose
product is -6. -1 and 6, 3 and -2

The middle term of the trinomial indicates that the sum of the last terms of
the binomials is 1. So, choose the pair whose sum is 1.

$$3 \text{ and } -2$$

Write the binomial factors. $(a + 3)(a - 2)$

Answer $(a + 3)(a - 2)$

Check $(a + 3)(a - 2) = a^2 - 2a + 3a - 6$
$\qquad\qquad\qquad = a^2 + a - 6$ It checks.

Example 3 Factor. $2x^2 - 22x - 24$

Solution Factor out the GCF. $2x^2 - 22x - 24 = 2(x^2 - 11x - 12)$
Factor the trinomial. $= 2(x - 12)(x + 1)$

Answer $2(x - 12)(x + 1)$

Check The check is left to the student.

Example 4 Factor. $x^2 + 8x + 6$

Solution Since the middle and last terms of the trinomial are positive, the last terms
of the binomials are positive. The choices are 1 and 6 or 2 and 3. The possible factorizations are

$$(x + 1)(x + 6)$$
$$(x + 2)(x + 3)$$

Mentally expand each possible factorization to see which one produces the
middle term $8x$.

	Middle term
$(x + 1)(x + 6)$	$+ 7x$
$(x + 2)(x + 3)$	$+ 5x$

In no case is there a middle term of $8x$. Thus, $x^2 + 8x + 6$ cannot be factored into the product of two binomials.

Answer $x^2 + 8x + 6$ is not factorable over the set of integers.

Example 5   Solve. $x^2 + 4x = 12$

  Solution   *Collect terms so that the right side is 0.* $x^2 + 4x - 12 = 0$
               Factor. $(x + 6)(x - 2) = 0$
               Use the zero-product property. $x + 6 = 0$ or $x - 2 = 0$
                                                          $x = -6$ or $x = 2$

  Answer $\{-6, 2\}$

  Check   The check is left to the student.

■ CLASSROOM EXERCISES

Factor.

1. $x^2 + 2x - 8$ **2.** $y^2 - y - 6$ **3.** $b^2 + b - 6$ **4.** $a^2 - 7ab - 18b^2$

Solve.

5. $x^2 - 3x - 4 = 0$ **6.** $x^2 = 2x + 15$ **7.** $x^2 + x = 12$ **8.** $-11x - 28 = x^2$

■ WRITTEN EXERCISES

A Factor.

1. $a^2 - 5a - 6$ **2.** $a^2 - a - 6$ **3.** $b^2 + b - 6$

4. $b^2 + 5b - 6$ **5.** $c^2 + c - 12$ **6.** $c^2 + 4c - 12$

7. $a^2 - 11a - 12$ **8.** $a^2 - a - 12$ **9.** $b^2 - 8b - 9$

10. $b^2 - 9$ **11.** $c^2 - 16$ **12.** $c^2 - 6c - 16$

13. $2x^2 + 12x - 14$ **14.** $3x^2 + 3x - 6$ **15.** $y^2 - 5y - 36$

16. $y^2 - 9y - 36$ **17.** $z^2 + 16z - 36$ **18.** $z^2 + 23z - 24$

19. $x^2 + 5x - 24$ **20.** $x^2 + 2x - 24$ **21.** $x^2 + 2xy - 3y^2$

22. $x^2 - 10xy - 11y^2$ **23.** $x^2 + xy - 2y^2$ **24.** $x^2 + 3xy - 4y^2$

Solve.

25. $x^2 - 3x - 4 = 0$ **26.** $x^2 - 4x - 5 = 0$ **27.** $y^2 - 6y - 7 = 0$

28. $y^2 - 2y - 8 = 0$ **29.** $a^2 + 3a - 10 = 0$ **30.** $a^2 + 3a - 18 = 0$

31. $b^2 + 4b = 12$ **32.** $b^2 + b = 12$ **33.** $t^2 = 11t + 12$

34. $t^2 = 5t + 6$ **35.** $v^2 + 9v + 18 = 0$ **36.** $v^2 + 17v - 18 = 0$

For each problem, let n represent the missing number, and write an equation that fits the problem. Then solve.

37. The square of a number increased by the number equals 20. What is the number?

38. The square of a number increased by the number equals 30. What is the number?

39. Twice a number subtracted from the square of the number equals 35. What is the number?

40. Twice a number subtracted from the square of the number equals 48. What is the number?

B List all integer values of k for which these polynomials can be factored.

41. $x^2 + kx - 7$ **42.** $x^2 + kx - 3$ **43.** $x^2 + kx - 4$

44. $x^2 + kx - 9$ **45.** $x^2 + kx - 12$ **46.** $x^2 + kx - 18$

Suppose the equation $x^2 + bx - 24 = 0$ has two solutions, r and s, with $r > s$.

47. Which of these expressions must represent positive numbers?

$$r \qquad s \qquad rs \qquad r + s$$

48. Which of these expressions must represent negative numbers?

$$r \qquad s \qquad rs \qquad r + s$$

Write an equation that fits the problem. Then solve.

49. The area of a rectangular lawn is 54 m². The lawn is 3 m longer than it is wide. What is the width of the lawn?

50. The area of a rectangular court is 140 m². The court is 4 m longer than it is wide. What is the width of the court?

51. The product of two consecutive negative integers is 272. What are the integers?

52. The product of two consecutive negative odd integers is 143. What are the integers?

C **53.** One number is 3 larger than another and the sum of their squares is 269. What are the numbers?

54. The difference between two numbers is 5 and the sum of the squares of the numbers is 193. What are the numbers?

55. The sum of two numbers is 15 and the sum of their squares is 117. What are the numbers?

56. An 8-in. by 10-in. picture has a border of uniform width. The total area of the border and picture is 224 in.² What is the width of the border?

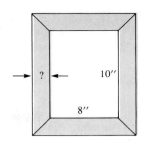

Solve.

57. $(x + 4)(x - 3) = 5x$

58. $(x + 4)(x - 3) = 12x$

59. $(2x + 1)(x - 3) = x^2 + 11$

60. $(2x + 3)(3x - 2) = 5x^2 + 9x + 6$

61. $(x + 1)(x + 3) = x + 13$

62. $(2x - 3)(x - 4) = (x - 4)^2$

■ REVIEW EXERCISES

Expand and simplify.

1. $(2x + 3)(3x + 1)$

2. $2(3a + b)(a - 2b)$ [7–5]

3. State the slope and the y-intercept of the line for the equation $y = \frac{2}{3}x + 4$. [5–6]

4. Graph the line with slope $-\frac{1}{2}$ and y-intercept 1. [5–6]

5. State whether the relation is a function. [5–8]

$$\{(1, 4), (2, 5), (1, 10)\}$$

6. State whether the equation represents a function. [5–8]

$$y^2 = x$$

Self-Quiz 3

Factor.

7–7   **1.** $x^2 + 8x + 16$ **2.** $y^2 - 14y + 49$ **3.** $3w^2 - 6w + 3$

7–8   **4.** $a^2 + 5a + 6$ **5.** $t^2 - 11t + 28$ **6.** $r^2 - 10r + 16$

7–9   **7.** $x^2 - 5x - 6$ **8.** $b^2 + 12b - 28$ **9.** $y^2 + 15y - 16$

Solve.

7–8,
7–9   **10.** $x^2 - 16x + 63 = 0$ **11.** $y^2 + y = 20$

7-10 Factoring $ax^2 + bxy + cy^2$

Preview

The Great Gustav posed this puzzle.

I am thinking of four integers. The product of the first and third is 12. The product of the second and fourth is -2. The product of the first and fourth added to the product of the second and third is 5. What are the integers?

This puzzle is quite difficult. There are many possibilities to check. In this lesson you will learn a method of factoring trinomials that could be used to solve the puzzle.

■ LESSON

The technique for factoring a trinomial whose first coefficient is different from 1 is the same as the technique that we used in earlier lessons. But there are more choices. Consider the expression

$$6x^2 + 11x + 3$$

The possible first terms of the binomials are

$$(1x \qquad)(6x \qquad)$$
$$(2x \qquad)(3x \qquad)$$

Since the middle and last terms of the trinomials are positive, the signs of the last terms of the binomial factors must be positive.

$$(1x + \quad)(6x + \quad)$$
$$(2x + \quad)(3x + \quad)$$

The factors of the last term of the trinomial are 1 and 3. These are all of the possible factorizations.

$$(1x + 1)(6x + 3)$$
$$(1x + 3)(6x + 1)$$
$$(2x + 1)(3x + 3)$$
$$(2x + 3)(3x + 1)$$

Next find the middle terms of the expansions.

$$9x \qquad 19x \qquad 9x \qquad 11x$$

The last expression is the correct middle term. Therefore,
$6x^2 + 11x + 3 = (2x + 3)(3x + 1)$.

Example 1 Factor. $6a^2 - 10ab - 4b^2$

Solution

Factor out the GCF of the three terms.	$2(3a^2 - 5ab - 2b^2)$
List possible first terms of the binomials.	$(3a \quad)(a \quad)$
List possible second terms of the binomials.	b and $-2b$ or $-b$ and $2b$

These are all the possible factorizations and their expansions.

$(3a + 2b)(a - b) = 3a^2 - ab - 2b^2$ Wrong middle term
$(3a - 2b)(a + b) = 3a^2 + ab - 2b^2$ Wrong middle term
$(3a - b)(a + 2b) = 3a^2 + 5ab - 2b^2$ Wrong middle term
$(3a + b)(a - 2b) = 3a^2 - 5ab - 2b^2$ Correct middle term!

Answer $2(3a + b)(a - 2b)$

If the first and last terms of the trinomial have several factors, the number of combinations to check may be great. However, the general technique is the same.

To factor trinomials of the form $ax^2 + bxy + cy^2$

1. Factor out the GCF of the terms.

2. List the possible first terms of the binomials.

3. List the possible last terms of the binomials.

4. Expand the possible products until the one that produces the correct middle term is found.

Example 2 Solve. $2x^2 + 7x = -3$

Solution $2x^2 + 7x = -3$

Collect terms on the left side of the equation. $2x^2 + 7x + 3 = 0$
Factor the trinomial. $(2x + 1)(x + 3) = 0$
Use the zero-product property. $2x + 1 = 0$ or $x + 3 = 0$
                                     $2x = -1$ or $x = -3$

$$x = -\frac{1}{2} \quad \text{or} \quad x = -3$$

Answer $\left\{ -\frac{1}{2}, -3 \right\}$

Check   The check is left to the student.

■ CLASSROOM EXERCISES

Given the trinomial $4x^2 - 4x - 3$, answer the following questions.

1. What are the possible first terms of the binomial factors?

2. What are the possible second terms of the binomial factors?

3. What do you know about the signs of the last terms of the binomial factors?

4. Factor the trinomial.

Factor.

5. $6a^2 + 7a + 2$ **6.** $2x^2 - 5x + 3$ **7.** $10m^2 + 13mn - 3n^2$

Solve.

8. $2x^2 - 3x - 2 = 0$ **9.** $2x^2 = x + 3$ **10.** $6x^2 - 22x - 8 = 0$

■ WRITTEN EXERCISES

Factor.

A   **1.** $2x^2 + 5x + 3$ **2.** $2x^2 + 7x + 3$ **3.** $3x^2 - 7x + 2$

     4. $3x^2 - 5x + 2$ **5.** $5a^2 - 13a - 6$ **6.** $5a^2 - 7a - 6$

     7. $5a^2 + 7a - 6$ **8.** $5a^2 + 13a - 6$ **9.** $5n^2 - 29n - 6$

   10. $5n^2 + 29n - 6$ **11.** $6n^2 - 13n - 5$ **12.** $6n^2 - 7n - 5$

   13. $7y^2 + 30y + 8$ **14.** $7y^2 + 18y + 8$ **15.** $7y^2 - 57y + 8$

   16. $7y^2 - 15y + 8$ **17.** $14a^2 - 20a + 6$ **18.** $14a^2 - 8a - 6$

   19. $8x^2 + 23xy - 3y^2$ **20.** $8x^2 - 5xy - 3y^2$ **21.** $4a^2 + ab - 5b^2$

   22. $4a^2 - 19ab - 5b^2$ **23.** $6x^2 - 11xy - 10y^2$ **24.** $6x^2 + 11xy - 10y^2$

Solve.

   25. $2a^2 - 7a + 5 = 0$ **26.** $2a^2 - 5a + 3 = 0$

   27. $3a^2 + 2a - 8 = 0$ **28.** $3a^2 - 10a - 8 = 0$

   29. $4a^2 - 4a - 15 = 0$ **30.** $4a^2 - 8a - 21 = 0$

   31. $6x^2 + 23x + 7 = 0$ **32.** $6x^2 - 17x + 7 = 0$

   33. $12x^2 - 28x + 11 = 0$ **34.** $12x^2 + 68x + 11 = 0$

   35. $12x^2 + x = 20$ **36.** $12x^2 - x = 20$

For each problem, let n represent the missing number, and write an equation that fits the problem. Then solve.

37. The sum of twice the square of a number and 7 times the number equals 15. What is the number?

38. Twice the square of a number added to the number equals 15. What is the number?

39. Twice the square of a number equals 3 more than the number. What is the number?

40. Twice the square of a number equals the sum of 5 times the number and 7. What is the number?

List all integer values of k for which these trinomials can be factored.

B   41. $2x^2 + kx + 3$ 42. $4x^2 + kx + 3$ 43. $6x^2 + kx + 3$

44. $2x^2 + kx - 3$ 45. $4x^2 + kx - 3$ 46. $6x^2 + kx - 3$

For each problem, write an equation that fits the problem. Then solve the equation and answer the question.

47. The length of a rectangle is 3 m more than twice the width. What is the width of the rectangle if its area is 9 m²?

48. A rectangular garden has a length that is 6 m greater than 4 times its width. What are the dimensions of the garden if its area is 40 m²?

49. The height of a triangle is 4 m less than the base. If the area of the triangle is 6 m², what is the height?

50. The length of a rectangular garden is 2 times the width. There is a cement walk around the garden that is 1 m wide. The area of the walk and the garden is 544 m². What are the dimensions of the garden?

51. Solve the problem in the Preview.

For how many different integer values of k are these trinomials factorable?

C   52. $10x^2 + kx + 21$ 53. $15x^2 + kx - 22$ 54. $4x^2 + kx + 9$

Solve.

55. $(x + 2)(x + 3) + (x + 5)(x - 3) = 0$ 56. $(x + 1)^2 + (x + 1)(x + 2) = 0$

57. The shorter base of a trapezoid is 1 m longer than the altitude, and the longer base is 5 m longer than the altitude. Find the altitude of the trapezoid if its area is 40 m². $\left[\text{Remember: } A = \dfrac{h(a + b)}{2}. \right]$

58. The height h, in feet, of an object projected upward with an initial velocity of v feet per second is given by the formula $h = vt - 16t^2$, where t represents the number of seconds. After how many seconds will an object projected vertically at 88 ft/s reach a height of 120 ft?

■ REVIEW EXERCISES

1. Use the graph to find the solution to this system.

$$x - y = 2$$
$$x + y = 0$$

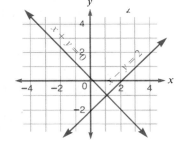

[6–1]

2. Solve by substitution.

$$y = 2x + 5$$
$$3x + 4y = -13$$

[6–2]

3. Solve this system by multiplication and addition.

$$3a + 2b = 5$$
$$5a + b = 27$$

[6–6]

4. Solve by using a system of equations.

[6–7]

The Drama Society sold 200 tickets and raised a total of $650. If student tickets cost $1.00 and adult tickets cost $4.00, how many of each kind of ticket were sold?

EXTENSION Compound interest _____

If $1000 is invested at a simple-interest rate of 6% per year, the value of the investment after 5 years is

$$1000 + 1000(0.06)5 = 1000(1 + (0.06)5).$$

↑ ↑   ↑ ↑
principal   rate time

If the same principal is invested at 6% per year compounded annually, the value of the investment after 5 years is

$$1000(1.06)^5.$$

1. Use a calculator to compute each amount.

    a. $1000[1 + (0.06)5]$
    b. $1000(1.06)^5$ [*Hint:* Use the power key if your calculator has one, or use the key sequence 1.06 × = = = = to evaluate $(1.06)^5$.]

2. Which investment is worth more?
3. How much more is it worth?

■ CHAPTER SUMMARY

- **Vocabulary**

factor	[page 305]	greatest common factor of two	[page 310]
prime number	[page 305]	or more integers [GCF]	
composite number	[page 305]	greatest common factor of two	[page 311]
to factor a number	[page 306]	or more monomials	
prime factorization	[page 306]	expanding	[page 318]
common factors	[page 310]	factoring	[page 318]

- *Factoring over a set of numbers* means that the factors must belong to the set. [7–1]

- The Zero-Product Property [7–3]

$$\text{For all numbers } a \text{ and } b,$$
$$\text{if } ab = 0, \text{ then } a = 0 \text{ or } b = 0.$$

- The FOIL method of expanding binomials [7–5]

    F   Multiply FIRST terms of the binomials.
    O   Multiply OUTER terms of the binomials.
    I   Multiply INNER terms of the binomials.
    L   Multiply LAST terms of the binomials.

- The sum of these products gives the expanded form of $(a + b)(c + d)$.

$$(a + b)(c + d) = ac + ad + bc + bd$$

    first   outer   inner   last
    terms   terms   terms   terms

- Factoring a difference of squares [7–6]

$$\text{For all numbers } a \text{ and } b,$$
$$a^2 - b^2 = (a + b)(a - b).$$

- Factoring a perfect square trinomial [7–7]

$$a^2 + 2ab + b^2 = (a + b)^2$$
$$\text{and}$$
$$a^2 - 2ab + b^2 = (a - b)^2$$

- To factor trinomials of the form $ax^2 + bxy + cy^2$ [7–10]

    1. Factor out the GCF of the terms.

    2. List the possible first terms of the binomials.

    3. List the possible last terms of the binomials.

    4. Expand the possible products until the one that produces the correct middle term is found.

■ CHAPTER REVIEW

7–1   **Objective:** To write the prime factorization of an integer.

  1. List the prime numbers between 1 and 10.

  2. a. Which of the following are composite numbers?

$$11, 12, 13, 14, 15$$

  b. Write a prime factorization of each composite number found in part (a), listing the factors in increasing order.

List all the positive integer factors.

  3. 30   **4.** 100

7–2   **Objective:** To find the greatest common factor of a pair of monomials.

  5. List the common factors of 20 and 30.

  6. List the greatest common factor of 18 and 27.

  7. List the common factors of $3a^2$ and $6ab$.

  8. List the greatest common factor of $6xy$ and $8y^2$.

7–3   **Objective:** To solve equations using the zero-product property.

Solve.

  9. $(x - 4)x = 0$ **10.** $(x + 3)(x - 7) = 0$ **11.** $6(2x - 1)(2x + 3) = 0$

7–4   **Objective:** To factor a monomial from a polynomial.

Factor completely.

  12. $3x + 6y$ **13.** $2x^2 - 10x$

Solve.

  14. $x^2 + 5x = 0$

  15. The square of a number is 3 times the number. What is the number?

7–5   **Objective:** To expand the product of two binomials.

Expand and simplify.

  16. $(x + 2)(x + 5)$ **17.** $(a + 10)(a - 10)$

  18. $(y - 4)^2$ **19.** $(2x + 3)(x - 4)$

7–6 Objective: To factor the difference of two squares.

Factor.

20. $x^2 - 25$ **21.** $9x^2 - 1$

Solve.

22. $x^2 - 100 = 0$ **23.** $2x^2 = 32$

7–7 Objective: To factor perfect square trinomials.

Factor.

24. $x^2 + 14x + 49$ **25.** $2x^2 - 12x + 18$

26. Solve $x^2 - 12x + 36 = 0$.

7–8 Objective: To factor trinomials of the form $x^2 - bxy + cy^2$ with c positive.

Factor.

27. $x^2 + 5x + 6$ **28.** $x^2 - 7xy + 12y^2$

29. Solve $x^2 - 12x + 20 = 0$.

7–9 Objective: To factor trinomials of the form $x^2 + bxy + cy^2$ with c negative.

Factor.

30. $b^2 - 2b - 8$ **31.** $a^2 + 7a - 8$

Solve.

32. $x^2 - 5x - 6 = 0$ **33.** $x^2 + x = 20$

7–10 Objective: To factor trinomials of the form $ax^2 + bxy + cy^2$.

Factor.

34. $2x^2 + 11x + 5$ **35.** $3x^2 + xy - 2y^2$

Solve.

36. $2x^2 - 3x + 1 = 0$ **37.** $2x^2 = 3x + 5$

Mathematics and Your Future

Strong preparation in mathematics is important for both boys and girls. Nine out of ten females will work fulltime at some point in their lives. Many will become heads of households or be major contributors in two-earner families. By continuing to study mathematics you will improve your earning capacity.

■ CHAPTER 7 SELF-TEST

7–1 State the prime factorization.

    1. 252   **2.** 600

7–2 Find the GCF.

    3. 252 and 600   **4.** $36a^2b^3c$ and $84ab^2c^2$

7–3 Solve.

    5. $y(y + 3) = 0$ **6.** $(2x - 1)(x + 5) = 0$

7–5,
7–6 Expand and simplify.

    7. $3x(5x - 9)$ **8.** $(2x + 7y)(2x - 7y)$

    9. $(3a - 11)(2a + 9)$ **10.** $(t - 9)^2$

7–4–
7–6 Factor completely.

    11. $5x^2 - 35x$ **12.** $9a^2 - 4$

    13. $y^2 + 10y + 25$ **14.** $2w^2 - 12w + 18$

7–7– **15.** $x^2 + 9x + 20$ **16.** $c^2 + 5c - 24$
7–10 **17.** $a^2 - ab - 6b^2$ **18.** $r^2 - 5r + 6$

    19. $2x^2 - 3x - 54$

7–4–
7–9 Solve.

    20. $5a^2 = 125$ **21.** $b^2 - 14b + 49 = 0$

    22. $x^2 - 7x + 10 = 0$ **23.** $w^2 + 7w = 78$

    24. The square of a number added to its double is 143. Find the number.

    25. A 5-inch by 8-inch picture has a border of uniform width. The total area of the border and picture is 130 in². What is the width of the border?

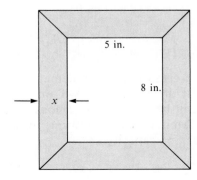

■ PRACTICE FOR COLLEGE ENTRANCE TESTS

1.

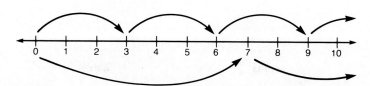

In the figure above, arcs are drawn above the number line every 3 units starting at 0 and arcs are drawn below the number line every 7 units starting at 0. At which of the following points on the number line will the arcs meet?

   A. 101   **B.** 102   **C.** 103   **D** 104   **E.** 105

2. If $a = 5 - t$, $b = 5 + t$, and $t^2 = 9$, then $ab =$ ___?___.

   A. 4   **B.** 7   **C.** 16   **D.** 34   **E.** 64

3. The number n is a positive whole number whose only positive whole number factors are 1 and n. If $15n$ is divisible by 10, then $n =$ ___?___.

   A. 2   **B.** 3   **C.** 9   **D.** 10   **E.** 13

4. If $xy = 3$ and $(x + y)^2 = 16$, then $x^2 + y^2 =$ ___?___.

   A. 6   **B.** 7   **C.** 9   **D.** 10   **E.** 13

5. If $x - 2 = 5$, then $(x - 1)^2 =$ ___?___.

   A. 4   **B.** 9   **C.** 16   **D.** 25   **E.** 36

6. If x is an even integer and n is the number of distinct prime factors of x, then the number of distinct prime factors of $2x$ is ___?___.

   A. $n - 1$ **B.** n **C.** $n + 1$ **D.** x **E.** $2x$

7. If n is a positive integer and $n^2 - n = k$, then which of the following could be the value of k?

   A. 42   **B.** 43   **C.** 44   **D.** 45   **E.** 46

8. $35^2 - 2(35)(15) + 15^2 =$ ___?___.

   A. 225   **B.** 400   **C.** 625   **D.** 1225   **E.** 2500

9. If $a + b = 4$ and $a - b = 2$, then $a^2 - b^2 =$ ___?___.

   A. 4   **B.** 8   **C.** 12   **D.** 16   **E.** 20

10. If $r = 3t + 2$ and $s = 9t^2$, then what is s in terms of r?

   A. $(r - 2)^2$ **B.** $(4 + 2)^2$ **C.** $3(r - 2)^2$ **D.** $(r + 2)^2$ **E.** $9(r - 2)^2$

11. What fraction of the number of integer multiples of 5 between 1 and 49 are also multiples of 3?

   A. $\dfrac{3}{49}$ **B.** $\dfrac{5}{49}$ **C.** $\dfrac{1}{3}$ **D.** $\dfrac{5}{9}$ **E.** $\dfrac{3}{5}$

8 Algebraic Fractions and Applications

By the late 1700s, French paper makers and scientists began experimenting with hot air balloons. The Montgolfier brothers discovered that hot air can cause a paper bag to rise. They were successful in launching the first hot air balloon on June 4, 1783.

Later that year, a French chemist, Jacques Charles, began experimenting with hydrogen and launched the first hydrogen balloon on August 27, 1783. Charles went on to discover nearly every essential of modern balloon design.

His work with balloons led to the discovery of the law that bears his name, the Charles Law, as follows:

The volume V of a gas under constant pressure is directly related to its temperature T in degrees Kelvin:

$$V = kT \quad \text{or} \quad \frac{V_1}{V_2} = \frac{T_1}{T_2}$$

8-1 Simplifying Algebraic Fractions

Preview

The Wizard of Odds performed number feats. In one trick the Wizard would ask a volunteer to do the following:

- Think of a number other than 1.

- Square the number and subtract 1.

- Divide this difference by 1 less than the original number.

When told the result, the Wizard could immediately give the original number. Copy and complete this table to determine how the Wizard was so quick with the answer.

Number	2	3	4	5	6	7	8	9	10
Result	?	?	?	?	?	?	?	?	?

The trick used by the Wizard can be explained by simplifying an algebraic fraction.

■ LESSON

Fractions that contain variables are called **algebraic fractions.** These are examples of algebraic fractions.

$$\frac{x}{5} \qquad \frac{7}{y} \qquad \frac{3+a}{2+a} \qquad \frac{6n}{7n}$$

Recall that division by 0 is not defined. Therefore, the values of the variables in algebraic fractions must be restricted so that the denominators of the fractions do not equal 0. Consider, for example, this fraction:

$$\frac{3}{x^2 + 2x}$$

To determine the restrictions on the value of x:

Solve the equation. $x^2 + 2x = 0$
Factor and use the $x(x + 2) = 0$
zero-product property. $x = 0$ or $x + 2 = 0$
                                     $x = 0$ or $x = -2$

In order for the denominator not to be 0, these restrictions must be placed on the values of x:

$$x \neq 0, \qquad x \neq -2$$

Values of the variables for which a fraction is defined are called **admissible values.** In the example above, the set of admissible values is the set of all real numbers x, except $x = 0$ and $x = -2$.

Example 1 State the restrictions on the variables in this algebraic fraction.

$$\frac{b}{a^2 - ab}$$

Solution Find the values of a and b that make the denominator 0.

$$a^2 - ab = 0$$
$$a(a - b) = 0$$
$$a = 0 \quad \text{or} \quad a = b$$

Answer These restrictions must be placed on a and b: $a \neq 0$, $a \neq b$.

If two fractions are not equivalent, we can show that fact by finding admissible values that give the fractions different values.

Example 2 Show that $\frac{x + 1}{3x + 1}$ and $\frac{x}{3x}$ are not equivalent.

Solution Substitute 2 for x and simplify.

$$\frac{x + 1}{3x + 1} \qquad\qquad \frac{x}{3x}$$

$$\frac{2 + 1}{3(2) + 1} \qquad\qquad \frac{2}{3 \cdot 2}$$

$$\frac{3}{7} \qquad\qquad \frac{2}{6} \text{ or } \frac{1}{3}$$

Answer Since $\frac{3}{7} \neq \frac{1}{3}$, $\frac{x + 1}{3x + 1}$ and $\frac{x}{3x}$ are not equivalent.

If two fractions are **equivalent fractions,** we can show that fact by transforming one fraction into the other, using the properties given below.

Multiplicaton and Division Properties for Equivalent Fractions

For all admissible values of a, b, and c,

$$\frac{a}{b} = \frac{ac}{bc} \quad \text{and} \quad \frac{a}{b} = \frac{a \div c}{b \div c}$$

These same properties can also be used to simplify algebraic fractions.

Example 3 Simplify. $\dfrac{3x + 6y}{21}$

Solution *Factor the numerator and denominator.*

$$\frac{3x + 6y}{21} = \frac{3(x + 2y)}{3 \cdot 7}$$

Divide both numerator and denominator by the common factor 3.

$$= \frac{\overset{1}{3}(x + 2y)}{\underset{1}{3} \cdot 7}$$

Answer $\dfrac{x + 2y}{7}$

$$= \frac{x + 2y}{7}$$

Example 4 Simplify. $\dfrac{2a - 2b}{6a^2 - 6b^2}$

Solution *Factor the numerator and denominator.* $\dfrac{2a - 2b}{6a^2 - 6b^2} = \dfrac{2(a - b)}{2 \cdot 3(a - b)(a + b)}$

Use the factored forms to determine the necessary restrictions on the values of the variables: $a \neq b, \ a \neq -b$

Eliminate the common factors.

$$= \frac{\overset{1}{2}(\overset{1}{a - b})}{\underset{1}{2} \cdot 3(\underset{1}{a - b})(a + b)}$$

$$= \frac{1}{3(a + b)}$$

Answer $\dfrac{1}{3(a + b)}, \ a \neq b, \ a \neq -b$

Note that factors that restrict the values of variables may be removed in the simplifying process. (For example, in Example 4, $a - b$ was removed from both numerator and denominator.) Therefore, restrictions on variables must be determined from the original factored expression, *not* the simplified result.

Example 5 Simplify. $\dfrac{x - 1}{1 - x}$

Solution Note that the denominator is the opposite of the numerator.

Factor. List the necessary restrictions on the value of x.

$$\frac{x - 1}{1 - x} = \frac{(x - 1)}{-1(x - 1)}, \ x \neq 1$$

Eliminate the common factor.

$$= \frac{\overset{1}{(x - 1)}}{-1\underset{1}{(x - 1)}}$$

$$= -1$$

Answer $-1, \ x \neq 1$

■ CLASSROOM EXERCISES

List any necessary restrictions on the value of the variable.

1. $\dfrac{7}{x}$

2. $\dfrac{3}{x+1}$

3. $\dfrac{5}{x(x-1)}$

4. $\dfrac{19+y}{x^2-4}$

5. $\dfrac{2a}{a-b}$

6. $\dfrac{17}{a^2+6a}$

7. $\dfrac{7b}{a^2+4}$

8. $\dfrac{2a-3}{5}$

State whether the following pairs of fractions are equivalent.

9. $\dfrac{x+1}{x+2}$ and $\dfrac{1}{2}$

10. $\dfrac{x}{2x}$ and $\dfrac{1}{2}$

11. $\dfrac{3x+2}{5x+2}$ and $\dfrac{3x}{5x}$

12. $\dfrac{3(x+2)}{5(x+2)}$ and $\dfrac{3}{5}$

Simplify. List any necessary restrictions on the variables.

13. $\dfrac{(a+b)(a-b)}{(a-b)(a-b)}$

14. $\dfrac{2c+2d}{5c+5d}$

15. $\dfrac{-(a-b)}{b-a}$

16. $\dfrac{x^2+2x+1}{2x+2}$

■ WRITTEN EXERCISES

List the values of x that are not admissible.

A **1.** $\dfrac{x-1}{x-3}$

2. $\dfrac{x-2}{x-4}$

3. $\dfrac{5}{2x-3}$

4. $\dfrac{4}{3x-5}$

5. $\dfrac{5x}{(x-2)(x-4)}$

6. $\dfrac{3x}{(x-1)(x+2)}$

7. $\dfrac{10}{x^2-5x+6}$

8. $\dfrac{8}{x^2-6x+8}$

List any necessary restrictions for the fractions to be defined.

9. $\dfrac{5}{a-5}$

10. $\dfrac{7}{7-a}$

11. $\dfrac{b}{b+3}$

12. $\dfrac{b}{b+4}$

13. $\dfrac{4}{a(b-2)}$

14. $\dfrac{8}{b(a-3)}$

15. $\dfrac{b}{a^2+3a}$

16. $\dfrac{a}{b^2+4b}$

State whether these pairs of fractions are equivalent for admissible values of x.

17. $\dfrac{3x}{4x}$ and $\dfrac{3}{4}$

18. $\dfrac{3x}{8x}$ and $\dfrac{3}{8}$

19. $\dfrac{5+x}{8+x}$ and $\dfrac{5}{8}$

20. $\dfrac{5+x}{6+x}$ and $\dfrac{5}{6}$

21. $\dfrac{1}{x-2}$ and $\dfrac{2}{2x-4}$

22. $\dfrac{1}{x-3}$ and $\dfrac{2}{2x-6}$

23. $\dfrac{1}{x+5}$ and $\dfrac{2}{x+10}$

24. $\dfrac{1}{x+4}$ and $\dfrac{3}{x+12}$

Simplify. List any necessary restrictions on the variables.

25. $\dfrac{5x}{15x}$

26. $\dfrac{6x}{16x}$

27. $\dfrac{4x^2}{15x^3}$

28. $\dfrac{12x}{15x^2}$

29. $\dfrac{(b+2)(b+3)}{(b+3)(b+4)}$

30. $\dfrac{(c+3)(c+4)}{(c+2)(c+3)}$

31. $\dfrac{2x+4}{5x+10}$

32. $\dfrac{6x+9}{9x+12}$

33. $\dfrac{x^2 + 5x}{2x^2 - 7x}$

34. $\dfrac{x^2 - 8x}{3x^2 + 5x}$

35. $\dfrac{5a^2}{a^3 + 2a^2 + a}$

36. $\dfrac{7a^2}{a^3 - 5a^2 + 6a}$

37. $\dfrac{c + 3}{c^2 + 5c + 6}$

38. $\dfrac{c - 2}{c^2 - 5c + 6}$

39. $\dfrac{3t - 12}{t^2 + t - 20}$

40. $\dfrac{2t + 10}{t^2 + t - 20}$

41. $\dfrac{a^2 + a - 2}{a^2 + 3a + 2}$

42. $\dfrac{a^2 - a - 2}{a^2 + 3a + 2}$

43. $\dfrac{2x^2 + 3x}{2x^2 + 5x + 3}$

44. $\dfrac{12x - 4}{3x^2 + 5x - 2}$

The values of the area and width of a rectangle are given. Express the value of the length of the rectangle in simplest form.

45. area: $x^2 + 3x$;
width: x

46. area: $5y + 20$;
width: 5

47. area: $a^2 - 4$;
width: $a - 2$

48. area: $b^2 - 9$;
width: $b - 3$

49. area: $a^2 + 4a - 12$;
width: $a - 2$

50. area: $a^2 + 3a - 18$;
width: $a - 3$

51. area: $a^2 + 8a + 16$;
width: $a + 4$

52. area: $a^2 + 6a + 9$;
width: $a + 3$

Evaluate the fractions for the given values of x and state whether the fractions are equivalent.

B　**53.**

x	$\dfrac{2 + x}{6 + x}$	$\dfrac{1 + x}{3 + x}$
2	?	?
1	?	?
0	?	?
-1	?	?

54.

x	$\dfrac{x - 3}{x^2 - 9}$	$\dfrac{1}{x + 3}$
2	?	?
1	?	?
0	?	?
-1	?	?

55.

x	$\dfrac{x + 3}{x^2 + x - 6}$	$\dfrac{x}{x^2 - 2x}$
-2	?	?
-1	?	?
1	?	?
3	?	?

56.

x	$\dfrac{x + 4}{x^2 + 2x - 8}$	$\dfrac{x}{x^2 - 2x}$
-3	?	?
-1	?	?
1	?	?
3	?	?

The following exercises list the formulas for the volume and surface area of some solids.

   a. Write a fraction for the ratio of the volume to the total surface area of each solid.

   b. Simplify the fractions.

57. Cube

$V = s^3$

$S = 6s^2$

58. Square pyramid

$V = \dfrac{1}{3}b^2h$

$S = b^2 + 2ab$

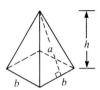

59. Square prism

$V = hs^2$

$S = 2s^2 + 4hs$

60. Cylinder

$V = \pi r^2 h$

$S = 2\pi r^2 + 2\pi rh$

61. Sphere

$V = \dfrac{4}{3}\pi r^3$

$S = 4\pi r^2$

62. Circular torus

$V = 2\pi^2 R r^2$

$S = 4\pi^2 Rr$

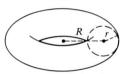

63. Write an algebraic fraction to explain the number trick presented in the Preview. Then write the fraction in simplest form.

For each of the following fractions, (a) state the values of the variable for which the fraction is undefined, (b) state the values of the variable that make the fraction equal to zero, and (c) simplify the fraction.

C **64.** $\dfrac{a^2 - a - 2}{a^2 + 3a + 2}$

65. $\dfrac{a^2 + 3a}{3a + 9}$

66. $\dfrac{b^2 + b - 6}{b^2 - b - 6}$

67. $\dfrac{a^2 - 16}{a^2 + 8a + 16}$

68. $\dfrac{3b^2 - 14b + 8}{3b^2 + 14b + 8}$

69. $\dfrac{2b^2 + 7b - 15}{2b^2 - b - 6}$

■ REVIEW EXERCISES

Substitute and simplify.

[1–2]

a	b	c	d
12	18	2	3

1. $\dfrac{a}{b} \cdot \dfrac{c}{d}$

2. $\dfrac{a}{b} \div \dfrac{c}{d}$

3. $\dfrac{a}{c} + \dfrac{b}{c}$

4. $\dfrac{a + b}{c}$

5. $\dfrac{a}{d} - \dfrac{b}{d}$

6. $\dfrac{a - b}{d}$

Preview

Simplify this product:

$$\frac{21}{22} \cdot \frac{33}{35}$$

In simplifying did you multiply first then remove the common facors? Or did you remove the common factors first and then multiply?

a. $\dfrac{21}{22} \cdot \dfrac{33}{35} = \dfrac{693}{770} = \dfrac{9 \cdot \overset{1}{\cancel{77}}}{10 \cdot \underset{1}{\cancel{77}}} = \dfrac{9}{10}$

b. $\dfrac{\overset{3}{\cancel{21}}}{\underset{2}{\cancel{22}}} \cdot \dfrac{\overset{3}{\cancel{33}}}{\underset{5}{\cancel{35}}} = \dfrac{9}{10}$

c. $\dfrac{21}{22} \cdot \dfrac{33}{35} = \dfrac{3 \cdot \overset{1}{\cancel{7}} \cdot 3 \cdot \overset{1}{\cancel{11}}}{2 \cdot \underset{1}{\cancel{11}} \cdot 5 \cdot \underset{1}{\cancel{7}}} = \dfrac{9}{10}$

In this lesson you will multiply fractions using these methods.

■ LESSON

Algebraic fractions are multiplied in the same way as arithmetic fractions.

Definition: Multiplication of Fractions

For all real numbers a and c, and all nonzero real numbers b and d,

$$\frac{a}{b} \cdot \frac{c}{d} = \frac{ac}{bd}, \ b \neq 0, \ d \neq 0$$

Suppose that we wish to simplify the product

$$\frac{5ab^2}{3} \cdot \frac{6}{a^2b}$$

Here are two methods of doing so.

Method 1. Multiply first and then eliminate common factors.

Multiply. State restrictions. $\quad \dfrac{5ab^2}{3} \cdot \dfrac{6}{a^2b} = \dfrac{30ab^2}{3a^2b}, \quad a \neq 0, \quad b \neq 0$

Factor the numerator and denominator and eliminate common factors. $\quad = \dfrac{2 \cdot \overset{1}{\cancel{3}} \cdot 5 \cdot \overset{1}{\cancel{a}} \cdot \overset{1}{\cancel{b}} \cdot b}{\underset{1}{\cancel{3}} \cdot \underset{1}{\cancel{a}} \cdot a \cdot \underset{1}{\cancel{b}}}$

The simplified product is as follows: $\quad = \dfrac{10b}{a}, \quad a \neq 0, \ b \neq 0$

Method 2. Eliminate common factors first and then multiply.

Factor the numerators and denominators.

$$\frac{5ab^2}{3} \cdot \frac{6}{a^2b} = \frac{5 \cdot a \cdot b \cdot b}{3} \cdot \frac{2 \cdot 3}{a \cdot a \cdot b}$$

State restrictions. $a \neq 0, \quad b \neq 0$

Eliminate common factors.

$$= \frac{5 \cdot \overset{1}{\cancel{a}} \cdot \overset{1}{\cancel{b}} \cdot b}{\cancel{3}} \cdot \frac{2 \cdot \cancel{3}}{\cancel{a} \cdot a \cdot \cancel{b}}$$

Multiply. $= \dfrac{10b}{a}, \quad a \neq 0, \quad b = 0$

Either of these methods can be used to simplify products.

> **Note:** From this point on in the text, you can assume that unless otherwise stated, the variables are restricted so that no denominator of an algebraic fraction equals zero.

Example 1   Simplify. $\dfrac{x^2}{y}(3y^2 + 2y)$

    Solution   Use the distributive property to expand.

$$\frac{x^2}{y}(3y^2 + 2y) = \frac{x^2}{y}(3y^2) + \frac{x^2}{y}(2y)$$

             Factor each term and eliminate common factors.

$$= \frac{x \cdot x}{\cancel{y}} \cdot \frac{3 \cdot y \cdot \overset{1}{\cancel{y}}}{1} + \frac{x \cdot x}{\cancel{y}} \cdot \frac{2 \cdot \overset{1}{\cancel{y}}}{1}$$

    Answer $3x^2y + 2x^2$

$$= 3x^2y + 2x^2$$

Example 2   Simplify. $\dfrac{x}{4y} \cdot \dfrac{y^2 - y}{xy - x}$

    Solution   Factor.

$$\frac{x}{4y} \cdot \frac{y^2 - y}{xy - x} = \frac{x}{4y} \cdot \frac{y(y - 1)}{x(y - 1)}$$

             Eliminate common factors.

$$= \frac{\cancel{x}}{4\cancel{y}} \cdot \frac{\cancel{y}(\cancel{y - 1})}{\cancel{x}(\cancel{y - 1})} = \frac{1}{4}$$

    Answer $\dfrac{1}{4}$

Example 3 Simplify. $\dfrac{x^2-1}{x^2+1} \cdot \dfrac{1}{x-1}$

Solution $\dfrac{x^2-1}{x^2+1} \cdot \dfrac{1}{x-1} = \dfrac{\overset{1}{\cancel{(x-1)}}(x+1)}{x^2+1} \cdot \dfrac{1}{\underset{1}{\cancel{x-1}}}$

$$= \dfrac{x+1}{x^2+1}$$

Note that the expressions $x+1$ and x^2+1 have no common factors other than 1. Therefore, $\dfrac{x+1}{x^2+1}$ cannot be simplified further.

Answer $\dfrac{x+1}{x^2+1}$

Dividing by a real number is equivalent to multiplying by its reciprocal. So we can write the following definition of division of fractions.

Definition: Division of Fractions

For all real numbers a, and all nonzero real numbers b, c, and d,

$$\frac{a}{b} \div \frac{c}{d} = \frac{a}{b} \cdot \frac{d}{c}, \quad b \neq 0, \quad c \neq 0, \quad d \neq 0$$

Any division expression can be transformed into an equivalent multiplication expression. Then that product can be simplified by the previous methods.

Example 4 Simplify. $\dfrac{3x}{5y} \div \dfrac{y}{15x}$

Solution *Rewrite as a multiplication expression.* $\qquad \dfrac{3x}{5y} \div \dfrac{y}{15x} = \dfrac{3x}{5y} \cdot \dfrac{15x}{y}$

Eliminate common factors and simplify. $\qquad = \dfrac{3x}{\underset{1}{\cancel{5}y}} \cdot \dfrac{\overset{3}{\cancel{15}x}}{y} = \dfrac{9x^2}{y^2}$

Answer $\dfrac{9x^2}{y^2}$

Example 5 Simplify. $\dfrac{2a+2b}{a^2+ab} \div \dfrac{4}{a}$

Solution *Rewrite as a multiplication expression.*

$$\dfrac{2a+2b}{a^2+ab} \div \dfrac{4}{a} = \dfrac{2a+2b}{a^2+ab} \cdot \dfrac{a}{4}$$

Factor and eliminate common factors.

$$= \dfrac{\overset{1}{\cancel{2}}\overset{1}{\cancel{(a+b)}}}{a\underset{1}{\cancel{(a+b)}}} \cdot \dfrac{\overset{1}{\cancel{a}}}{2 \cdot 2} = \dfrac{1}{2}$$

Answer $\dfrac{1}{2}$

■ CLASSROOM EXERCISES

Simplify.

1. $\dfrac{x}{y} \cdot \dfrac{y}{x}$

2. $2x\left(\dfrac{3}{x} + \dfrac{5}{2x}\right)$

3. $\dfrac{3(x-2)}{4} \cdot \dfrac{2}{5(x-2)}$

4. $\dfrac{c+3d}{4} \cdot \dfrac{12}{2c+6d}$

5. $\dfrac{6x^2}{y} \div \dfrac{2y}{3x}$

6. $\dfrac{ab+b^2}{4a+4b} \div \dfrac{b}{4}$

■ WRITTEN EXERCISES

Write each product in simplest form.

A

1. $\dfrac{a}{b} \cdot \dfrac{b}{a^2}$

2. $\dfrac{a}{b^2} \cdot \dfrac{b}{a}$

3. $\dfrac{x}{3} \cdot \dfrac{6}{x^2}$

4. $\dfrac{y}{2} \cdot \dfrac{8}{y^3}$

5. $\dfrac{a^2}{3} \cdot \dfrac{9}{a}$

6. $\dfrac{s^3}{2} \cdot \dfrac{4}{s}$

7. $\dfrac{2x}{5} \cdot \dfrac{3}{x}$

8. $\dfrac{4a}{3} \cdot \dfrac{2}{3a}$

9. $\dfrac{5x}{y^2} \cdot xy$

10. $\dfrac{3x^2}{y^2} \cdot xy$

11. $12\left(\dfrac{a}{3} + \dfrac{b}{4}\right)$

12. $16\left(\dfrac{x}{4} - \dfrac{y}{8}\right)$

13. $2x\left(\dfrac{3}{x} + \dfrac{5}{2}\right)$

14. $4a\left(\dfrac{3}{4a} + \dfrac{4}{a}\right)$

15. $\dfrac{x}{y}(3y^2 + y)$

16. $\dfrac{2x}{y}(y^2 + 3y)$

17. $\dfrac{x}{3}\left(\dfrac{2}{x} + \dfrac{6}{x^2}\right)$

18. $\dfrac{a^2}{4}\left(\dfrac{2}{a} + \dfrac{8}{a^3}\right)$

19. $\dfrac{a(a-b)}{b} \cdot \dfrac{b}{2(a-b)}$

20. $\dfrac{a}{b(a+b)} \cdot \dfrac{3(a+b)}{a}$

Write each quotient in simplest form.

21. $\dfrac{3a}{5b} \div \dfrac{9a}{10a}$

22. $\dfrac{3a}{4b} \div \dfrac{12a}{5b}$

23. $\dfrac{a^2}{2b} \div \dfrac{a}{b}$

24. $\dfrac{x^3}{3y} \div \dfrac{x^2}{y}$

25. $\dfrac{2x^2}{3y} \div \dfrac{2y}{3x^2}$

26. $\dfrac{3a^2}{2b} \div \dfrac{3b}{2a^2}$

27. $\dfrac{5c^2d}{2d^2} \div \dfrac{3cd}{c}$

28. $\dfrac{4a^2b}{3a} \div \dfrac{2a}{3b}$

29. $\dfrac{a(a-b)}{b(a+b)} \div \dfrac{a-b}{ab}$

30. $\dfrac{a(a+b)}{a-b} \div \dfrac{ab}{b(a-b)}$

31. $\dfrac{2(x+3)}{(x-4)^2} \div \dfrac{4}{x-4}$

32. $\dfrac{2(x+3)^2}{x-4} \div \dfrac{4(x+3)}{(x-4)^2}$

The values of the length and width of a rectangle are given. Find the area of each rectangle.

33. length: $\dfrac{2a}{b}$

width: $\dfrac{b}{a^2}$

34. length: $\dfrac{b}{3a}$

width: $\dfrac{a^2}{b^2}$

35. length: $\dfrac{2x^2}{y}$

width: $\dfrac{3y^2}{x}$

36. length: $\dfrac{a^2b}{3}$

width: $\dfrac{5}{ab}$

The values of the area and width of a rectangle are given. Find the length of each rectangle.

37. area: $\dfrac{2x}{y}$

width: $\dfrac{xy}{3}$

38. area: $\dfrac{3}{2}$

width: $\dfrac{3x}{y}$

39. area: $\dfrac{x}{2}$

width: $\dfrac{2x}{y}$

40. area: $\dfrac{5ab^2}{2}$

width: $\dfrac{10ab}{3}$

An object's average speed (in meters per second) and the time (in seconds) it travels are given. Find the distance traveled.

B **41.** speed: $\dfrac{t^2 - 4}{4}$

time: $\dfrac{12}{t - 2}$

42. speed: $(t + 3)(t + 4)$

time: $\dfrac{2}{t + 3} + \dfrac{5}{t + 4}$

An object's average speed (in meters per second) and the distance (in meters) it travels are given. Find the time traveled.

43. speed: $\dfrac{x + 5}{x + 3}$

distance: $\dfrac{x + 5}{x + 4}$

44. speed: $\dfrac{x + 2}{x - 3}$

distance: $\dfrac{x^2 - x - 6}{x^2 - 2x - 3}$

Find the area of each triangle. Assume that the dimensions are given in centimeters.

45. base: $\dfrac{4a + 12}{a - 2}$

altitude: $\dfrac{a^2 + 2a - 8}{2a + 6}$

46. base: $\dfrac{x + 2}{x + 1}$

altitude: $\dfrac{x^2 - 2x - 3}{x^2 + 5x + 6}$

47. base: $\dfrac{2x^2 + 7x + 6}{x + 6}$

altitude: $\dfrac{6x - 24}{2x^2 - 5x - 12}$

48. base: $\dfrac{2a + 2b}{a - b}$

altitude: $\dfrac{a - b}{a^2 + 3ab + 2b^2}$

Write each expression in simplest form.

49. $\dfrac{x^2 - 3x - 4}{4} \cdot \dfrac{8x - 16}{x^2 - x - 2}$

50. $\dfrac{y^2 - 4}{9} \div \dfrac{y^2 + 2y}{3y}$

51. $\dfrac{5r + 5}{r + 3} \cdot \dfrac{2r + 6}{r^2 + 2r + 1}$

52. $\dfrac{1}{x^2 + 2x - 3} \div \dfrac{x + 3}{2x - 2}$

53. $\dfrac{10}{a^2 + 2ab + b^2} \div \dfrac{5}{a^2 - b^2}$

54. $\dfrac{p^2 + 2p}{p^2 + 5p + 6} \cdot \dfrac{p^2 + 3p}{4}$

55. $\dfrac{s}{s - 3} \cdot \dfrac{s^2 - 9}{3s^2 + 9s}$

56. $\dfrac{2x^2y - 8xy}{y - 2} \div \dfrac{xy - 4y}{2y - 4}$

Find the volume of each prism. Assume that the dimensions are given in centimeters.

C **57.**

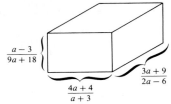

$\dfrac{a-3}{9a+18}$

$\dfrac{3a+9}{2a-6}$

$\dfrac{4a+4}{a+3}$

58.

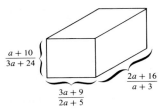

$\dfrac{a+10}{3a+24}$

$\dfrac{2a+16}{a+3}$

$\dfrac{3a+9}{2a+5}$

59.

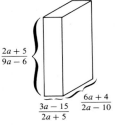

$\dfrac{2a+5}{9a-6}$

$\dfrac{3a-15}{2a+5}$ $\dfrac{6a+4}{2a-10}$

60.

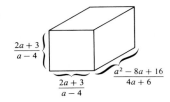

$\dfrac{2a+3}{a-4}$

$\dfrac{2a+3}{a-4}$ $\dfrac{a^2-8a+16}{4a+6}$

The values of the volume, width, and height of a prism are given. Find the length of each prism.

61. volume: $\dfrac{2a+3}{a-4}$

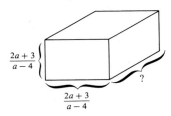

$\dfrac{2a+3}{a-4}$

$\dfrac{2a+3}{a-4}$?

62. volume: $\dfrac{a}{a+5}$

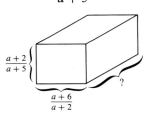

$\dfrac{a+2}{a+5}$

$\dfrac{a+6}{a+2}$?

63. volume: $\dfrac{x^2+7x+10}{x^2-2x}$

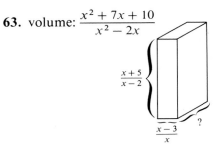

$\dfrac{x+5}{x-2}$

$\dfrac{x-3}{x}$?

64. volume: $\dfrac{x+1}{x^2-7x+6}$

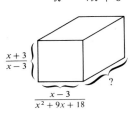

$\dfrac{x+3}{x-3}$

$\dfrac{x-3}{x^2+9x+18}$?

■ REVIEW EXERCISES

Compute.

1. $\dfrac{3}{4} \cdot \dfrac{4}{5}$

2. $\dfrac{3}{4} \div \dfrac{4}{5}$

3. $\dfrac{3}{4} + \dfrac{4}{5}$

4. $\dfrac{3}{4} - \dfrac{4}{5}$

5. $4\left(\dfrac{1}{3}\right) + 5\left(\dfrac{1}{3}\right)$

6. $\dfrac{4+5}{\dfrac{1}{3}}$

Preview Fractional notation

The evolution of terms and symbols for fractions is an interesting account of how such numbers were considered by early mathematicians. The Arabic word for fraction, "al-kasr," comes from the verb "to break." Early English writers referred to fractions as "broken numbers." Authors of eighteenth-century American arithmetics called fractions "vulgar" or "common."

In the sixth century A.D., fractions were written with the numerator above the denominator but without a dividing bar. The vinculum (or fraction bar) was used by the Arab writers as early as the year 1000 in the forms

$$a/b \text{ and } \frac{a}{b}$$

but the modern notation was not commonly accepted until the thirteenth century.

■ LESSON

Algebraic fractions are added or subtracted in the same way as arithmetic fractions. If the fractions have the same denominators, the following definition is used.

> ### Definition: Addition and Subtraction of Fractions with Like Denominators
>
> For all real numbers a, b, and c ($c \neq 0$)
>
> $$\frac{a}{c} + \frac{b}{c} = \frac{a+b}{c} \text{ and } \frac{a}{c} - \frac{b}{c} = \frac{a-b}{c}$$

For example, consider adding $\frac{x}{10}$ and $\frac{3x}{10}$.

Write the sum as a single fraction.
$$\frac{x}{10} + \frac{3x}{10} = \frac{x+3x}{10}$$

Simplify.
$$= \frac{\overset{2}{4x}}{\underset{5}{10}} = \frac{2x}{5}$$

The simplified sum is $\frac{2x}{5}$.

Example 1 Simplify. $\frac{a}{6} - \frac{b}{6}$

Solution Use the definition of subtraction to write the difference as a single fraction.

$$\frac{a}{6} - \frac{b}{6} = \frac{a-b}{6}$$

Answer $\frac{a-b}{6}$

Example 2 Simplify. $\dfrac{x^2}{x-1} - \dfrac{x}{x-1}$

　　　Solution　Write the difference as a single fraction.

$$\dfrac{x^2}{x-1} - \dfrac{x}{x-1} = \dfrac{x^2 - x}{x-1}$$

　　　　　　Simplify.

$$= \dfrac{x(\cancel{x-1})^{1}}{\cancel{x-1}_{1}}$$

$$= x$$

　　　Answer　x

Example 3 Simplify. $\dfrac{a(a+4)}{a+2} + \dfrac{4}{a+2}$

　　　Solution　$\dfrac{a(a+4)}{a+2} + \dfrac{4}{a+2} = \dfrac{a^2 + 4a + 4}{a+2}$

$$= \dfrac{\cancel{(a+2)^2}^{(a+2)}}{\cancel{a+2}_{1}}$$

$$= a + 2$$

　　　Answer　$a + 2$

■ CLASSROOM EXERCISES

Simplify.

1. $\dfrac{7}{8}b + \dfrac{1}{8}b$　　　　　**2.** $\dfrac{5x}{9} - \dfrac{2x}{9}$　　　　　**3.** $\dfrac{4}{a} + \dfrac{1}{a}$

4. $\dfrac{2c}{2c+4} + \dfrac{4}{2c+4}$　　　**5.** $\dfrac{x}{x^2-1} - \dfrac{1}{x^2-1}$　　　**6.** $\dfrac{a}{a^2-1} + \dfrac{1}{a^2-1}$

■ WRITTEN EXERCISES

Simplify.

A　**1.** $\dfrac{3x}{2} + \dfrac{9x}{2}$　　　　　**2.** $\dfrac{x}{2} + \dfrac{5x}{2}$　　　　　**3.** $\dfrac{5}{a} + \dfrac{1}{a}$

4. $\dfrac{2}{a} + \dfrac{3}{a}$　　　　　**5.** $\dfrac{11x}{3} - \dfrac{2x}{3}$　　　　**6.** $\dfrac{10x}{3} - \dfrac{4x}{3}$

7. $\dfrac{7}{2y} - \dfrac{4}{2y}$　　　　**8.** $\dfrac{10}{2y} - \dfrac{3}{2y}$　　　　**9.** $\dfrac{b}{b+2} + \dfrac{1}{b+2}$

10. $\dfrac{b}{b-2} + \dfrac{2}{b-2}$　　　**11.** $\dfrac{c}{c-7} - \dfrac{3}{c-7}$　　　**12.** $\dfrac{c}{c-5} - \dfrac{8}{c-5}$

13. $\dfrac{2s}{s-1} - \dfrac{2}{s-1}$　　　**14.** $\dfrac{3s}{s-4} - \dfrac{12}{s-4}$　　　**15.** $\dfrac{4x}{x+1} + \dfrac{4}{x+1}$

16. $\dfrac{5x}{x+2} + \dfrac{10}{x+2}$

17. $\dfrac{a+1}{2a+6} + \dfrac{a+7}{2a+6}$

18. $\dfrac{2a+1}{3a+6} + \dfrac{a+8}{3a+6}$

19. $\dfrac{5a+11}{3a+6} - \dfrac{a+3}{3a+6}$

20. $\dfrac{6a+13}{2a+6} - \dfrac{3a+4}{2a+6}$

21. $\dfrac{c^2+7c}{c^2-4c} - \dfrac{c^2+2c}{c^2-4c}$

22. $\dfrac{r^2+5r}{r^2-3r} - \dfrac{r^2+r}{r^2-3r}$

23. $\dfrac{s^2+2s}{s^2-9} + \dfrac{2s+3}{s^2-9}$

24. $\dfrac{s^2+s}{s^2-4} + \dfrac{2s+2}{s^2-4}$

Simplify. State the values of the variables that are not admissible.

B **25.** $\dfrac{2c}{c+3} + \dfrac{6}{c+3}$

26. $\dfrac{3h}{h-4} - \dfrac{12}{h-4}$

27. $\dfrac{6h}{h^2-1} - \dfrac{6}{h^2-1}$

28. $\dfrac{2k+3}{k^2-k} + \dfrac{5k-3}{k^2-k}$

29. $\dfrac{3d+7}{d-4} - \dfrac{3d+7}{d-4}$

30. $\dfrac{7p+5}{(p+2)^2} + \dfrac{p+11}{(p+2)^2}$

31. $\dfrac{q^2-3q}{4q} + \dfrac{q^2+6q}{4q}$

32. $\dfrac{v^2-6}{3v^2} - \dfrac{2v^2-6}{3v^2}$

33. $\dfrac{x^2}{x^2-4} - \dfrac{x}{x^2-4} - \dfrac{6}{x^2-4}$

34. $\dfrac{x(x-4)}{x-2} + \dfrac{4}{x-2}$

35. $\dfrac{a(a+3)}{a+1} + \dfrac{2}{a+1}$

36. $\dfrac{x}{x-1} + \dfrac{1}{1-x}$

C **37.** $\dfrac{x+8}{x^2-5x+6} + \dfrac{x^2-7x}{x^2-5x+6}$

38. $\dfrac{x^2-3}{x^2+2x-8} + \dfrac{x-9}{x^2+2x-8}$

39. $\dfrac{a^2+8a+5}{2a^2+5a-12} - \dfrac{4a+1}{2a^2+5a-12}$

40. $\dfrac{8a^2+3a+3}{6a^2+a-2} - \dfrac{2a^2-4a+1}{6a^2+a-2}$

41. $\dfrac{6}{2a-3} - \dfrac{4a}{3-2a}$

42. $\dfrac{b^2}{2-b} + \dfrac{2b}{b-2}$

■ REVIEW EXERCISES

Compute.

1. $12\left(\dfrac{1}{4} + \dfrac{1}{3}\right)$

2. $10\left(\dfrac{3}{5} - \dfrac{1}{2}\right)$

3. What is the least common denominator of $\dfrac{7}{8}$ and $\dfrac{5}{12}$?

Solve.

4. $\dfrac{1}{3}x - 5 = \dfrac{3}{4}$

5. $\dfrac{x}{100} = \dfrac{4}{5}$

[3–7]

Substitute and simplify.

	a	b	c	d
	100	-50	-5	10

6. $\dfrac{a}{b}$ **7.** $\dfrac{ac}{bc}$ **8.** $\dfrac{(a+c)}{(b+c)}$ [2–5]

Self-Quiz 1

8–1 List any restrictions on the variables.

1. $\dfrac{2}{ab}$ **2.** $\dfrac{x(x-4)}{(x-2)(x+2)}$

Simplify. Assume that no denominators are equal to zero.

3. $\dfrac{3xy}{12x^2y}$ **4.** $\dfrac{5(a+2b)}{(a+2b)^2}$

8–2 **5.** $\dfrac{12a^2b}{5} \cdot \dfrac{25}{3ab^2}$ **6.** $\dfrac{x(y-1)}{xy} \cdot \dfrac{y^2}{y(y-1)}$

7. $\dfrac{2m}{7n^2} \div \dfrac{4m^2}{21n}$ **8.** $\dfrac{9(r+2)}{2} \div \dfrac{3(r+2)}{8r}$

8–3 **9.** $\dfrac{6}{x} - \dfrac{3}{x} + \dfrac{2}{x}$ **10.** $\dfrac{c}{c-2} - \dfrac{2}{c-2}$

11. $\dfrac{2b}{b+4} + \dfrac{2-b}{b+4}$ **12.** $\dfrac{x+7}{2x+6} - \dfrac{x+1}{2x+6}$

EXTENSION   Using a calculator to determine equivalent fractions

You can use a calculator to determine whether two fractions $\dfrac{a}{b}$ and $\dfrac{c}{d}$ are equivalent by following these steps.

1. Use the key sequence $a \ \times \ d \ \div \ b \ = \ $.
2. Look at the display. If the display shows the value of c, then the fractions are equivalent.

Use the steps above to decide whether these fractions are equivalent.

1. $\dfrac{117}{149}, \ \dfrac{585}{755}$ **2.** $\dfrac{221}{718}, \ \dfrac{884}{2872}$ **3.** $\dfrac{118}{127}, \ \dfrac{2714}{2921}$

Suppose the fractions in each pair are equivalent. Find the value of x.

4. $\dfrac{97}{107}, \ \dfrac{x}{3103}$ **5.** $\dfrac{37}{83}, \ \dfrac{x}{18426}$ **6.** $\dfrac{53530}{67670}, \ \dfrac{x}{67}$

8–4 Adding and Subtracting Fractions with Unlike Denominators

Preview

How would you simplify the following expression?

$$\frac{7}{12} + \frac{4}{15}$$

Most students use one of these methods to simplify this sum.

Use $12 \cdot 15$ as the common denominator.	Use 60 as the common denominator.

$$\frac{7}{12} + \frac{4}{15} = \frac{7 \cdot 15 + 4 \cdot 12}{12 \cdot 15} \qquad \frac{7}{12} + \frac{4}{15} = \frac{7}{12} \cdot \frac{5}{5} + \frac{4}{15} \cdot \frac{4}{4}$$

$$= \frac{105 + 48}{180} \qquad\qquad = \frac{35}{60} + \frac{16}{60}$$

$$= \frac{153}{180} \qquad\qquad = \frac{35 + 16}{60}$$

$$= \frac{17}{20} \qquad\qquad = \frac{51}{60}$$

$$\qquad\qquad\qquad\qquad = \frac{17}{20}$$

Which method would you use? Which method is simpler?

In this lesson you will use one of the above methods to add or subtract algebraic fractions.

■ LESSON

If fractions to be added or subtracted do not have common denominators, the following definition can be used.

Definition: Addition and Subtraction of Fractions with Unlike Denominators

For all real numbers a, b, c, and d ($b \neq 0$ and $d \neq 0$),

$$\frac{a}{b} + \frac{c}{d} = \frac{ad + bc}{bd} \quad \text{and} \quad \frac{a}{b} - \frac{c}{d} = \frac{ad - bc}{bd}, \; b \neq 0, \text{ and } d \neq 0$$

For example, consider this difference:

$$\frac{2x}{3y} - \frac{5}{3}$$

Write the difference as a single fraction.

$$\frac{2x}{3y} - \frac{5}{3} = \frac{2x(3) - 5(3y)}{3y(3)}$$

Simplify the numerator and denominator.

$$= \frac{6x - 15y}{9y}$$

Factor the numerator and denominator and simplify.

$$= \frac{\overset{1}{\cancel{3}}(2x - 5y)}{\underset{1}{\cancel{3}} \cdot 3y} = \frac{2x - 5y}{3y}$$

The difference is $\dfrac{2x - 5y}{3y}$.

The definition for adding or subtracting fractions with unlike denominators works for all fractions. However, the work can be done more efficiently if the **least common denominator,** rather than the product of the denominators, is used.

Definition: Least Common Denominator

The least common denominator of two or more algebraic fractions is the least common multiple of the denominators of the fractions.

To find the least common multiple of the denominators of two algebraic fractions, write the prime factorization of each denominator. Then identify the highest power of each prime factor and write the product of the highest powers.

Example 1 Write the least common denominator of $\dfrac{3}{8}$ and $\dfrac{5}{12}$.

Solution *Write the prime factorization of the denominators.*

$$8: 2^3$$
$$12: 2^2 \cdot 3$$

Identify the highest power of each prime factor.

$$8: 2^3$$
$$12: 2^2 \cdot 3$$

Write the product of these highest powers. $\quad 2^3 \cdot 3 = 24$

Answer 24

The same process can be used with algebraic fractions.

Example 2 Write the least common denominator of $\frac{5}{6x^2}$ and $\frac{2}{9x}$.

Solution

$$6x^2 = 2 \cdot 3 \cdot x^2$$
$$9x = 3^2 \cdot x$$

Write the product of the highest powers of the prime factors.

$$2 \cdot 3^2 x^2 = 18x^2$$

Answer $18x^2$

Fractions with unlike denominators can be added or subtracted by changing them to equivalent fractions that have the least common denominator as their denominators.

Example 3 Simplify. $\frac{5}{6x^2} - \frac{2}{9x}$

Solution *Find the least common denominator.* The least common denominator of $6x^2$ and $9x$ is $18x^2$.

Write equivalent fractions with $18x^2$ as a common denominator.

$$\frac{5}{6x^2} = \frac{5}{6x^2} \cdot \frac{3}{3} = \frac{15}{18x^2}$$

$$\frac{2}{9x} = \frac{2}{9x} \cdot \frac{2x}{2x} = \frac{4x}{18x^2}$$

Substitute and simplify.

$$\frac{5}{6x^2} - \frac{2}{9x} = \frac{15}{18x^2} - \frac{4x}{18x^2}$$

$$= \frac{15 - 4x}{18x^2}$$

Answer $\dfrac{15 - 4x}{18x^2}$

Example 4 Simplify. $x + \frac{5x}{3y}$

Solution $x + \dfrac{5x}{3y} = \dfrac{x}{1} + \dfrac{5x}{3y}$

$$= \frac{x(3y)}{3y} + \frac{5x}{3y}$$

$$= \frac{3xy + 5x}{3y}$$

Answer $\dfrac{3xy + 5x}{3y}$

Example 5 Simplify. $\dfrac{3}{a+b} + \dfrac{4}{a-b}$

Solution The least common denominator is $(a+b)(a-b)$.

$$\frac{3}{a+b} + \frac{4}{a-b} = \frac{3}{a+b} \cdot \frac{a-b}{a-b} + \frac{4}{a-b} \cdot \frac{a+b}{a+b}$$

$$= \frac{3(a-b) + 4(a+b)}{(a+b)(a-b)}$$

$$= \frac{3a - 3b + 4a + 4b}{(a+b)(a-b)}$$

$$= \frac{7a+b}{(a^2 - b^2)}$$

Answer $\dfrac{7a+b}{a^2 - b^2}$

■ CLASSROOM EXERCISES

State the least common denominator of the fractions.

1. $\dfrac{1}{x}, \dfrac{1}{2x}$ **2.** $\dfrac{3}{2a^2}, \dfrac{5}{6a}$ **3.** $\dfrac{1}{4b}, \dfrac{1}{6ab}$ **4.** $\dfrac{2}{2x^2y}, \dfrac{4}{2xy^2}$ **5.** $\dfrac{2}{x+1}, \dfrac{3}{x^2-1}$

Simplify.

6. $\dfrac{3}{2x} + \dfrac{2y}{x}$ **7.** $\dfrac{3}{x} + \dfrac{2}{x^2}$ **8.** $\dfrac{2x}{y} + \dfrac{2y}{x}$ **9.** $\dfrac{9}{5xy^2} - \dfrac{3}{2x^2y}$ **10.** $\dfrac{4}{a+3} - \dfrac{12}{(a+3)^2}$

■ WRITTEN EXERCISES

State the least common denominator of the fractions.

A   **1.** $\dfrac{5}{6}, \dfrac{2}{9}$ **2.** $\dfrac{1}{4}, \dfrac{5}{6}$ **3.** $\dfrac{2}{3}, \dfrac{1}{3x}$ **4.** $\dfrac{4}{5}, \dfrac{3}{5x}$

5. $\dfrac{3}{a}, \dfrac{2}{b}$ **6.** $\dfrac{4}{x}, \dfrac{1}{y}$ **7.** $\dfrac{1}{a^2}, \dfrac{2}{a}$ **8.** $\dfrac{3}{b^2}, \dfrac{5}{b}$

9. $\dfrac{x}{a^2b}, \dfrac{x}{ab^2}$ **10.** $\dfrac{a}{x^2y}, \dfrac{b}{xy^2}$

11. $\dfrac{5}{2a^3b}, \dfrac{7}{6ab^2}$ **12.** $\dfrac{6}{5a^2b^2}, \dfrac{1}{15a^2b}$

13. $\dfrac{x}{x+2}, \dfrac{x}{x^2-4}$ **14.** $\dfrac{5}{x-5}, \dfrac{5}{x^2-25}$

15. $\dfrac{3}{x^2+3x}, \dfrac{4}{2x}$ **16.** $\dfrac{5}{x^2-4x}, \dfrac{2}{3x}$

Simplify.

17. $\dfrac{5}{6} + \dfrac{2}{9}$

18. $\dfrac{1}{6} + \dfrac{3}{8}$

19. $\dfrac{5}{a} - \dfrac{2}{3a}$

20. $\dfrac{4}{a} - \dfrac{1}{2a}$

21. $\dfrac{5}{a} + \dfrac{2}{3}$

22. $\dfrac{4}{a} + \dfrac{1}{2}$

23. $\dfrac{4}{x} - \dfrac{3}{x^2}$

24. $\dfrac{7}{x^2} - \dfrac{5}{x}$

25. $\dfrac{x}{3} + \dfrac{x}{2}$

26. $\dfrac{x}{4} + \dfrac{x}{8}$

27. $\dfrac{x}{y} - \dfrac{y}{x}$

28. $\dfrac{a}{b} - \dfrac{b}{a}$

29. $\dfrac{3b}{a^2} - \dfrac{2b}{a}$

30. $\dfrac{2y}{x^2} - \dfrac{3y}{x}$

31. $\dfrac{6y}{5x} + \dfrac{7y}{10x^2}$

32. $\dfrac{3z}{7y^2} - \dfrac{2z}{y}$

Find the perimeter of each figure. Assume that the dimensions are given in centimeters.

33.
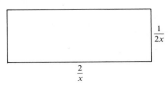

34.
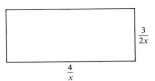

B **35.**
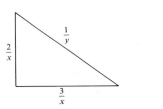

36.
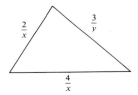

37.
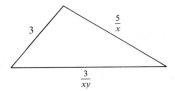

38.
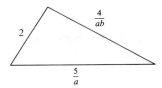

Simplify. State the values of the variable that are not admissible.

39. $\dfrac{3}{x+2} + \dfrac{2}{x}$

40. $\dfrac{3}{x-2} + \dfrac{3}{x+2}$

41. $\dfrac{x}{x+2} + \dfrac{3}{x+3}$

42. $\dfrac{3x+2}{x} + \dfrac{2x+4}{x+2}$

43. $\dfrac{5}{x} - \dfrac{3}{x+6}$

44. $\dfrac{2x}{x-2} - \dfrac{3}{x-3}$

45. $\dfrac{7}{x-5} - \dfrac{3}{x-2}$

46. $\dfrac{2x}{x+5} - \dfrac{2x}{x+4}$

47. $\dfrac{3}{2x} + \dfrac{2}{x+1}$

48. $\dfrac{5}{1-x} - \dfrac{4}{x}$

49. $\dfrac{x}{2x-1} + \dfrac{3}{x}$

50. $\dfrac{x}{x+1} - \dfrac{x+1}{x}$

C **51.** $\dfrac{3}{x^2-4}+\dfrac{4}{x+2}$

52. $\dfrac{x}{x+3}+\dfrac{x}{x^2+x-6}$

53. $\dfrac{1}{x^2+5x+6}+\dfrac{1}{x^2+3x+2}$

54. $\dfrac{2}{x^2+2x-8}-\dfrac{1}{x^2-x-2}$

55. $\dfrac{1}{x^2+7x+12}-\dfrac{2}{x^2+6x+8}$

56. $\dfrac{3}{x^2-x-2}+\dfrac{1}{x^2-5x+6}$

Find the perimeter of each polygon. Assume that the dimensions are given in centimeters.

57.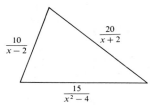

58.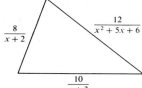

59.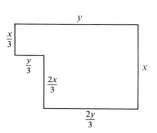

60.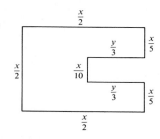

■ REVIEW EXERCISES

1. State which of the numbers -2, -1, 0, 1, and 2 are solutions to the equation. [3–1]

$$x^2=\dfrac{1}{x^2}$$

Solve. [3–6]

2. $3(x-2)=2-x$

3. $\dfrac{2(x+4)}{3}=12$

4. Solve $P=2l+2w$, for w.

5. Solve $A=\dfrac{1}{2}bh$, for h. [3–8]

Solve by writing an equation.

6. There are 20 bills in a stack of five and ten dollar bills worth $140. How many of each kind are there? [3–9]

7. A class of 30 students went on a field trip and stopped at a refreshment stand. Each of the boys had a regular ice cream cone that cost 60 cents and each of the girls had a large ice cream cone that cost 90 cents. If the total bill (not including sales tax) was $21, then how many of these students were boys and how many were girls?

380   Chapter 8   Algebraic Fractions and Applications

8-5 Solving Fractional Equations

Preview

Tom Sawyer can whitewash the back fence in 6 h, while Ben Rogers can whitewash it in only 4 h. How long would it take them to whitewash it if they work together?

- Make a rough estimate of the answer. Will it take them longer than 6 h, between 4 and 6, or less than 4 h?
- What fraction of the fence can Tom whitewash in 1 h?
- What fraction of the fence can Ben whitewash in 1 h?
- Together what fraction of the fence can they whitewash in 1 h?
- If t is the number of hours required to whitewash the fence together, what fraction of the fence can they whitewash in 1 h?

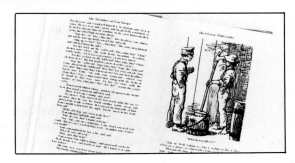

One equation that can be used to solve this problem is:

$$\frac{1}{6} + \frac{1}{4} = \frac{1}{t}$$

In this lesson you will learn to solve equations that contain fractions.

■ LESSON

Equations with fractions are usually more complicated than equations without fractions. Therefore, the first step in solving the equation is to rid the equation of fractions by multiplying both sides of the equation by the least common denominator of the fractions. For example, to solve the problem given in the Preview we must solve the equation:

$$\frac{1}{6} + \frac{1}{4} = \frac{1}{t}$$

Find the least common denominator of the fractions.

The least common denominator is $12t$.

Eliminate the fractions by multiplying both sides of the equation by the least common denominator. Simplify.

$$12t\left(\frac{1}{6} + \frac{1}{4}\right) = 12t\left(\frac{1}{t}\right)$$
$$2t + 3t = 12$$
$$5t = 12$$

Solve the new equation.

$$t = \frac{12}{5}$$

The solution of the equation is $\left\{\frac{12}{5}\right\}$.

Since $t = \frac{12}{5} = 2\frac{2}{5}$, Tom and Ben working together can whitewash the fence in

$2\frac{2}{5}$ h (2 h and 24 min).

Example 1   Solve. $\dfrac{1}{x} + 3 = \dfrac{2}{x}$

Solution

$$\frac{1}{x} + 3 = \frac{2}{x}$$

Multiply both sides by the least common denominator x.

$$x\left(\frac{1}{x} + 3\right) = x\left(\frac{2}{x}\right)$$

$$1 + 3x = 2$$

$$3x = 1$$

$$x = \frac{1}{3}$$

Note that multiplying both sides of an equation by the same number and then applying the distributive property has the effect of multiplying each term on both sides by that number.

Answer $\left\{\dfrac{1}{3}\right\}$

Check $\dfrac{1}{x} + 3 = \dfrac{2}{x}$

$$\frac{1}{\frac{1}{3}} + 3 \overset{?}{=} \frac{2}{\frac{1}{3}}$$

$$3 + 3 = 6 \qquad \text{It checks.}$$

Remember that division by 0 is not allowed. When there are variables in the denominators of fractions, the variables must be restricted to admissible values so that the denominators do not equal 0.

Example 2   Solve. $1 + \dfrac{5}{2x} = \dfrac{2.5}{x}$

Solution

$$1 + \frac{5}{2x} = \frac{2.5}{x}$$

First, list any restrictions.   Restriction: $x \neq 0$

Multiply both sides by 2x. $2x\left(1 + \dfrac{5}{2x}\right) = 2x\left(\dfrac{2.5}{x}\right)$

$$2x + 5 = 5$$

$$2x = 0$$

$$x = 0$$

We know that x cannot equal 0 in the original equation. Therefore, even though 0 is a solution of the last equation, it is not a solution of the original equation.

Answer $\emptyset$ (The equation has *no* solution.)

Example 2 (continued)

Check Note what happens if we check whether 0 is a solution.

$$1 + \frac{5}{2x} = \frac{2.5}{x}$$

$$1 + \frac{5}{0} \stackrel{?}{=} \frac{2.5}{0}$$

Since $\frac{5}{0}$ and $\frac{2.5}{0}$ are not real numbers, 0 is not a solution.

Example 3 Solve. $x - \frac{6}{x} = 1$

Solution

$$x - \frac{6}{x} = 1$$

Restriction: $x \neq 0$

Multiply both sides by x.

$$x\left(x - \frac{6}{x}\right) = x \cdot 1$$

$$x^2 - 6 = x$$
$$x^2 - x - 6 = 0$$

Factor and solve.

$$x + 2 = 0 \quad \text{or} \quad x - 3 = 0$$
$$x = -2 \quad \text{or} \qquad x = 3$$

Answer $\{-2, 3\}$

Check Check *each* solution in the original equation.

$$x - \frac{6}{x} = 1 \qquad\qquad x - \frac{6}{x} = 1$$

$$3 - \frac{6}{3} \stackrel{?}{=} 1 \qquad\qquad -2 - \frac{6}{-2} \stackrel{?}{=} 1$$

$$3 - 2 \stackrel{?}{=} 1 \qquad\qquad -2 - (-3) \stackrel{?}{=} 1$$

$$1 = 1 \qquad\qquad\qquad 1 = 1 \qquad \text{The answer checks.}$$

Equations that contain decimal fractions are sometimes difficult to solve. One way to eliminate decimal fractions is to multiply both sides of an equation by the power of 10 that makes each term in the equation a whole number.

Example 5 Solve. $0.5x = 5 + 0.25x$

Solution

$$0.5x = 5 + 0.25x$$

Multiply both sides by 100.
Simplify and solve.

$$100(0.5x) = 100(5 + 0.25x)$$
$$50x = 500 + 25x$$
$$25x = 500$$
$$x = 20$$

Answer $\{20\}$

Check The check is left to the student.

■ CLASSROOM EXERCISES

Solve.

1. $\dfrac{2x}{3} + \dfrac{5}{3} = 8$ 2. $\dfrac{3}{5} = \dfrac{x}{100}$ 3. $\dfrac{8}{x} + 1 = \dfrac{1}{x}$ 4. $\dfrac{45}{x} = \dfrac{60}{100}$

5. $x - \dfrac{4}{x} = 3$ 6. $0.5x = 7 - 0.2x$ 7. $0.87x - 0.41(x - 1) = 1.79$

■ WRITTEN EXERCISES

State the least common denominator of each of these pairs of fractions.

A 1. $\dfrac{5}{6}, \dfrac{3}{4}$ 2. $\dfrac{1}{6}, \dfrac{7}{8}$ 3. $\dfrac{2}{3}, \dfrac{5}{x}$ 4. $\dfrac{3}{4}, \dfrac{2}{2}$

5. $\dfrac{x}{7}, \dfrac{5}{14x}$ 6. $\dfrac{3}{4x}, \dfrac{x}{6}$ 7. $\dfrac{8}{3x}, \dfrac{7}{2x}$ 8. $\dfrac{5}{4x}, \dfrac{8}{3x}$

Solve. Write ∅ if there is no solution.

9. $\dfrac{2}{3} = \dfrac{x}{24}$ 10. $\dfrac{x}{12} = \dfrac{3}{4}$ 11. $\dfrac{a}{30} = \dfrac{5}{6}$ 12. $\dfrac{9}{6} = \dfrac{a}{12}$

13. $\dfrac{12}{b} = \dfrac{3}{2}$ 14. $\dfrac{24}{b} = \dfrac{4}{3}$ 15. $\dfrac{x}{3} + \dfrac{1}{2} = 10$ 16. $\dfrac{x}{4} + \dfrac{1}{3} = 3$

17. $\dfrac{x}{5} - 4 = \dfrac{9}{2}$ 18. $\dfrac{x}{4} - 3 = \dfrac{8}{5}$ 19. $\dfrac{1}{2}x + \dfrac{3}{4}x = 10$ 20. $\dfrac{1}{3}x + \dfrac{3}{4}x = 13$

21. $\dfrac{1}{5}x - \dfrac{7}{10}x = 5$ 22. $\dfrac{5}{12}x - \dfrac{7}{12}x = 1$ 23. $\dfrac{2}{3}x - \dfrac{1}{6}x = 9$

24. $\dfrac{4}{5}x - \dfrac{1}{5}x = 9$ 25. $\dfrac{1}{3}x - \dfrac{3}{4}x = 10$ 26. $\dfrac{2}{3}x - \dfrac{8}{9}x = 2$

27. $\dfrac{3}{x} + \dfrac{1}{2} = \dfrac{5}{x}$ 28. $\dfrac{3}{x} + \dfrac{5}{2} = \dfrac{8}{x}$ 29. $\dfrac{3}{2x} + \dfrac{1}{x} = 5$

30. $\dfrac{2}{x} + \dfrac{1}{x} = 9$ 31. $\dfrac{6}{x} - \dfrac{16}{x} = 5$ 32. $\dfrac{6}{x} - \dfrac{20}{2x} = 1$

B 33. $2.4x - 0.8 = 0.4$ 34. $0.3x - 0.12 = 0.06$ 35. $0.05x + 0.73 = 0.8$

36. $0.05x + 0.84 = 1$ 37. $x + \dfrac{6}{x} = 5$ 38. $x + \dfrac{4}{x} = 5$

39. $x + 2 = \dfrac{8}{x}$ 40. $x + 1 = \dfrac{6}{x}$ 41. $x + 3 = \dfrac{10}{x}$

42. $x + 4 = \dfrac{5}{x}$ 43. $4 + \dfrac{3}{x} = \dfrac{6}{2x}$ 44. $5 + \dfrac{2}{x} = \dfrac{6}{3x}$

45. $x = 10 - \dfrac{24}{x}$

46. $x = 12 - \dfrac{27}{x}$

47. $1 = \dfrac{2}{x} + \dfrac{3}{x^2}$

48. $1 = \dfrac{1}{x} + \dfrac{12}{x^2}$

49. $\dfrac{5}{x^2} + 1 = \dfrac{6}{x}$

50. $\dfrac{x}{4} + \dfrac{1}{x} = 1$

C **51.** $2x = 1 + \dfrac{3}{x}$

52. $3x + \dfrac{4}{x} = 8$

53. $6x + \dfrac{3}{x} = 11$

54. $7 = \dfrac{4}{x} - 2x$

55. $\dfrac{12}{x} = 23 - 5x$

56. $\dfrac{3x}{2} + \dfrac{4}{x} = 5$

57. $3x = \dfrac{1}{2} + \dfrac{1}{x}$

58. $1 = \dfrac{3}{2x} - 4x$

59. $\dfrac{5}{x} = 7 - \dfrac{6x}{5}$

■ REVIEW EXERCISES

Simplify.

1. $(x^2 + 4x - 3) + (x^2 - 2x) - (3x + 5)$ [4–1]

2. $(2ab^2) \cdot (-3ab)$

3. $(-2xy)(-3y^2)$ [4–3]

4. $(3x^3)^2$

5. $(-2ab^2)^3$ [4–5]

6. $\dfrac{3x^2}{9x^4} \ (x \neq 0)$

7. $\dfrac{8x^4}{4x^3} \ (x \neq 0)$ [4–8]

8. Write $4.32 \cdot 10^6$ in standard decimal notation. [4–4]

9. Simplify and write the quotient in scientific notation. [4–8]

$$\dfrac{6.4 \cdot 10^7}{1.6 \cdot 10^3}$$

Solve.

10. Genevieve and Helen like to run on a path around a lake with a circumference of 5 km. Genevieve runs at a rate of 12 km/h and Helen runs at a rate of 8 km/h. If they start at the same point and run in opposite directions, how long will it take until they meet? When they meet, how far will each have run?

Mathematics and Your Future

Today in many colleges and universities, at least three years of precollege high school mathematics, including algebra, geometry, and advanced algebra is required for many majors. Four years of mathematics, including trigonometry, is necessary for majors in mathematics, science, computer science, or engineering. Take all the mathematics in high school that you will need to prepare you for college. Do not limit your options or delay your college education while you take enough mathematics courses to catch up.

Preview Applications: Electrical circuits

All electrical circuits have parts that resist the flow of electricity. The filament of a lightbulb and the heating element of a toaster are examples of resistors. The ohm (Ω) is a unit of electrical resistance.

Two resistors may be connected in a series so that all of the current must pass through each one. The total resistance (R) of the two resistors in series is the sum of the individual resistances, r_1 and r_2.

$$R = r_1 + r_2$$

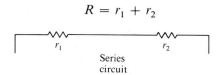

Series circuit

For two resistors connected in a parallel circuit, the current "splits up," with part of it going through one resistor and part going through the other. The total resistance R is given by the formula

$$\frac{1}{R} = \frac{1}{r_1} + \frac{1}{r_2}$$

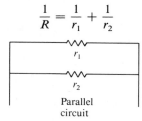

Parallel circuit

In this lesson you will use this formula and other formulas that contain algebraic fractions.

■ LESSON

The formula

$$\frac{1}{R} = \frac{1}{r_1} + \frac{1}{r_2}$$

can be used to solve problems about resistances in parallel circuits. For example, suppose that we wish to know the total resistance of a 3-ohm (3Ω) resistance and a 4-ohm (4Ω) resistance connected in parallel.

Substitute 3 for r_1 and 4 for r_2 in the formula.

$$\frac{1}{R} = \frac{1}{r_1} + \frac{1}{r_2}$$

$$\frac{1}{R} = \frac{1}{3} + \frac{1}{4}$$

Multiply both sides by the least common denominator, 12R.
Simplify and solve.

$$12R\left(\frac{1}{R}\right) = 12R\left(\frac{1}{3} + \frac{1}{4}\right)$$

$$12 = 4R + 3R$$

$$12 = 7R$$

$$R = \frac{12}{7}$$

The total resistance of the circuit is $\frac{12}{7}$ ohms. Note that the resistance of the parallel circuit is less than either of the individual resistances.

Fractional equations are also used in problems involving work rates, since fractions are used to express the rate of work. For example, if a farmer can mow a hay field in 5 h, the farmer's rate of work is $\frac{1}{5}$ of a field per hour. In general, if a job can be done in n hours, the rate is $\frac{1}{n}$ of the job per hour.

The relationship between the amount (w) of work, the rate (r) of work, and the time (t) worked is

$$w = rt$$

If the farmer mows at the rate of $\frac{1}{5}$ of a field per hour, then in 3 h he can mow $3\left(\frac{1}{5}\right)$ of the field.

Example 1   A master painter can paint a house in 12 h. An assistant can paint the house in 16 h. How long will it take the two painters working together to paint the house?

Solution   Let t = the number of hours for the painters to paint the house together.

We know three pieces of information for each painter and for the two painters together: (1) the rate worked, (2) the time worked, and (3) the amount of work done. A table can be used to organize the information.

	Rate (houses per hour)	Time (hours)	Amount of work (houses painted)
Painter	$\frac{1}{12}$	t	$\left(\frac{1}{12}\right)t$
Assistant	$\frac{1}{16}$	t	$\left(\frac{1}{16}\right)t$
Together	$\frac{1}{t}$	t	$\left(\frac{1}{t}\right)t$ or 1

Word equation: Amount of work done by master painter   plus   Amount of work done by assistant   equals   Amount of work done by pair

Equation:
$$\frac{1}{12}t \quad + \quad \frac{1}{16}t \quad = \quad 1$$

Multiply both sides by 48.
$$48\left(\frac{1}{12}t + \frac{1}{16}t\right) = 48(1)$$

$$4t + 3t = 48$$

$$7t = 48$$

$$t = \frac{48}{7}$$

Answer   The two painters can paint the house in $6\frac{6}{7}$ h (about 6 h and 51 min).

Example 1 (continued)

Check (Check by estimating.) Two fast painters working at the rate of the master painter (one house in 12 h) could paint the house in 6 h. Two slower painters working at the rate of the assistant (one house in 16 h) could paint the house in 8 h. Therefore, it will take the painter and his assistant between 6 and 8 h to paint the house. The answer ($6\frac{6}{7}$ h) is reasonable.

Example 2 Mark uses his tractor mower to mow his neighbor's 2-acre lawn. He can mow the lawn by himself in 3 h. One day, after Mark had mowed for a half hour, his younger sister began helping him with a small mower. Together they finished the job in 2 more hours. How long would it have taken Mark's sister to have done the job alone using the small mower?

Solution Let $x =$ the number of hours it would take Mark's sister to mow the lawn alone.

	Rate (lawns per hour)	Time worked (hours)	Amount of work (lawns mowed)
Mark	$\dfrac{1}{3}$	$\dfrac{5}{2}$	$\left(\dfrac{1}{3}\right)\left(\dfrac{5}{2}\right)$
Sister	$\dfrac{1}{x}$	2	$\left(\dfrac{1}{x}\right)(2)$
			Total: 1 lawn

Word equation:
Amount of work done by Mark plus Amount of work done by sister equals Total amount of work

Equation:
$$\left(\frac{1}{3}\right)\left(\frac{5}{2}\right) \quad + \quad \left(\frac{1}{x}\right)(2) \quad = \quad 1$$

Multiply both sides by 6x.
$$6x\left[\left(\frac{1}{3}\right)\left(\frac{5}{2}\right) + \left(\frac{1}{x}\right)(2)\right] = 6x(1)$$
$$5x + 12 = 6x$$
$$x = 12$$

Answer Mark's sister could mow the lawn in 12 h.

Check Mark's sister mowed $\dfrac{1}{12}$ of the lawn per hour. In the 2 hours she worked, she mowed $2\left(\dfrac{1}{12}\right)$, or $\dfrac{1}{6}$ of the lawn. Mark mowed $\left(\dfrac{1}{3}\right)\left(\dfrac{5}{2}\right)$, or $\dfrac{5}{6}$ of the lawn in the $2\dfrac{1}{2}$ hours he worked. Since $\dfrac{1}{6} + \dfrac{5}{6} = 1$, the answer checks.

Example 3 Solve this literal equation for b. $\dfrac{1}{a} + \dfrac{1}{2a} = \dfrac{1}{b}$

Solution

The least common denominator is $2ab$.

$$\dfrac{1}{a} + \dfrac{1}{2a} = \dfrac{1}{b}$$

Multiply both sides by $2ab$.

$$2ab\left(\dfrac{1}{a} + \dfrac{1}{2a}\right) = 2ab\left(\dfrac{1}{b}\right)$$

Simplify.

$$2b + b = 2a$$

$$3b = 2a$$

Divide both sides by 3.

$$b = \dfrac{2a}{3}$$

Answer $\quad b = \dfrac{2a}{3}$

Check $\quad$ The check is left to the student.

■ CLASSROOM EXERCISES

1. If two 10-ohm resistors are connected in parallel, what is the total resistance of the circuit?

2. Cindy works 2 times as fast as her brother. Together they cleaned the house in 4 h. How long would it take each of them alone to clean the house?

3. Find the total resistance of a circuit made up of a 5-ohm resistor and a 10-ohm resistor connected in parallel.

4. If the total resistance of a circuit with two parallel resistances is 100 ohms and one resistance is 300 ohms, then what is the other resistance?

5. If $\dfrac{1}{a} = \dfrac{1}{b} - \dfrac{1}{c}$, solve for b in terms of a and c.

■ WRITTEN EXERCISES

Find the total resistance of a circuit if the two given resistors are connected in

parallel. Use the formula $\dfrac{1}{R} = \dfrac{1}{r_1} + \dfrac{1}{r_2}$.

A 1. 6 ohms and 6 ohms $\qquad$ 2. 4 ohms and 4 ohms $\qquad$ 3. 2 ohms and 4 ohms

4. 3 ohms and 6 ohms $\qquad$ 5. 12 ohms and 6 ohms $\qquad$ 6. 12 ohms and 4 ohms

Two resistors are connected in parallel. The total resistance of the circuit and the resistance of one of the resistors is given. Find the resistance of the other resistor. Use the formula $\frac{1}{R} = \frac{1}{r_1} + \frac{1}{r_2}$.

7. $R = 4$ ohms

$r_1 = 20$ ohms

8. $R = 8$ ohms

$r_1 = 24$ ohms

9. $R = 6$ ohms

$r_1 = 9$ ohms

10. $R = 12$ ohms

$r_1 = 18$ ohms

11. $R = 14$ ohms

$r_1 = 42$ ohms

12. $R = 1$ ohm

$r_1 = 2$ ohms

Solve these work problems. Write an equation that fits the problem. Solve the equation and answer the question.

13. Kim can paint a room in 4 h. Kristen can paint the same room in 6 h. How long does it take Kim and Kristen to paint the room if they work together?

14. Chris can address all the envelopes for a publicity mailing in 8 h. Mike can address all the envelopes in 7 h. How long does it take Chris and Mike to address the envelopes if they work together?

15. Amy delivered an advertisement to the residents in her section of town in 6 h. Last week, when Amy and Adam worked together, they completed the deliveries in 4 h. How long would it take Adam working alone to make the deliveries?

16. Jon can mow a lawn in 4 h. When Jon and Alexander work together, they mow the same lawn in 1 h. How long does it take Alexander to mow the lawn alone?

17. An old machine and a new machine produce the same item. The new machine produces twice as many items as the old machine in the same time. When both machines work, they produce the day's quota in 3 h. How long would it take the old machine alone to produce the quota? How long would it take the new machine alone to produce the quota?

18. A small pipe and a large pipe both carry water to a storage tank. The large pipe fills the tank twice as fast as the small pipe. When both pipes are used, the tank is filled in 8 h. How long would it take the small pipe to fill the tank by itself? How long would it take the large pipe to fill the tank by itself?

Solve.

19. Solve for r_1.

$$\frac{1}{R} = \frac{1}{r_1} + \frac{1}{r_2}$$

20. Solve for r_2.

$$\frac{1}{R} = \frac{1}{r_1} + \frac{1}{r_2}$$

The total resistance R of a circuit made up of three resistors, r_1, r_2, and r_3, connected in parallel is given by the formula

$$\frac{1}{R} = \frac{1}{r_1} + \frac{1}{r_2} + \frac{1}{r_3}$$

B **21.** Individual resistors of 2 ohms, 3 ohms, and 6 ohms are connected in parallel. What is the total resistance?

22. Find the total resistance when individual resistors of 2 ohms, 5 ohms, and 10 ohms are connected in parallel. What is the total resistance?

23. Three identical resistors connected in parallel have a total resistance of 6 ohms. What is the resistance of each resistor?

24. Three resistors are connected in parallel. The resistance of the second is twice the first and the resistance of the third is twice the second. The total resistance is 4 ohms. What is the resistance of each resistor? [*Hint:* The three resistances could be called x, $2x$, and $4x$.]

Solve.

25. A large water tank has a leak that completely drained the tank in 20 hours. It takes 4 hours to fill the tank when there is no leak. How long will it take to fill the leaking tank?

26. Three machines working alone can produce a day's quota in 2 h, 3h, and 6 h, respectively. How long does it take to produce the quota when all three machines work together?

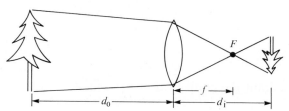

When parallel light rays pass through a convex lens, they converge at a point beyond the lens. The distance from the center of the lens to this focal point F is called the focal length f. An object at distance d_o from the lens projects an image on the other side of the lens at distance d_i from the lens. The relationship between d_o, d_i, and f is given by this formula:

$$\frac{1}{d_o} + \frac{1}{d_i} = \frac{1}{f}.$$

27. An object is 30 cm from a lens that has a focal length of 10 cm. How far is the image from the lens?

28. An object is 24 cm from a lens that has a focal length of 6 cm. How far is the image from the lens?

29. An object is 12 cm from a lens and its image is 4 cm from the lens. What is the focal length of the lens?

30. Solve the formula for f. (Express f in terms of d_o and d_i.)

31. Solve the formula for d_i. (Express d_i in terms of f and d_o.)

The *harmonic mean* h of two numbers a and b is given by the formula

$$\frac{1}{h} = \frac{\frac{1}{a} + \frac{1}{b}}{2}$$

If a and b are rates, then h is the average rate.

C 32. Solve the formula for h.

33. A salesperson drives to an office at 60 km/h and returns home by the same route at 40 km/h. What was the average speed for the total trip? (If $a = 60$ and $b = 40$, find h.)

34. A passenger plane flies into the wind to its destination and flies with the wind on the return trip. The ground speed is 450 mph going and 550 mph returning. What is the average speed for the round trip?

35. Lori paddled a canoe from the dock to an island at 4 km/h going with the current. On her return trip, against the current, Lori averaged 1 km/h. What was her average speed for the round trip?

■ REVIEW EXERCISES

1. For each point below, either state the quadrant the point is in or the axis it is on. [5–1]

   a. $(0, -5)$ **b.** $(4, -3)$ **c.** $(-3, 4)$ **d.** $(1, 7)$

2. The table defines a relation. [5–2]

x	0	1	1	2	2
y	0	1	-1	2	-2

   a. List the domain of the relation.

   b. List the range of the relation.

   c. Graph the relation.

3. Indicate whether the ordered pair is a solution of the equation $3x + y = 12$. [5–3]

   a. $(0, 4)$ **b.** $(2, 6)$ **c.** $(-2, 18)$

4. Find the missing number for each solution pair of the equation
$2a - b = 10$. [5–3]

a	0	?	2	?
b	?	0	?	4

5. Select three values of x, complete the table, plot the points, and then graph
the equation $y = \frac{1}{2}x - 2$. [5–4]

x	?	?	?
y	?	?	?

6. Complete the table and then use the ordered pairs to graph the equation
$y = x^2 - 4$. [5–4]

x	-3	-2	-1	$-\frac{1}{2}$	0	$\frac{1}{2}$	1	2	3
y	?	?	?	?	?	?	?	?	?

Self-Quiz 2

8–4 Simplify.

1. $\dfrac{2}{x} - \dfrac{5}{3x}$

2. $\dfrac{8}{y} - \dfrac{y}{4}$

3. $\dfrac{6}{x^2} + \dfrac{7}{x}$

4. $\dfrac{3}{x-2} + \dfrac{1}{x}$

8–5 Solve.

5. $\dfrac{s}{4} - 1 = \dfrac{3}{2}$

6. $x - 3 = \dfrac{18}{x}$

7. $\dfrac{1}{3x} + \dfrac{1}{2x} = \dfrac{5}{6}$

8–6 Solve.

8. Using the formula $\dfrac{1}{R} = \dfrac{1}{r_1} + \dfrac{1}{r_2}$, find the resistance r_1 when $r_2 = 3$ ohms and
the total resistance $R = 2$ ohms.

9. A new letter-sorting machine A can complete a task in $3\frac{1}{2}$ h. An older machine
can do the same job in 5 h. How long will it take both machines to do the
task together?

8–7 Ratio and Proportion

Preview   **Historical note: Agreeable sounds**

In experimenting with the relative lengths of strings needed to produce musical notes, Pythagoras (pih-THAG-oh-ras), a Greek mathematician who lived more than 2500 years ago, found that string lengths in ratios of small integers, such as 2 to 1, 3 to 2, and 4 to 3, produced especially agreeable sounds. Two strings of a more complicated ratio of length, such as 23 to 13, produced an unpleasant sound combination.

Determine whether Pythagoras would have thought the sound pleasant if the string-lengths were in these ratios:

a. 24 to 12   **b.** 24 to 16

c. 24 to 18   **d.** $11\frac{1}{2}$ to $6\frac{1}{2}$

e. 1.5 to 1   **f.** 2.3 to 1.3

In this lesson you will work with ratios. Many people consider the ability to use ratios as the most practical problem-solving skill studied in school.

■ LESSON

The **ratio** of a number a to a number b ($b \neq 0$) is the quotient of a divided by b. The ratio can be written in several ways.

$$a \text{ to } b \qquad a{:}b \qquad \frac{a}{b}$$

For example, the ratio of the number of inches in a foot to the number of inches in a yard can be expressed as 12 to 36 or 12:36 or $\frac{12}{36}$.

In this text, ratios will most often be written using fraction notation.

Ratios are simplified in the same way that fractions are simplified. For example, the above ratio of the number of inches in a foot to the number of inches in a yard $\left(\frac{12}{36}\right)$ can be simplified to $\frac{1}{3}$.

An equation that states that two ratios are equal is called a **proportion.**

For example, $\frac{12}{36} = \frac{1}{3}$.

Consider how fractions can be eliminated from a proportion. Let a, b, c, and d be real numbers ($b \neq 0$ and $d \neq 0$).

$$\frac{a}{b} = \frac{c}{d}$$

Multiply both sides by bd. $bd \cdot \dfrac{a}{b} = bd \cdot \dfrac{c}{d}$

Simplify. $ad = bc$

We have just proved the **cross-multiplication** property.

Cross-Multiplication Property

For all numbers a, b, c, and d ($b \neq 0$, $d = 0$),

$$\text{if } \frac{a}{b} = \frac{c}{d}, \text{ then } ad = bc.$$

The operation is called *cross multiplying* because the numbers that are multiplied (a and d, b and c) are related by a cross.

$$\frac{a}{b} \diagdown\!\!\!\!\diagup \frac{c}{d}$$

The cross-multiplication property can be used to determine whether a proportion is true or false. For example, consider the proportions

$$\frac{57}{38} = \frac{3}{2} \quad \text{and} \quad \frac{27}{81} = \frac{1}{4}.$$

$\dfrac{57}{38} = \dfrac{3}{2}$	$\dfrac{27}{81} = \dfrac{1}{4}$
$57(2) = 38(3)$	$27(4) = 81(1)$
$114 = 114$ True.	$108 = 81$ False.

The first proportion is true because the cross products are equal. The second proportion is false because the cross products are not equal.

The cross-multiplication property can also be used to solve proportions.

Mathematics and Your Future

As the use of technology in vocational fields increases, so too do the mathematics requirements for vocational and technical programs. Most of these programs require a knowledge of algebra and geometry. Many programs in the agricultural, business. health, trade, and industrial areas also require advanced algebra, while some specialized fields require trigonometry. If you want to keep your options open for enrolling in various vocational programs, be sure to take enough mathematics.

Example 1 Solve. $\frac{x}{6} = \frac{5}{8}$

Solution

$$\frac{x}{6} = \frac{5}{8}$$

Cross multiply.

$$8x = 6 \cdot 5$$

$$x = \frac{30}{8}$$

$$= \frac{15}{4}$$

Answer $\left\{\frac{15}{4}\right\}$

Check The check is left to the student.

Example 2 Solve. $\frac{2}{3} = \frac{x+4}{x+13}$

Solution

$$\frac{2}{3} = \frac{x+4}{x+13}$$

Cross multiply.

$$2(x + 13) = 3(x + 4)$$

$$2x + 26 = 3x + 12$$

$$-x = -14$$

$$x = 14$$

Answer $\{14\}$

Check The check is left to the student.

Example 3 The ratio of the length of a rectangle to the width is 5 to 2. Find the dimensions of the rectangle if its perimeter is 70 cm.

Solution Let $5x$ = the length and $2x$ = the width.
(Note that $5x$ and $2x$ are in the ratio 5 to 2.)

$$5x + 2x + 5x + 2x = 70$$

$$14x = 70$$

$$x = 5$$

Answer The length is $5 \cdot 5$ cm, or 25 cm. The width is $2 \cdot 5$ cm, or 10 cm.

Check $\frac{25}{10} = \frac{5}{2}$ The perimeter is 25 cm + 10 cm + 25 cm + 10 cm = 70 cm.
It checks.

First write the two measurements using the same unit. Then write the ratio in simplest form.

1. 1 in. to 1 ft

2. 1 year to 5 weeks

3. 1 cm to 1 m

True or false?

4. $\dfrac{3}{4} = \dfrac{15}{20}$

5. $\dfrac{5}{8} = \dfrac{35}{54}$

6. $\dfrac{7}{4} = \dfrac{3.5}{2}$

Solve.

7. $\dfrac{x}{5} = \dfrac{12}{15}$

8. $\dfrac{12}{a} = 6$

9. $\dfrac{3}{4} = \dfrac{y}{27}$

10. If there are 10 boys and 20 girls in an algebra class, simplify the ratio of the number of boys to the number of girls.

11. In Exercise 10, simplify the ratio of the number of students to the number of girls.

■ WRITTEN EXERCISES

Write the two measurements using the same units. Then write the ratios in simplest form.

A

1. 2 feet to 1 yard

2. 1 yard to 2 feet

3. 1 kilometer to 1 meter

4. 1 meter to 1 kilometer

5. 10 seconds to 1 minute

6. 1 day to 32 hours

7. 8 ounces to 2 pounds

8. 1 pound to 12 ounces

A bicycle shop has 20 five-speeds, 34 ten-speeds, and 45 twelve-speeds. Simplify the ratio of bikes for the given speeds.

9. five-speeds to ten-speeds

10. ten-speeds to twelve-speeds

11. ten-speeds to total

12. twelve-speeds to total

Simplify these ratios.

13. $\dfrac{36}{48}$

14. $\dfrac{40}{48}$

15. $\dfrac{28}{32}$

16. $\dfrac{20}{32}$

17. $\dfrac{80}{100}$

18. $\dfrac{75}{100}$

19. $\dfrac{34}{51}$

20. $\dfrac{38}{57}$

True or false?

21. $\dfrac{30}{80} = \dfrac{12}{32}$

22. $\dfrac{50}{70} = \dfrac{15}{21}$

23. $\dfrac{12}{15} = \dfrac{20}{25}$

24. $\dfrac{6}{15} = \dfrac{8}{20}$

25. $\dfrac{18}{30} = \dfrac{14}{20}$

26. $\dfrac{9}{12} = \dfrac{18}{21}$

27. $\dfrac{26}{34} = \dfrac{38}{52}$

28. $\dfrac{46}{38} = \dfrac{68}{59}$

Solve these proportions.

29. $\dfrac{2}{3} = \dfrac{x}{54}$

30. $\dfrac{3}{4} = \dfrac{x}{68}$

31. $\dfrac{48}{42} = \dfrac{56}{x}$

32. $\dfrac{63}{56} = \dfrac{36}{x}$

33. $\dfrac{5}{8} = \dfrac{7.5}{x}$

34. $\dfrac{5}{6} = \dfrac{7.5}{x}$

35. $\dfrac{x}{6} = \dfrac{9}{4}$

36. $\dfrac{x}{10} = \dfrac{11}{4}$

Solve.

37. If 60 of the 300 students at Grant School are left-handed, what is the ratio of left-handed students to right-handed students?

38. The ratio of the sides of two squares is 3 to 1. What is the ratio of their perimeters?

39. The ratio of the length to the width of a rectangle is 4 to 3. What are the dimensions of the rectangle if its perimeter is 84 cm?

40. An 81-cm cable is cut into two pieces whose lengths are in the ratio 4 to 5. How long are the pieces?

41. An inheritance of $50,000 is divided between two people in the ratio 7 to 3. How much does each person get?

Solve these proportions.

B **42.** $\dfrac{3}{4} = \dfrac{x+5}{x+10}$

43. $\dfrac{2}{3} = \dfrac{x+4}{x+10}$

44. $\dfrac{5}{2} = \dfrac{x+6}{x-6}$

45. $\dfrac{7}{3} = \dfrac{x+6}{x-6}$

46. $\dfrac{5}{7} = \dfrac{3x+6}{5x+2}$

47. $\dfrac{4}{9} = \dfrac{3x-1}{6x+3}$

48. $\dfrac{1}{x-3} = \dfrac{x-2}{2}$

49. $\dfrac{x-2}{5x-16} = \dfrac{1}{x-2}$

50. $\dfrac{2x-1}{x+1} = \dfrac{x-2}{x-3}$

51. The ratio of the sides of two squares is 3 to 1. What is the ratio of their areas?

52. The ratio of the radii of two circles is 4 to 1. What is the ratio of their circumferences?

53. The radii of two circles are in the ratio 10 to 1. What is the ratio of their areas?

54. The edges of two cubes are in the ratio $\dfrac{1}{2}$. What is the ratio of their surface areas?

55. What is the ratio of the volumes of two cubes if their edges are in the ratio $\dfrac{2}{3}$?

56. The ratio of the lengths, of the widths, and of the heights of two rectangular prisms is $\frac{3}{4}$. Is the volume of the smaller prism more than half, half, or less than half the volume of the larger prism? Explain.

57. The ratio of adult to student tickets for a concert is $\frac{3}{2}$. How many of each kind were sold if 300 tickets were sold in all?

58. Three members are in the ratio 2:3:4. What are the numbers if their sum is 30? [*Hint:* Let $2x$, $3x$, and $4x$ represent the numbers.]

59. Three numbers are in the ratio 3:4:5. What are the numbers if their sum is 240?

60. Three numbers are in the ratio 5:7:9. What are the numbers if their sum is 105?

Suppose that $\frac{a}{b} = \frac{c}{d}$ and $a \neq 0$, $b \neq 0$, $c \neq 0$, and $d \neq 0$. Show that these statements are true.

C **61.** $\dfrac{a}{c} = \dfrac{b}{d}$ **62.** $\dfrac{a}{b} = \dfrac{a+c}{b+d}$ **63.** $\dfrac{a}{b} = \dfrac{a+2c}{b+2d}$

■ REVIEW EXERCISES

1. Draw these three lines through the point $(2, 0)$.   [5–5]
Label the lines a, b, and c.

   a. Line with slope 2

   b. Line with slope $-\dfrac{2}{3}$

   c. Line with no slope

2. State the slopes of the lines.   [5–5]

   a. $\overleftrightarrow{AB}$

   b. $\overleftrightarrow{AC}$

   c. $\overleftrightarrow{BD}$

   d. $\overleftrightarrow{BC}$

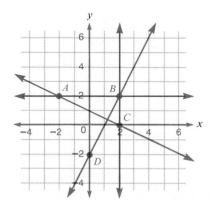

3. State the slope and y-intercept of the line whose equation is $y = 2x - 5$.   [5–6]

4. Write this equation in slope–intercept form.   [5–6]

$$3x + 5y = 15$$

8-8 Percent

Preview

A student noticed that some words containing "cent" have similar meanings. For example,

cent: A monetary unit equal to $\frac{1}{100}$ of the dollar of the United States, Australia, Canada, Ethiopia, Guyana, Liberia, Malaysia, New Zealand, Hong Kong, and Singapore.

centenary: Pertaining to a 100-year period.

centigram: $\frac{1}{100}$ of a gram.

centuplicate: To multiply by 100.

century: 100 years.

Can you guess the meaning of the following words that contain "cent"?

- A souvenir in Mexico City costs 37 pesos and 50 centavos. What is one centavo worth?

- Payette, Idaho, had a centennial celebration. How old is the town?

- The Roman centurion led his soldiers in battle. How many soldiers did he command?

In this lesson, you will learn to use an important ratio called percent.

■ LESSON

The ratio of one quantity to another is sometimes written as a percent. **Percent** means "hundredth" and is usually denoted by the symbol %. For example,

$$5\% = \frac{5}{100} = 0.05 \qquad 100\% = \frac{100}{100} = 1$$

$$6.7\% = \frac{6.7}{100} = 0.067 \qquad 143\% = \frac{143}{100} = 1.43$$

If the ratio is given as a fraction, it can be changed to a percent by

1. changing to an equivalent fraction with denominator of 100 and then writing the numerator of the fraction followed by a percent symbol, or

2. changing the fraction to a decimal, multiplying the decimal by 100, and writing that number followed by a percent symbol.

For example, to change $\frac{3}{20}$ to a percent notation:

Method 1	*Method 2*
Change the fraction to an equivalent fraction with a denominator of 100.	Divide the numerator by the denominator to change the fraction to a decimal.
$$\frac{3}{20} = \frac{15}{100} = 15\%$$	$$\frac{3}{20} = 0.15 = 15\%$$

Proportions can be used to solve percent problems.

Example 1 About 17% of the 30 students in an algebra class earned an A during the second quarter. About how many students earned an A?

Solution Let P = the number of students who earned an A. Then the ratio of P to 30 must be equal to the ratio of 17 to 100.

$$\frac{P}{30} = \frac{17}{100}$$

$$100P = 510$$

$$P = 5.1$$

Answer About 5 students earned an A.

Example 2 In one high school about 53% of the students are girls. If there are 332 girls in the school, how many students are there in all?

Solution Let n = the number of students in the school. Then the ratio of 332 to n must equal the ratio of 53 to 100.

$$\frac{332}{n} = \frac{53}{100}$$

$$53n = 33{,}200$$

$$n \approx 626.42$$

Answer There are about 626 students in the school.

In general, if $a\%$ of b equals c, then $\dfrac{a}{100} \cdot b = c$ or $\dfrac{a}{100} = \dfrac{c}{b}$.

Example 3 A few years ago, the best time for the women's marathon was about 111% of the best time for the men's marathon. If the women's best time was about 2 h 22 min, approximately what was the men's best time?

Solution We know that 2h 22 min = 142 min. Let m = the men's best time.

$$111\% \text{ of } m = 142$$

$$\frac{111}{100} \cdot m = 142$$

$$\frac{111}{100} = \frac{142}{m}$$

$$111m = 14{,}200$$

$$m \approx 127.93$$

Answer The men's best time was about 128 min, or 2h 8 min.

Another important use of percents is in interest problems. The relationship between **interest** i (in dollars), **principal** p (in dollars), **rate** r (percent per year), and time t (in years) is given by the formula $i = prt$.

Example 4 Barbara Lynch borrowed $5200 for 2 years to pay for her new car. The interest rate was 9.9% per year. How much did she pay each month for the two years?

Solution She paid interest on the entire $5200 even though she paid some of the principal each month.

p	r	t
5200	9.9%, or 0.099	2

$$i = prt$$
$$i = 5200 \cdot 0.099 \cdot 2$$
$$i = 1029.6$$

She paid $1029.60 in interest and $5200 in principal for a period of 24 months.

Answer The monthly payment was
$$\frac{\$5200 + \$1029.60}{24}, \quad \text{or} \quad \$259.57.$$

■ CLASSROOM EXERCISES

Write these ratios as percents.

1. $\dfrac{1}{2}$ **2.** $\dfrac{7}{10}$ **3.** $\dfrac{5}{8}$ **4.** $\dfrac{1}{3}$ **5.** $\dfrac{5}{4}$

Write these percents as simplified ratios.

6. 45%   **7.** 5.5%   **8.** 150%

Solve.

9. 15% of $95 = x$ **10.** 130% of $45 = n$ **11.** 25% of $y = 40$

12. 1.4% of $m = 0.7$ **13.** x% of $20 = 15$ **14.** z% of $60 = 70$

Solve.

15. Ms. Carson requires students to score at least 92% on a test in order to get an A. Carolyn scored 43 out of 46 possible points. Did she get an A on the test?

16. Mr. Crandal borrowed $2500 for 6 months. He paid interest at the rate of 11.5% per year. How much did he have to repay at the end of 6 months?

■ WRITTEN EXERCISES

Write these ratios as percents.

A 1. $\dfrac{3}{8}$ 2. $\dfrac{7}{8}$ 3. $\dfrac{6}{5}$ 4. $\dfrac{3}{2}$

5. $\dfrac{1}{400}$ 6. $\dfrac{1}{200}$ 7. $\dfrac{2.5}{50}$ 8. $\dfrac{3.5}{25}$

Write these percents as simplified ratios.

9. 32%   10. 44%   11. 125%   12. 175%

13. 0.5%   14. 0.4%   15. $1\dfrac{1}{2}\%$ 16. $2\dfrac{1}{2}\%$

Solve these proportions.

17. $25\% = \dfrac{x}{12}$ 18. $20\% = \dfrac{x}{40}$ 19. $30\% = \dfrac{60}{x}$

20. $10\% = \dfrac{30}{x}$ 21. $x\% = \dfrac{12}{15}$ 22. $x\% = \dfrac{9}{15}$

Solve.

23. 5% of 80 = x 24. 8% of 50 = x 25. 150% of x = 9

26. 140% of x = 7   27. x% of 500 = 40   28. x% of 400 = 60

29. 150% of 50 = x 30. 125% of 40 = x 31. 12% of x = 9.6

32. 15% of x = 7.2   33. x% of 250 = 35   34. x% of 180 = 27

Solve. Use the formula $i = prt$, where i = interest in dollars, p = principal in dollars, r = annual rate as a percent, and t = time in years.

35. Mr. Banks invested $4000 for 2 years at an annual rate of 9.5%. How much interest did he earn?

36. Ms. Cash invested $800 for 5 years at an annual rate of 7.5%. How much interest did she earn?

37. Sarah Blank borrowed $6000 for 3 years at an annual rate of 12%. How much interest did she pay on her loan?

38. Bud Nichols borrowed $1500 for 2 years at an annual rate of 18%. How much interest did he pay on his loan?

Solve.

39. In a basketball game, the Lakers made 36 out of 45 free throws. What percent of their free throws were made?

40. A record store reported that only 18 out of 90 records sold were instrumentals. What percent of the records sold were instrumentals?

41. A newspaper reported that 30% of the registered voters questioned in a telephone poll favored candidate Robertson. If 66 voters questioned favored Robertson, how many voters were questioned in all?

42. The Rattlers' leadoff batter got on base 40% of the time in the month of July. If the batter was on base 38 times, how many times was the batter at bat during July?

Solve these proportions.

B 43. $\frac{1}{2}\% = \frac{x}{300}$

44. $\frac{3}{4}\% = \frac{x}{200}$

45. $\frac{117}{60} = x\%$

46. $\frac{126}{90} = x\%$

47. $0.35\% = \frac{28}{x}$

48. $0.18\% = \frac{9}{x}$

Solve.

49. Burt Corning borrowed $300 for 6 months and was charged $30 interest. What was the annual interest rate of the loan?

50. Carole Ziegler invested $15,000 for 6 months and earned $900 interest. What was the annual interest rate?

51. The Bailey family budgets 30% of its income for food. If they spent $5700 on food in one year, what was their family income?

52. The annual budget of a large city department was $1,200,000. The next year the budget was 5% greater and the following year it was decreased by 5%. What was the budget after the decrease?

Solve.

C 53. $\frac{x+2}{x+4} = 20\%$

54. $\frac{x-2}{x-4} = 20\%$

55. $\frac{x-2}{80} = (x+4)\%$

56. $\frac{2x-3}{x+6} = 40\%$

57. $\frac{2x-3}{20} = (3x-10)\%$

58. $\frac{3x+4}{40} = (5x-3)\%$

■ REVIEW EXERCISES

1. Use the graph to find the solution
   of this system.

   $$x - 3y = 1$$
   $$2x + y = -5$$

[6–1]

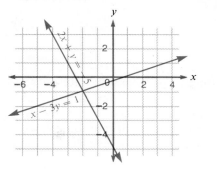

Solve each system by substitution.

[6–2]

2. $y = 2x - 2$

   $2x + 3y = 10$

3. $y = 2x + 2$

   $3x + 2y = 32$

State whether the system has one solution, no solution, or infinitely many
solutions.

[6–3]

4. $x + y = 2$

   $3x + 3y = 6$

5. $2x + y = 3$

   $2x + y = 4$

Solve each problem by writing a system of equations.

[6–4]

6. One number is 4 more than 3 times another. If the sum of the two numbers
   is 100, what are the numbers?

7. A pile of 50 coins consists of nickels and dimes. If its total value is four
   dollars, how many coins are nickels and how many are dimes?

EXTENSION   The "waste factor" in computing prices

Some products that appear to be similar have different prices. One reason for
the difference in price is the varying amount of "waste" included in the product. For example, a ham with the bone in costs less *per pound* than boned ham,
because the bone is waste and because extra labor is needed to remove the bone.
The following problem involves differences in the amounts of a "waste"
element.

Solve.

Regular ground beef that contains 28% fat sells for $1.29 per lb. Lean
ground beef that contains 24% fat sells for $1.89 per lb. Extra-lean ground
beef that contains 20% fat sells for $2.19 per lb. Diet ground beef that
contains 15% fat sells for $2.49 per lb. For which type is the cost per
pound of the nonfat beef the least? For which type is it the greatest?

8-9 Problem Solving—Using Proportions

Preview A sampling technique

The Department of Natural Resources wanted to estimate the number of fish in a pond. They caught and tagged 30 fish and returned them to the pond. Several days later, they caught a sample of 40 fish and found that 2 of them had tags. They organized the information in a table.

	In the sample	In the pond
Number of fish with tags	2	30
Total number of fish	40	?

Can you find an estimate of the total number of fish in the pond?

In this lesson you will learn how to solve problems using proportions.

■ LESSON

There are many practical situations in which the ratio of two quantities remains constant or is assumed to remain constant. For example, suppose that a school cafeteria serves 9 students in 2 minutes. It is reasonable to assume that all students will be served at the same rate. That is, the ratio of the number of students served to the number of minutes it takes to serve them is $\frac{9}{2}$. If we wish to determine how long it will take to serve 225 students, we set up and solve this proportion.

$$\frac{9}{2} = \frac{225}{x}$$

$$9x = 450$$

$$x = 50$$

It will take 50 minutes to serve 225 students.

Example 1   Solve.

Cathy scored 115 points on a 130-point test. What percent of the possible points did she score?

Solution   A table can help you write the proportion.

Let x be the percent of points scored by Cathy.

	Ratios	
Number scored	115	x
Number possible	130	100

$$\frac{115}{130} = \frac{x}{100}$$

$$130x = (115)(100)$$

$$x \approx 88.46$$

Answer   Cathy scored about 88%.

Example 2   Mary and Cherie shared the costs of buying a car. The ratio of the amount Mary paid to the amount Cherie paid was $\frac{4}{3}$. If the car cost $8400, how much did each pay?

Solution   Let m = the amount Mary paid. Then $(8400 - m)$ is the amount Cherie paid.

$$\frac{m}{8400 - m} = \frac{4}{3}$$

$$3m = 33,600 - 4m$$

$$7m = 33,600$$

$$m = 4800$$

Answer   Mary paid $4800 and Cherie paid $3600.

Check $4800 + $3600 = $8400; \quad \dfrac{4800}{3600} = \dfrac{4 \cdot 1200}{3 \cdot 1200} = \dfrac{4}{3}$ The answer checks.

■ CLASSROOM EXERCISES

State a proportion for the problem. Solve the proportion and answer the question.

1. If 8 cm on a map represents 96 km, how many kilometers are represented by 12 cm on the map?

2. Human hair grows about 1.95 in. in 2 mo. How long will it take hair to grow 8 in.?

■ WRITTEN EXERCISES

Write a proportion that fits the problem. Do not solve.

A 1. The cashier at a supermarket checked out 3 customers every 10 minutes. At that rate, how many customers would be checked out in one hour?

2. Twenty-four persons go on one of the rides at an amusement park every 3 minutes. At that rate, how long would it take 120 people to go on the ride?

3. Cheese is selling for $3.20 per lb. What does 6 oz of cheese cost at that rate?

4. A 3-oz package of oriental noodle soup mix costs 36¢. What does 1 lb of mix cost at that rate?

5. Ms. Reed drove at a constant rate of 54 mph. How far did she go in 20 min?

6. Mr. Wright drove at a constant rate of 52 mph. How long did it take him to go 65 mi?

7. Mr. Jefferson received 45% of the votes in a recent election. If 1460 persons voted, how many votes did he get?

8. Ms. Monroe received 40% of the votes in a recent election. If she got 672 votes, how many people voted in the election?

9. Kathy made 12 of 16 free throws in a basketball game. What percent of her free throws did she make?

10. Brad made 8 of 12 field goal attempts in a basketball game. What percent of his field goal attempts did he make?

Write a proportion for each problem. Solve the proportion and answer the question.

11. On the map of Antarctica, 1 in. represents 1200 mi. How far is Palmer from the South Pole if the distance is $1\frac{1}{2}$ in. on the map?

12. On the map of Colorado, 1.5 cm represents 40 km. How many kilometers is Cortez from Pueblo if the distance is 14.1 cm on the map?

13. Ms. Maple drove 756 km and used 56 L of gasoline. What was her average fuel consumption (km/L)?

14. Mr. Pine drove 588 mi and used 21 gal of gasoline. What was his average fuel consumption (mi/gal)?

15. The sales tax in Davis City is 5%. What is the tax on a purchase that cost $22.84? (Round your answer to the nearest cent.)

16. The hotel tax in Benton is 8%. What is the tax on a hotel bill of $55.50?

17. The Mustangs' quarterback completed 80% of his passes. How many passes did he throw if he completed 12?

18. The catcher on the Panthers softball team was hitting .350 (350 hits per 1000 at bat). If she had 56 hits, how many times was she at bat?

19. Three of the 25 students in a class were absent on Monday. What percent of the students were absent?

20. Tony scored 38 out of a possible 40 points on a test. What percent did he get wrong?

21. Marie is paid $6.25 per hour. How much is she paid for working 4 h, 15 min?

22. Crystal is paid $6.30 per hour. How many minutes did she work if she was paid $8.19?

23. Bars of soap are selling at the price of 3 for $1.17. What will 5 bars of soap cost?

24. Cans of soup are selling at the price of 2 for 65¢. What will 10 cans of soup cost?

25. At a constant speed of 55 mph, how many minutes will it take to drive 20 mi?

26. At a constant rate of 4 mph, how many minutes will it take to walk $\frac{2}{3}$ mi?

B 27. The width and length of a rectangle are in the ratio $\frac{5}{7}$. How wide is the rectangle if the length is 34 cm more than the width?

28. Approximately 9,912,500 people live in Belgium, which has an area of 30,500 km^2. If Iceland, with an area of approximately 100,000 km^2, had the same population density (population per square kilometer) as Belgium, what would Iceland's population be? (Iceland's actual population is approximately 230,000.)

29. Alberts topped Barlow in a recent election by a 6-to-5 margin. (The ratio was $\frac{6}{5}$.) How many votes did each get if a total of 4191 votes were cast for the two candidates?

30. Approximately 127,500,000 people live in Brazil, which has an area of 8,500,000 km^2. India has an area of about 3,187,500 km^2. If India had the same population density as Brazil, what would India's population be? (The actual population of India is more than 670,000,000.)

31. The Ace Financial Company charges 18% on loans for one year. If $2000 is borrowed, how much must be repaid, including the interest charges?

32. If an investment earns $9\frac{1}{2}$% interest for one year, what is the value of a $500 investment after one year, including the interest earned?

33. During their careers in the National Football League, which of these quarterbacks completed the greatest percent of his passes?

Player	Passes attempted	Passes completed
Sonny Jurgenson	4262	2433
Bart Starr	3149	1808
Roger Staubach	2958	1685
Fran Tarkenton	6467	3686
Johnny Unitas	5186	2830

34. A $32 sweater is marked "20% off." What is the sale price?

35. A clerk earns $90 a week plus a commission of 10% on sales. What were the clerk's sales if her total wage for a week was $202.55?

36. The Tylers drove 341 mi, getting 27.5-mi/gal fuel consumption. Then they drove 315 mi, getting 25-mi/gal fuel consumption. What was their average fuel consumption for the entire trip?

37. An investment of $1280 was worth $1382.40 at the end of one year. What was the annual interest rate?

38. A jacket marked $48.50 was on sale for $39.77. What percent was the discount?

39. On a map of Switzerland, 4 cm represents 50 km. How many miles is it from Zurich to Berne if it is 7.4 cm on the map? (1 mi = 1.6 km) Round your answer to the nearest mile.

40. In Germany the price of beef is given in marks per kilogram. Suppose that the monetary exchange rate is 2.5 marks per dollar. What is the cost in marks per kilogram of beef that costs $2.40 per pound? (1 kg = 2.2 lb)

41. A jogging suit was on sale for 25% off. When it didn't sell, the price was reduced an additional 20% and marked $19.20. What was the original price?

■ REVIEW EXERCISES

Solve each system by multiplication and addition.

1. $2x - 3y = -7$

$4x + 2y = 10$

2. $2x + 3y = 4$

$3x + 5y = 10$ [6–6]

Solve.

3. We want to create a 10-lb assortment of nuts costing $4.25/lb. In order to do this, we mix nuts worth $4.00/lb with nuts worth $5.00/lb. How many pounds of each will we use?   [6–7]

4. List all of the positive integer factors of 18.   [7–1]

5. Write the prime factorization of 60.   [7–1]

6. State the greatest common factor of 40 and 60.   [7–2]

8–10 Problem Solving—Direct Variation

Preview

The distance traveled by a person walking at a rate of 5 km/h is shown in the equation, table, and graph.

$$d = 5t$$

Time (h)	0	0.5	1	2	3
Distance (km)	0	2.5	5	10	15

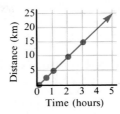

- If the time is doubled, what happens to the distance?
- If the time is tripled, what happens to the distance?
- If the time is halved, what happens to the distance?

In the above example one quantity varies proportionately as another. In this lesson you will study this type of function.

■ LESSON

Suppose that someone is paid $4 per hour. If x is the number of hours worked and y is the total wage paid, then these formulas show the relationship between x and y:

$$y = 4x \quad \text{and} \quad \frac{y}{x} = 4$$

This relationship is called **direct variation,** and 4 is called the **constant of variation.** We say that y varies directly as x. That is, the wage paid varies directly as the number of hours worked. Here are some values of x and y.

x	1	2	3	4	5	6
y	4	8	12	16	20	24

Note that if the number of hours is doubled, the wage is doubled, and if the number of hours is divided by three, the total wage is also divided by three. That is what is meant by direct variation—if one value increases or decreases, the other value increases or decreases proportionately.

Both the equation $y = 4x$ and the graph show that when two quantities vary directly, they are components of a linear function in which the y-intercept is 0.

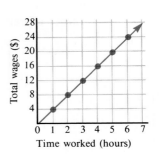

Definition: Direct Variation

A direct variation is a function defined by an equation of the form

$$y = kx \quad \text{or} \quad \frac{y}{x} = k$$

where k is a constant not equal to zero. y is said to *vary directly* as x, and k is called the *constant of variation*.

In this text, we will consider only direct variation in which k is positive.

Example 1 A certain brand of taco sauce costs $.99 per bottle. Does the total cost vary directly as the number of bottles purchased?

Solution If b is the number of bottles purchased and C is the total cost, then this equation shows the relationship:

$$C = 0.99b \quad (b \text{ is a positive integer})$$

Answer The equation is in the form $y = kx$ ($k \neq 0$). Therefore, the total cost varies directly as the number of bottles purchased.

Example 2 A class made arrangements to go on a field trip by having some parents drive their cars. One car would be needed for every 5 students. Does the number of cars vary directly as the number of students?

Solution Here are some values of the numbers of cars and students.

Number of students	1	2	5	6	7	11	16	20
Number of cars	1	1	1	2	2	3	4	4

The number of cars does not increase in the same way that the number of students increases.

Answer The number of cars needed does not vary directly as the number of students.

This graph shows that the relationship between the number of students and the number of cars needed is not linear.

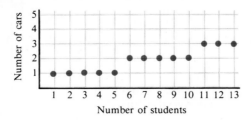

If (x_1, y_1) and (x_2, y_2) are ordered pairs of a direct variation, there is a number k such that

$$y_1 = kx_1 \quad \text{and} \quad y_2 = kx_2,$$

$$\frac{y_1}{x_1} = k \quad \text{and} \quad \frac{y_2}{x_2} = k.$$

So,
$$\frac{y_1}{x_1} = \frac{y_2}{x_2} \quad \text{and} \quad \frac{x_1}{y_1} = \frac{x_2}{y_2}.$$

The following example shows how to use proportions to solve direct variation problems.

Example 3 If y varies directly as x, and $x = 2$ when $y = 7$, then what is the value of x when $y = 16$?

Solution Since y varies directly as x, we can write this proportion:

$$\frac{2}{7} = \frac{x}{16}$$

$$7x = 32$$

$$x = \frac{32}{7}$$

Answer When $y = 16$, $x = \frac{32}{7}$.

■ CLASSROOM EXERCISES

Does y vary directly as x?

1. $y = 7x$ **2.** $y = 3x + 1$ **3.** $y = x^2$ **4.** $\dfrac{y}{x} = 10$

5.

x	1	2	3	4	5
y	1.5	3	4.5	6	7.5

6.

x	1	2	3	4	5
y	3	6	7	8	9

7.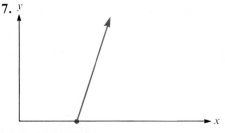

8.

9. Suppose that x and y vary directly. If x is 5 when y is 3, what is the value of x when y is 15?

10. A bicyclist rides at the constant rate of 25 km/h. Do the distance ridden and the time ridden vary directly?

■ WRITTEN EXERCISES

Determine whether y varies directly as x.

A **1.** $y = 3x + 2$ **2.** $y = \frac{1}{3}x + 6$ **3.** $y = \frac{1}{2}x$

     4. $y = 4x$ **5.** $\frac{y}{x} = 6$ **6.** $\frac{y}{x} = 5$

7.

x	2	4	6	8
y	1	2	3	4

8.

x	3	6	9	12
y	3	6	7	8

9.

x	4	8	12	16
y	2	6	10	14

10.

x	1	2	3	4
y	12	6	4	3

11.

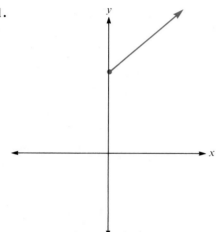

12.

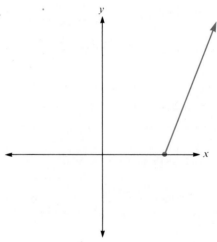

13.

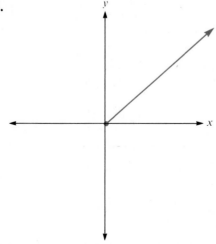

14.

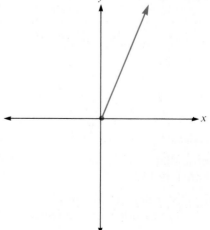

15. First-class stamps cost 22¢. Let x = the number of stamps purchased and y = the total cost of the stamps.

16. Shawn charges $1 plus $1.50 per hour for babysitting. Let x = the number of hours babysitting and y = Shawn's total charge.

17. The width of a rectangle is 5 cm. Let x = the length in centimeters and y = the area in square centimeters.

18. The base of a triangle is 10 cm. Let x = the altitude in centimeters and y = the area in square centimeters.

19. The area of a rectangle is 24 cm². Let x = the length in centimeters and y = the width in centimeters.

20. A walker traveled 6 miles. Let x = the walker's rate (in miles per hour) and y = the time (in hours).

Suppose y varies directly as x. Find the constant of variation, k.

21. $y = 8$ when $x = 2$ **22.** $y = 8$ when $x = 4$ **23.** $y = 2$ when $x = 8$

24. $y = 4$ when $x = 8$ **25.** $y = 13$ when $x = 7$ **26.** $y = 17$ when $x = 11$

Suppose y varies directly as x in the following exercises.

27. $y = 2$ when $x = 3$. Find y when $x = 24$.

28. $y = 2$ when $x = 3$. Find y when $x = 18$.

29. $y = 2$ when $x = 3$. Find x when $y = 24$.

30. $y = 2$ when $x = 3$. Find x when $y = 18$.

31. $y = 15$ when $x = 12$. Find y when $x = 40$.

32. $y = 15$ when $x = 12$. Find y when $x = 100$.

33. $y = 15$ when $x = 12$. Find x when $y = 40$.

34. $y = 15$ when $x = 12$. Find x when $y = 100$.

Suppose the quantities in the following exercises vary directly. Write an equation that fits the problem. Solve the equation and answer the question.

35. Jenny worked 5 h and was paid $18.75. How much is she paid for working 8 h?

36. A 10-m length of string weighs 350 g. How long is a string that weighs 875 g?

37. When the population of Falls City was 25,000, the residents used 40,000,000 gal of water in one year. Estimate the amount of water which will be used by the residents of Falls City when the population grows to 35,000.

38. Eight slices of white enriched bread yield 18 g of protein. How many grams of protein are in 3 slices of white enriched bread?

39. A gallon of paint will cover 450 ft^2 of a previously painted smooth surface. How many gallons of paint are needed to cover 2000 ft^2? (Round your answer to the nearest whole number of gallons.)

40. A gallon of paint will cover 300 ft^2 of an unpainted rough surface. How many gallons of paint are needed to cover 2000 ft^2? (Round your answer to the nearest whole number of gallons.)

Which of the following relations are examples of direct variation?

B **41.** $y - 2x = 0$ **42.** $y - 8 = x + 8$ **43.** $\dfrac{y + 14}{2} = x + 7$

44. The amount of gasoline purchased and its cost. (Number of gallons, number of dollars.)

45. The weight of an object in grams and its weight in kilograms. (Number of grams, number of kilograms.)

46. A student's age and the student's grade level. (Number of years, number of the grade level.)

The formula $V = kT$ (Charles' Law) expresses the relationship between the volume of a gas V under constant pressure and its temperature T measured in degrees Kelvin. Use the formula to solve these problems.

47. Nine cubic meters of oxygen are kept under constant pressure while the oxygen's temperature is raised from 300° Kelvin to 350° Kelvin. What is the new volume?

48. Forty liters of helium, kept under constant pressure, are cooled from 373° Kelvin (the boiling point of water) to 273° Kelvin (the freezing point of water). What is the new volume? (Round your answer to the nearest whole number of liters.)

If there is a constant k such that $y = kx^2$ (or $y = kx^3$), then we say that y varies directly as the square (or cube) of x. Write an equation that fits the problem. Solve the equation and answer the question.

C **49.** The surface area of a cube varies directly as the square of the length of an edge. If the edge is 3 cm and the surface area is 54 cm^2, what is the surface area of a cube when the edge is 6 cm? [*Hint:* To find k, solve $54 = k(3^2)$.]

50. The area of a circle varies directly as the square of its diameter. If the diameter is 100 m, the area is 7854 m^2. What is the area of a circle when the diameter is 1000 m?

51. The volume of a tetrahedron varies directly as the cube of the length of one of its edges. The volume is 117.85 cm^3 when the edge is 10 cm. What is the volume when the edge is 100 cm?

52. The surface area of a sphere varies directly as the square of its diameter. The surface area is 28.278 cm^2 when the diameter is 3 cm. What is the radius of a sphere if its surface area is 113.112 cm^2?

Expand and simplify.

1. $(x + 3)(x - 5)$ **2.** $(2x + 3)(x + 4)$ [7–5]

Factor completely.

3. $2a^2 - 6a$ [7–4]

4. $x^2 - 49$ [7–6]

5. $x^2 + 10x + 25$ [7–7]

6. $x^2 - 7x + 12$ [7–8]

7. $x^2 + 2x - 8$ [7–9]

8. $2x^2 + 5x + 3$ [7–10]

a	b	c	d
-6	12	-4	3

Substitute and simplify.

9. $\dfrac{a}{b} + \dfrac{c}{d}$ **10.** $\dfrac{(a + c)}{(b + d)}$ **11.** $\dfrac{a}{b} \cdot \dfrac{c}{d}$ **12.** $\dfrac{ac}{bd}$ [2–5]

Self-Quiz 3

**8–7,
8–8**   Solve.

1. $\dfrac{16}{x} = \dfrac{4}{3}$ **2.** $\dfrac{y - 1}{y + 2} = \dfrac{2}{5}$

3. $15\% = \dfrac{45}{x}$ **4.** $\dfrac{w}{w + 17} = 32\%$

**8–8,
8–9**   **5.** The interest charged on a $4500 loan over a period of 30 months was $1575. What was the annual interest rate?

6. The width and height of a rectangular window are in the ratio $\dfrac{5}{9}$. What is the height if it is 36 cm more than the width?

7. The Rowdies basketball team was successful on 45% of their shots in one game. How many shots did they shoot if they made 36 baskets?

8–10   **8.** Suppose that y varies directly as x. If $y = 48$ when $x = 15$, find y when $x = 35$.

8-11 Problem Solving—Inverse Variation

Preview

Compare the *x*- and *y*-values in these functions. In which functions is the product of the *x*- and *y*-values constant?

In which ones is the difference or quotient constant?

A:

x	30	60	45	15	900
y	1	2	1.5	0.5	30

B:

x	30	15	10	7.5	6
y	1	2	3	4	5

C:

x	10	12	14	16	18
y	14	15	16	17	18

D:

x	10	12	14	16	18
y	12	14	16	18	20

E:

x	0.5	1	2	4	8
y	32	16	8	4	2

F:

x	6	10	14	18	22
y	24	20	16	12	8

Did you find that two functions have constant products?

For each statement below, identify the function above (A, B, C, D, E, or F) that is being described.

- *x* and *y* are the length and width of a rectangle that has an area of 30 cm².
- *x* and *y* are the area and width of a rectangle that is 30 cm long.
- *x* and *y* are the width and length of a rectangle in which the length is 2 cm more than the width.

In this lesson you will study functions involving constant products.

■ LESSON

Suppose that you have to drive 100 mi. This table shows the time that would be required to complete the trip at various speeds.

Rate (mph)	2	5	10	20	25	50
Time (h)	50	20	10	5	4	2

Note that as one quantity gets larger, the other gets proportionately smaller. For example, if the speed is doubled from 5 mph to 10 mph, the time is halved from 20 h to 10 h. If the speed is quartered from 20 mph to 5 mph, the time is quadrupled from 5 h to 20 h. The product of the rate *r* and the time *t* is always equal to the constant distance 100.

$$rt = 100$$

The graph of the equation shows that as r gets larger, t gets smaller, and as r gets smaller, t gets larger. The graph is not a straight line, so clearly this is not a linear function. The equation is a second-degree equation and its graph is part of a curve called a **hyperbola**.

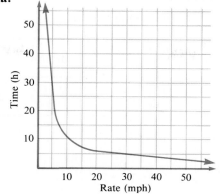

We say that time **varies inversely** as the rate.

Definition: Inverse Variation

An inverse variation is a function defined by an equation of the form

$$xy = k \quad \text{or} \quad y = \frac{k}{x},$$

where k is a constant greater than 0. y is said to *vary inversely* as x, and k is called the *constant of variation.*

Example 1   Consider all rectangles with area equal to 12 cm². Do their lengths and widths vary inversely?

Solution   The formula for the area of a rectangle is $lw = A$. In this case the area is 12.

$$lw = 12$$

This equation has the required form, $xy = k$ where $k = 12$.

Answer   The length and width of rectangles with areas equal to 12 cm² vary inversely.

Example 2 Draw the graph of the equation $lw = 12$.

Solution Since it would not make sense to have negative values for the length and width, $l > 0$ and $w > 0$.

Make a table of some values of l and w.

Length (cm)	$\frac{1}{2}$	1	2	3	4	6	12	24
Width (cm)	24	12	6	4	3	2	1	$\frac{1}{2}$

Answer

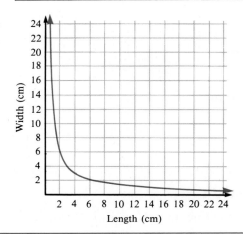

Example 3 Suppose that y varies inversely as x. If $x = 3$ when $y = 12$, what is the value of y when $x = 9$?

Solution Since y varies inversely as x, there is a k such that

$$x_1y_1 = k \quad \text{and} \quad x_2y_2 = k$$

so

$$x_1y_1 = x_2y_2.$$

Substitute 3 for x_1, 12 for y_1, and 9 for x_2.

$$3(12) = 9y_2$$

Solve for y_2. $\qquad 36 = 9y_2$

$$4 = y_2$$

Answer When $x = 9$, $y = 4$.

CLASSROOM EXERCISES

Does y vary inversely as x? If it does, state the value of the constant of variation.

1. $xy = 4$

2. $2xy = 8$

3. $xy - 1 = 9$

4. $3x = y$

5. $\dfrac{xy}{2} = 3$

6. $y = \dfrac{2}{x}$

7.

x	1	2	3	6	9	18
y	18	9	6	3	2	1

8.

x	1	2	3	4	5	6	7	8	9
y	9	8	7	6	5	4	3	2	1

9.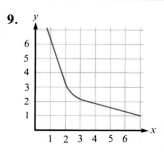

10.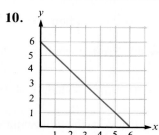

11. Suppose that y varies inversely as x. If $x = 5$ when $y = 10$, what is the value of x when $y = 4$?

12. Graph $xy = 6$ for $x > 0$.

WRITTEN EXERCISES

Indicate how x and y vary—inversely, directly, or "other."

A

1. $xy = 18$

2. $xy = 24$

3. $y = \dfrac{2}{3}x$

4. $y = \dfrac{3}{4}x$

5. $xy = \dfrac{4}{3}$

6. $xy = \dfrac{6}{5}$

7. $x + y = 24$

8. $x + y = 18$

9. $y = \dfrac{5}{x}$

10. $y = \dfrac{8}{x}$

11. $xy + 2 = 2xy - 2$

12. $xy - 3 = 3xy - 30$

Find the constant of variation, k.

13. $xy = 20$

14. $xy = \dfrac{1}{20}$

15. $y = \dfrac{0.5}{x}$

16. $y = \dfrac{10}{x}$

17. $3xy - 6 = 15$

18. $2xy - 4 = 16$

Graph each equation for $x > 0$.

19. $xy = 18$ **20.** $xy = 15$ **21.** $y = \dfrac{1}{x}$

22. $y = \dfrac{4}{x}$ **23.** $xy + 2 = 8$ **24.** $xy - 3 = 6$

Assume that x and y vary inversely in the following exercises.

25. $y = 5$ when $x = 6$. Find y when $x = 12$.

26. $y = 8$ when $x = 5$. Find y when $x = 2.5$.

27. $y = 5$ when $x = 6$. Find x when $y = 10$.

28. $y = 8$ when $x = 5$. Find x when $y = 4$.

29. $y = \dfrac{1}{2}$ when $x = 40$. Find y when $x = 20$.

30. $y = \dfrac{1}{3}$ when $x = 30$. Find y when $x = 10$.

Assume that the quantities in the following exercises vary inversely. Write an equation that fits the problem. Solve the equation and answer the question.

31. A distance was traveled in 4 h at an average speed of 50 mph. The return trip was made in 5 h. What was the average speed on the return trip?

32. A trip was made in 4 h at an average speed of 55 mph. The return trip was made in 5 h. What was the average speed on the return trip?

33. A canoeist paddled downstream for 5 h, averaging 6 mph. On the return trip the average speed against the current was 3 mph. How long did it take the canoeist to make the return trip?

34. Flying into a headwind, a light plane, averaging 180 mph, took 2 h 20 min to reach its destination. The return trip, flying with a tailwind, took only 2 h. What was the average speed on the return trip?

35. The average score for the starting five players on a basketball team was 12 points each. If one of the players made no points, how many points would each of the other 4 players have to average in order for the starting team to maintain its average score?

36. Twenty-five students were asked to contribute $4 each for a class project. If 5 of the students did not contribute, how much must each of the other students contribute to achieve the same total?

37. A gallon of paint will cover a length of 50 ft of a wall that is 9 ft high. What length of a 10-ft high wall would be covered by this gallon of paint?

38. A bag of lawn food covers a rectangular lawn that is 40 ft wide and 125 ft long. A golf course has a strip of lawn that is 8 ft wide along a path. How long a strip could be covered by a bag of lawn food?

39. A regular hexagon and a regular octagon have the same perimeter. The hexagon's sides are 12 cm long. How long is each of the octagon's sides?

40. An equilateral triangle and a square have the same perimeter. How long is a side of the square if each side of the triangle is 90 cm?

If two objects on a lever are balanced, their mass and their distance from the *fulcrum* (point of balance) vary inversely. For example, a 60-g object that is 8 cm from the fulcrum balances a 40-g object that is 12 cm from the fulcrum (since $60 \cdot 8 = 40 \cdot 12$).

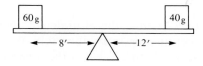

B **41.** If a 50-kg object 2 m from a fulcrum balances a 40-kg object, how far is the 40-kg object from the fulcrum?

42. A 100-kg object 1.8 m from a fulcrum balances an object that is 1.2 m from the fulcrum. What is the mass of the other object?

43. How much mass must be placed 2 m from the fulcrum to balance a 74-kg object that is 2.2 m from the fulcrum?

44. A 40-kg object and a 60-kg object are placed on opposite ends of a board that is 3 m long. How far from the 40-kg object is the fulcrum if the objects are balanced?

45. A 174-lb person and 114-lb person sitting on opposite ends of a 12-ft seesaw are in balance. How far is the fulcrum from the 174-lb person?

The formula $PV = k$ (Boyle's Law) expresses the relationship between the volume of a gas V at a constant temperature and its pressure P. Use the formula to solve these problems.

46. The pressure acting on 10 m³ of air is raised from 1 atmosphere to 2 atmospheres. The temperature is kept constant. What is the volume of the air after the pressure is applied?

47. A helium-filled balloon has a volume of 20 m³ at sea level (where the pressure is 1 atmosphere). What is the volume of the balloon after it rises to an altitude where the pressure is 0.8 atmosphere? (Assume its temperature is kept constant.)

48. At a depth of 21 m under water, the pressure is 3 atmospheres. At the surface the pressure is 1 atmosphere. A bubble of air with a volume of 6 cm³ escapes from the mouthpiece of a diver working 21 m below the surface. What is the volume of the bubble as it reaches the surface?

Solve.

C **49.** In ellipses that have an area of 100 cm², the values of a and b vary inversely. If $a = 3$ cm when $b = 10.61$ cm, what does b equal when $a = 1.5$ cm?

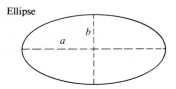
Ellipse

50. In pyramids that have a volume of 720 cm³, the altitude (in centimeters) and the area of the base (in square centimeters) vary inversely. When the altitude is 30 cm, the area of the base is 72 cm². What is the area of the base if the altitude is 20 cm?

51. In pyramids with square bases that have a volume of 540 cm³, the altitude (in centimeters) and the square of the length of a side of the square vary inversely. When the altitude is 20 cm, the side of the square is 9 cm. What is the side of the square when the altitude is 5 cm?

■ REVIEW EXERCISES

Solve.

1. $x(x + 5) = 0$ **2.** $(x + 4)(2x - 1) = 0$ [7–3]

3. $x^2 + 3x = 0$ **4.** $2x^2 = -6x$ [7–4]

5. $x^2 + 12x + 36 = 0$ **6.** $x^2 + 16x + 64 = 0$ [7–7]

7. $2x^2 + 7x - 4 = 0$ **8.** $3x^2 + 4x - 4 = 0$ [7–10]

■ CHAPTER SUMMARY

• **Vocabulary**

algebraic fractions	[page 359]	rate	[page 402]
admissible value	[page 359]	principal	[page 402]
equivalent fractions	[page 360]	interest	[page 402]
least common denominator	[page 376]	direct variation	[page 411]
ratio	[page 394]	constant of variation	[page 411]
proportion	[page 394]	hyperbola	[page 419]
cross multiply	[page 395]	inverse variation	[page 419]
percent	[page 400]		

- Two fractions are equivalent if one of them can be transformed into the other by multiplication or division. [8–1]

  For all numbers a, b, and c ($b \neq 0$ and $c \neq 0$),

  $$\frac{a}{b} = \frac{ac}{bc} \quad \text{and} \quad \frac{a}{b} = \frac{a \div c}{b \div c}$$

- Multiplying fractions [8–2]

  For all numbers a, b, c, and d ($b \neq 0$ and $d \neq 0$),

  $$\frac{a}{b} \cdot \frac{c}{d} = \frac{ac}{bd}$$

- Dividing fractions [8–2]

  For all numbers a, b, c, and d ($b \neq 0$, $c \neq 0$, and $d \neq 0$),

  $$\frac{a}{b} \div \frac{c}{d} = \frac{a}{b} \cdot \frac{d}{c}$$

- Adding or subtracting fractions with like denominators [8–3]

  For all numbers a, b, and c ($c \neq 0$),

  $$\frac{a}{c} + \frac{b}{c} = \frac{a+b}{c} \quad \text{and} \quad \frac{a}{c} - \frac{b}{c} = \frac{a-b}{c}$$

- To add or subtract fractions with unlike denominators, transform the fractions into equivalent fractions with a common denominator. [8–4]
- Adding or subtracting fractions with unlike denominators

  For all numbers a, b, c, and d ($c \neq 0$ and $d \neq 0$),

  $$\frac{a}{c} + \frac{b}{d} = \frac{ad + bc}{cd} \quad \text{and} \quad \frac{a}{c} - \frac{b}{d} = \frac{ad - bc}{cd}$$

- To find the least common multiple of the denominators of two algebraic fractions, write the prime factorization of each denominator. Then identify the highest power of each prime factor and write the product of the highest powers. [8–5]
- To solve an equation containing fractions, multiply both sides of the equation by the least common denominator of the fractions to clear the equation of fractions.
- Cross-multiplication property [8–7]

  For all real numbers a, b, c, and d ($b \neq 0$ and $d \neq 0$),

  $$\text{if } \frac{a}{b} = \frac{c}{d}, \text{ then } ad = bc.$$

- A ratio written as a fraction can be changed to a percent by: [8–8]

  1. changing to an equivalent fraction with a denominator of 100 and then writing the numerator of that fraction followed by a percent symbol, or

  2. changing the fraction to a decimal, multiplying the decimal by 100, and writing that number followed by a percent symbol.

- If $a\%$ of $b = c$, then

  $$\frac{a}{100} \cdot b = c \text{ and } \frac{a}{100} = \frac{c}{b}$$

- The relationship between interest (i), principal (p), rate (r), and time (t) is given by the formula $i = prt$.   [8–10]

- A direct variation is a function defined by an equation of the form   [8–11]

$$y = kx \quad \text{or} \quad \frac{y}{x} = k, \; k \neq 0$$

- An indirect variation is a function defined by an equation of the form   [8–11]

$$xy = k \quad \text{or} \quad y = \frac{k}{x}, \; k > 0$$

■ CHAPTER REVIEW

8–1 Objectives: To determine admissible values of variables in algebraic fractions.
To simplify algebraic fractions.

1. Which values of x are not admissible for the fraction $\dfrac{x}{x-1}$ to be defined?

2. State whether the fractions $\dfrac{2x+1}{3x+1}$ and $\dfrac{2x}{3x}$ are equivalent.

Simplify. List any necessary restrictions on the variables.

3. $\dfrac{2x^2 y}{4xy}$

4. $\dfrac{(x+2)(x-1)}{(x-1)(x-2)}$

8–2 Objective: To multiply and divide algebraic fractions.

Simplify. Assume that no denominator is zero.

5. $\dfrac{6x}{y} \cdot \dfrac{2y}{3x^2}$

6. $\dfrac{(a+b)^2}{(a+b)(a-b)} \cdot \dfrac{a-b}{a+b}$

7. $\dfrac{16a^2}{9b} \div \dfrac{4a}{3b}$

8. $\dfrac{x(x-y)}{xy} \div \dfrac{x-y}{y(x-y)}$

8–3 Objective: To add and subtract fractions with like denominators.

Simplify. Assume that no denominator is zero.

9. $\dfrac{2c}{c+4} + \dfrac{8}{c+4}$

10. $\dfrac{4h}{h^2-16} - \dfrac{16}{h^2-16}$

Simplify. List any necessary restrictions on the variables.

11. $\dfrac{b}{b+2} - \dfrac{4}{b+2}$

12. $\dfrac{2a+4}{a-1} + \dfrac{a-7}{a-1}$

8–4 Objective: To add and subtract fractions with unlike denominators.

13. State the least common denominator for $\dfrac{3}{2ab^2}$ and $\dfrac{5}{12a^2}$.

Simplify. Assume that no denominator is zero.

14. $\dfrac{3}{x} - \dfrac{2}{3x}$ 15. $\dfrac{1}{mn} + \dfrac{3}{n^2}$ 16. $\dfrac{6}{y^2 + 2} - \dfrac{4}{y}$

8–5 Objective: To solve equations that contain algebraic fractions.

Solve.

17. $\dfrac{x}{36} = \dfrac{4}{9}$ 18. $\dfrac{3}{y} = \dfrac{2}{3} - \dfrac{5}{y}$ 19. $x + \dfrac{5}{x} = 6$

8–6 Objective: To use fractional equations to solve problems.

Solve.

20. A circuit is made up of two resistors in parallel. The total resistance is 4 ohms. If one of the resistors has a resistance of 12 ohms, what is the resistance of the other resistor? (Use the equation $\dfrac{1}{R} = \dfrac{1}{r_1} + \dfrac{1}{r_2}$).

21. If one printing press can complete a job in a 4 h and another press can complete the job in 6 h, how long will it take both machines working together to complete the job?

8–7 Objective: To simplify ratios and to solve proportions.

22. For the ratio 14 in. to 2 ft, write the measurements using the same units. Then simplify the ratio.

23. Simplify this ratio. $\dfrac{45}{100}$

24. Solve this proportion. $\dfrac{2}{3} = \dfrac{x+4}{x+9}$

25. The ratio of the adjacent sides of a rectangle is 5 to 2. What are the length and width if the perimeter is 21 cm?

8–8 Objective: To solve percent and simple interest problems.

26. Write $\dfrac{2}{5}$ as a percent.

27. Write 65% as a simplified ratio.

Solve.

28. $75\% = \dfrac{300}{x}$ 29. 6% of $x = 15$

30. Ms. Salgado invested $2000 for 3 years at an annual rate of 9%. How much interest did she earn?

8–9   Objective: To solve problems using proportions.

Solve.

31. John rode his bicycle at the rate of 24 km/h. How far did he go in 25 min?

32. When Mr. Kozlak's home room raised $200 in a student fundraising drive, he told them that they had achieved 80% of their goal. What was their goal?

8–10   Objective: To solve direct variation problems.

33. Does y vary directly as x when $y = \frac{2}{3}(x + 6)$?

34. If y varies directly as x and $y = 15$ when $x = 10$, find y when $x = 24$.

35. If Suzanne earns $7.00 in 2 h, how much will she earn at the same rate in 5 h?

8–11   Objective: To solve inverse variation problems.

36. Graph the equation $xy = 6$ for $x > 0$.

37. Suppose that x and y vary inversely and that $y = 8$ when $x = 3$. Find y when $x = 6$.

38. Averaging 15 km/h, June and Sue took 2 h to ride their bikes to Long Lake. If their return trip along the same route took an hour and a half, what was their average speed returning?

■ CHAPTER 8 SELF-TEST

Simplify. List any restrictions on the variables.

8–1   **1.** $\dfrac{6x^2y}{3xy^2}$ **2.** $\dfrac{(y-2)(y+3)}{5(y-2)}$

Simplify. Assume that no divisor is zero.

8–2,
8–3,
8–4

3. $\dfrac{9a}{7b^2} \cdot \dfrac{14b}{3a}$ **4.** $\dfrac{6x}{25y} \div \dfrac{2}{5xy}$ **5.** $\dfrac{5m}{m+2n} \cdot \dfrac{6(m+2n)}{10m^2}$

6. $\dfrac{(x-y)(x+y)}{3(x+2y)} \div \dfrac{x-y}{9}$ **7.** $\dfrac{2w}{w-5} - \dfrac{10}{w-5}$ **8.** $\dfrac{12}{a^2} - \dfrac{3}{a^2}$

9. $\dfrac{1}{y} + \dfrac{3}{4}$ **10.** $\dfrac{1}{x^2} - \dfrac{2}{xy}$ **11.** $\dfrac{5}{x-2} - \dfrac{4}{x}$

Solve.

8–5   **12.** $\dfrac{y}{32} = \dfrac{3}{4}$ **13.** $\dfrac{5}{x} + \dfrac{4}{3} = \dfrac{9}{x}$ **14.** $t + 1 = \dfrac{2}{t}$

8–7,
8–8   **15.** $\dfrac{2}{5} = \dfrac{x + 1}{x - 2}$ **16.** 125% of $x = 15$ **17.** $\dfrac{35}{125} = y\,\%$

8–6   **18.** Use the formula $\dfrac{1}{R} = \dfrac{1}{r_1} + \dfrac{1}{r_2}$ to find r_2 when $R = 3\ \Omega$ and $r_1 = 12\ \Omega$.

8–7   **19.** The ratio of the sides of two squares is 2 to 5. What is the ratio of their areas?

8–10   **20.** Suppose y varies directly as x. If $y = 48$ when $x = 30$, find y when $x = 105$.

8–11   **21.** Suppose y varies inversely as x. If $y = 15$ when $x = 56$, find x when $y = 35$.

8–8   **22.** In a six-month period, Mr. Q earned $450 interest on $15,000. What annual rate of interest did Mr. Q earn?

8–9   **23.** If one side of a square is decreased by 2 m and another side increased by 2 m, the sides will be in the ratio 5 to 7. Find the original length of a side of the square.

8–10   **24.** Ms. Melville drove 288 mi on 14 gal of gasoline. How far can she drive on 35 gal of gasoline?

8–11   **25.** Twelve boxes of oranges contain 8 oranges each. Hoping to boost sales, the grocer repackages the oranges 6 to a package. How many smaller packages can he make?

■ PRACTICE FOR COLLEGE ENTRANCE TESTS

Each question consists of two quantities, one in column A and one in column B. Compare the two quantities and select one of the following answers.

A if the quantity in column A is greater;
B if the quantity in column B is greater;
C if the two quantities are equal;
D if the relationship cannot be determined from the information given.

Comments:

- Letters such as a, b, x, and y are variables that can be replaced by real numbers.
- A symbol that appears in both columns in a question stands for the same thing in column A as in column B.
- In some questions information that applies to quantities in both columns is centered above the two columns.

	Column A	Column B	Answers
	$3 + 4$	$3 \cdot 4$	B
	$a = 5$		
	$a + b$	$b + 5$	C
	$2x + y$	$x + 2y$	D

	Column A	Column B
1.	20% of 300	300% of 20
2.	x is 50% of y	
	The percent that y is of x	500%
3.	$\dfrac{1}{2} + \dfrac{1}{4} + \dfrac{1}{8}$	$\dfrac{1}{16} + \dfrac{1}{8} + \dfrac{1}{4}$
4.	$2a - 4$	$\dfrac{6a - 12}{3}$
5.	$x \neq 2$	
	0	$\dfrac{1}{(x - 2)^2}$
6.	The ratio of $\dfrac{2}{3}$ to $\dfrac{1}{4}$	The ratio of $\dfrac{3}{4}$ to $\dfrac{1}{3}$
7.	a and b are positive and $\dfrac{a}{b} = 3$	
	$a + b$	4
8.	The price of a stereo increases by 20% and then decreases by 20%.	
	The original price of the stereo	The final price of the stereo
9.	a and b are positive integers	
	40% of a + 20% of b	30% of $(a + b)$
10.	12% of $3x = 45$	
	12% of x	15
11.	$\dfrac{0.2}{0.5} = \dfrac{0.5}{x}$	
	x	1

■ CUMULATIVE REVIEW (*Chapters 7–8*)

7–1 List all positive-integer factors of the given number.

    1. 39   **2.** 92

Write the prime factorization of each number.

    3. 48   **4.** 60

7–2 State the greatest common factor of each pair of numbers or monomials.

    5. 24, 56   **6.** 45, 72

    7. $9xy^2$, $12x^2y$ **8.** $6a^2b$, $4a^2b^3$

7–3 Solve.

    9. $x(x - 2) = 0$ **10.** $(a - 7)(a + 7) = 0$ **11.** $(2y - 1)(3y + 1) = 0$

7–4   **12.** Find the greatest common factor of the terms of the polynomial $8x + 10y$.

Factor.

    13. $14y - 7x$ **14.** $ab - b^2$

Solve.

    15. $x^2 + 4x = 0$ **16.** $y^2 = 5y$

7–5 Expand and simplify.

    17. $(a + 2)(a + 7)$ **18.** $(c - 5)(c + 5)$

    19. $(2n + 4)(n - 3)$ **20.** $(x - 8)^2$

7–6   **21.** Write $49b^2$ as a square.

Factor.

    22. $x^2 - 9$ **23.** $4x^2 - 1$

Solve.

    24. $x^2 - 49 = 0$ **25.** $x^2 = 81$

7–7 Expand and simplify.

    26. $(x + 10)^2$ **27.** $(2b - 1)^2$

7–7, Factor.
7–8

    28. $a^2 + 12a + 36$ **29.** $x^2 - 10x + 25$

    30. $p^2 - 6p + 5$ **31.** $a^2 + 7a + 12$

Solve.

32. $x^2 - 3x + 2 = 0$ **33.** $y^2 + 6y + 9 = 0$

Factor.

34. $a^2 - 11a - 12$ **35.** $y^2 + 3y - 18$

36. $2x^2 + 7x + 3$ **37.** $6n^2 - n - 1$

Solve.

38. $x^2 - x - 30 = 0$ **39.** $b^2 - b = 6$

40. Twice the square of a number minus 3 times the number equals 2. What is the number?

41. State whether this pair of fractions is equivalent for admissible values of x.

$$\frac{1}{x - 4} \quad \text{and} \quad \frac{3}{x - 12}$$

42. What value(s) of b are not admissible for the fraction to be defined?

$$\frac{b}{b + 2}$$

Simplify. Assume that no denominators are zero.

43. $\dfrac{6x}{18x}$ **44.** $\dfrac{2x - 4}{5x - 10}$ **45.** $\dfrac{b + 2}{b^2 + 5b + 6}$

46. $\dfrac{a}{b^2} \cdot \dfrac{b}{a}$ **47.** $\dfrac{16x}{y} \cdot \dfrac{3y}{4x}$ **48.** $\dfrac{2a}{b} \cdot (b^2 + 4b)$

49. $\dfrac{6x}{y} \div \dfrac{2x}{y}$ **50.** $\dfrac{5a}{c^2} \div \dfrac{10a}{c}$ **51.** $\dfrac{12}{a - 2} \div \dfrac{a}{a^2 - 4}$

Simplify. Assume that no denominators are zero.

52. $\dfrac{4b}{3} + \dfrac{2b}{3}$ **53.** $\dfrac{3}{a - 2} + \dfrac{4}{a - 2}$ **54.** $\dfrac{c}{c - 6} - \dfrac{6}{c - 6}$

55. $\dfrac{x}{4} + \dfrac{x}{3}$ **56.** $\dfrac{3}{4y} - \dfrac{1}{2y}$ **57.** $\dfrac{5}{x - 1} - \dfrac{1}{x}$

Solve.

58. $\dfrac{16}{12} = \dfrac{2a}{3}$ **59.** $\dfrac{x}{3} - \dfrac{1}{2} = \dfrac{7}{6}$

60. $\dfrac{4}{y} + \dfrac{5}{3} = \dfrac{9}{y}$ **61.** $\dfrac{5}{x} - \dfrac{1}{2x} = 3$

62. Use the formula $\dfrac{1}{R} = \dfrac{1}{r_1} + \dfrac{1}{r_2}$ to find the total resistance R when $r_1 = 4$ ohms and $r_2 = 6$ ohms.

8–6 Write an equation to solve this problem.

63. Lucinda can make the pep-rally posters in 5 h. Gary can make the same number of posters in 6 h. How long does it take Lucinda and Gary to make posters for the pep rally if they work together?

8–7 **64.** Write the measurement in the same units, and then write the ratio in simplest form.   50 minutes to 2 hours

65. Simplify the ratio. $\dfrac{26}{39}$

8–8 **66.** Write $\dfrac{7}{20}$ as a percent.

67. Write 48% as a simplified ratio.

Solve.

68. x% of $40 = 32$ **69.** 16% of $x = 9.6$

70. Claude Rich invested $2000 at 8% for 3 years. How much did he earn in interest? (Use $i = prt$.)

8–9 **71.** Mattie stocked four shelves at Foodland in 30 minutes. At that rate, how long would it take her to stock 14 shelves?

72. Jerry drove 135 mi on 6 gal of gasoline. How far can he drive on a full tank of 15 gal?

73. The ratio of ninth-grade students who can curl their tongues to those who cannot is 4 to 3. How many students have this genetic trait if there are 56 ninth-grade students?

8–10 **74.** Suppose y varies directly as x. If $y = 24$ when $x = 15$, find y when $x = 40$.

75. Does y vary directly as x if $y = 4x + 2$?

76. Suppose y varies directly as x. If $y = 16$ when $x = 12$, find the constant of variation and write the direct-variation equation.

8–11 **77.** Suppose y varies inversely as x. If $y = 12$ when $x = 3$, find x when $y = 2$.

78. If y varies inversely as x and k is the constant of variation, which of these equations is true?

a. $x + y = k$ **b.** $xy = k$ **c.** $\dfrac{y}{x} = k$

d. $y = x - k$ **e.** $y = x^k$

79. A trucker can drive from Baltimore to Indianapolis in 11 h at 50 mph. How fast must he drive in order to complete the return trip in 10 h?

9 Inequalities and Their Graphs

A manufacturer makes a cattle feed from various grains. Rather than requiring exact amounts of the grains, the recipe for the feed specifies ranges of the amounts to be used. The amount of each grain used also depends on the amounts of the other grains used. As costs of the various grains fluctuate, the manufacturer can vary the amount of each grain in order to minimize the cost of the feed and maximize the profits.

The relative amounts of the grains and their ranges are stated as inequalities. The graphs of the inequalities help to determine the optimal amount of each grain to use for maximum profit.

$$c + t \geq 60$$
$$c + t \leq 100$$
$$c \geq 2t$$
$$c \leq 5t$$

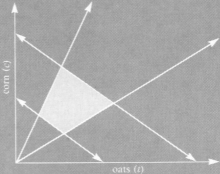

9-1 Graphing Inequalities

Preview

The qualifying standards for the 1984 Olympic trials were as follows.

Running Events	Time (h:min:s)	Field Events	Distance
100 meters	10:35	High jump	2.24 m
200 meters	20:74	Long jump	7.85 m
400 meters	46:00	Triple jump	16.20 m
800 meters	1:47:44	Shot put	19.81 m
1,500 meters	3:42:20	Discus	61.26 m
5,000 meters	13:49:00	Pole vault	5.40 m
10,000 meters	28:46:00	Javelin	78.80 m
		Hammer throw	63.00 m

In order to compete for a berth on the 1984 U.S. Olympic track and field team, an athlete must have had a performance in 1982–83 that equaled or bettered the qualifying standards. To qualify for a running event, the athlete's time must have been equal to or less than these standards. To compete in a field event, the athlete's performance must have been equal to or greater than the standard height or distance.

■ LESSON

Mathematical sentences that state that two quantities are not equal are called **inequalities.** These symbols are used in inequalities:

$<$ is less than
$>$ is greater than
$\neq$ is not equal to

Here are some examples of inequalities. These inequalities are true.

$6 < 8$ 6 is less than 8
$8 > 6$ 8 is greater than 6
$6 \neq 8$ 6 is not equal to 8

The two other inequality symbols are $\leq$ and $\geq$.

$\leq$ is less than or equal to
$\geq$ is greater than or equal to

Inequalities using the $\leq$ or $\geq$ symbols are **compound inequalities.** The sentence

$$6 \leq 8$$

is a short way of writing

$$6 < 8 \quad \text{or} \quad 6 = 8$$

A compound sentence using "or" is true if at least one of the two conditions is true. It is false if both of the conditions are false.

Example 1   True or false? $12 \leq 13$

Solution $12 \leq 13$ says that "$12 < 13$ or $12 = 13$." Since $12 < 13$ is true, the compound inequality is true.

Example 2   True or false? $12 \leq 12$

Solution $12 \leq$ says that "$12 < 12$ or $12 = 12$." Since $12 = 12$ is true, the compound inequality is true.

Inequalities with variables are open sentences just as equations with variables are open sentences. Open sentences are neither true nor false. The **solutions of an inequality** are those numbers that make the inequality true when they are substituted for the variable. Inequalities in one variable usually have infinitely many numbers as solutions. Number-line graphs make the solutions easier to understand.

Example 3   Graph on a number line. $x < 4$

Solution   Any number less than 4 is a solution. This is shown by coloring red points on the number line that have coordinates less than 4. The red arrowhead indicates that the graph extends infinitely far to the left. The open circle indicates that the graph begins at 4 but does not include 4.

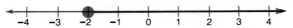

Example 4   Graph on a number line. $y \geq -2$

Solution -2 and any number that is greater than -2 is a solution. The solid circle indicates that the graph starts at -2 and includes -2. The red arrowhead indicates that the graph extends infinitely far to the right.

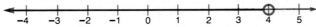

Example 5   Graph on a number line. $b \neq 2.5$

Solution   Every number except 2.5 makes the inequality true. The red arrowheads indicate that the graph extends infinitely far in both directions.

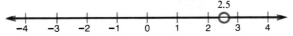

■ CLASSROOM EXERCISES

True or false?

1. $5 < 12$

2. $5 < -12$

3. $5 \leq 12$

4. $5 \leq -12$

5. $5 \leq 5$

6. $-3 > -6$

7. $-2 \geq 0$

8. $7 \geq -13$

9. $3.5 \neq -3.5$

Let x represent the number. Write an inequality that fits the situation.

10. The number is greater than 15.

11. The number is at least 20.

12.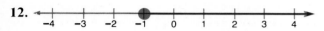

13.

Graph on a number line.

14. $x < 5$

15. $y \geq -3$

16. $a \neq 0$

■ WRITTEN EXERCISES

True or false?

1. $3 < 5$

2. $4 < -6$

3. $2 > -7$

4. $0.3 > 0.5$

5. $-8 > 1$

6. $9 > -3$

7. $0.5 < -0.3$

8. $-500 < 300$

9. $-\dfrac{1}{5} \leq -\dfrac{1}{5}$

10. $-\dfrac{1}{3} \leq -\dfrac{1}{3}$

11. $-3\dfrac{1}{2} \geq -5\dfrac{1}{2}$

12. $-\dfrac{81}{2} \geq 10$

13. $-2\dfrac{1}{2} > -2\dfrac{1}{4}$

14. $2\dfrac{1}{2} > 2\dfrac{1}{4}$

15. $3\dfrac{1}{5} < 3\dfrac{1}{10}$

16. $-3\dfrac{1}{5} < -3\dfrac{1}{10}$

Let x represent the number. Write an inequality that fits the description.

17. The number is less than 100.

18. The number is greater than 500.

19. The number is at least 70.

20. The number is not more than 30.

21. The number is greater than -90.

22. The number is less than -10.

23. The number is not 2.

24. The number is not 0.

Graph on a number line.

25. $x > 3$

26. $x > 5$

27. $x < 5$

28. $x < 3$

29. $x \neq 2$

30. $x \neq -2$

31. $x \geq -3$

32. $x \geq -5$

33. $x \leq -5$

34. $x \leq -3$

35. $x < \dfrac{1}{2}$

36. $x > -\dfrac{1}{2}$

Write an inequality using x for each graph.

37.

38.

39.

40.

41.

42.

43.

44.

List all the numbers in $\{-100, -20, 0, 20, 100\}$ that are solutions of the inequality.

45. $x < 35$　　　　**46.** $x < -35$　　　　**47.** $x \neq -20$　　　　**48.** $x \neq 100$

49. $x \geq 0$　　　　**50.** $x \geq -15$　　　　**51.** $x \leq 15$　　　　**52.** $x \leq 20$

List all the numbers in $\{-5, -0.55, 0.505, 0.55, 5.5\}$ that are solutions of the inequality.

53. $x \geq 0.5$　　　　**54.** $x \geq -0.5$　　　　**55.** $x \leq -0.5$　　　　**56.** $x \leq -0.55$

57. $x \neq 5$　　　　**58.** $x \neq -5$　　　　**59.** $x < 5(0.11)$　　　　**60.** $x < \dfrac{1.01}{2}$

Write an inequality for each situation. Then graph the inequality on a number line.

B 61. To fit in the trash can, the sticks cannot be more than 90 cm long.
(Let l represent the length in centimeters.)

62. A delivery service will accept packages up to and including 70 lb.
(Let w represent the weight in pounds.)

63. The tomato plants were at least 10 cm tall before they were put on sale.

64. The diameter of each tree in the park was at least 40 cm.

65. Each coin in the collection was at least 7 years old.

66. No one had a score of 5 on the test.

67. The value of the purchase was under 50¢.

68. The purchase was at least $2.25.

List the numbers in $\{-3, -2, -1, 0, 1, 2, 3\}$ that satisfy the inequality.

C   **69.** $5x + 2 < -6$ **70.** $4x - 3 \leq -7$ **71.** $x + x + x \geq 2$

72. $2 + x > 0$ **73.** $1 - x > 0$ **74.** $-2x \geq -1$

75. For what values of b is it true that $b + \frac{1}{b} > 2?$

76. For what values of a is it true that $a - \frac{2}{a} > 1?$

[*Hint:* Some values may be negative.]

77. For what values of x is $x^2 > x?$

78. For what values of y is $y \leq \sqrt{y}?$

Graph.

79. $x^2 > 1$ **80.** $x^2 \leq 4$ **81.** $x^2 > x$ **82.** $x^2 \neq x$

■ REVIEW EXERCISES

a	b	x	y
-3	2	-5	-4

Substitute and simplify.

1. $(a - b) - (x - y)$ **2.** $x - (a + y)$ [2–3]

3. $|ax| + |by|$ **4.** $|ab + xy|$ [2–4]

Solve.

5. $3x + 8 = 5 - 2x$ **6.** $2x - 5 = 4(x + 3)$ [3–6]

7. Solve $P = 2\pi r$ for r.   **8.** Solve $ax + by = c$ for y.   [3–8]

EXTENSION   Extra costs _____

Increasing the number of steps used in producing an item usually increases the cost of the product. For example, a frozen TV dinner costs more than the same items of food purchased fresh and cooked at home. The extra costs reflect the labor, equipment, and packaging materials needed to prepare and package the dinner.

Solve the following problem to find the additional cost of extra steps in production.

Thrift-T mending tape in a dispenser costs $.82 for 950 ft of tape. The same kind of tape without the dispenser costs $.82 for 1020 feet. How much does the dispenser cost?

9–2 Compound Inequalities and Graphs

Preview Discovery

The Wizard of Odds operates two games of skill at the amusement park.

DART THROW

1	19	6	25	84
27	55	48	3	16
0	72	8	30	24
13	42	27	69	36
64	20	15	81	7

WHEEL OF FORTUNE

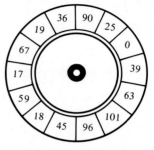

To win on the dart throw, a player has to hit a number that is a square *or* a cube. On which of these numbers does the Wizard have to award a prize?

69   36   15   27   64

In order for a player to win on the wheel of fortune, a number has to be odd *and* a multiple of 9. On which of these numbers does the Wizard award a prize?

25   36   96   18   45

In this lesson you will find numbers that satisfy two conditions joined by the word *or* or *and*.

■ LESSON

In the preceding lesson we saw that the symbols $\geq$ and $\leq$ are short ways to state that two conditions are joined by the word *or*. For example, $x \leq 4$ means "$x < 4$ or $x = 4$." Sentences that state two or more conditions are called compound sentences.

Consider this compound sentence in which the two conditions are joined by the word *or*.

$$a < -1 \quad \text{or} \quad a > 2$$

To be a solution of the sentence, a number must satisfy one condition or the other, but not necessarily both conditions. To determine whether a number is a solution, substitute it into each part of the sentence.

$$Substitute -4: \quad \begin{array}{ccc} a < -1 & or & a > 2 \\ -4 < -1 & or & -4 > 2 \\ \text{True} & & \text{False} \end{array}$$

-4 is a solution since it satisfies one of the conditions.

$$Substitute \ 0: \quad \begin{array}{ccc} a < -1 & or & a > 2 \\ 0 < -1 & or & 0 > 2 \\ \text{False} & & \text{False} \end{array}$$

0 is not a solution since it does *not* satisfy *either* condition.

$$Substitute \ 5: \quad \begin{array}{ccc} a < -1 & or & a > 2 \\ 5 < -1 & or & 5 > 2 \\ \text{False} & & \text{True} \end{array}$$

5 is a solution since it satisfies one of the conditions.

Numbers less than -1 are solutions of the compound sentence since they satisfy the first condition. Numbers between -1 and 2 and the numbers -1 and 2 are not solutions since they do not satisfy either condition. Numbers greater than 2 are solutions since they satisfy the second condition. The solution sets of compound sentences can be described using graphs. The following graph shows all the solutions of the compound sentence $a < -1$ or $a > 2$.

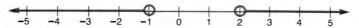

Note that -1 and 2 are key numbers. They are the boundaries of the parts of the number line that are solutions.

Some compound sentences contain the word *and*. When two conditions of a compound sentence are joined by "and," *both* conditions must be satisfied to make the sentence true.

Example 1 Graph on a number line. $x > -1$ and $x \le 2$

Solution $x > -1$ and $x \le 2$

The numbers -1 and 2 are key numbers. They divide the number line into three parts. It is important to test numbers in each of the three parts to determine whether they satisfy both conditions of the inequality.

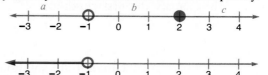

Test part a.

-1 and each number less than -1 make the compound inequality false because those numbers do not satisfy the first condition.

$$\begin{array}{ccc} -1 > -1 & and & -1 \le 2 \\ \text{False} & & \text{True} \end{array}$$

Example 1 (continued)

Test part b.

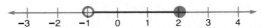

Numbers between − 1 and 2, and 2, satisfy both conditions and are, therefore, solutions of the inequality.

$$1 > -1 \quad \text{and} \quad 1 \le 2$$
$$\text{True} \qquad\qquad \text{True}$$

Test part c.

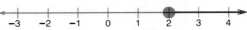

Numbers greater than 2 are not solutions of the inequality because those numbers do not satisfy the second condition.

$$5 > -1 \quad \text{and} \quad 5 \le 2$$
$$\text{True} \qquad\qquad \text{False}$$

Answer

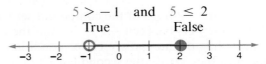

Example 2 Graph. $x < 2$ and $x > 3$

Solution There are no numbers that are both less than 2 and greater than 3. Therefore, there are no solutions to the compound sentence, and the graph contains no points.

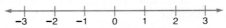

Example 3 Graph. $a \ge 3$ or $a > 4$

Solution Numbers less than 3 are not solutions since they do not satisfy either condition.

$$1 \ge 3 \quad \text{or} \quad 1 > 4$$
$$\text{False} \qquad\quad \text{False}$$

The numbers 3 and 4, and numbers between 3 and 4 are solutions since they satisfy the first condition

$$3.5 \ge 3 \quad \text{or} \quad 3.5 > 4$$
$$\text{True} \qquad\quad \text{False}$$

Numbers greater than 4 are solutions since they satisfy both conditions.

$$6 \ge 3 \quad \text{or} \quad 6 > 4$$
$$\text{True} \qquad\quad \text{True}$$

Answer

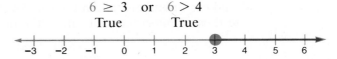

Example 4 Graph. $b \leq 0$ and $b \leq 3$

Solution Numbers less than or equal to 0 are solutions since they satisfy both conditions.

$$-1 \leq 0 \quad \text{and} \quad -1 \leq 3$$
$$\text{True} \qquad\qquad \text{True}$$

Numbers between 0 and 3, and 3, are not solutions since they do not satisfy the first condition.

$$2 \leq 0 \quad \text{and} \quad 2 \leq 3$$
$$\text{False} \qquad\qquad \text{True}$$

Numbers greater than 3 are not solutions since they do not satisfy either condition.

$$5 \leq 0 \quad \text{and} \quad 5 \leq 3$$
$$\text{False} \qquad\qquad \text{False}$$

Answer

CLASSROOM EXERCISES

True or false?

1. $2 < 4$ and $2 < 6$ **2.** $2 < 4$ or $2 < 6$ **3.** $6 < 8$ and $6 < -4$

4. $9 < 12$ or $9 < -4$ **5.** $6 < 5$ or $6 < 1$ **6.** $8 \leq 8$ or $8 > 8$

Graph.

7. $a < -1$ or $a > 3$ **8.** $x \leq -1$ and $x > 3$ **9.** $y > -2$ or $y > 0$

10. $y \geq -2$ and $y \geq 0$ **11.** $-2 < x$ and $x \leq 2$ **12.** $-2 < x$ or $x \leq 2$

Write a compound inequality for the graph.

13.

WRITTEN EXERCISES

True or false?

A **1.** $10 < 20$ or $10 < 30$ **2.** $10 < 15$ or $10 < 25$ **3.** $10 < 20$ and $10 < 30$

4. $10 < 15$ and $10 < 25$ **5.** $5 < 8$ or $5 < 0$ **6.** $5 < 8$ or $5 < 3$

7. $5 < 8$ and $5 < 0$ **8.** $5 < 8$ and $5 < 3$ **9.** $2 < -20$ or $2 < -30$

True or false?

10. $3 < -15$ or $3 < -25$

11. $6 < -12$ and $6 < -18$

12. $7 < -5$ and $7 < -15$

13. $5 \geq 5$ or $5 \geq 3$

14. $3 \geq 2$ or $3 \geq 3$

15. $4 > 2$ or $3 < 1$

16. $6 \geq 5$ and $5 \leq 6$

17. $7 \leq 8$ and $8 > -10$

18. $7 < 10$ or $7 > 10$

Graph on a number line.

19. $a < 2$ or $a > 3$

20. $a > 4$ or $a < 1$

21. $b \geq 4$ or $b \leq 0$

22. $b \leq 3$ or $b \geq 5$

23. $x < -2$ or $x \geq -1$

24. $x \leq -3$ or $x > 0$

25. $y > 2$ and $y < 5$

26. $y < 10$ and $y > 7$

27. $f < 5$ and $f > -2$

28. $f > -3$ and $f < 4$

29. $q < 3$ or $q > 2$

30. $q < 4$ or $q > 1$

31. $n < 3$ and $n > 2$

32. $n < 4$ and $n > 1$

33. $r = 2$ or $r < 1$

34. $r = 3$ or $r < -1$

35. $s \neq 3$ and $s > 0$

36. $s \neq 2$ and $s > 0$

Write a compound sentence for each graph.

37.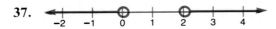

38.

39.

40.

41.

42.

43.

44.

List all the numbers in $\left\{2, 2\frac{1}{2}, 3, 3\frac{1}{2}, 4\right\}$ that are solutions of the compound sentence.

B **45.** $x < 3\frac{7}{8}$ and $x > 2\frac{1}{2}$

46. $x > 2$ and $x \leq 3\frac{1}{4}$

47. $x < 2\frac{3}{4}$ or $x > 3\frac{3}{4}$

48. $x > 3\frac{1}{4}$ or $x < 2\frac{1}{4}$

49. $x \neq 3\frac{1}{2}$ and $x \geq 3$

50. $x \neq 3$ and $x < 3\frac{1}{2}$

51. $x > 3$ and $x > 2\frac{1}{2}$

52. $x < 3$ and $x < 3\frac{1}{2}$

Write a compound sentence for each situation. Then graph the sentence on a number line.

53. An applicant's height must be between 157 cm and 173 cm. (Let h represent the height in centimeters.)

54. An applicant's mass must be between 45 kg and 79 kg. (Let m represent the mass in kilograms.)

55. The rod must be at least 6.5 cm long but it cannot be more than 6.7 cm long.

56. The food should be heated at 325°, plus or minus 10°. ("Plus or minus" 10° means the temperature could be as great as 10° more than 325° or as little as 10° less than 325°.)

57. The food should be heated for 12 min, plus or minus 2 min.

58. The denominator could be any number greater than -2 but it cannot be 0.

59. The numerator is either less than -5 or greater than 5.

60. The absolute value of a number is greater than 4.

61. The absolute value of a number is less than 4.

List all the numbers in $\{-2, -1, 0, 1, 2\}$ that are solutions of the compound sentence.

C **62.** $x + 2 > 1$ and $2x < 3$

63. $2x > 1$ or $x + 3 < 1\frac{1}{2}$

64. $3x + 1 \leq 5$ and $3x - 1 \geq 1$

65. $x^2 > 0.5$ and $x \neq -1$

66. $|x| > 1.5$ and $x \leq 1.5$

67. $|x + 1| > 1.5$ or $x < -1.5$

68. $x^2 + 3 > 5$ or $(x + 3)^2 < 5$

69. $x^2 - 2x < 0$ or $x^2 + 2x < 0$

■ REVIEW EXERCISES

Simplify.

1. $(x^2 + 3x - 4) + (x - 5) - (x^2 - 2)$

2. $2x^2 \cdot xy^2$ [4–1, 4–3]

3. $(2x^2)^3$ **4.** $2x \cdot (x^2 - 3x + 4)$

5. $\dfrac{10^3}{10^5}$ [4–5, 4–6, 4–8]

6. Change 463,000 to scientific notation. [4-4]

7. Change $4.65 \cdot 10^6$ to standard decimal notation. [4-4]

Solve.

8. Ms. Jones can grade 25 mathematics tests per hour and Ms. Smith can grade 20 mathematics tests per hour. If Ms. Smith starts at 3:00 P.M. and Ms. Jones starts at 3:30 P.M., at what time will they finish grading a total of 100 tests? [4-10]

Self-Quiz 1

9-1 True or false?

1. $6 < -4$ **2.** $5 \geq 5$

Graph on a number line.

3. $x < 2\frac{1}{2}$ **4.** $x \geq -4$

Write an inequality using x for this graph.

5.

9-2 True or false?

6. $-7 < -8$ and $5 > 4\frac{1}{2}$ **7.** $15 \geq 18$ or $-3 < 0$

Graph on a number line.

8. $x > 5$ or $x < 1$ **9.** $x \neq 3$ and $x \geq -1$

Write a compound inequality using x for this graph.

10.

9-1,
9-2 Write an inequality or compound inequality for each situation.

11. No student in the class is taller than 135 cm. (Let h represent the height in centimeters.)

12. The cooking time should be between 3 and 5 min. (Let t represent the cooking time in minutes.)

List the numbers in $\left\{3, 3\frac{1}{2}, 4, 4\frac{1}{2}, 5\right\}$ that are solutions of the inequality.

13. $x > 3\frac{1}{2}$ **14.** $x < 4$ or $x \geq 4\frac{1}{2}$

Preview   A puzzle

Graybeard the pirate and his 13 crew members captured 975 gold pieces and a solid gold box. Graybeard wanted the gold box for himself, but promised to buy it by putting an equal weight of gold pieces into the pile of loot. He said that the bag of gold pieces on this balance matched the box in weight. Was Graybeard lying or telling the truth? How do you know?

In this lesson you will learn a property that can be used to solve some inequalities.

■ LESSON

The basic strategy used to solve equations is to change them to simpler equivalent equations until the variable is alone on one side. We use the same technique to solve an inequality — changing to simpler equivalent inequalities until the variable is alone on one side. The addition property of inequalities permits us to change from one inequality to an equivalent inequality (one with the same set of solutions).

The Addition Property of Inequalities

For all numbers a, b, and c,

$$\text{if } a < b, \text{ then } a + c < b + c$$

and

$$\text{if } a > b, \text{ then } a + c > b + c.$$

For example, the inequality $x - 5 < 2$ can be solved as follows.

$$x - 5 < 2$$
Add 5 to both sides. $\quad x - 5 + 5 < 2 + 5$
Simplify. $\quad\quad\quad\quad x < 7$

The solution set is the set of real numbers less than 7. The graph is as shown here.

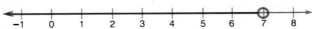

Example 1 Solve. $y + 4 \geq 3.2$

Solution

Add -4 to both sides.

$$y + 4 \geq 3.2$$
$$y \geq -0.8$$

Answer The set of all numbers greater than or equal to -0.8

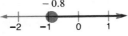

Example 2 Solve. $3x - 6 < 2x + 3$

Solution

Add $-2x$ to both sides.
Add 6 to both sides.

$$3x - 6 < 2x + 3$$
$$x - 6 < 3$$
$$x < 9$$

Answer The set of all numbers less than 9

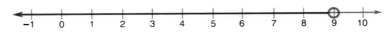

Example 3 Solve and graph. $x - 3 \neq 2$

Solution

Add 3 to both sides.

$$x - 3 \neq 2$$
$$x \neq 5$$

Answer The set of all numbers except 5

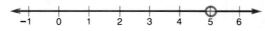

Example 4 Solve by writing and solving an inequality: After finding a quarter, Beverly knew that she had at least $2.15. How much did she have before she found the quarter?

Solution "At least" means the amount of money is greater than or equal to $2.15. Let x = the amount of money (in dollars) that Beverly started with.

$$x + 0.25 \geq 2.15$$
$$x \geq 1.90$$

Answer Beverly started with at least $1.90.

■ CLASSROOM EXERCISES

Solve.

1. $a + 6 > -2$

2. $x - 2 \leq -3$

3. $5 + x < 3$

4. $4x + 3 \geq 3x + 1$

5. $4 - 3x < 6 - 4x$

6. $2x + 5 \neq x - 3$

■ WRITTEN EXERCISES

Solve.

A

1. $a + 5 > -1$

2. $x + 4 > -3$

3. $x - 3 \geq 6$

4. $x - 2 \geq 5$

5. $3x \leq 2x + 4$

6. $4x \leq 3x + 2$

7. $x - 6 < -2$

8. $x - 4 < -3$

9. $5x + 2 \geq 4x + 3$

10. $3x + 6 \geq 2x + 10$

11. $2x + 6 < x - 4$

12. $3x + 2 < 2x - 5$

Solve. Graph the solutions on a number line.

13. $6x \leq 5x - 2$

14. $5x \leq 4x - 3$

15. $2x + 3 > 3x + 2$

16. $4x + 5 < 5x + 3$

17. $x + 2 > 2x + 5$

18. $3x + 4 > 4x + 1$

Solve the compound sentence. Graph the solutions on a number line.

19. $x + 2 < 5$ or $x + 1 > 6$

20. $x + 1 < 3$ or $x + 2 > 6$

21. $x + 2 > 5$ and $x + 1 < 6$

22. $x + 1 > 3$ and $x + 2 < 6$

23. $2x \geq x + 5$ or $x + 6 < 7$

24. $2x < x + 2$ or $x + 2 \geq 6$

25. $2x \geq 3x + 5$ and $3x + 2 \geq 2x - 4$

26. $3x + 5 \geq 2x$ and $2x \geq 3x + 2$

Write an inequality for each situation. Then solve the inequality.

27. If Dustin grew 2 cm, he would be at least 172 cm tall. How tall is Dustin now?

28. If Nicole gained 3 kg, she would weigh at least 45 kg. What is Nicole's present weight?

29. After Kara spent $2.50 she had less than $5.50 left. How much money did Kara have originally?

30. Jeremy had less than $2.50 left after paying a $4 bill. How much money did Jeremy have before he paid the bill?

31. If the music shop sold 25 records it would still have more than 850 records. How many records does it have now?

32. The video store had more than 635 tapes left after selling 35 tapes. How many tapes did the store have originally?

From these unbalanced situations, find the number of unit weights (O) that would balance a block (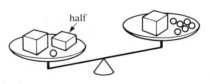). Assume that each block weighs the same and each rod (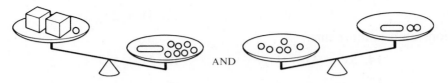) weighs the same.

33.

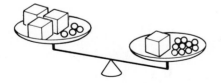

34.

35.

AND

Solve.

B **36.** $3(x - 3) > 2x + 1$

37. $2(x - 3) > 3x + 1$

38. $2(x + 4) < 3x + 4$

39. $4(x + 3) < 3x - 4$

40. $2(x + 3) + 5 \geq x + 8$

41. $5x - 3 \geq 6(x - 1) + 2$

42. $x^2 + 3x + 4 \leq x^2 + 4x$

43. $x^2 + 3x + 4 \leq x^2 + 2x$

44. $x(x + 6) + 1 > x(x + 5) + 3$

45. $x(x + 6) + 5 > x(x + 7) + 8$

Solve. Graph the solutions on a number line.

C **46.** $x + 5 < 2(x + 4)$ and $x + 5 > 2(x - 4)$

47. $x + 5 > 2(x + 4)$ or $x + 5 < 2(x - 4)$

48. $4.5x + 3.2 \geq 3.5x + 5.7$ and $2.3x - 1.7 \leq 1.3x + 1.8$

49. $4.5x + 3.2 \leq 3.5x + 5.7$ or $2.3x - 1.7 \geq 1.3x + 1.8$

List all the numbers in $\{-2, -1, 0, 1, 2\}$ that are solutions of these compound sentences.

50. $x^2 + 2x < x^2 + x$ and $5x + 2 > 4x$

51. $x^2 + 3x < x^2 + 2x$ or $3x - 1 > 2x$

52. $2(x - 1) \le 3x - 1$ and $3(x - 1) \le 2x - 2$

53. $x(x + 1) \ge x^2 + 1$ or $3(x + 2) \le 2x + 4$

■ REVIEW EXERCISES

1. In which quadrant or on which axis do the following points lie? [5–1]

   a. (3, 0)   **b.** (4, −3)   **c.** (−1, 5)

2. The set {(0, 4), (1, 4), (1, 5), (2, 5)} describes a relation. [5–2]

   a. List the domain.   **b.** List the range.   **c.** Graph the relation.

3. Select three values of x, complete the table, plot the points, and then graph [5–4]
the equation $3x + 3y = 12$.

x	?	?	?
y	?	?	?

4. Copy and complete the table. Then graph the equation $y = (x - 1)^2$. [5–4]

x	−3	−2	−1	0	1	2	3
y	?	?	?	?	?	?	?

5. Draw these three lines through the point (0, −2) labeling the lines a, b, and c. [5–5]

   a. line with slope $\dfrac{2}{3}$ **b.** line with slope −1   **c.** line with slope 0

6. State the slopes of lines a, b, and c. [5–5]

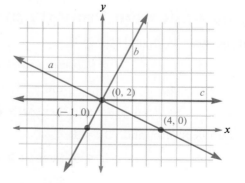

7. State the slope and y-intercept of the equation $2x - 3y = 6$. [5–6]

9–4 Solving Inequalities by Multiplication

Preview

The multiplication property of equality states that multiplying both sides of an equation by a nonzero number results in an equivalent equation. What can you discover about multiplication of inequalities from these examples?

True or false?

1. a. $3 < 7$
  b. $6(3) < 6(7)$
  c. $0.2(3) < 0.2(7)$
  d. $0(3) < 0(7)$
  e. $-2(3) < -2(7)$
  f. $-2(3) > -2(7)$

2. a. $4 \geq -2$
  b. $3(4) \geq 3(-2)$
  c. $0.5(4) \geq 0.5(-2)$
  d. $0(4) \geq 0(-2)$
  e. $-1(4) \geq -1(-2)$
  f. $-3(4) \leq -3(-2)$

3. a. $-5 < -2$
  b. $2(-5) < 2(-2)$
  c. $0.1(-5) < 0.1(-2)$
  d. $0(-5) < 0(-2)$
  e. $-10(-5) < -10(-2)$
  f. $-2(-5) > -2(-2)$

The multiplication property of inequalities has two cases, one for positive multipliers and one for negative multipliers. Try to state both cases.

In this lesson you will use the multiplication property of inequalities to solve inequalities.

■ LESSON

The inequalities $a > b$ and $c > d$ are said to be inequalities in the **same sense,** whereas $a < b$ and $c > d$ are said to be inequalities in the **opposite sense** (that is, the inequality signs are reversed).

Multiplying both sides of an inequality by a positive number results in an equivalent inequality in the same sense. Multiplying both sides of an inequality by a negative number results in an equivalent inequality in the opposite sense.

The Multiplication Property of Inequalities

For all numbers a, b, and c, $c > 0$,

$$\text{if } a < b, \text{ then } ca < cb$$

$$\text{and}$$

$$\text{if } a > b, \text{ then } ca > cb.$$

For all numbers a, b, and c, $c < 0$,

$$\text{if } a < b, \text{ then } ca > cb$$

$$\text{and}$$

$$\text{if } a > b, \text{ then } ca < cb.$$

These examples show how to use the multiplication property to solve inequalities.

Example 1   Solve and graph on a number line. $4x > 9$

Solution $4x > 9$

Multiply both sides by $\frac{1}{4}$. $\frac{1}{4}(4x) > \frac{1}{4}(9)$ (The sense of the inequality is maintained.)

$$x > \frac{9}{4}$$

Answer   The set of all numbers greater than $\frac{9}{4}$

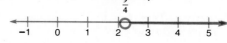

Example 2   Solve and graph on a number line. $-2x \geq 5$

Solution $-2x \geq 5$

Multiply both sides by $-\frac{1}{2}$. $-\frac{1}{2}(-2x) \leq -\frac{1}{2}(5)$ (The sense of the inequality is reversed.)

$$x \leq -\frac{5}{2}$$

Answer   The set of all numbers less than or equal to $-\frac{5}{2}$

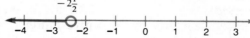

Example 3   Solve and graph. $-\frac{3x}{8} \leq 24$

Solution $-\frac{3x}{8} \leq 24$

Multiply both sides by 8 and simplify. $8\left(-\frac{3x}{8}\right) \leq 8(24)$

$$-3x \leq 192$$

Multiply both sides by $-\frac{1}{3}$ and simplify. $-\frac{1}{3}(-3x) \geq -\frac{1}{3}(192)$

$$x \geq -64$$

Answer   The set of all numbers greater than or equal to -64

Example 4 A 12-oz bottle of liquid soap sells for $0.82. At what price is a 15-oz bottle more economical?

Solution One bottle is more economical than another if its unit price is less. The unit price can be expressed as the ratio of cost to weight. Let c be the cost of the 15-oz bottle. The unit price of the 12-oz bottle is $\frac{0.82}{12}$ and the unit price of the 15-oz bottle is $\frac{c}{15}$.

$$\frac{0.82}{12} > \frac{c}{15}$$

Multiply by 60 to eliminate fractions. $\quad 60\left(\frac{0.82}{12}\right) > 60\left(\frac{c}{15}\right)$

$$4.1 > 4c$$

$$\frac{1}{4}(4.1) > \frac{1}{4}(4c)$$

$$1.025 > c$$

Answer The cost of a 15-oz bottle must be less than $1.025. That means that the cost must be $1.02 or less.

■ CLASSROOM EXERCISES

What is the missing inequality sign: $<$, $>$, $\leq$, or $\geq$?

1. If $x < 9$, then $2x$ ⑦ $2(9)$.

2. If $y < 4$, then $-3y$ ⑦ $-3(4)$.

3. If $a > 2$, then $-0.2a$ ⑦ $-0.2(2)$.

4. If $b > 5$, then $4b$ ⑦ $4(5)$.

Solve and graph.

5. $\frac{x}{3} < 6$ **6.** $-5y > 5$ **7.** $-2a \leq -4$ **8.** $\frac{3x}{4} \geq -2$

■ WRITTEN EXERCISES

What is the missing inequality sign, $<$ or $>$?

A **1.** If $x < 3$, then $4x$ ⑦ $4(3)$.

2. If $x < 2$, then $3x$ ⑦ $3(2)$.

3. If $x > 5$, then $4x$ ⑦ $4(5)$.

4. If $x > 12$, then $3x$ ⑦ $3(12)$.

5. If $x < 8$, then $-5x$ ⑦ $-5(8)$.

6. If $x < 4$, then $-6x$ ⑦ $-6(4)$.

7. If $x > 7$, then $-10x$ ⑦ $-10(7)$.

8. If $x > 6$, then $-8x$ ⑦ $-8(6)$.

True or false?

9. If $x \geq -2$, then $5x \geq 5(-2)$.

10. If $x \geq -5$, then $3x \geq 3(-5)$.

True or false?

11. If $x \le -6$, then $\frac{1}{2}x \ge \frac{1}{2}(-6)$.

12. If $x \le -6$, then $\frac{1}{3}x \ge \frac{1}{3}(-6)$.

13. If $2x \ge 10$, then $\frac{1}{2}(2x) \ge \frac{1}{2}(10)$.

14. If $3x \le 12$, then $\frac{1}{3}(3x) \le \frac{1}{3}(12)$.

15. If $-2x \le 10$, then $-\frac{1}{2}(-2x) \le -\frac{1}{2}(10)$.

16. If $-3x \ge 12$, then $-\frac{1}{3}(-3x) \ge -\frac{1}{3}(12)$.

Solve and graph on a number line.

17. $3a < 6$

18. $2a < 6$

19. $\frac{1}{2}b > 6$

20. $\frac{1}{3}b > 6$

21. $-4c < 12$

22. $-6c < 12$

23. $-\frac{1}{6}d > 12$

24. $-\frac{1}{4}d > 12$

25. $2r \ge -5$

26. $2r \ge -3$

27. $-\frac{1}{2}s \le -9$

28. $-\frac{1}{2}s \le -7$

Write an inequality for each situation. Then solve the inequality and answer the question.

29. Mr. Bradley paid more than $7.20 for 3 lb of meat. What was the cost per pound? (Let c represent the cost per pound.)

30. Two pounds of cheese cost Mrs. Corning more than $5.50. What was the cost per pound?

31. Denise Dravell bought $\frac{1}{2}$ lb of spices and paid less than $4.00. What was the cost per pound?

32. Ernie Ersland bought $\frac{1}{4}$ lb of chemicals and paid less than $2.00. What was the cost per pound?

33. The Zelco Corporation made a profit of less than $-\$10,000$ (actually a loss) during the first 5 months of the year. What was their average monthly profit?

34. The Velnor Company made a profit of less than $-\$1800$ (actually a loss) during the first 3 months of the year. What was their average monthly profit?

B **35.** A 6-pack of juice costs $1.00. At what price is an 8-pack of juice more economical?

36. A $2\frac{1}{2}$-oz jar of onion salt sells for 75¢. The $9\frac{1}{4}$-oz jar has a smaller unit selling price. What is the selling price of the $9\frac{1}{4}$-oz jar?

37. The Wildcats basketball team made 12 field goals in 25 attempts in the first half of a game. In the second half of the game they took only 20 shots but had a higher shooting percentage than in the first half. How many field goals did the Wildcats make in the second half?

38. Denise Waller used 16 gal of gasoline driving 480 mi on the first day of a trip. On the second day she drove 500 mi and had a better fuel consumption rate. How many gallons of gasoline did Denise use on the second day?

Solve and graph on a number line.

39. $\frac{2}{3}x \leq -12$

40. $-\frac{2}{3}x \leq -12$

41. $\frac{5}{8}x \leq 40$

42. $-\frac{5}{4}x \leq 20$

43. $-\frac{3}{4}y \geq -12$

44. $\frac{3}{4}y \geq -12$

45. $\frac{9}{4}y \geq 72$

46. $-\frac{2}{9}y \geq 90$

47. $\frac{x}{5} > 10$

48. $\frac{x}{5} > -10$

49. $-\frac{x}{3} > -60$

50. $-\frac{x}{4} > 20$

Solve and graph these compound inequalities.

C **51.** $-2x \leq -6$ and $x - 2 \leq 4$

52. $-2x \geq -6$ or $x - 2 \geq 4$

53. $-\frac{1}{2}x \geq 4$ or $x + 2 \geq 12$

54. $\frac{1}{2}x \leq -4$ and $x + 12 \geq 2$

55. $\frac{3}{8}x \geq 24$ or $x + 8 \leq 24$

56. $-\frac{3}{8}x \geq -24$ and $x - 8 \geq 24$

Solve and graph on a number line. [*Remember:* The values for a variable may be positive or negative.]

57. $\frac{1}{y} \leq 3$

58. $\frac{1}{2}y \leq 10$

59. $\frac{4}{x} > \frac{1}{2}$

60. $-\frac{6}{x} < 3$

■ REVIEW EXERCISES

1. Solve this system by graphing.   [6–1]
$$x + y = 6$$
$$x - 2y = -6$$

2. Solve this system by substitution.   [6–2]
$$x + 2y = 12$$
$$x = 2y$$

3. State whether the following system has zero, one, or infinitely many solutions.   [6–3]
$$2x + y = 4$$
$$6x + 3y = 12$$

4. Solve this system by multiplication and addition.   [6–6]
$$3x + y = 5$$
$$x - 2y = 4$$

Solve.

5. One number is 10 more than twice another. If the sum of the two numbers is 100, what are the numbers? [6-4]

6. We need to make 20 L of a solution that is 50% alcohol. If we mix a solution of 40% alcohol with a solution of 80% alcohol, how many liters of each solution should we use? [6-7]

EXTENSION   Solving direct variation problems on a computer

The electrical resistance of a wire is directly proportional to the length of the wire. If 200 m (X1) of an 18-gauge wire has a resistance of 4.3 ohms (Y1), we can compute the length (X2) of this wire for a given resistance (Y2). We can also compute its resistance (Y2) for any length (X2). The following program can be used to solve problems like this that involve direct variation.

```
10   PRINT "WHAT ARE THE VALUES OF X1 AND Y1";
20   INPUT X1, Y1
30   PRINT "IS THE VALUE OF X2 KNOWN (Y OR N)";
40   INPUT A$
50   IF A$ = "Y" THEN 110
60   PRINT "WHAT IS THE VALUE OF Y2";
70   INPUT Y2
80   X2 = X1 * Y2/Y1
90   PRINT "X2 =     "; X2
100  GOTO 150
110  PRINT "WHAT IS THE VALUE OF X2";
120  INPUT X2
130  Y2 = Y1 * X2/X1
140  PRINT "Y2 =     "; Y2
150  END
```

[The dollar sign indicates to the computer that the input is a letter (or any other symbol) rather than a number.]

What is the resistance of the following lengths of the wire described above?

1. 500 m   2. 10 m   3. 1 m   4. 2 km

What length of this wire can be used if its resistance can be no more than the following?

5. 10 ohms   6. 450 ohms   7. 1 ohm   8. 0.3 ohm

Preview Average gain

Ms. Armand's students are running a carnival booth as a money-raising project. The class developed a game in which the probability of winning for each play is 0.06. The average amount a player can expect to gain per play is given by this formula:

$$\text{Player's average gain in one play} = \text{Probability of winning} \times \text{Value of prize} - \text{Cost of one play}$$

A positive average gain means that the player can expect to win and the class can expect to lose. A negative average gain means that the player can expect to lose and the class can expect to win.

What should be the value of each prize if the class charges $0.25 for each play and wants its profit to be at least $0.13 on each play?

In this lesson you will learn to solve inequalities such as the following to help you solve problems like the one above.

$$0.06x - 0.25 \le -0.13$$

■ LESSON

The addition and multiplication properties of inequalities can be used together to solve inequalities. The general approach to solving inequalities is the same as that for solving equations. However, care must be taken to reverse the sense of the inequality when multiplying or dividing by a negative number.

Example 1   Solve and graph. $4x - 2 > x + 1$

Solution

$$4x - 2 > x + 1$$

Add $-x$ *to both sides.* $3x - 2 > 1$

Add 2 to both sides. $3x > 3$

Multiply both sides by $\frac{1}{3}$. $x > 1$

Answer   The set of all numbers greater than 1

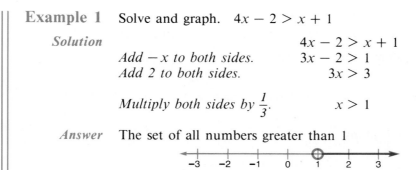

Example 2 Solve and graph. $2(x - 3) \le 5x - 12$

Solution

$$2(x - 3) \le 5x - 12$$

Expand. $2x - 6 \le 5x - 12$

Add $-5x$ to both sides. $-3x - 6 \le -12$

Add 6 to both sides. $-3x \le -6$

Divide both sides by -3. $x \ge 2$ (The sense of the inequality is reversed.)

Answer The set of all numbers greater than or equal to 2

Example 3 In one city a taxi ride costs \$0.50 for the first quarter-mile and \$0.20 for each additional quarter-mile or part of a quarter-mile. How far can you ride at a cost of \$5.00 or less?

Solution Setting up a table can make the problem easier to understand. Let q be the number of quarter-miles ridden.

Number of quarter-miles ridden		Cost in dollars
More than	Less than or equal to	
0	1	0.50
1	2	$0.50 + 0.20(1)$
2	3	$0.50 + 0.20(2)$
3	4	$0.50 + 0.20(3)$
.	.	.
.	.	.
.	.	.
$q - 1$	q	$0.50 + 0.20(q - 1)$

Inequality: $0.50 + 0.20(q - 1) \le 5.00$

Multiply both sides by 10 to eliminate $5 + 2(q - 1) \le 50$

decimals. $5 + 2q - 2 \le 50$

$$2q + 3 \le 50$$

$$2q \le 47$$

$$q \le 23.5$$

Answer The number of quarter-miles that can be ridden for \$5.00 or less is less than or equal to 23.5. However, the inequality indicates that the cost increases smoothly rather than jumping \$.20 for each additional quarter-mile. Therefore, the number of quarter-miles must be less than or equal to 23, and the number of miles is less than or equal to $5\frac{3}{4}$.

Example 3 (continued)

Check A trip of exactly 23 quarter-miles will cost $.50 + $.20(22) = $4.90. A trip of a little more than 23 quarter-miles will cost $.50 + $.20(23) = $5.10. The answer checks.

■ CLASSROOM EXERCISES

Solve and graph on a number line.

1. $5x + 5 > 15$
2. $9m - 9 < 45$
3. $6z - 4 \leq 20$

4. $5y + 16 \geq 11$
5. $3 - 2x < 11$
6. $5a - 7 > 2a + 8$

■ WRITTEN EXERCISES

Solve and graph on a number line.

A
1. $4x + 8 > 20$
2. $3x + 6 > 15$
3. $5y - 10 < 20$

4. $2y - 6 < 10$
5. $3n - 6 \leq 0$
6. $4n - 8 \leq 12$

7. $3a + 2 \geq a + 6$
8. $4a + 6 \geq 2a + 14$
9. $4(t - 1) > 3t + 5$

10. $3(t - 2) > 2t + 1$
11. $2(x + 3) < 5x - 3$
12. $4(x + 2) < 6x - 2$

13. $-2a + 3 \leq 8$
14. $-3a + 3 \leq 7$
15. $-\frac{1}{2}c + 5 > 7$

16. $-\frac{1}{3}c + 1 > 4$
17. $\frac{1}{2}x + 5 \geq x - 1$
18. $\frac{1}{2}x + 3 \geq x - 2$

Write an inequality. Solve the inequality and answer the question.

19. The length of a rectangular lot is 2 m more than the width. What is the width of the lot if the perimeter is less than 28 m? (Let w be the width in meters and $(w + 2)$ be the length in meters.)

20. The length of a rectangle is 3 cm more than the width. What is the width of the rectangle if the perimeter is less than 52 cm? (Let w be the width in centimeters and $(w + 3)$ be the length in centimeters.)

21. Tina is half as old as her uncle. The sum of their ages is at least 45 years old. How old is Tina? (Let t represent Tina's age.)

22. Tim is one third as old as his aunt. The sum of their ages is at least 64 years. How old is Tim? (Let t represent Tim's age.)

23. The lengths of the sides of a triangle are in the ratio 5:6:7. What is the length of the longest side if the perimeter is not more than 54 cm? (Let the sides be $5x$, $6x$, and $7x$.)

24. The lengths of the sides of a triangle are in the ratio 4:5:6. What is the length of the shortest side if the perimeter is not more than 60 cm? (Let the sides be $4x$, $5x$, and $6x$.)

25. The sum of two consecutive integers is greater than 16. What is the smaller integer?

26. The sum of two consecutive integers is greater than 22. What is the smaller integer?

27. The sum of two consecutive even integers is less than -13. What is the smaller integer?

28. The sum of two consecutive odd integers is less than -13. What is the smaller integer?

Solve and graph on a number line.

B   **29.** $5(x + 3) \leq 2(x - 6)$ **30.** $3x + 2 \geq 2\frac{1}{2}x - 6$

31. $3(x + 2) \geq 2\frac{1}{2}(x - 6)$ **32.** $\frac{1}{2}(x + 2) > \frac{1}{3}(x + 6)$

33. $x(x + 3) < x\left(x - 2 + \dfrac{10}{x}\right)$ **34.** $0.2x + 1.3 < x - 0.3$

35. $0.5(x + 3) > 0.6 - 0.7x$ **36.** $0.25x - 0.2 > 0.15(x + 2)$

Write an inequality. Solve the inequality and answer the question.

37. A salesperson earns a salary of $500 per month plus 2% of the sales. What must the sales be if the salesperson is to have a monthly income of at least $1600?

38. The owner of an apartment house must get at least $100,000 in rent each year to meet expenses and make a reasonable profit. On the average, 80% of the 25 apartments are occupied. What should the monthly rent be for one of the apartments if the same rent is charged for all the occupied apartments?

39. A manufacturer of components for motorcycles estimates that 1% of the parts are defective. How many parts must be made in order to have at least 4500 nondefective parts?

40. A worker is paid $9.00 an hour but 30% of the earnings is deducted for taxes, social security, and insurance. What whole number of hours must be worked to have at least $250 take-home pay (after deductions)?

41. A band agrees to play for $200 plus 25% of the ticket sales. How much must the ticket sales be if the band is to make at least $500?

42. A car repair costs $20 for the replacement part plus $25 per hour for labor. The owner has $100. How many hours of labor can she afford before she needs additional money?

Solve and graph these compound sentences on a number line.

C **43.** $2x + 3 < 6x - 1$ and $3(x + 2) < x + 10$ **44.** $2(x + 3) \leq 5x - 6$ or $3x + 2 \geq 5x + 6$

45. $-\frac{1}{2}x + 2 > 5$ or $-\frac{2}{3}x + 2 < -4$ **46.** $\frac{3}{4}x \geq x + 2$ and $\frac{1}{4}x \leq x + 12$

■ REVIEW EXERCISES

1. Write the prime factorization of 28. [7–1]

2. Give the greatest common factor of 12 and 18. [7–2]

3. Write the greatest common factor of $3ab$ and $6b^2$. [7–2]

4. Expand and simplify $(2x + 1)(x - 2)$. [7–5]

Self-Quiz 2

True or false?

9–3   **1.** If $x \leq -2$, then $3(x) \leq 3(-2)$.   **2.** If $-\frac{1}{2}x > 5$, then $-2\left(-\frac{1}{2}x\right) < -2(5)$.

Solve the compound inequality and graph the solution on a number line.

3. $x - 1 < 3$ or $x + 3 > 8$ **4.** $2x > x + 6$ and $x - 7 < 2$

Graph on a number line.

9–4   **5.** $-\frac{1}{3}x > 2$ **6.** $2x + 3 \leq -1$

Solve.

9–4,   **7.** $3x + 1 \leq 2x + 4$ **8.** $4(x - 2) > 3(2x - 1)$
9–5

Write and solve an inequality for the situation.

9. In 3 min, there will be less than 1 h left. How much time is left now? (Let t represent the time left in minutes.)

10. A buyer's share of $\frac{1}{4}$ of the boat's price was more than $1200. What was the boat's price? (Let b represent the boat's price.)

11. The width of a rectangle is half its length. The perimeter must be no more than 60 cm. What is the rectangle's length? (Let l be the rectangle's length in centimeters.)

9–6 Absolute Values and Inequalities

The absolute value of a number can be defined as the distance of that number from 0. Therefore, the solution of the inequality

$$|x| < 2$$

is the set of all numbers that are less than 2 units from 0.

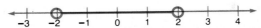

Here is another inequality involving absolute value, a table of some values, and a graph of all solutions.

$$|x - 1| < 2$$

solutions

x	-3	-2	-1	-0.9	0	0.5	1	2	2.9	3	4
$|x - 1|$	4	3	2	1.9	1	0.5	0	1	1.9	2	3

less than 2

- The solutions of $|x - 1| < 2$ are all numbers that are less than 2 units from __?__ .
- Can you write a similar sentence for this inequality? $|x - 3| \le 2$

In this lesson you will learn to solve inequalities that involve absolute values.

■ LESSON

Inequalities that contain absolute values are equivalent to compound inequalities. For example, we saw in the Preview that the solutions of $|x| < 2$ are numbers between -2 and 2. These numbers can also be described by this compound inequality:

$$-2 < x \quad \text{and} \quad x < 2$$

A short way of writing this compound inequality is $-2 < x < 2$.

The Absolute Value Property of Inequalities

For all real numbers x and a,

if a is 0 or positive
$|x| < a$ is equivalent to $-a < x$ *and* $x < a$
which is equivalent to $-a < x < a$.

Now consider this inequality:

$$|x| > 3$$

Solutions of this inequality are numbers that are greater than 3 and numbers that are less than -3. Therefore, the inequality $|x| > 3$ is equivalent to the compound inequality

$$x < -3 \quad \text{or} \quad x > 3.$$

The Absolute Value Property of Inequalities

For all real numbers x and a,

if a is 0 or positive,
$|x| > a$ is equivalent to $x < -a$ or $x > a$.

Study these examples to see how to solve inequalities that involve absolute values.

Example 1 Graph on a number line. $|x| < 3$

Solution $|x| < 3$
$-3 < x < 3$

Answer

Example 2 Graph. $|2r| \geq 6$

Solution $|2r| \geq 6$
$2r \leq -6 \quad \text{or} \quad 2r \geq 6$
$r \leq -3 \quad \text{or} \quad r \geq 3$

Example 2 (continued)

Answer

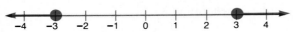

Example 3 Graph. $|x + 1| > 2$

Solution
$$|x + 1| > 2$$
$$x + 1 < -2 \quad \text{or} \quad x + 1 > 2$$
$$x < -3 \quad \text{or} \quad x > 1$$

Answer

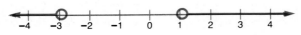

Example 4 Graph. $|y - 3| \leq 1$

Solution
$$|y - 3| \leq 1$$
$$y - 3 \geq -1 \quad \text{and} \quad y - 3 \leq 1$$
$$y \geq 2 \quad \text{and} \quad y \leq 4$$
$$2 \leq y \leq 4$$

Answer

We can think of $|y - 3|$ on the number line as the distance between a number y and the number 3. Then the inequality in Example 4 describes all numbers y that are no more than 1 unit away from 3.

Example 5 Write an inequality involving absolute value that describes this graph.

Solution Note that the graph includes all numbers that are 1 unit or more from 3. An inequality that describes those numbers is:

$$|x - 3| \geq 1$$

■ CLASSROOM EXERCISES

Write as compound sentences without absolute values.

1. $|x| < 5$ **2.** $|x| > 3$ **3.** $|d - 4| \leq 7$ **4.** $|b + 3| \geq 1$

Graph.

5. $|x| < 1$ **6.** $|y| > 1$ **7.** $|c| \geq 4$ **8.** $|b| \leq 0$

9. $|x + 1| < 4$ **10.** $|t - 2| > 2$ **11.** $|3x| < 6$ **12.** $|y - 3| \geq 0$

Use absolute value to write an inequality that describes the graph.

13.

14.

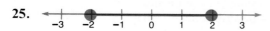

■ WRITTEN EXERCISES

Write as compound sentences without absolute values.

A   **1.** $|x| < 10$ **2.** $|x| < 20$ **3.** $|x| > 15$ **4.** $|x| > 5$

    5. $|b + 3| \leq 8$ **6.** $|b + 5| \leq 12$ **7.** $|a - 6| \geq 8$ **8.** $|a - 4| \geq 7$

Graph on a number line.

    9. $|x| < 4$ **10.** $|x| < 6$ **11.** $|x| > 7$ **12.** $|x| > 5$

13. $|x + 4| \leq 6$ **14.** $|x + 2| \leq 4$ **15.** $|x - 1| \geq 3$ **16.** $|x - 5| \geq 2$

17. $|2x| \geq 5$ **18.** $|4x| \geq 6$ **19.** $|-3x| < 12$ **20.** $|-5x| < 15$

21. $\left|\dfrac{x}{3}\right| > 6$ **21.** $\left|\dfrac{x}{2}\right| > 8$ **23.** $\left|-\dfrac{x}{4}\right| \leq 1$ **24.** $\left|-\dfrac{x}{5}\right| \leq 2$

Use absolute value to write an inequality that describes the graph.

25.

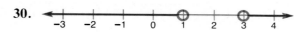

26.

27.

28.

29.

30.

31.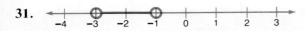

32.

Use absolute value to write an inequality for each problem. Then graph the inequality. (Use x for the variable.)

33. The temperature was within 5 degrees of 0.

34. The temperature was within 10 degrees of 0.

35. The production goal for the week was 2500 cars. The number of cars produced was within 100 of the goal.

36. The charity campaign goal was $50,000. The fund raisers came within $3000 of their goal.

37. The rod was supposed to be 25 cm long, plus or minus 0.5 cm.

38. The temperature of the oven was supposed to be 160°C, plus or minus 10°C.

Solve.

B **39.** $|2x + 1| > 4$ **40.** $|2x - 1| < 4$ **41.** $|3x + 2| \leq 5$

42. $3|x + 2| \geq 6$ **43.** $|-2x + 3| \geq 11$ **44.** $|-3x + 6| \leq 9$

45. $|4x - 2| \leq 10$ **46.** $2|3x - 9| \geq 12$ **47.** $-2|x - 3| \leq 8$

Graph.

C **48.** $|x| \geq 5$ and $|x| \leq 6$ **49.** $|x| \geq 6$ or $|x| < 5$ **50.** $|x - 5| \leq 3$ or $|x + 5| \leq 3$

51. $|x - 5| \leq x$ **52.** $|x| \leq x + 5$ **53.** $|x + 10| \leq |x|$

54. $|x - 3| \neq |3 - x|$ **55.** $|x| \cdot |x| \leq |x|$ **56.** $|x| \cdot |x| \geq |x|$

■ REVIEW EXERCISES

Solve.

1. $(x + 3) \cdot x = 0$ **2.** $(x + 2)(x - 5) = 0$ [7–3]

3. $x^2 - 4x = 0$ **4.** $2x^2 - 5x = 0$ [7–4]

5. $x^2 - 81 = 0$ **6.** $16x^2 - 25 = 0$ [7–6]

Factor.

7. $x^2 + 10x + 25$ **8.** $x^2 - 9x + 18$ [7–7, 7–8]

9. $x^2 + 4x - 5$ **10.** $2x^2 + x - 1$ [7–9, 7–10]

9–7 Inequalities in Two Variables

Preview   Application: Highway expansion

Highways sometimes buckle on very hot days. Construction engineers know they must provide expansion joints between sections of the highway to allow for the expansion. The following formula can be used to determine the expansion E (in inches) of one mile of a two-lane highway that was built when the temperature was 60°F, where T is the present temperature in degrees Fahrenheit.

$$E = 0.75T - 45$$

To prevent buckling, the total length of the expansion joints in each mile of highway must be greater than or equal to E.

- If the temperature reaches 80°F, which of these total lengths for the joints will be enough?

  a. 10 in.   **b.** 15 in.
  c. 20 in.   **d.** 25 in.

- If the temperature reaches 100° F, which of these total lengths for the expansion joints will be enough?

  a. 15 in.   **b.** 25 in.
  c. 35 in.   **d.** 45 in.

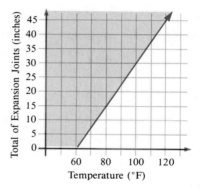

In this lesson you will make graphs of inequalities like the one above.

■ LESSON

Since the solutions of inequalities in two variables are ordered pairs of numbers, the solutions of the inequalities can be graphed in the coordinate plane. For example, suppose we wish to graph the solutions of this inequality:

$$y \geq x$$

This sentence is equivalent to:

$$y = x \quad \text{or} \quad y > x$$

Solutions of the compound sentence must satisfy one condition or the other.

First, graph the ordered pairs that satisfy the equation $y = x$.

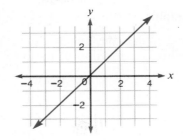

Next, find some ordered pairs that satisfy the condition $y > x$ and graph them.

x	2	1	1	0	-1	-3	-3
y	3	1.5	3	1	-0.5	-2	3

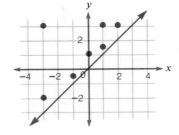

It seems that every point "above" the line is in the graph. To test that guess, find more solutions.

x	2	0	-2	-2
y	2.5	3	0	-1.5

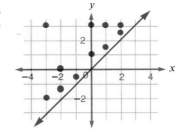

Then, test some points "below" the line.

(1, 0) is not a solution, because $0 > 1$ is false.
(-1, -3) is not a solution, because $-3 > -1$ is false.

It seems that no point below the line is a solution. The graph includes the line $y = x$ and all points above it.

We shade the region above the line and use a solid line to show that the boundary is part of the graph. If the boundary is not part of the graph, a dashed line is used.

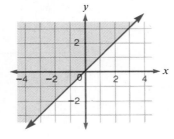

Example 1 Graph. $x + y < 3$

Solution First, graph the line $x + y = 3$. Use a dashed line because none of those points is a solution of the inequality.

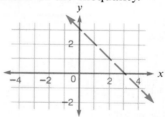

Next, find some solutions of the inequality $x + y < 3$ and graph them.

x	y
1	1
4	-2
2	-2
0	2
-1	3

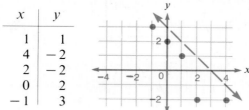

Answer The graph includes all points below the line $x + y = 3$.

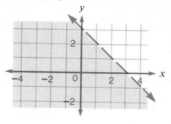

Example 2 Graph. $y \le x^2$

Solution Graph $y = x^2$.

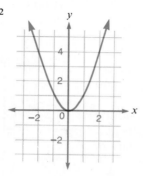

Example 2 (continued)

Find some solutions of the inequality $y < x^2$.

x	y
3	4
0	-2
-2	3
-1	-3

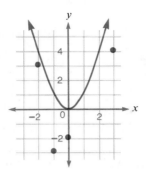

Answer   The graph includes all points on the curve and below it.

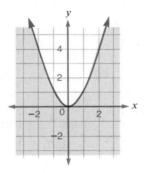

These steps should be used to graph an inequality in two variables.

1. Graph the corresponding equation to get the boundary of the graph. The boundary should be a dashed line if the boundary points are not included in the graph. The boundary should be a solid line if the boundary points are included in the graph.
2. The boundary divides the plane into regions. Test points in all regions to find which regions are to be included in the graph.
3. Shade the points that are included in the graph.

■ CLASSROOM EXERCISES

State whether the boundary should be dashed or solid for these graphs.

1. $y \geq 2x + 1$

2. $y < 3x - 5$

State whether the given point is in the graph of the inequality.

3. $(0, 0)$; $y > x - 1$

4. $(1, 10)$; $y \geq 2x + 5$

Graph.

5. $y < x$ **6.** $y \geq 2x$ **7.** $y > 2x + 1$ **8.** $y > x^2 - 1$

9. Write an inequality for this graph.

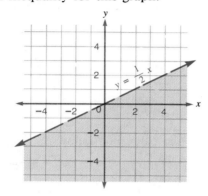

■ WRITTEN EXERCISES

State whether the boundary should be dashed or solid for these graphs.

A **1.** $y \leq 3x - 1$ **2.** $y \geq 2x + 3$ **3.** $y > \frac{1}{2}x + 4$ **4.** $y < \frac{1}{3}x + 2$

State whether the given point is in the graph of the inequality.

5. $(0, 0)$; $y < x + 3$ **6.** $(0, 0)$; $y < x + 5$

7. $(0, 0)$; $y \leq x - 4$ **8.** $(0, 0)$; $y \leq x - 1$

9. $(1, 10)$; $y \geq 2x - 3$ **10.** $(10, 1)$; $y \geq 3x - 2$

11. $(100, 1)$; $y > \frac{1}{2}x + 5$ **12.** $(1, 100)$; $y > \frac{1}{3}x + 1$

13. $(0, 1)$; $y > \frac{1}{2}x - 2$ **14.** $(2, 0)$; $y \geq \frac{1}{2}x - 3$

Graph.

15. $y \geq 2x - 3$ **16.** $y \geq \frac{1}{2}x + 2$ **17.** $y > \frac{1}{3}x$

18. $y > 3x$ **19.** $y < x + 3$ **20.** $y < x - 3$

21. $y \leq -2x + 2$ **22.** $y \leq -3x + 3$ **23.** $y \leq \frac{1}{2}x^2$

24. $y \leq \frac{1}{4}x^2$ **25.** $x + y \geq 6$ **26.** $x + y \geq 3$

Write an inequality for each graph.

27.

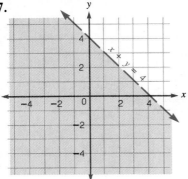

28.

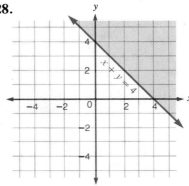

B 29.

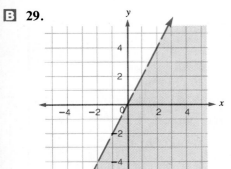

30.

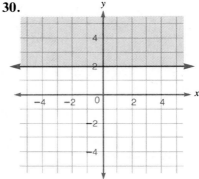

Graph.

31. $2x + 3y \leq 6$ **32.** $2x - 3y \leq 6$ **33.** $-2x + 3y \leq 6$ **34.** $-2x - 3y \leq 6$

35. $y \geq 2$ **36.** $x \leq 5$ **37.** $xy > 12$ **38.** $xy > -12$

C 39. $x \geq 2$ and $y \geq 3$ **40.** $x \geq 2$ or $y \leq 3$

41. $|x| + |y| \leq 5$ **42.** $|x + y| < 5$

43. $|y| < |x|$ [*Hint:* What are the boundaries?]   **44.** $|y| > x^2$ [*Hint:* What are the boundaries?]

■ REVIEW EXERCISES

1. Which values of x are not admissible?   [8–1]

$$\frac{3}{x^2 - 5x}$$

Simplify.

2. $\dfrac{6x}{9x}$ **3.** $\dfrac{6x^2}{5} \cdot \dfrac{3}{4x}$ **4.** $\dfrac{1}{x} + \dfrac{1}{2x}$ [8–2, 8–3, 8–4]

Solve.

5. $\dfrac{3}{x} = \dfrac{60}{80}$ **6.** $\dfrac{1}{x} = \dfrac{1}{3} + \dfrac{1}{5}$ [8–5]

7. $24 = x\%$ of 80   **8.** $80\% = \dfrac{48}{x}$ [8–8]

9. George can stock the grocery shelves in 8 h and Bill can stock them in 10 h.   [8–6]
How long does it take George and Bill working together to stock the shelves?

10. A resistance of 10 ohms is placed in parallel with a resistance of 20 ohms. What is the total resistance of the circuit?

$\left(\text{Use the formula } \dfrac{1}{R} = \dfrac{1}{r_1} + \dfrac{1}{r_2}.\right)$

Self-Quiz 3

9–6 Graph on a number line.

    1. $|x| \geq 3$ **2.** $|x + 4| < 2$ **3.** $|2x + 1| \leq 5$

Use absolute value to write an inequality to describe the graph.

    4.

    5.

9–7 State whether the given number pair is a solution of the inequality $y \leq x + 3$.

    6. $(-2, 4)$ **7.** $(-3, 0)$ **8.** $(-1, -5)$

Graph.

    9. $y \leq x - 4$ **10.** $y > x^2$

Write an inequality for each condition.

11. Maureen decided to add a minimum of $5 each week to her existing savings account of $315. (Let x represent the number of weeks later and y the amount of money in Maureen's account.)

12. A carpenter must construct this month's quantity of 3-legged stools and 4-legged stools. His supply of wooden legs for the month will be no more than 196 legs. (Let x represent the number of 3-legged stools and y the number of 4-legged stools.)

9-8 Compound Inequalities in Two Variables

Preview

The April Frost Show is a 30-minute television show that divides its time among three types of activity: (1) April Frost sings, dances, and tells jokes, (2) guest stars perform, and (3) sponsors of the show have commercial messages. If e represents the number of minutes April entertains and c represents the number of minutes for commercials, match an inequality to each of these conditions.

1. The sponsors insist on at least 3 min of commercials.
2. The network limits the total commercial time to 6 min.
3. April's agent insists that she perform at least 10 min.
4. April's writer says that she cannot perform more than 25 min.
5. It is a half-hour show.
6. Guest stars cannot perform more than 16 minutes.

a. $e + c \geq 14$ **b.** $e + c \leq 30$
c. $c \leq 6$ **d.** $c \geq 3$
e. $e \leq 25$ **f.** $e \geq 10$

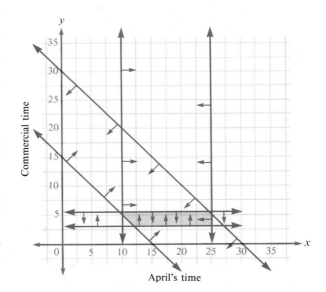

Commercial time / April's time

The graph shows all the conditions at once. The shaded region is where all six conditions are satisfied.

The graph helps answer such questions as, "What combination of times is most pleasing to viewers?"

In this lesson you will study systems of linear inequalities and their applications to problems such as the one above.

■ LESSON

Solutions to compound sentences involving "and" must satisfy all the conditions. Solutions to compound sentences involving "or" must satisfy at least one of the conditions. Consider the following compound inequality in two variables.

$$x + y \leq 6 \quad \text{and} \quad x > 3$$

To graph this compound inequality, we first graph the condition $x + y \leq 6$.

Draw the boundary with a solid line and use arrows to show which side of the boundary is included in the graph.

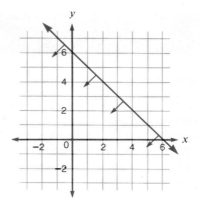

Next, graph the second condition, $x > 3$. Draw the boundary with a dashed line, and use arrows to show which side of the boundary is included in the graph.

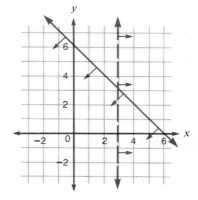

Finally, shade the region that lies within the graphs of *both* conditions. This is the graph of the compound inequality.

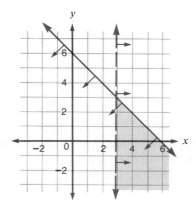

Example 1   Graph. $y \leq x + 2$ and $x \leq 3$ and $y \geq -1$

Solution   First graph $y \leq x + 2$.

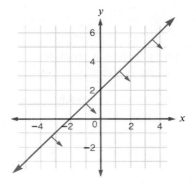

Then graph $x \leq 3$.

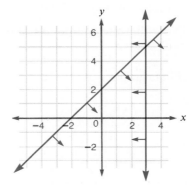

Finally, graph $y \geq -1$.

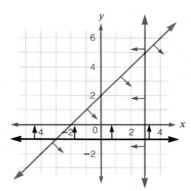

Example 1 (continued)

Shade the region that satisfies all three conditions.

Answer   The graph is a triangle and its interior.

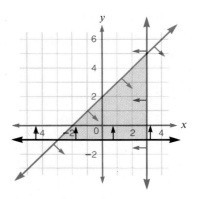

Example 2   Graph. $y \leq x - 2$ or $y < -x + 1$

Solution   Note that the compound sentence joins the conditions with "or." The graph will consist of points that satisfy either condition.

First, graph $y \leq x - 2$.

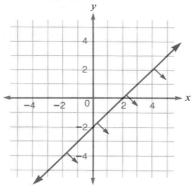

Then graph $y < -x + 1$.

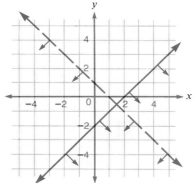

Shade the region that satisfies either condition.

Answer

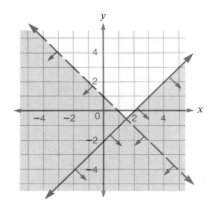

These steps should be used to graph compound inequalities in two variables.

1. Graph each of the conditions, using arrows to indicate which side of the boundary is included.
2. If the compound sentence involves "and," shade the region that satisfies all the conditions.
3. If the compound sentence involves "or," shade the region that satisfies one or more of the conditions.

■ CLASSROOM EXERCISES

State whether the boundaries of these graphs will be solid lines or dashed lines.

1. $x + y < 5$ and $x - y < 6$

2. $x - 2y \le 6$ and $y \le 2x - 6$

3. $y \ge 3$ or $x \ge 4$

4. $y > \frac{1}{2}x - 2$ or $y > 2x + 2$

Match the graphs in Exercises 5–8 with these compound inequalities.

a. $y \le 2$ or $x \le 4$

b. $x + y \le 4$ and $y \ge -x$

c. $y \le 2$ and $x \le 4$

d. $x + y \ge 4$ or $y \le -x$

5.

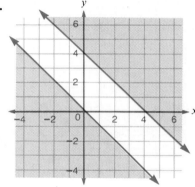

6.

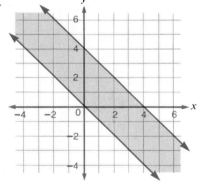

7.

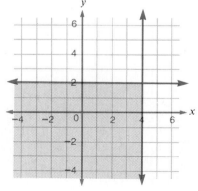

8.

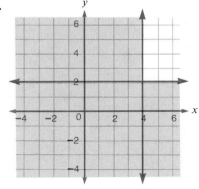

Graph.

9. $x + y \geq -2$ and $y \leq 3$

10. $y > 2x + 3$ or $x > 3$

■ WRITTEN EXERCISES

Match the graphs in Exercises 1–8 with the compound inequalities.

A **a.** $x \leq -2$ and $y \leq 3$

b. $x \leq -2$ or $y \leq 3$

c. $x \leq -2$ and $y \geq 3$

d. $x \leq -2$ or $y \geq 3$

e. $x \geq -2$ and $y \leq 3$

f. $x \geq -2$ or $y \leq 3$

g. $x \geq -2$ and $y \geq 3$

h. $x \geq -2$ or $y \geq 3$

1.

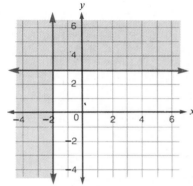

2.

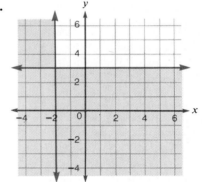

3.

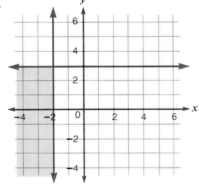

4.

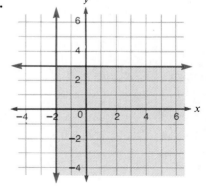

5.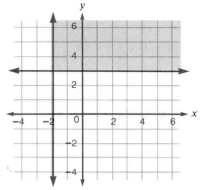

6.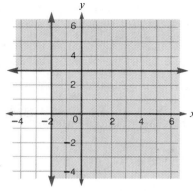

7.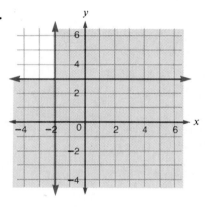

8.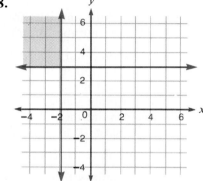

Graph.

9. $y \geq 2x$ and $x < 0$

10. $y \geq \frac{1}{2}x$ and $x < 2$

11. $y < 3$ or $y > x$

12. $y < 4$ or $y < x$

13. $y \leq 5$ and $y \geq -2$

14. $x \geq -2$ and $x \leq 3$

15. $y \geq 5$ or $y \leq -2$

16. $x \leq -2$ or $x \geq 3$

17. $y \geq x + 2$ and $y \leq 2x$

18. $y \leq x + 2$ and $y \geq 2x$

19. $y < 2x - 3$ and $y > x - 3$

20. $y > 2x - 3$ and $y < x - 3$

In each situation, let x represent the mass in kilograms of the flour and y the mass in kilograms of the sugar. Write an inequality for each condition. Graph both conditions.

21. The total mass of a flour–sugar mixture is less than 5 kg. There is at least 1 more kg of flour than sugar.

22. The total mass of a flour–sugar mixture is at least 5 kg. There is more flour than sugar.

23. There is at least twice as much flour as sugar. There are less than 3 kilograms of sugar.

24. There are at least 2 more kg of flour than sugar. The amount of flour exceeds the amount of sugar by less than 4 kg.

25. The total mass of the flour–sugar mixture is more than 6 kg. The mass of the flour exceeds the mass of the sugar by less than 3 kg.

26. The mass of the sugar is at least half the mass of the flour. There are at least 3 kg of flour.

State whether the point of intersection of the two boundaries is part of the graph.

B **27.** $y > x$ and $x \leq 4$

28. $y \leq 2x$ or $y > 5$

29. $y > \frac{1}{2}x$ or $y < 4x$

30. $x \leq 5$ and $y \geq -3$

A farmer wants to mix corn and oats for feed. Let x represent the number of pounds of corn and y the number of pounds of oats.

a. The total mixture should not exceed 100 lb.
b. There should be at least 50 lb of the mixture.
c. There should be at least 40 lb of corn.
d. There should be no more than 70 lb of corn.
e. There should be no more than 50 lb of oats.
f. There should be at least 10 lb of oats.

31. Write an inequality for each condition.

32. Graph the six inequalities from Exercise 31.

33. Which of the six conditions is not needed because it is covered by the other conditions?

34. Which of these (corn, oats) combinations satisfy all the conditions?

    a. (40, 10)   **b.** (30, 20)   **c.** (40, 50)   **d.** (50, 50)   **e.** (70, 30)
    f. (70, 10)   **g.** (60, 30)   **h.** (80, 20)   **i.** (40, 60)

35. Which of the acceptable (corn, oats) combinations has the largest corn/oats ratio?

36. Which of the acceptable (corn, oats) combinations has the largest oats/corn ratio?

Two classes decide to raise money by running a chili supper.

a. They have 200 tickets to sell.
b. Class A agrees to sell at least 80 tickets.
c. Class B agrees to sell at least 60 tickets.
d. They must sell at least 160 tickets.
e. Class B will sell no more than 20 more tickets than class A.

C **37.** Let x represent the number of tickets sold by class A and y the number of tickets sold by class B. Write an inequality for each condition.

38. Graph the five inequalities.

39. Which of these (class A, class B) combinations satisfy all the conditions?

    a. (80, 120)   **b.** (120, 80)   **c.** (140, 60)   **d.** (60, 140)   **e.** (100, 60)
    f. (80, 80)   **g.** (80, 100)   **h.** (90, 110)   **i.** (100, 80)

40. In which of the acceptable combinations does class A sell the greatest number of tickets?

41. In which of the acceptable combinations does class B sell the greatest number of tickets?

42. If 160 tickets were sold and class A sold as many as possible, how many tickets did each class sell?

43. If class A made a profit of 50¢ on each ticket it sold and class B made a 40¢ profit on each ticket it sold, what is the maximum profit possible?

■ REVIEW EXERCISES

1. What is the ratio of the width to the length of a 60-ft by 150-ft rectangular lot?   [8–7]

2. What is the interest on $300 invested for three years at an annual rate of 10%?   [8–8]

3. If Caryl's car travels 300 mi on 12 gal of gasoline, how far should it be able to travel on 10 gal?   [8–9]

4. A recipe for 8 people calls for a half-dozen eggs. How many eggs should be used if 12 people are to be served?   [8–9]

5. Heidi's mother earns three times as much as Heidi does. How much does her mother earn if Heidi earns $150.00?   [8–10]

6. If y varies directly as x, and $y = 12$ when $x = 5$, find y when $x = 100$.   [8–10]

7. Suppose x and y vary inversely and $y = 10$ when $x = 2$. Find y when $x = 5$.   [8–11]

8. Graph the equation $xy = 12$ for $x > 0$.   [8–11]

■ CHAPTER SUMMARY

● **Vocabulary**

inequality	[page 435]	equivalent inequalities	[page 447]
compound inequality	[page 435]	solve an inequality	[page 447]
$<, >, \neq, \leq, \geq$	[page 435]	sense of an inequality	[page 452]
solutions of inequalities	[page 436]		

● The Addition Property of Inequalities   [9–3]

    For all numbers a, b, and c,
$$\text{if } a < b, \text{ then } a + c < b + c$$
$$\text{and}$$
$$\text{if } a > b, \text{ then } a + c > b + c.$$

- The Multiplication Property of Inequalities   [9–4]

  For all numbers a, b, and c, $c > 0$,

  $$\text{if } a < b, \text{ then } ca < cb$$
  $$\text{and}$$
  $$\text{if } a > b, \text{ then } ca > cb.$$

  For all numbers a, b, and c, $c < 0$,

  $$\text{if } a < b, \text{ then } ca > cb$$
  $$\text{and}$$
  $$\text{if } a > b, \text{ then } ca < cb.$$

- The Absolute Value Property of Inequalities   [9–6]

  If $a \geq 0$, then

  $$|x| < a \text{ is equivalent to } -a < x \text{ and } x < a,$$
  $$\text{which is equivalent to } -a < x < a;$$
  $$|x| > a \text{ is equivalent to } x < -a \text{ or } x > a.$$

- Solutions to compound sentences involving "and" must satisfy all the   [9–8]
  conditions.

- Solutions to compound sentences involving "or" must satisfy at least one of   [9–8]
  the conditions.

■ CHAPTER REVIEW

9–1 Objective: To determine whether inequalities are true or false, and to graph inequalities
on a number line.

    1. Letting x be a number, write an inequality that states that the number is not less
than 21.

    2. Graph on a number line $x > -2$.

    3. Write the inequality that is graphed on this number line.

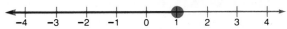

9–2 Objective: To graph compound inequalities on a number line.

Graph on a number line.

    4. $x \geq -2$ and $x \leq 1$ **5.** $x < -2$ or $x > 2$ **6.** $x > -3$ and $x \neq 0$

9–3 Objective: To solve inequalities by using the addition property of inequalities.

Solve.

    7. $x + 5 < 12$ **8.** $3x - 5 \geq 2x + 3$

Graph the solution on a number line.

    9. $2x + 4 < 3x + 2$

9–4 Objective: To solve inequalities by using the multiplication property of inequalities.

10. What is the missing inequality sign, $<$ or $>$?
If $x < 5$, then $-2x \;\boxed{?}\; -2 \cdot 5$.

Graph on a number line.

11. $4x \geq -12$ **12.** $-3x < 6$

9–5 Objective: To solve inequalities by using both the addition and multiplication properties of inequalities.

Solve and graph on a number line.

13. $3x + 5 > 17$ **14.** $3 - x \leq 2(x - 3)$

Write an inequality. Solve the inequality and answer the question.

15. If Sally begins with $50, what is the least amount she must save per week to buy a $199 stereo in 10 weeks?

9–6 Objective: To solve inequalities that contain absolute values.

16. Write $|x - 3| > 5$ as a compound sentence without absolute values.

17. Graph on a number line $|x + 3| \leq 2$.

18. Use absolute values to write an inequality that describes this graph.

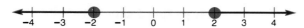

9–7 Objective: To graph inequalities in two variables.

Graph.

19. $y \geq \frac{1}{2}x - 2$ **20.** $x + y < 4$ **21.** $2x > 5$

9–8 Objective: To graph compound inequalities in two variables.

Graph.

22. $y > x + 1$ and $x > 0$ **23.** $x + y \leq 4$ or $y \geq x$ **24.** $y \leq 5$ and $x \geq 3$ and $y \geq x - 2$

■ CHAPTER 9 SELF-TEST

9–1 Write an inequality for the expression. Let n represent the number.

1. The number is at least -8. **2.** One-fourth the number is less than 17.

9–2 Graph on a number line.

3. $x \geq \frac{7}{2}$ **4.** $y < 1$ or $y > 4$ **5.** $w \geq -\frac{1}{2}$ and $w \leq \frac{5}{2}$

Write a compound inequality for these graphs.

6.

7.

9–3 Solve.

8. $2w - 3 > w + 5$

9. $4(x - 5) \leq 5(x + 3)$

9–4, Graph the solution of the inequality.
9–5

10. $\frac{5}{4}t - 1 \neq 4$

11. $-6s < 15$

12. $3(3 - x) \geq 2x + 14$

13. $\frac{1}{4}y + 8 < \frac{1}{2}y + 6$

9–5 Write and solve an inequality for each situation.

14. The area of a rectangular flower bed must be at least 42 ft². The width must be 3 ft. (Use x to express possible lengths.)

15. The perimeter of an isosceles triangle can be no more than 25 cm. The length of an equal side is twice the base. (Use x to express possible lengths for the base.)

9–2, True or false?
9–5

16. $-13 < -11$ or $4 \geq 5$

17. $|-3| > 2$ and $-3 < -5$

9–6 Write as a compound inequality.

18. $|s - 1| > 3$

19. $|2w| < 6$

9–7 State whether the given point is part of the graph of the inequality.

20. $(3, -1); 2x - y < 6$

21. $(2, 5); x \geq y - 4$

9–7, Graph.
9–8

22. $y > 2x - 6$

23. $x + y \geq 4$ and $x - y \leq 1$

■ PRACTICE FOR COLLEGE ENTRANCE TESTS

Each question consists of two quantities, one in column A and one in column B. Compare the two quantities and select one of the following answers.

A if the quantity in column A is greater
B if the quantity in column B is greater
C if the two quantities are equal
D if the relationship cannot be determined from the information given

Comments

- Letters such as a, b, x, and y are variables that can be replaced by real numbers.
- A symbol that appears in both columns in a question stands for the same thing in column A as in column B.
- In some questions information that applies to quantities in both columns is centered above the two columns.

EXAMPLES

Column A		Column B	Answers
$3 + 4$		$3 \cdot 4$	B
$a + b$	$a = 5$	$b + 5$	C
$2x + y$		$x + 2y$	D

1.

Column A		Column B
$\dfrac{a}{b}$	$a > 1$ $b < -1$	-1

2.

Column A		Column B
y	y and $\dfrac{49}{y}$ are both prime numbers.	49

3.

Column A		Column B
$a - b$	$a < b$	$b - a$

4.

$$\boxed{x} = |x - 6|$$

Column A	Column B
$\boxed{2}$	$\boxed{10}$

5.

Column A		Column B
$a^2 + b^2$	$a < 0 < b$	$(a + b)^2$

6.

Column A		Column B
x	$4x^2 > y^2$	y

7.

In 30 years Margaret will be 3 times as old as she is now. Seven years ago Louis was half as old as he is now.

Column A	Column B
Margaret's age now	Louis' age now

8.

Column A		Column B
x	$3x + 5 < 2$	0

9.

Column A		Column B
x	$2x > y > 0$	y

10.

Column A		Column B
$\dfrac{1}{a}$	$a < b < 0$	$\dfrac{1}{b}$

10 Rational and Irrational Numbers

If a point C divides a line segment AB into two parts so that $\dfrac{AC}{CB} = \dfrac{AB}{AC}$, each fraction is equal to the Golden Ratio, $\dfrac{1 + \sqrt{5}}{2}$.

In architecture and nature, rectangles in which the ratio of length to width equals the Golden Ratio are considered most pleasing to the eye. Many blossoms have the shape of a pentagon, which also illustrates the Golden Ratio.

$$\frac{PR}{PQ} = \frac{1 + \sqrt{5}}{2}$$

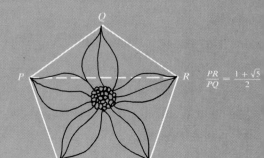

$\dfrac{PR}{PQ} = \dfrac{1 + \sqrt{5}}{2}$

10-1 Square Roots

Preview

A manufacturer of card tables makes one model that is 3 ft long on each side. A new model is to have a playing surface that is $\frac{1}{3}$ greater than the 3-ft model.

- What is the surface area of the top of the new model?
- How long must each side of the new model be?

■ LESSON

The number 25 is called the **square** of 5 because $5^2 = 25$. The number 25 is also the square of -5 because $(-5)^2 = 25$.

For the same reason, we say that 5 and -5 are **square roots** of 25.

> ### Definition: Square Root
>
> If $x^2 = y$, then x is a *square root* of y.

The positive square root of 25 is written $\sqrt{25}$.

$$\sqrt{25} = 5$$

The negative square root of 25 is written $-\sqrt{25}$.

$$-\sqrt{25} = -5$$

For example,

$$\sqrt{16} = 4 \qquad -\sqrt{16} = -4$$
$$\sqrt{49} = 7 \qquad -\sqrt{49} = -7$$
$$\sqrt{144} = 12 \qquad -\sqrt{144} = -12$$

The symbol $\sqrt{}$ is called a **radical sign**. The expression under the radical sign is the **radicand**. The entire expression is called a **radical**.

$$\sqrt{25} \text{ is a radical} \qquad 25 \text{ is the radicand}$$

Note that $\sqrt{16}$, $\sqrt{49}$, $\sqrt{144}$, and $\sqrt{0}$ are positive numbers or zero. The radical sign is used to indicate the **principal square root** of a number. The principal square root is never negative.

For the set of real numbers, a radicand cannot be a negative number. No real number squared is equal to a negative number. In the expression $\sqrt{x}$, x represents a number greater than or equal to 0.

If x is positive, then $\sqrt{x}$ is positive.
If x is positive, then $-\sqrt{x}$ is negative.
If x is 0, then $\sqrt{x}$ is 0.

The expression $\sqrt{25}$ can be read "the principal square root of 25," "the square root of 25," or "radical 25." The expression $-\sqrt{25}$ is read "the negative square root of 25" or "negative radical 25."

Example 1 Simplify. $\sqrt{100}$

Solution The radical sign identifies the principal square root, a positive number.

Answer $\sqrt{100} = 10$, because $10^2 = 100$.

Example 2 Simplify. $-\sqrt{64}$

Solution The expression $-\sqrt{64}$ identifies the opposite of the principal square root, a negative number.

Answer $-\sqrt{64} = -8$

Example 3 Simplify. $\sqrt{225}$

Solution *Find the prime factorization of the radicand.* $\quad \sqrt{225} = \sqrt{3^2 \cdot 5^2}$
Write the radicand as a square. $\qquad\qquad\qquad = \sqrt{(3 \cdot 5)^2}$
Simplify. $\qquad\qquad\qquad\qquad\qquad\qquad = 3 \cdot 5 = 15$

Answer $\sqrt{225} = 15$

Example 4 Simplify. $\sqrt{\dfrac{4}{9}}$

Solution *Factor the numerator and denominator.* $\quad \sqrt{\dfrac{4}{9}} = \sqrt{\dfrac{2^2}{3^2}}$

Write the radicand as a square. $\qquad\qquad = \sqrt{\left(\dfrac{2}{3}\right)^2}$

Simplify. $\qquad\qquad\qquad\qquad\qquad\quad = \dfrac{2}{3}$

Answer $\sqrt{\dfrac{4}{9}} = \dfrac{2}{3}$

Sometimes we wish to refer to both the positive square root and the negative square root in the same equation. We can write this as follows:

$$\pm \sqrt{25} = \pm 5$$

The sentence above is a short way of writing

$$+\sqrt{25} = +5 \quad \text{and} \quad -\sqrt{25} = -5$$

The symbol $\pm$ is read "positive or negative."

Example 5 Simplify. $\pm\sqrt{81}$

Solution $\pm\sqrt{81} = \pm 9$ because $+\sqrt{81} = +9$ and $-\sqrt{81} = -9$.

Example 6 Simplify. $(\sqrt{36})^2$

Solution $\sqrt{36}$ is the principal square root of 36. That is, $\sqrt{36}$ is the nonnegative number whose square is 36.

Answer $(\sqrt{36})^2 = 36$

Check $(\sqrt{36})^2 = 6^2 = 36$

■ CLASSROOM EXERCISES

Complete by choosing the correct word.

1. $\sqrt{49}$ is ___?___ (positive/negative).   **2.** $-\sqrt{36}$ is ___?___ (positive/negative).

3. For all positive values of a, $\sqrt{a}$ is ___?___ (positive/negative).

Simplify.

4. $\sqrt{4}$ **5.** $-\sqrt{9}$ **6.** $\sqrt{1}$ **7.** $\sqrt{0}$

8. $\sqrt{196}$ **9.** $\sqrt{1.96}$ **10.** $-\sqrt{441}$ **11.** $-\sqrt{4.41}$

12. $\sqrt{\dfrac{16}{25}}$ **13.** $\dfrac{\sqrt{16}}{\sqrt{25}}$ **14.** $\sqrt{\dfrac{49}{36}}$ **15.** $\dfrac{\sqrt{49}}{\sqrt{36}}$

16. Answer both questions and state how the answers differ.

    a. What does $\sqrt{100}$ equal?   **b.** What are the square roots of 100?

17. How are the solutions to these equations different?

    a. $x^2 = 64$ **b.** $x = \sqrt{64}$

■ WRITTEN EXERCISES

Identify whether the expression represents a positive number, a negative number, or zero, or does not represent a real number.

A **1.** $-\sqrt{49}$ **2.** $-\sqrt{16}$ **3.** $\sqrt{81}$ **4.** $\sqrt{121}$

    5. $\sqrt{-25}$ **6.** $\sqrt{-36}$ **7.** $\sqrt{2.56}$ **8.** $\sqrt{1.96}$

    9. $\sqrt{0.64}$ **10.** $\sqrt{0.49}$ **11.** $-\sqrt{1.21}$ **12.** $-\sqrt{1.69}$

Simplify.

13. $\sqrt{10,000}$ **14.** $\sqrt{400}$ **15.** $\sqrt{2500}$ **16.** $\sqrt{1600}$

17. $-\sqrt{40,000}$ **18.** $-\sqrt{100}$ **19.** $-\sqrt{1.44}$ **20.** $-\sqrt{2.25}$

Simplify.

21. $\dfrac{\sqrt{9}}{\sqrt{16}}$ **22.** $\dfrac{\sqrt{4}}{\sqrt{25}}$ **23.** $\sqrt{\dfrac{9}{16}}$ **24.** $\sqrt{\dfrac{4}{25}}$

Find the square roots of these numbers.

25. 400 **26.** 2500 **27.** 1600 **28.** 100

Solve these equations. The equations may have two solutions, one solution, or no solution. Write $\emptyset$ if the equation has no solution.

29. $x^2 = 16$ **30.** $x^2 = 36$ **31.** $x = \sqrt{25}$ **32.** $x = \sqrt{16}$

33. $x = -\sqrt{64}$ **34.** $x = -\sqrt{25}$ **35.** $x^2 = 64$ **36.** $x^2 = 25$

37. $x = \sqrt{-36}$ **38.** $x = \sqrt{-64}$ **39.** $\sqrt{x} = 4$ **40.** $\sqrt{x} = 9$

The area A of a square with side s is given by the formula $A = s^2$. The formula $s = \sqrt{A}$ expresses the length of the side of a square when its area is known. Find the length of the side of the square with the given area.

41. 100 cm^2 **42.** 64 km^2 **43.** 36 km^2 **44.** 49 m^2

Solve.

B **45.** Jacksonville, Florida, is one of the largest cities in the United States. Its area is approximately 2025 km^2. The area of Los Angeles, California is approximately 1225 km^2. If both cities were in the shape of squares, how much longer would a side of the Jacksonville square be.?

46. The area of Oklahoma City is approximately 1681 km^2. The area of San Francisco is approximately 121 km^2. If both cities were in the shape of squares, how much longer would a side of the Oklahoma City square be?

Simplify.

47. $(\sqrt{0.36})^2$ **48.** $(\sqrt{0.49})^2$ **49.** $\sqrt{1.6^2}$ **50.** $\sqrt{2.5^2}$

51. $\sqrt{3^2 \cdot 4^2}$ **52.** $3^2 \cdot \sqrt{4^2}$ **53.** $3 \cdot \sqrt{4}$ **54.** $3 \cdot (\sqrt{4})^2$

55. $\sqrt{\dfrac{0}{9}}$ **56.** $\sqrt{\dfrac{81}{100}}$ **57.** $\dfrac{\sqrt{81}}{100}$ **58.** $\dfrac{81}{\sqrt{100}}$

59. $(\sqrt{0.64})^2$ **60.** $\sqrt{0.64^2}$ **61.** $(\sqrt{0.0001})^2$ **62.** $\sqrt{0.01^2}$

C An old "rule of thumb" for sailors is that the maximum speed of a sailboat in knots (nautical miles per hour) is 1.35 times the square root of the length (in feet) of the waterline ($K = 1.35\sqrt{l}$). Find the maximum speed of a sailboat with each of these waterline lengths.

63. 25 ft **64.** 36 ft

65. 49 ft **66.** 64 ft

The time in seconds for an object h meters above the ground to fall to the ground is given by the formula $t = \sqrt{\dfrac{h}{5}}$. How long will it take an object to fall to the ground from each of these heights?

67. 5 m   **68.** 20 m   **69.** 80 m   **70.** 500 m

The time in seconds for a pendulum to swing back and forth is given by the formula $t = 2\pi\sqrt{\dfrac{l}{10}}$, where l is the length of the pendulum in meters. (Use 3.14 as an approximation for π.) For pendulums of the following lengths, find the time for a full swing.

71. 10 m   **72.** 2.5 m   **73.** 40 m   **74.** 0.1 m

A *geometric mean* of two numbers a and b is a number x such that $\dfrac{a}{x} = \dfrac{x}{b}$. Find the positive geometric mean for each pair of numbers.

75. 12; 48   **76.** 75; 12   **77.** 45; 20   **78.** 7; 0.28

79. A square swimming pool is bounded by a concrete walk that is 3 ft wide. The area of the pool plus the walk is 2304 ft². Find the dimensions of the pool.

■ REVIEW EXERCISES

State the property that is illustrated in each equation.   [1–5]

1. $4x + 5x = (4 + 5)x$ **2.** $(37 \cdot 25)4 = 37(25 \cdot 4)$

Solve.   [3–7]

3. $3x = 4(x - 3)$ **4.** $3(x - 2) = 5x + 4$

Solve each equation for r.   [3–9]

5. $L = 2\pi rh$ **6.** $A = p + prt$

7. Write 93,000,000 in scientific notation.   [4–4]

EXTENSION   Evaluating square roots on a calculator

You can use the square root key ($\sqrt{\ }$) on a calculator to compute the square root of a number. For example, if you enter 9 and then press the square root key, you should see a calculator display of 3, the principal square root of 9.
    Follow these steps.

1. Enter any number greater than 1.
2. Press $\sqrt{\ }$ $\sqrt{\ }$ $\sqrt{\ }$ $\sqrt{\ }$ $\sqrt{\ }$, noting the display after each time you press the key.
3. Repeat with other numbers greater than 1.
4. Repeat with numbers between 0 and 1.

Can you explain why, in all cases, the sequence of displayed numbers approaches 1?

10–2 Approximations of Square Roots

The ancient Mesopotamians were skilled at finding square roots. They used the symbol  to represent the square root of a number. Using a base-60 numeration system, they achieved amazing accuracy in determining approximations of common square roots. An ancient Babylonian tablet lists the square root of 2 as 〈symbols〉 or 1.414222, a result that was not exceeded in accuracy until the Renaissance.

Babylonian table of squares and square roots

The Mesopotamians, however, believed that all square roots could be expressed as rational numbers, the quotient of two whole numbers. It was not until about 450 B.C. that the Pythagorean Society of Greece discovered that $\sqrt{2}$ is not a rational number, and that some square roots can only be approximated by a decimal representation.

In this lesson you will learn how to find approximations of square roots.

■ LESSON

If you wish to build a square storage-shed floor with an area of 70 ft², the length of each side must be $\sqrt{70}$ ft. However, since measuring tapes are not marked with symbols such as $\sqrt{70}$, we must find another way of writing that number.

We know that $\sqrt{70}$ is not an integer or even a rational number. This means that it cannot be expressed exactly as a common fraction or as a decimal. Thus, $\sqrt{70}$ is an example of an **irrational number**.

> ## Definition: Irrational Number
>
> An *irrational number* is a real number that cannot be expressed as the quotient of two integers.

Although we cannot express $\sqrt{70}$ as an exact decimal fraction, we can find a decimal approximation to any degree of accuracy required by using systematic trials. For example, consider finding a decimal approximation of $\sqrt{70}$ accurate to the nearest hundredth.

First, determine which consecutive integers $\sqrt{70}$ lies between.

$$8^2 = 64$$
$$(\sqrt{70})^2 = 70$$
$$9^2 = 81$$

So, $8 < \sqrt{70} < 9$.

Next, improve the approximation to tenths.

$$8.3^2 = 68.89$$
$$(\sqrt{70})^2 = 70$$
$$8.4^2 = 70.56$$

Thus, $8.3 < \sqrt{70} < 8.4$.

Continue to hundredths.

$8.36^2 \approx 69.89$ [*Remember:* $\approx$ means "is approximately equal to."]
$(\sqrt{70})^2 = 70$
$8.37^2 \approx 70.06$

So, $8.36 < \sqrt{70} < 8.37$ ($\sqrt{70}$ is between 8.36 and 8.37, but it is closer to 8.37). Thus, the decimal approximation of $\sqrt{70}$ to the nearest hundredth is 8.37.

The decimal approximations of square roots of numbers can be determined by using a calculator or by using the table on page 641.

Example 1   Use this table to find a decimal approximation of $\sqrt{20}$ to the nearest hundredth.

Number	Square of number	Square root of number
n	n^2	$\sqrt{n}$
18	324	4.243
19	361	4.359
20	400	4.472
21	441	4.583

Solution   Find 20 in the n-column. Then find a decimal approximation of $\sqrt{20}$ in the $\sqrt{n}$-column.

Answer $\sqrt{20} \approx 4.47$

The same table can be used to find the squares of integers.

Example 2   Use this table to simplify 53^2.

n	n^2	$\sqrt{n}$
51	2601	7.141
52	2704	7.211
53	2809	7.280
54	2916	7.348

Solution   Find 53 in the n-column. Then find 53^2 in the n^2-column.

Answer $53^2 = 2809$

The solutions of some equations are square roots. We can use a table or a calculator to find decimal approximations for irrational numbers.

Example 3   Solve $x^2 = 37$, giving a decimal approximation of the solution to the nearest hundredth.

Solution
$$x^2 = 37$$
$$x = \pm\,\sqrt{37}$$

The exact solution is $\{\sqrt{37}, -\sqrt{37}\}$.

Answer   The approximate solution is $\{6.08, -6.08\}$.

■ CLASSROOM EXERCISES

State the consecutive integers between which the given number lies.

1. $\sqrt{39}$ **2.** $\sqrt{75}$ **3.** $\sqrt{14.4}$

Use the table on page 641 to find an approximation to the nearest hundredth.

4. $\sqrt{23}$ **5.** $\sqrt{68}$ **6.** $\sqrt{85}$

Solve, giving approximations of the solutions to the nearest hundredth.

7. $x^2 = 45$ **8.** $x^2 = 59$ **9.** $x^2 = 76$

■ WRITTEN EXERCISES

State the consecutive integers between which the given number lies.

A   **1.** $\sqrt{57}$ **2.** $\sqrt{95}$ **3.** $\sqrt{20.3}$ **4.** $\sqrt{27.9}$

Use the table on page 641 to find a decimal approximation to the nearest hundredth.

5. $\sqrt{2}$ **6.** $\sqrt{3}$ **7.** $\sqrt{30}$ **8.** $\sqrt{20}$

9. $-\sqrt{18}$ **10.** $-\sqrt{43}$ **11.** $-\sqrt{72}$ **12.** $-\sqrt{92}$

The number given is the square root of an integer accurate to the nearest tenth. What is the integer?

13. 3.6 **14.** 6.4 **15.** 7.8 **16.** 7.2

Use the table on page 641 to simplify.

17. 75^2 **18.** 88^2 **19.** 44^2 **20.** 76^2

21. $\sqrt{8281}$ **22.** $\sqrt{6889}$ **23.** $\sqrt{1444}$ **24.** $\sqrt{3844}$

Solve. List the solutions exactly, using radicals, not decimals.

25. $x^2 = 10$ **26.** $x^2 = 21$ **27.** $x^2 = 19$ **28.** $x^2 = 91$

Solve. List approximations for the solutions to the nearest hundredth.

29. $x^2 = 10$ **30.** $x^2 = 21$ **31.** $x^2 = 19$ **32.** $x^2 = 91$

33. The area of a square is 57cm². What is the length of each side of the square to the nearest hundredth of a centimeter?

34. The area of a square is 95 cm². What is the length of each side of the square to the nearest hundredth of a centimeter?

B The time in seconds for an object to fall h meters is given by the formula $t = \sqrt{\dfrac{h}{5}}$.
Find to the nearest tenth of a second the time for an object to fall to the ground from these heights.

35. 25 m **36.** 50 m

37. 100 m **38.** 200 m

The electrical current in amperes (I) that flows through an appliance is given by the formula $I = \sqrt{\dfrac{W}{R}}$, where W is the power of the appliance in watts and R is the resistance of the appliance in ohms. Find the current through these appliances.

39. A 50-watt lightbulb with a resistance of 200 ohms

40. A 500-watt iron with a resistance of 20 ohms

41. The surface area of a cube is 240 cm². Find the length of an edge to the nearest thousandth of a centimeter. (The six faces of a cube are squares.)

Use systematic trials to calculate the principal square root to the nearest tenth. Show your work.

42. $\sqrt{22}$ **43.** $\sqrt{45}$ **44.** $\sqrt{92}$ **45.** $\sqrt{13}$ **46.** $\sqrt{231}$ **47.** $\sqrt{185}$

Use systematic trials to calculate the principal square root to the nearest hundredth. Show your work.

C **48.** $\sqrt{300}$ **49.** $\sqrt{417}$ **50.** $\sqrt{631}$ **51.** $\sqrt{722}$ **52.** $\sqrt{200}$ **53.** $\sqrt{156}$

54. True or false?

$$\sqrt{9 + 16} = \sqrt{9} + \sqrt{16}$$

What do you conclude from your answer?

55. True or false?

$$\sqrt{100 - 36} = \sqrt{100} - \sqrt{36}$$

What do you conclude from your answer?

56. True or false?

$$\sqrt{4 \cdot 49} = \sqrt{4} \cdot \sqrt{49}$$

What do you conclude from your answer?

57. True or false?

$$\sqrt{\frac{25}{9}} = \frac{\sqrt{25}}{\sqrt{9}}$$

What do you conclude from your answer?

■ REVIEW EXERCISES

1. Find the missing number for each solution pair of the equation $d = 10t + 40$. [5–3]

t	0	2	?	?	?	0.2
d	?	?	50	100	20	?

2. Select three values of x, complete the table, plot the points, and then graph the equation $3x + y = 2$. [5–4]

x	?	?	?
y	?	?	?

3. Copy and complete the table and then use the ordered pairs to graph the equation $y = -x^3$. [5–4]

x	-3	-2	-1	$-\frac{1}{2}$	0	$\frac{1}{2}$	1	2	3
y	?	?	?	?	?	?	?	?	?

4. Draw these three lines through point $(2, 1)$. Label the lines a, b, and c. [5–5]

a. line with slope $\frac{3}{2}$

b. line with slope $-\frac{1}{3}$

c. line with no slope

5. Graph the line with a slope of $-\frac{1}{2}$ and y-intercept of -1. [5–6]

6. Write the equation $3x - 2y = 6$ in slope–intercept form. [5–6]

7. State whether the relation $\{(1, 2), (1, 3), (2, 4)\}$ represents a function. [5–8]

8. State whether the equation $y = \frac{4}{x}$ represents a linear function. [5–8]

EXTENSION A computer program for square root

The computer program below takes any positive number and any estimate of its square root to compute better and better approximations of the principal square root of the number. Lines 70 and 80 compute these approximations. A new value of the estimate (E) is computed by averaging the old estimate and the quotient of the number divided by the old estimate (A). The new estimate becomes the old estimate for the next approximation. Lines 60 and 90 count the number of times that lines 70 and 80 are used. Line 120 rounds the answer to the nearest tenth.

```
 10  PRINT "ENTER A POSITIVE NUMBER."
 20  INPUT P
 30  IF P < = 0 THEN 10
 40  PRINT "TYPE ANY POSITIVE ESTIMATE OF THE
      SQUARE ROOT OF YOUR NUMBER."
 50  INPUT E
 60  X = 0
 70  A = P/E
 80  E = (A + E)/2
 90  X = X + 1
100  PRINT X
110  IF ABS(A - E) > = .1 THEN 70
120  E = INT(10 * E + .5)/10
130  PRINT "SQUARE ROOT OF    "; P; "     =    "; E
140  PRINT "ANY MORE SQUARE ROOTS? (Y/N)"
150  INPUT Q$
160  IF Q$ = "N" THEN 180
170  GOTO 10
180  END
```

Use the program to find the principal square root of each number to the nearest tenth.

1. 29 **2.** 53 **3.** 82 **4.** 14

Change .1 to .01 in line 110 and change (10 * E + .5)/10 to (100 * E + .5)/100 in line 120 to find the principal square root of each number to the nearest hundredth.

5. 29 **6.** 53 **7.** 82 **8.** 14

How many times are steps 70 and 80 used (with new line 110) if P is 200 and E is each of the following?

9. 10 **10.** 50 **11.** 14 **12.** 15

10-3 The Pythagorean Theorem

Preview

More than 2500 years ago Egyptian and Greek mathematicians made an interesting and useful discovery about the sides of a right triangle. It is possible that the relationship among the sides was discovered while the mathematicians were studying patterns in floor tiles.

Triangle ABC is a right triangle. Each floor tile consists of four triangular parts.

- How many triangular parts cover the square on side AB?
- How many triangular parts cover the square on side AC?
- How many triangular parts cover the square on side BC?
- What relationship do you see among the numbers of parts in the three squares?

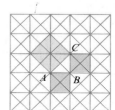

In this lesson you will study the relationship among the areas of the three squares. This relationship, named for the Greek mathematician Pythagoras, has many uses.

■ LESSON

Ancient mathematicians discovered that the area of the square on the longest side (**hypotenuse**) of a right triangle is equal to the sum of the areas of the squares on the two shorter sides (**legs**).

That property can be stated in algebraic terms as follows.

The Pythagorean Theorem

For a right triangle, the square of the hypotenuse is equal to the sum of the squares of the legs.

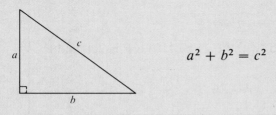

$$a^2 + b^2 = c^2$$

Right triangles are the only triangles for which this special relationship is true.

Converse of the Pythagorean Theorem

If sides a, b, and c of a triangle are such that $a^2 + b^2 = c^2$, then the triangle is a right triangle. The right angle is opposite the longest side.

These examples show some uses of the Pythagorean theorem and its converse.

Example 1   Are these triangles right triangles?

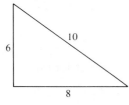   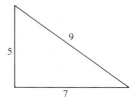

Solutions   Let $a = 6$, $b = 8$, and $c = 10$.   Let $a = 5$, $b = 7$, and $c = 9$.
Then $a^2 + b^2 = 36 + 64 = 100$ Then $a^2 + b^2 = 25 + 49 = 74$
$c^2 = 100$ $c^2 = 81$
$a^2 + b^2 = c^2$ $a^2 + b^2 \neq c^2$

Answer   The triangle is a right triangle.   The triangle is not a right triangle.

Example 2   How long is the third side of this right triangle?

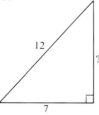

Solution $a^2 + b^2 = c^2$
$7^2 + b^2 = 12^2$
$49 + b^2 = 144$
$b^2 = 95$
$b = \pm \sqrt{95} \approx \pm 9.75$

There are two solutions of the algebraic equation. However, only the positive solution makes sense as an answer to the problem. The length of the third side is $\sqrt{95}$.

Answer   The length is 9.75 to the nearest hundredth. (See the table on page 641.)

Example 3   A baseball diamond is a square 90 ft long on each side. How far is home plate from second base?

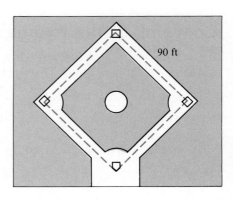

90 ft

Solution
$$a^2 + b^2 = c^2$$
$$90^2 + 90^2 = c^2$$
$$c^2 = 16{,}200$$
$$c = \pm\sqrt{16{,}200}$$

The negative solution of the equation is not an answer to the problem. The distance from home plate to second base is exactly $\sqrt{16{,}200}$. Using a calculator, we find an approximation to the nearest tenth:

$$c \approx 127.3$$

Answer   The length is 127.3 ft to the nearest tenth.

■ CLASSROOM EXERCISES

State whether the triangle is a right angle.

1.
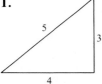
5
3
4

2.
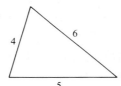
4
6
5

3.
$\sqrt{7}$
3
4

Find the missing lengths.

4.
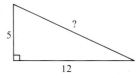
5
?
12

5.

1
2
?

6.
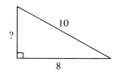
10
?
8

■ WRITTEN EXERCISES

State whether a triangle with these sides is a right triangle.

A **1.** 20 cm, 21 cm, 29 cm

2. 7 cm, 24 cm, 25 cm

3. 42 m, 40 m, 52 m

4. 28 m, 45 m, 63 m

5. 16 m, 30 m, 34 m

6. 12 m, 35 m, 37 m

7. 28 km, 96 km, 100 km

8. 32 m, 32 m, 40 m

Find the missing lengths.

9.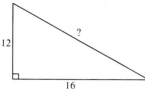

10.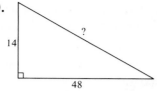

11.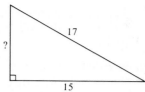

12.

13.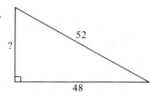

14.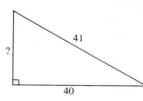

15.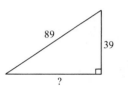

16.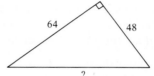

17.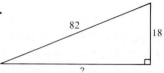

18.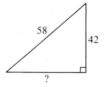

19.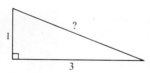

20.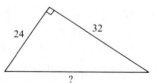

Find the missing lengths. Use the table on page 641 to approximate to the nearest tenth.

21.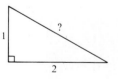

22.

23.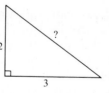

24.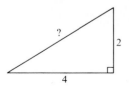

25.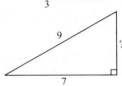

26.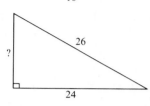

Find the length of the diagonal of a rectangle with the given length and width.
Use the table on page 641 to approximate to the nearest tenth.

27. length: 8 cm
width: 2 cm

28. length: 6 cm
width: 2 cm

29. length: 30 km
width: 20 km

30. length: 35 km
width: 27 km

31. length: 48 mm
width: 36 mm

32. length: 62 mm
width: 43 mm

Solve.

33. The highway route from Cedar Falls to Cedar Rapids is 40 mi east, then 30 mi south. What is the airline distance from Cedar Falls to Cedar Rapids?

34. Pearl City is approximately 8 mi west and 6 mi north of Honolulu. What is the straight-line distance between the two cities?

B **35.** A vacant lot is 36 m long and 27 m wide. How many meters shorter is the direct (diagonal) route from one corner to the opposite corner than the route that follows the boundary of the lot?

36. Cities A and B are joined by a direct highway. They are also connected by a highway that runs 24 km east from city A and then 70 km north to city B. How much shorter is the direct route than the east–north route?

37. An antenna is 60 m tall. A supporting cable is attached to the top of the antenna and to an anchor 11 m from the foot of the antenna. How long is the cable?

38. What is the longest line segment that can be drawn on a rectangular sheet of wrapping paper that is 28 in. wide and 45 in. long?

39. The diagonals of two different rectangles are 85 cm long. The length of one rectangle is 77 cm and the length of the other rectangle is 84 cm. How much wider is one rectangle than the other rectangle?

40. The diagonal brace of a rectangular door is 106 in. long. If the width of the door is 56 in., how high is the door?

41. A square field is 1 km long on each side. How many meters shorter is the diagonal distance across the field than the distance along two sides?

42. A hiker can walk on a road around the field described in Exercise 41 at the rate of 6 km/h. Walking diagonally across the field, the hiker walks at a rate of 4 km/h. Which route from one corner of the field to the opposite corner is faster? How much faster?

Triangles *ACD* and *BCD* are formed by joining points *A* and *B* on circle *O* to the ends of diameter *DC*. The angles at *A* and *B* are right angles. The diameter of the circle is 20 cm. The length of side *AD* is 10 cm and the length of side *BD* is 14 cm.

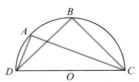

43. Find the perimeter and area of triangle *ACD*.

44. Find the perimeter and area of triangle *BCD*.

45. A room is 12 m long and 4 m wide. The ceiling is 3 m above the floor. How far is one corner of the floor from the opposite corner of the ceiling?

46. Can a 1-m stick be put in a box that measures 33 cm by 56 cm by 72 cm? What is the length of the longest stick that can fit in the box?

47. The sides of an acute triangle (each angle less than $90°$) are a, b, and c with c being the longest side. How is $a^2 + b^2$ related to c^2?

48. The sides of an obtuse triangle (one angle greater than $90°$) are a, b, and c with c being the longest side. How is $a^2 + b^2$ related to c^2?

■ REVIEW EXERCISES

1. Solve this system by graphing.
$x + y = 2$
$x - 2y = -4$

2. Solve this system by substitution.   [6–1, 6–2]
$y = 5x$
$25x + 10y = 300$

3. State whether this system has 0, 1, or infinitely many solutions.   [6–3]
$y = 2x + 2$
$y = 3x + 2$

Self-Quiz 1

10–1 Simplify.

1. $\sqrt{\dfrac{49}{25}}$ **2.** $-\sqrt{8100}$ **3.** $\pm\sqrt{0.36}$

Solve these equations.

4. $x^2 = 144$ **5.** $x^2 = 225$ **6.** $x^2 = -16$

10–2 Use the table on page 641 to find these square roots to the nearest tenth.

7. $\sqrt{12}$ **8.** $\sqrt{19,600}$ **9.** $\sqrt{169}$

10–3 Find the missing lengths.

10. 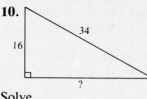   **11.** 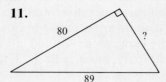   **12.**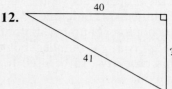

Solve.

13. What length of wire is needed to reach from the top of a 52-ft antenna to a point on the ground 20 ft from its base?

10–4 Simplifying Square Roots

Preview

- Which of these equivalent algebraic expressions is simplest?

$$5x - 3x - x \qquad \frac{10x}{10} \qquad x \qquad 1x$$

Most people would agree that x is simpler than the others.

- Which of these expressions is simplest?

$$\sqrt{8} \qquad 2\sqrt{2} \qquad \sqrt{\frac{24}{3}} \qquad \frac{4}{\sqrt{2}}$$

Although both $\sqrt{8}$ and $2\sqrt{2}$ are simple expressions, mathematicians have generally agreed that $2\sqrt{2}$ is "simpler" than $\sqrt{8}$.

In this lesson you will study how to simplify expressions that include radicals.

■ LESSON

Notice that

$$\sqrt{4} \cdot \sqrt{9} = 2 \cdot 3 = 6$$

and that

$$\sqrt{4 \cdot 9} = \sqrt{36} = 6;$$

therefore,

$$\sqrt{4} \cdot \sqrt{9} = \sqrt{4 \cdot 9}.$$

The last equation is an example of the multiplication property for radicals.

The Multiplication Property for Radicals

For all nonnegative numbers a and b,

$$\sqrt{ab} = \sqrt{a} \cdot \sqrt{b}$$

In the property above, the values of a and b must be restricted to 0 or the positive numbers. If only one of a and b is a negative number, then the expression $\sqrt{ab}$ is not a real number. If both a and b are negative numbers, then the expression $\sqrt{ab}$ is a real number, but the expressions $\sqrt{a}$ and $\sqrt{b}$ are not.

This property can be used to simplify some radicals. For example, consider simplifying $\sqrt{18}$.

Factor the radicand.	$\sqrt{18} = \sqrt{2 \cdot 3^2}$
Use the multiplication property for radicals.	$= \sqrt{2} \cdot \sqrt{3^2}$
Simplify.	$= 3\sqrt{2}$

When the radicand does not contain any perfect-square factors, we say that it is in **simplest form**.

The simplest form of $\sqrt{18}$ is $3\sqrt{2}$.

Example 1   Simplify. $\sqrt{48}$

   Solution

$$\begin{aligned}\sqrt{48} &= \sqrt{2^4 \cdot 3} \\ &= \sqrt{(2^2)^2} \cdot \sqrt{3} \\ &= 2^2 \cdot \sqrt{3} \\ &= 4\sqrt{3}\end{aligned}$$

   Answer $\sqrt{48} = 4\sqrt{3}$

Example 2   Simplify. $\sqrt{360}$

   Solution

$$\begin{aligned}\sqrt{360} &= \sqrt{2^3 \cdot 3^2 \cdot 5} \\ &= \sqrt{2^2 \cdot 3^2 \cdot 2 \cdot 5} \\ &= \sqrt{(2 \cdot 3)^2} \cdot \sqrt{2 \cdot 5} \\ &= 6\sqrt{10}\end{aligned}$$

   Answer $\sqrt{360} = 6\sqrt{10}$

The multiplication property for radicals is also used to simplify expressions involving the product of radicals.

Example 3   Simplify. $-\sqrt{24} \cdot \sqrt{30}$

   Solution

$$\begin{aligned}-\sqrt{24} \cdot \sqrt{30} &= -\sqrt{24 \cdot 30} \\ &= -\sqrt{2^3 \cdot 3 \cdot 2 \cdot 3 \cdot 5} \\ &= -\sqrt{2^4 \cdot 3^2 \cdot 5} \\ &= -\sqrt{(2^2 \cdot 3)^2} \cdot \sqrt{5} \\ &= -(2^2 \cdot 3) \cdot \sqrt{5} \\ &= -12\sqrt{5}\end{aligned}$$

   Answer $-\sqrt{24} \cdot \sqrt{30} = -12\sqrt{5}$

Example 4   Simplify. $\sqrt{3}(\sqrt{3} + \sqrt{8})$

   Solution

$$\begin{aligned}\sqrt{3}(\sqrt{3} + \sqrt{8}) &= \sqrt{3} \cdot \sqrt{3} + \sqrt{3} \cdot \sqrt{8} \\ &= \sqrt{3^2} + \sqrt{3} \cdot 2\sqrt{2} \\ &= 3 + 2\sqrt{6}\end{aligned}$$

   Answer $\sqrt{3}(\sqrt{3} + \sqrt{8}) = 3 + 2\sqrt{6}$

■ CLASSROOM EXERCISES

Write the prime factorization.

1. 18

2. 150

Simplify.

3. $\sqrt{8}$

4. $\sqrt{12}$

5. $\sqrt{16}$

6. $\sqrt{24}$

7. $\sqrt{40}$

8. $\sqrt{160}$

9. $\sqrt{2}(\sqrt{8} - 2\sqrt{3})$

10. $(\sqrt{10} + \sqrt{20})\sqrt{5}$

■ WRITTEN EXERCISES

Write the prime factorization.

A **1.** 20

2. 50

3. 300

4. 450

5. 75

6. 45

7. 200

8. 72

Simplify.

9. $\sqrt{50}$

10. $\sqrt{20}$

11. $-\sqrt{450}$

12. $-\sqrt{300}$

13. $\sqrt{15} \cdot \sqrt{3}$

14. $\sqrt{15} \cdot \sqrt{5}$

15. $\sqrt{24} \cdot \sqrt{3}$

16. $-\sqrt{10} \cdot \sqrt{20}$

17. $\sqrt{2}(\sqrt{2} + \sqrt{3})$

18. $\sqrt{3}(\sqrt{2} + \sqrt{3})$

19. $\sqrt{5}(2 + \sqrt{5})$

20. $\sqrt{7}(1 + \sqrt{7})$

21. $\sqrt{3}(\sqrt{3} - 1)$

22. $\sqrt{5}(\sqrt{5} - 1)$

23. $\sqrt{2}(2\sqrt{2} + \sqrt{6})$

24. $\sqrt{3}(\sqrt{6} + 2\sqrt{3})$

25. $\sqrt{5}(\sqrt{20} + \sqrt{45})$

26. $\sqrt{2}(\sqrt{50} + \sqrt{18})$

27. $(2 + \sqrt{3})(2 - \sqrt{3})$

28. $(3 + \sqrt{2})(3 - \sqrt{2})$

29. $(4 + \sqrt{5})^2$

30. $(5 - \sqrt{3})^2$

a	b
3	$\sqrt{3}$

Substitute and simplify.

31. $\sqrt{3 + a^2}$

32. $\sqrt{a^2 - 1}$

33. $\sqrt{15 + b^2}$

34. $\sqrt{25 + b^2}$

35. $(ab)^2$

36. ab^2

37. $(a - b)(a + b)$

38. $(a + b)^2$

39. $(a - b)^2$

B Use the Pythagorean theorem to find the length x. Write the answer in simplest form.

40.

41.

42.

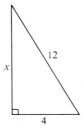

Use the Pythagorean theorem to find the length x. Write the answer in simplest form.

43.

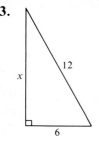

44.

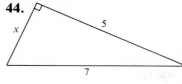

45.

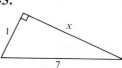

46.

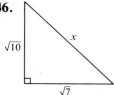

47.

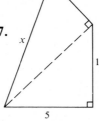

48.

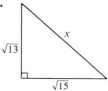

Approximate these radicals to the nearest hundredth. Use the table on page 641.

Sample   Approximate the square root of 300 correct to the nearest hundredth.

Solution   Simplify radicals before finding decimal approximations.

$$\sqrt{300} = 10\sqrt{3}$$
$$\approx 10(1.732) = 17.32$$

C **49.** $\sqrt{316}$ **50.** $\sqrt{320}$ **51.** $\sqrt{500}$ **52.** $\sqrt{775}$

53. Explain what is wrong with this demonstration that $-4 = 4$.

$$-4 = \sqrt{-4} \cdot \sqrt{-4}$$
$$= \sqrt{(-4)(-4)}$$
$$= \sqrt{16}$$
$$= 4$$

■ REVIEW EXERCISES

1. List all the positive integer factors of 18. [7–1]

2. Write the prime factorization of 42. [7–1]

3. State the greatest common factor of 18 and 36. [7–2]

4. State the greatest common factor of $8ab$ and $12a^2$. [7–2]

Solve.

5. $x(x + 5) = 0$ **6.** $3(2x + 4)(x - 5) = 0$ [7–3]

7. Factor completely. $4a^2 - 6a$ [7–4]

8. The square of a number is 9 times the number. What is the number? [7–4]

10–5 Simplifying Quotients of Square Roots

Preview

Before the invention of calculators and computers, most computations were done using paper and pencil. Thus it was important to use efficient techniques.

If we know that $\sqrt{6} \approx 2.449489743$, how can we find an approximation for $\dfrac{1}{\sqrt{6}}$ to the nearest hundredth?

We could divide 1 by 2.449489743, but there is a method that is less cumbersome.

In this lesson we will study how to simplify expressions with radicals in denominators. The decimal approximation for expressions like $\dfrac{1}{\sqrt{6}}$ can then be found efficiently.

■ LESSON

Notice that

$$\sqrt{\frac{36}{9}} = \sqrt{4} = 2$$

and that

$$\frac{\sqrt{36}}{\sqrt{9}} = \frac{6}{3} = 2.$$

Therefore,

$$\sqrt{\frac{36}{9}} = \frac{\sqrt{36}}{\sqrt{9}}.$$

The last equation is an example of the division property for radicals.

The Division Property for Radicals

For all nonnegative numbers a and b, $b \neq 0$,

$$\sqrt{\frac{a}{b}} = \frac{\sqrt{a}}{\sqrt{b}}$$

The division property for radicals can be used to simplify radical expressions that contain fractions. For example, consider simplifying $\sqrt{\dfrac{27}{4}}$.

Use the division property for radicals. $\qquad \sqrt{\dfrac{27}{4}} = \dfrac{\sqrt{27}}{\sqrt{4}}$

Simplify each radical. $\qquad\qquad\qquad\qquad = \dfrac{3\sqrt{3}}{2}$

The simplest form of $\sqrt{\dfrac{27}{4}}$ is $\dfrac{3\sqrt{3}}{2}$.

Example 1 Simplify. $\sqrt{\dfrac{1}{3}}$

Solution

$$\sqrt{\frac{1}{3}} = \frac{\sqrt{1}}{\sqrt{3}}$$

$$= \frac{1}{\sqrt{3}}$$

Multiply the numerator and denominator by $\sqrt{3}$ to make the denominator an integer.

$$= \frac{1}{\sqrt{3}} \cdot \frac{\sqrt{3}}{\sqrt{3}}$$

$$= \frac{\sqrt{3}}{3}$$

Answer $\sqrt{\dfrac{1}{3}} = \dfrac{\sqrt{3}}{3}$

The process of changing the form of a fraction from one with a radical in the denominator to an equivalent fraction without a radical in the denominator is called **rationalizing the denominator**. To rationalize a denominator, first write the denominator in simplest radical form. Then multiply the numerator and denominator by the remaining radical.

Example 2 Rationalize the denominator and simplify. $\dfrac{4}{\sqrt{80}}$

Solution *Simplify the denominator.*

$$\frac{4}{\sqrt{80}} = \frac{4}{4\sqrt{5}}$$

Simplify the fraction.

$$= \frac{1}{\sqrt{5}}$$

Multiply the numerator and denominator by $\sqrt{5}$.

$$= \frac{1}{\sqrt{5}} \cdot \frac{\sqrt{5}}{\sqrt{5}}$$

Simplify the numerator and denominator.

$$= \frac{\sqrt{5}}{5}$$

Answer $\dfrac{4}{\sqrt{80}} = \dfrac{\sqrt{5}}{5}$

Example 3 Write a decimal approximation of $\dfrac{\sqrt{27}}{\sqrt{2}}$ to the nearest hundredth.

Solution

$$\dfrac{\sqrt{27}}{\sqrt{2}} = \dfrac{3\sqrt{3}}{\sqrt{2}}$$

Rationalize the denominator.

$$= \dfrac{3\sqrt{3}}{\sqrt{2}} \cdot \dfrac{\sqrt{2}}{\sqrt{2}}$$

$$= \dfrac{3\sqrt{6}}{2}$$

Use the table of square roots.

$$\approx \dfrac{3(2.449)}{2}$$

$$= \dfrac{7.347}{2}$$

$$= 3.674$$

Answer $\dfrac{\sqrt{27}}{\sqrt{2}} = 3.67$ to the nearest hundredth

Example 4 Simplify. $\sqrt{15\dfrac{1}{2}}$

Solution *Write the radicand as a fraction.* $\sqrt{15\dfrac{1}{2}} = \sqrt{\dfrac{31}{2}}$

Use the division property for radicals. $= \dfrac{\sqrt{31}}{\sqrt{2}}$

Rationalize the denominaor. $= \dfrac{\sqrt{31}}{\sqrt{2}} \cdot \dfrac{\sqrt{2}}{\sqrt{2}}$

$$= \dfrac{\sqrt{62}}{2}$$

Answer $\sqrt{15\dfrac{1}{2}} = \dfrac{\sqrt{62}}{2}$

■ CLASSROOM EXERCISES

Simplify.

1. $\sqrt{\dfrac{8}{25}}$ 　　 **2.** $\sqrt{\dfrac{3}{5}}$ 　　 **3.** $\dfrac{\sqrt{5}}{\sqrt{2}}$ 　　 **4.** $\dfrac{\sqrt{6}}{\sqrt{8}}$ 　　 **5.** $\sqrt{\dfrac{5}{12}}$

6. Give a decimal approximation to the nearest thousandth of $\dfrac{\sqrt{19}}{\sqrt{3}}$.

Simplify.

7. $\sqrt{3\dfrac{1}{2}}$ 　　 **8.** $\sqrt{2\dfrac{1}{2}}$ 　　 **9.** $\sqrt{1\dfrac{1}{5}}$ 　　 **10.** $\sqrt{1\dfrac{2}{3}}$ 　　 **11.** $\sqrt{6\dfrac{2}{3}}$

■ WRITTEN EXERCISES

Simplify.

A 1. $\sqrt{\dfrac{6}{25}}$ 2. $\sqrt{\dfrac{3}{16}}$ 3. $\sqrt{\dfrac{7}{9}}$ 4. $\sqrt{\dfrac{3}{4}}$

5. $\dfrac{2}{\sqrt{7}}$ 6. $\dfrac{3}{\sqrt{10}}$ 7. $\dfrac{\sqrt{2}}{\sqrt{5}}$ 8. $\dfrac{\sqrt{2}}{\sqrt{3}}$

9. $\dfrac{\sqrt{7}}{\sqrt{2}}$ 10. $\dfrac{\sqrt{10}}{\sqrt{3}}$ 11. $\sqrt{\dfrac{7}{3}}$ 12. $\sqrt{\dfrac{10}{7}}$

13. $\dfrac{\sqrt{12}}{\sqrt{10}}$ 14. $\dfrac{\sqrt{6}}{\sqrt{15}}$ 15. $\dfrac{\sqrt{8}}{\sqrt{27}}$ 16. $\dfrac{\sqrt{18}}{\sqrt{125}}$

17. $\dfrac{\sqrt{12}}{\sqrt{27}}$ 18. $\dfrac{\sqrt{18}}{\sqrt{50}}$ 19. $\dfrac{\sqrt{50}}{\sqrt{72}}$ 20. $\dfrac{\sqrt{45}}{\sqrt{20}}$

Give a decimal approximation to the nearest hundredth.

21. $\dfrac{1}{\sqrt{3}}$ 22. $\dfrac{1}{\sqrt{2}}$ 23. $\dfrac{1}{\sqrt{10}}$ 24. $\dfrac{1}{\sqrt{6}}$

25. $\dfrac{\sqrt{21}}{\sqrt{2}}$ 26. $\dfrac{\sqrt{17}}{\sqrt{5}}$ 27. $\sqrt{\dfrac{13}{6}}$ 28. $\sqrt{\dfrac{23}{3}}$

29. $\dfrac{\sqrt{21}}{\sqrt{3}}$ 30. $\dfrac{\sqrt{21}}{\sqrt{7}}$ 31. $\dfrac{\sqrt{500}}{\sqrt{3}}$ 32. $\dfrac{\sqrt{700}}{\sqrt{11}}$

33. $\sqrt{8\dfrac{1}{3}}$ 34. $\sqrt{5\dfrac{1}{3}}$ 35. $\sqrt{6\dfrac{2}{5}}$ 36. $\sqrt{8\dfrac{4}{5}}$

Find the length of the rectangle having the given area and width. Write the answer in simplest form.

37. area: 7 cm²
   width: $\sqrt{2}$ cm

38. area: 8 cm²
   width: $\sqrt{3}$ cm

39. area: $\sqrt{15}$ m²
   width: $\sqrt{2}$ m

40. area: $\sqrt{40}$ m²
   width: $\sqrt{3}$ m

41. area: 12 km²
   width: $\sqrt{18}$ km

42. area: 24 km²
   width: $\sqrt{20}$ km

a	b	c
2	3	5

Substitute and simplify.

B 43. $\sqrt{\dfrac{a}{b}}$ 44. $\sqrt{\dfrac{b}{a}}$ 45. $\dfrac{\sqrt{a^2}}{\sqrt{b}}$ 46. $\dfrac{\sqrt{b^2}}{\sqrt{a}}$

47. $\sqrt{\dfrac{a^2b}{ab^2}}$ 48. $\sqrt{\dfrac{ab^2}{a^2b}}$ 49. $\dfrac{\sqrt{a^2c}}{\sqrt{ab}}$ 50. $\dfrac{\sqrt{b^2}}{\sqrt{abc}}$

Find the missing lengths.

51.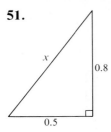

52.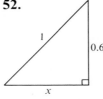

53.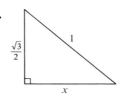

54.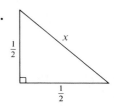

If $\sqrt{y}$ is an irrational number, then the two binomials of the form $x + \sqrt{y}$ and $x - \sqrt{y}$ are **conjugates** of each other. A conjugate can be used to rationalize a binomial denominator.

Example 5 Simplify. $\dfrac{5}{4 + \sqrt{3}}$

Solution *Multiply the numerator and denominator by the conjugate of the denominator.*

$$\frac{5}{4 + \sqrt{3}} = \frac{5}{4 + \sqrt{3}} \cdot \frac{4 - \sqrt{3}}{4 - \sqrt{3}}$$

Simplify.

$$= \frac{20 - 5\sqrt{3}}{16 - 3}$$

$$= \frac{20 - 5\sqrt{3}}{13}$$

Answer $\dfrac{5}{4 + \sqrt{3}} = \dfrac{20 - 5\sqrt{3}}{13}$

Use conjugates to simplify.

C **55.** $\dfrac{6}{3 - \sqrt{2}}$ **56.** $\dfrac{5}{2 + \sqrt{3}}$ **57.** $\dfrac{10}{4 + \sqrt{6}}$ **58.** $\dfrac{2}{5 - \sqrt{5}}$

Give a decimal approximation to the nearest hundredth.

59. $\dfrac{2}{2 - \sqrt{2}}$ **60.** $\dfrac{2}{2 + \sqrt{2}}$ **61.** $\dfrac{\sqrt{2}}{\sqrt{2} - 2}$ **62.** $\dfrac{\sqrt{2}}{2 + \sqrt{2}}$

■ REVIEW EXERCISES

Factor.

1. $x^2 - 100$ **2.** $x^2 + 12x + 36$ [7–6, 7–7]

3. $x^2 - 7x + 12$ **4.** $x^2 - x - 42$ [7–8, 7–9]

Solve.

5. $2x^2 - 50 = 0$ **6.** $x^2 + 10x + 25 = 0$ [7–6, 7–7]

7. $x^2 + 12x + 20 = 0$ **8.** $x^2 + 4x - 12 = 0$ [7–8, 7–9]

Preview A model for adding radicals

What is the total area of these two rectangles?

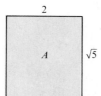
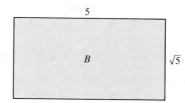

The area of rectangle A is $2\sqrt{5}$ and the area of rectangle B is $5\sqrt{5}$. The total area can be written as

$$2\sqrt{5} + 5\sqrt{5}$$

The diagram below suggests another way to express the area.

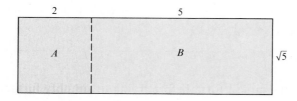

Find the combined area of rectangles C and D.

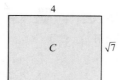
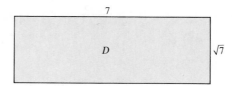

How much greater is the area of rectangle D than the area of rectangle C?

In this lesson we will study how to add, subtract, and simplify expressions containing radicals.

■ LESSON

Here are two sums that involve radicals:

$$3\sqrt{2} + 5\sqrt{2} \qquad 4\sqrt{7} + 2\sqrt{5}$$

The first can be simplified using the distributive property.

$$3\sqrt{2} + 5\sqrt{2} = (3 + 5)\sqrt{2}$$
$$= 8\sqrt{2}$$

The second sum $(4\sqrt{7} + 2\sqrt{5})$ cannot be simplified. In order to combine radical terms by adding or subtracting, we must be sure that the radical factors of the terms are the same.

> To simplify sums or differences of square roots:
>
> 1. Write each radical in simplest form.
> 2. Use the distributive property to add or subtract terms with like radicands.

Example 1 Simplify. $8\sqrt{5} - 2\sqrt{5} + 4$

Solution
$$8\sqrt{5} - 2\sqrt{5} + 4 = (8 - 2)\sqrt{5} + 4$$
$$= 6\sqrt{5} + 4$$

Answer $8\sqrt{5} - 2\sqrt{5} + 4 = 6\sqrt{5} + 4$

Be sure to write all radicals in simplest form before trying to simplify further.

Example 2 Simplify. $7\sqrt{2} - 4\sqrt{8} + \sqrt{18}$

Solution
$$7\sqrt{2} - 4\sqrt{8} + \sqrt{18} = 7\sqrt{2} - 4(2\sqrt{2}) + 3\sqrt{2}$$
$$= 7\sqrt{2} - 8\sqrt{2} + 3\sqrt{2}$$
$$= 2\sqrt{2}$$

Answer $7\sqrt{2} - 4\sqrt{8} + \sqrt{18} = 2\sqrt{2}$

Example 3 Simplify. $2\sqrt{27} + \sqrt{50} + 4\sqrt{75}$

Solution
$$2\sqrt{27} + \sqrt{50} + 4\sqrt{75} = 2 \cdot 3\sqrt{3} + 5\sqrt{2} + 4 \cdot 5\sqrt{3}$$
$$= 6\sqrt{3} + 20\sqrt{3} + 5\sqrt{2}$$
$$= 26\sqrt{3} + 5\sqrt{2}$$

Answer $2\sqrt{27} + \sqrt{50} + 4\sqrt{75} = 26\sqrt{3} + 5\sqrt{2}$

Example 4 Simplify. $\sqrt{\dfrac{2}{3}} + \sqrt{\dfrac{3}{8}}$

Solution

$$\sqrt{\dfrac{2}{3}} + \sqrt{\dfrac{3}{8}} = \dfrac{\sqrt{2}}{\sqrt{3}} + \dfrac{\sqrt{3}}{\sqrt{8}}$$

$$= \dfrac{\sqrt{2}}{\sqrt{3}} \cdot \dfrac{\sqrt{3}}{\sqrt{3}} + \dfrac{\sqrt{3}}{2\sqrt{2}} \cdot \dfrac{\sqrt{2}}{\sqrt{2}}$$

$$= \dfrac{\sqrt{6}}{3} + \dfrac{\sqrt{6}}{4}$$

$$= \dfrac{4\sqrt{6}}{12} + \dfrac{3\sqrt{6}}{12}$$

$$= \dfrac{7\sqrt{6}}{12}$$

Answer $\sqrt{\dfrac{2}{3}} + \sqrt{\dfrac{3}{8}} = \dfrac{7\sqrt{6}}{12}$

■ CLASSROOM EXERCISES

Simplify. Write "simplest form" if the expression cannot be simplified.

1. $5\sqrt{3} + 2\sqrt{3}$ **2.** $4\sqrt{2} - 9\sqrt{2}$ **3.** $6\sqrt{3} - 7\sqrt{2}$

4. $2\sqrt{27} + 3\sqrt{12} - \sqrt{24}$ **5.** $\sqrt{8} + \sqrt{75} - \sqrt{48} - \sqrt{50}$ **6.** $\sqrt{\dfrac{1}{5}} + \dfrac{\sqrt{5}}{3}$

■ WRITTEN EXERCISES

Simplify. Write "simplest form" if the expression cannot be simplified.

A   **1.** $3\sqrt{2} + 7\sqrt{2}$ **2.** $5\sqrt{2} + 6\sqrt{2}$ **3.** $6\sqrt{3} + \sqrt{3}$ **4.** $5\sqrt{7} + \sqrt{7}$

    5. $4\sqrt{7} - 6\sqrt{7}$ **6.** $7\sqrt{3} - 10\sqrt{3}$ **7.** $8\sqrt{5} - \sqrt{5}$ **8.** $4\sqrt{6} - \sqrt{6}$

    9. $5\sqrt{3} + 3\sqrt{2} + 2\sqrt{3}$ **10.** $3\sqrt{2} + 2\sqrt{3}$ **11.** $\sqrt{21} - \sqrt{4}$ **12.** $2\sqrt{3} + 5\sqrt{2} + 3\sqrt{2}$

  13. $\sqrt{8} + \sqrt{18}$ **14.** $\sqrt{12} + \sqrt{27}$ **15.** $\sqrt{10} + \sqrt{21}$ **16.** $\sqrt{15} + \sqrt{14}$

  17. $\sqrt{45} - \sqrt{20}$ **18.** $\sqrt{125} - \sqrt{80}$ **19.** $\sqrt{48} - \sqrt{75}$ **20.** $\sqrt{50} - \sqrt{98}$

Find the perimeter of the triangle. Simplify the answer.

21.

22.

23.

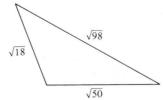

24.

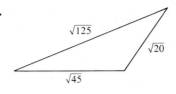

Find the perimeter and area of the rectangle. Simplify the answers.

25.

26.

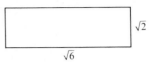

27.

28.

Simplify.

B **29.** $\sqrt{2} + \dfrac{3}{\sqrt{2}}$

30. $\dfrac{\sqrt{3}}{6} + \dfrac{2}{\sqrt{3}}$

31. $\dfrac{4}{\sqrt{6}} + \dfrac{\sqrt{2}}{\sqrt{3}}$

32. $5\sqrt{\dfrac{2}{3}} - 4\sqrt{\dfrac{3}{2}}$

33. $\dfrac{2\sqrt{5}}{12} - \sqrt{\dfrac{3}{20}}$

34. $\dfrac{6}{\sqrt{6}} + 4\sqrt{\dfrac{5}{6}}$

Find the length of the hypotenuse of the triangle.

35.

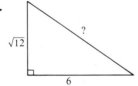

36.

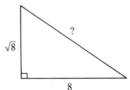

37. What is the perimeter of the triangle in Exercise 35?

38. What is the perimeter of the triangle in Exercise 36?

Find the perimeter of the triangle.

39.

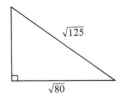

40.

41.

42.

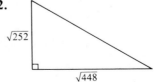

43. Find the perimeter of this parallelogram.

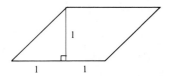

44. Find the area of this square.

In each of the following, is the equation true for all positive values of a and b? If not, write a counterexample to show that it is false.

C **45.** $\sqrt{a^2 - b^2} = a - b$

46. $\sqrt{a} + \sqrt{b} = \sqrt{a + b}$

47. $(\sqrt{a} - \sqrt{b})^2 = a - b$

48. $\dfrac{1}{\sqrt{a}} \cdot \dfrac{1}{\sqrt{b}} = \dfrac{1}{\sqrt{ab}}$

■ REVIEW EXERCISES

1. Are these fractions equivalent for admissible values of x? 　　　　[8–1]

$$\frac{3x + 2}{4x + 2} \text{ and } \frac{3x}{4x}$$

2. Which values of x are not admissible? 　$\dfrac{x(x + 3)}{(x - 1)(x + 2)}$ 　　[8–1]

Simplify. Assume that no denominators are zero.

3. $\dfrac{3a}{4b} \cdot \dfrac{6ab}{c}$

4. $\dfrac{x(x + 1)}{x(x + 2)} \div \dfrac{x + 1}{2}$ 　　[8–3]

5. $\dfrac{2}{x} + \dfrac{3}{x2}$

6. $\dfrac{1}{2x} + \dfrac{2}{3x}$ 　　[8–4]

Solve.

7. $\dfrac{24}{18} = \dfrac{36}{x}$

8. $\dfrac{1}{12} = \dfrac{1}{r} + \dfrac{1}{20}$ 　　[8–5]

9. It takes one machine 10 h to complete a job and it takes another machine 15 h to complete the same job. How long will it take both machines working together to complete the job?

10. What is the resistance in a circuit that contains a 10-ohm resistance and a 20-ohm resistance connected in parallel? 　　[8–6]

$\left[\text{Use the equation } \dfrac{1}{R} = \dfrac{1}{r_1} + \dfrac{1}{r_2}.\right]$

Simplify. 　　[4–6]

11. $3x(x^2 - 5x + 4)$

12. $(2xy^2)^3(x + y)$

Solve by making and using a table. 　　[2–9]

13. What number can be added to the numerator and denominator of $\dfrac{1}{4}$ so that the new fraction will equal $\dfrac{4}{5}$?

Self-Quiz 2

10–4, Simplify.
10–5,
10–6   **1.** $\sqrt{72}$ **2.** $\sqrt{5}(\sqrt{10} - \sqrt{15})$ **3.** $\sqrt{42} \cdot \sqrt{56}$ **4.** $\sqrt{\dfrac{4}{7}}$

5. $\dfrac{\sqrt{96}}{\sqrt{18}}$ **6.** $\dfrac{2}{\sqrt{6}}$ **7.** $5\sqrt{8} - \sqrt{18}$ **8.** $\dfrac{2}{\sqrt{12}} + \dfrac{1}{\sqrt{3}}$

10–4,
10–5

Substitute and simplify.

a	b	c
$\sqrt{6}$	$\sqrt{18}$	$\sqrt{24}$

9. $a^2 + b^2$ **10.** $\dfrac{1}{a + c}$ **11.** $\sqrt{b^2 - 2}$

12. $-\sqrt{a^2 - 2}$ **13.** $\dfrac{1}{c - a}$ **14.** $a(c - b)$

10–6   Find the perimeter of each figure.

15.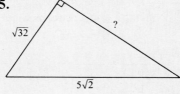

16.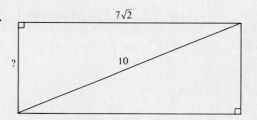

Mathematics and Your Future

Today students rarely decide on a career by their first year in high school. In fact, college students typically change majors several times. Therefore, even if you have a career in mind, you should take enough mathematics to keep your options open for other careers. For example, with fewer than four years of precollege high school mathematics you eliminate your options for college-level programs in science or engineering. With fewer than three years you eliminate your options for most college majors, while with fewer than two years you eliminate your options for many vocational and technical school programs. The best way to keep your options open is to take as many mathematics courses as you can.

Preview

Consider this equation.

$$\sqrt{x^2} = x$$

- Substitute 6 for x. Is the sentence that you get true or false?

- Substitute 0 for x. Is the sentence that you get true or false?

- Substitute -3 for x. Is the sentence that you get true or false?

- Which of the following statements is always true?

  a. For all real numbers a, $\sqrt{a^2} = a$.

  b. For all real numbers a, $\sqrt{a^2} = -a$.

  c. For all nonnegative real numbers a, $\sqrt{a^2} = a$.

- Complete this equation to obtain a true sentence.

$$\text{For all real numbers } a, \ \sqrt{a^2} = ?$$

In this lesson you will learn to simplify a radical when the radicand contains a variable.

■ LESSON

Radicals that contain variables are simplified by the same techniques used to simplify other radicals. For example, consider simplifying $\sqrt{54x^2y^5}$. Since it is difficult to simplify when the values of variables in radicals are negative numbers, we will assume at first that domains of the variables are restricted to the set of nonnegative numbers.

First write the radicand as two factors, one of which contains all square factors.

$$\sqrt{54x^2y^5} = \sqrt{9x^2y^4} \cdot \sqrt{6y}$$

Then simplify.
$$= 3xy^2\sqrt{6y}$$

	Example 1 Simplify. $\sqrt{25a^2b^3}$, $a \geq 0$, $b \geq 0$

Solution
$$\sqrt{25a^2b^3} = \sqrt{25a^2b^2} \cdot \sqrt{b}$$

$$= 5ab\sqrt{b}$$

Answer $\sqrt{25a^2b^3} = 5ab\sqrt{b}$, $a \geq 0$, $b \geq 0$

Example 2 Simplify. $\sqrt{\dfrac{48}{x^3}}$, $x > 0$

Solution Note that x is restricted to positive values because the denominator cannot be 0.

$$\sqrt{\frac{48}{x^3}} = \frac{4\sqrt{3}}{x\sqrt{x}}$$

Rationalize the denominator. $= \dfrac{4\sqrt{3}\sqrt{x}}{x\sqrt{x}\sqrt{x}}$

$$= \frac{4\sqrt{3x}}{x^2}$$

Answer $\sqrt{\dfrac{48}{x^3}} = \dfrac{4\sqrt{3x}}{x^2}$, $x > 0$

Now consider the difficulties caused when the values of variables in radicands are negative or zero.

1. The value of a square-root radicand cannot be negative. The domains of variables must be restricted to values that make the radicand positive or zero; for example,

$$\sqrt{3x^2y}, \qquad y \geq 0.$$

The domain of x is not restricted since x^2 is nonnegative for all values of x.

2. The principal square root of a quantity cannot be negative, so absolute value must be used whenever necessary to ensure that the principal square root is not negative.

$$\sqrt{x^2} = |x|, \quad \text{where } x \text{ is a real number}$$

Note that without using absolute value we would have a false statement if a negative number is substituted for x. For example, the principal square root of $\sqrt{(-3)^2}$ is 3, not -3.

$$\sqrt{(-3)^2} = |-3| = 3$$

3. Division by zero is never allowed. Thus, values of variables that make denominators in radicands zero are not allowed; for example,

$$\sqrt{\frac{5}{y^2}}, \qquad y \neq 0.$$

Example 3 Simplify, stating any necessary restrictions. $\sqrt{x^2y}$

Solution First state the restrictions. The restriction is $y \geq 0$.

$$\sqrt{x^2y} = \sqrt{x^2}\sqrt{y}$$
$$= |x|\sqrt{y}$$

Answer $\sqrt{x^2y} = |x|\sqrt{y}$, $y \geq 0$

Example 2   Solve. $\sqrt{y+2} = y - 4$

Solution

$$\sqrt{y+2} = y - 4$$
$$y + 2 = y^2 - 8y + 16$$
$$0 = y^2 - 9y + 14$$
$$0 = (y - 2)(y - 7)$$
$$y - 2 = 0 \quad \text{or} \quad y - 7 = 0$$
$$y = 2 \quad \text{or} \quad y = 7$$

Check   Check each solution in the original equation.

$\sqrt{y+2} = y - 4$	$\sqrt{y+2} = y - 4$
$\sqrt{2+2} \stackrel{?}{=} 2 - 4$	$\sqrt{7+2} \stackrel{?}{=} 7 - 4$
$\sqrt{4} \stackrel{?}{=} -2$	$\sqrt{9} \stackrel{?}{=} 3$
$2 = -2$ False	$3 = 3$ True

Answer   {7}

Note how important it is to test possible answers in the original equation. In Example 2, the solution 2 satisfies the equation $y + 2 = y^2 - 8y + 16$, but it does not satisfy the original equation.

Example 3   Solve. $\sqrt{x+5} + 16 = 4$

Solution

$$\sqrt{x+5} + 16 = 4$$

Isolate the radical on one side of the equation.
Square both sides.

$$\sqrt{x+5} = -12$$
$$x + 5 = 144$$
$$x = 139$$

Check

$$\sqrt{x+5} + 16 = 4$$
$$\sqrt{139+5} + 16 \stackrel{?}{=} 4$$
$$\sqrt{144} + 16 \stackrel{?}{=} 4$$
$$12 + 16 = 4 \quad \text{False}$$

Answer   ∅ (No number satisfies the equation.)

The following steps can be used to solve a radical equation.

1. Rewrite the equation with the radical isolated on one side.
2. Square both sides of the equation to eliminate the radical.
3. Solve the new equation.
4. Check possible answers in the original equation.

■ CLASSROOM EXERCISES

Solve.

1. $\sqrt{x} = 4$

2. $\sqrt{n} = 9$

3. $\sqrt{a - 7} = 0$

4. $\sqrt{5m - 1} = 7$

5. $\sqrt{x + 8} = 2$

6. $\sqrt{4n + 4} = 8$

7. $\sqrt{2n + 2} = 4$

8. $\sqrt{5n + 1} = 6$

9. $2\sqrt{x - 1} = x - 1$

10. $\sqrt{x + 10} + 2 = x$

■ WRITTEN EXERCISES

Solve.

A 1. $\sqrt{t} = 5$

2. $\sqrt{t} = 7$

3. $\sqrt{3x - 2} = 4$

4. $\sqrt{2x - 3} = 5$

5. $\sqrt{3y + 9} = 6$

6. $\sqrt{4y + 1} = 3$

7. $\sqrt{c + 6} = c$

8. $\sqrt{c + 12} = c$

9. $\sqrt{3x - 8} = x - 2$

10. $\sqrt{3x - 5} = x - 1$

11. $\sqrt{r + 14} = r + 2$

12. $\sqrt{9r + 19} = r + 3$

13. $\sqrt{x + 6} = x$

14. $\sqrt{x} + 2 = x$

15. $\sqrt{2x} + 3 = 8$

16. $\sqrt{3x} + 1 = 7$

17. $\sqrt{5p - 1} + p = 3$

18. $\sqrt{3p + 1} + p = 9$

19. $\sqrt{a + 1} = a - 5$

20. $\sqrt{5a + 6} = a + 2$

21. $x + \sqrt{5} = 4$

22. $x - \sqrt{3} = 6$

23. $\sqrt{10q + 4} - q = 2$

24. $\sqrt{10q - 6} - q = 1$

Sometimes it is helpful to examine the original radical equation before solving. Use your knowledge of principal square roots to solve these equations easily.

25. $\sqrt{x + 4} = -3$

26. $\sqrt{y} - \sqrt{2y} = 5$

27. $\sqrt{a} + \sqrt{-a} = 0$

28. Why is it inappropriate to square both sides as the first step in solving this equation?

$$2\sqrt{x} - 1 = x$$

Solve these equations containing two radicals. If there is no solution, write ∅.

B 29. $\sqrt{x} = \sqrt{5x - 8}$

30. $\sqrt{5x} = \sqrt{3x + 6}$

31. $\sqrt{6x} = \sqrt{2x - 8}$

32. $\sqrt{7x} = \sqrt{8x + 6}$

33. $\sqrt{x^2 + 2} = \sqrt{3x}$

34. $\sqrt{x^2 + 4} = \sqrt{5x}$

35. $\sqrt{x^2 - 10} = \sqrt{3x}$

36. $\sqrt{2x} = \sqrt{15 - x^2}$

37. $\sqrt{5x} = \sqrt{2x^2 + 3}$

38. $\sqrt{13x} = \sqrt{3x^2 + 4}$

39. $\sqrt{5x^2 + x} = \sqrt{3 - x}$

40. $\sqrt{5x^2 + x} = \sqrt{2 - x^2}$

528 Chapter 10 Rational and Irrational Numbers

To solve some equations containing two radicals, we must square both sides of the equation twice.

Example 4 Solve. $\sqrt{x} + 1 = \sqrt{2x - 2}$

> *Solution*
>
> $$\sqrt{x} + 1 = \sqrt{2x - 2}$$
>
> *Square both sides.* $$(\sqrt{x} + 1)^2 = (\sqrt{2x - 2})^2$$
> $$x + 2\sqrt{x} + 1 = 2x - 2$$
> $$2\sqrt{x} = x - 3$$
> *Square both sides.* $$4x = x^2 - 6x + 9$$
> $$0 = x^2 - 10x + 9$$
> $$0 = (x - 9)(x - 1)$$
> $$x - 9 = 0 \quad \text{or} \quad x - 1 = 0$$
> $$x = 9 \quad \text{or} \qquad x = 1$$
>
> *Check* $\sqrt{x} + 1 = \sqrt{2x - 2}$ $\sqrt{x} + 1 = \sqrt{2x - 2}$
>
> $\sqrt{9} + 1 \overset{?}{=} \sqrt{2(9) - 2}$ $\sqrt{1} + 1 \overset{?}{=} \sqrt{2(1) - 2}$
>
> $3 + 1 \overset{?}{=} \sqrt{16}$ $1 + 1 \overset{?}{=} \sqrt{0}$
>
> $\quad 4 = 4 \qquad$ True $\quad 2 = 0 \qquad$ False
>
> *Answer* $\{9\}$

Solve these equations containing two radicals. If there is no solution, write ∅.

C **41.** $\sqrt{x - 1} + 3 = \sqrt{4x + 5}$ **42.** $\sqrt{x + 10} + 2 = \sqrt{x + 26}$

43. $\sqrt{2x - 7} - 1 = \sqrt{x - 4}$ **44.** $\sqrt{x} + \sqrt{x + 5} = 5$

45. $\sqrt{x + 2} + \sqrt{x - 3} = -5$ **46.** $\sqrt{2x + 4} - \sqrt{x - 2} = 2$

■ REVIEW EXERCISES

Graph these inequalities on a number line.

1. $2x + 3 > x + 5$ **2.** $3x + 5 < 2x + 6$ **3.** $4 - 2x > 10$ [9-3, 9-5]

Self-Quiz 3

10–7 Simplify each expression. Assume that the variables are all positive.

   1. $\sqrt{121y^3z^2}$ **2.** $\sqrt{\dfrac{a^4}{b}}$ **3.** $\dfrac{\sqrt{y^5}}{\sqrt{x^3}}$

State restrictions, if any, on the values of the variables. Then simplify.

   4. $\sqrt{x^3y^4}$ **5.** $\sqrt{\dfrac{36}{z}}$ **6.** $\sqrt{-2x^5}$

10–8 Solve each equation for x.

   7. $\sqrt{2x} = \sqrt{x - 4}$ **8.** $\sqrt{y} + 2 = y$ **9.** $\sqrt{x^2 - 5} = 2\sqrt{x}$

10–9 The Distance Formula

Preview

What is the distance between the two points?

A and C

B and C

A and B

F and D

E and F

D and E

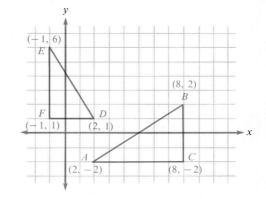

In this lesson you will learn to use a formula to find the distance between any two points whose coordinates are known.

■ LESSON

The distance between two points on the same horizontal line or the same vertical line is easy to compute.

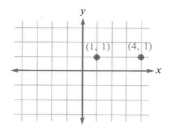

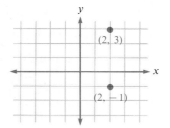

The distance d is the difference between the x-coordinates.

$$d = 4 - 1 = 3$$

In general, for points (x_1, y_1) and (x_2, y_2), when $y_1 = y_2$, $d = |x_2 - x_1|$.

The distance d is the difference between the y-coordinates.

$$d = 3 - (-1) = 4$$

In general, for points (x_1, y_1) and (x_2, y_2), when $x_1 = x_2$, $d = |y_2 - y_1|$.

The absolute value ensures that the positive difference is used to denote the distance.

When two points are not on the same horizontal or vertical line, the Pythagorean theorem is used to find the distance between them.

530   Chapter 10   Rational and Irrational Numbers

To find the distance d between points (2, 3) and (4, 7), draw a horizontal line through one point and a vertical line through the other. Note that a right triangle is formed, and the Pythagorean theorem can now be used.

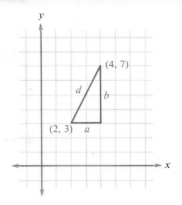

$d^2 = a^2 + b^2$
$d^2 = |4 - 2|^2 + |7 - 3|^2$
$d^2 = 2^2 + 4^2$
$d^2 = 20$
$d = \sqrt{20}$ [$d \neq -\sqrt{20}$ because the distance cannot be negative.]
$d = 2\sqrt{5}$

Since the differences above are squared, parentheses can be used instead of absolute value signs.

The distance between two points can be found by using this formula.

The Distance Formula

The distance d between two points

$$(x_1, y_1) \text{ and } (x_2, y_2)$$

is

$$d = \sqrt{(x_2 - x_1)^2 + (y_2 - y_1)^2}.$$

Example 1 Find the distance between points (2, −3) and (5, 1).

Solution
$$\begin{aligned}
d &= \sqrt{(x_2 - x_1)^2 + (y_2 - y_1)^2} \\
&= \sqrt{(5 - 2)^2 + [1 - (-3)]^2} \\
&= \sqrt{3^2 + 4^2} \\
&= \sqrt{9 + 16} \\
&= \sqrt{25} \\
&= 5
\end{aligned}$$

Answer The distance between (2, −3) and (5, 1) is 5.

Example 2 Find the value of y such that the distance between point A and point B is $2\sqrt{5}$.

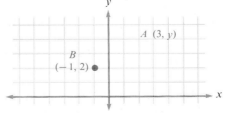

Example 2 (continued)

Solution

$$d = \sqrt{(x_2 - x_1)^2 + (y_2 - y_1)^2}$$
$$2\sqrt{5} = \sqrt{[3 - (-1)]^2 + (y - 2)^2}$$
$$2\sqrt{5} = \sqrt{16 + (y^2 - 4y + 4)}$$
$$2\sqrt{5} = \sqrt{y^2 - 4y + 20}$$
$$(2\sqrt{5})^2 = (\sqrt{y^2 - 4y + 20})^2$$
$$20 = y^2 - 4y + 20$$
$$0 = y^2 - 4y$$
$$0 = y(y - 4)$$
$$y = 0 \quad \text{or} \quad y = 4$$

Check   The check is left to the student.

Answer   {0, 4}

■ CLASSROOM EXERCISES

State the values of x_1, x_2, y_1, and y_2.

1. (4, 2), (5, 1) **2.** (−3, 4), (2, 8) **3.** (−1, −3), (−5, −6)

State the values of $|x_2 - x_1|$ and $|y_2 - y_1|$.

4. (−8, 1), (0, −5) **5.** (1, 4), (6, −8) **6.** (3, 7), (1, −9)

Find the distance between the two points.

7. (2, 4), (2, −5) **8.** (3, 3), (5, 3) **9.** (−3, −2), (3, −10)

■ WRITTEN EXERCISES

Find $|x_2 - x_1|$ and $|y_2 - y_1|$ for each pair of points.

A
1. (2, 2), (6, 8) **2.** (3, 3), (9, 7)

3. (10, 3), (4, 6) **4.** (9, 1), (2, 10)

5. (−2, 3), (4, −10) **6.** (5, −3), (−1, 7)

7. (0, 0), (−2, −10) **8.** (−3, −15), (0, 0)

Find the distance between the two points. Simplify the distances.

9. (2, −3), (2, 7) **10.** (−3, 5), (−3, 1)

11. (4, −2), (−1, −2) **12.** (−3, 6), (2, 6)

13. (2, 4), (6, 7) **14.** (3, 6), (7, 9)

15. (3, 7), (15, 2) **16.** (3, 9), (9, 1)

17. (10, 10), (−6, −2) **18.** (4, 4), (−11, −4)

19. (4, 3), (4, 10) **20.** (6, 2), (6, 11)

Find the distance between the two points. Simplify the distances.

21. $(5, -2), (6, -2)$ **22.** $(7, -3), (4, -3)$

23. $(100, 100), (114, 52)$ **24.** $(50, 50), (74, 18)$

Find the value of x or y given the distance d between the two points.

B   **25.** $(11, 7), (3, y); d = 10$ **26.** $(7, 10), (x, 4); d = 10$

    27. $(5, -5), (x, 7); d = 13$ **28.** $(10, 10), (-5, y); d = 17$

    29. $(2, 3), (4, y); d = \sqrt{5}$ **30.** $(5, 1), (x, 5); d = 4\sqrt{2}$

    31. $(x, 4), (5, 9); d = 5\sqrt{2}$ **32.** $(3, y), (7, 1); d = 2\sqrt{5}$

Find the distance between the two points. Write the approximation of the distance to tenths.

33. $(5, 4), (7, 9)$ **34.** $(-3, 7), (2, 4)$

35. $(-2, -3), (5, 3)$ **36.** $(-2, -3), (-6, 2)$

37. Find the length of side BC.   **38.** Find the length of the longer diagonal.

                                             39. Find the length of the shorter diagonal.

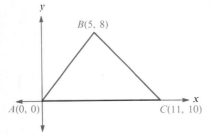

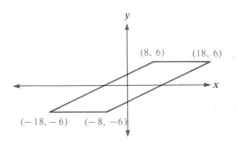

40. Use the distance formula to show that the triangle with the following vertices is isosceles: $A(3, 1)$, $B(3, 9)$, and $C(8, 5)$.

41. Use the distance formula to show that the triangle with the following vertices is isosceles: $A(-4, 3)$, $B(5, 1)$, and $C(-2, -5)$.

Find the perimeter of triangle ABC given the coordinates of A, B, and C.

42. $A(7, 10)$ **43.** $A(-3, 5)$ **44.** $A(10, 6)$
      $B(10, 14)$ $B(2, 17)$ $B(10, 12)$
      $C(10, 10)$ $C(2, 5)$ $C(7, 6)$

Find the length of each diagonal of quadrilateral $ABCD$.

45. $A(10, 0)$ **46.** $A(-10, -10)$ **47.** $A(35, 12)$
      $B(2, 15)$ $B(-10, 11)$ $B(21, 0)$
      $C(17, 23)$ $C(10, 11)$ $C(0, 0)$
      $D(25, 8)$ $D(10, -10)$ $D(16, 12)$

C **48.** Show that ABC is a right triangle.

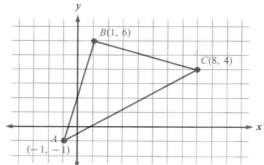

49. Find the area of rectangle $MNOP$.

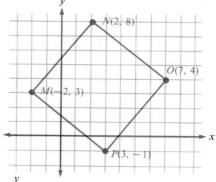

50. Show that the lengths of the diagonals of the square $PQRS$ are equal.

51. The point M is on a coordinate axis, a distance of 5 units from $N(-3, 4)$. What are the possible coordinates of M?

■ **REVIEW EXERCISES**

1. Use absolute values to write an inequality that describes this graph.   [9–6]

Graph on a number line.

2. $|2 + x| < 4$ **3.** $|x - 2| > 3$ [9–6]

Graph on a coordinate plane.

4. $x + y \geq 2$ **5.** $y < \frac{1}{2}x + 3$ [9–7]

6. $y < x$ or $y \geq 0$ **7.** $y > x - 1$ and $x \geq 0$ [9–8]

EXTENSION   Using the distance formula on a computer

Number pairs (x,y) that satisfy the equation $4 = \sqrt{(1-x)^2 + (2-y)^2}$ are coordinates of points that are a distance of 4 units from the point $(1, 2)$. The following program takes any value of x between -3 and 5 and computes the corresponding value(s) of y. The BASIC function SQR instructs the computer to take a square root.

```
 10   PRINT "ENTER A NUMBER BETWEEN -3 AND 5."
 20   INPUT X
 30   Y = 2 + SQR (16 - (1 - X) ↑ 2)
 40   Y = INT(Y * 10 + .5)/10
 50   PRINT "X =     "; X, "Y =     "; Y
 60   Y = 2 - SQR (16 - (1 - X) ↑ 2)
 70   Y = INT(Y * 10 + .5)/10
 80   PRINT "X =     "; X, "Y =     "; Y
 90   PRINT "ANY MORE NUMBERS? (Y/N)"
100   INPUT P$
110   IF P$ = "N" THEN 130
120   GOTO 10
130   END
```

1. Use the program to find many (x, y) pairs that satisfy the given equation.

2. Graph the equation.

3. Enter a value of x that is either less than -3 or greater than 5. What happens? Why?

4. How can you change lines 10, 30, and 60 to define a circle with radius 4 centered at $(-2, 3)$?

Mathematics and Your Future

Counselors have information about admissions and scholarships to colleges and universities. They also know about vocational schools and employment opportunities. Ask your counselor for information about entrance requirements for the schools you may be considering. Be sure to find out the mathematics requirements for major fields that you may wish to study. The amount of mathematics needed for a specific major is frequently greater than the amount of mathematics you will need to get admitted to the institution. You will find that if you take plenty of mathematics, you will keep your options open.

■ CHAPTER SUMMARY

• Vocabulary

• If $x^2 = y$, then x is a *square root* of y. [10–1]

• An *irrational number* is a real number that cannot be expressed as a quotient of two integers. [10–2]

• The Pythagorean Theorem [10–3]

For a right triangle, the square of the hypotenuse is equal to the sum of the squares of the legs.

$$a^2 + b^2 = c^2$$

• Converse of the Pythagorean Theorem [10–3]

If sides a, b, and c of a triangle are such that $a^2 + b^2 = c^2$, then the triangle is a right triangle. The right angle is opposite the longest side.

• The Multiplication Property for Radicals [10–4]

For all nonnegative numbers a and b,

$$\sqrt{ab} = \sqrt{a} \cdot \sqrt{b}.$$

• The Division Property for Radicals [10–5]

For all nonnegative numbers a and b, $b \neq 0$,

$$\sqrt{\frac{a}{b}} = \frac{\sqrt{a}}{\sqrt{b}}.$$

• Simplest Form of a Radical Expression [10–5]

A radical expression is in simplest form if:

1. the radicand does not contain a fraction;
2. the radicand has no perfect-square factor other than 1;
3. no radicals appear in the denominator.

• For all real numbers x, [10–7]

$$\sqrt{x^2} = |x|.$$

- When working with square roots and variables: [10–7]

  1. The value of the radicand cannot be negative, so the domains of the variables must be properly restricted.
  2. The principal square root of a quantity cannot be negative. Absolute value must be used whenever necessary to be sure that the principal square root is not negative.

  $$\sqrt{x^2} = |x|, \text{ where } x \text{ is a real number}$$

  3. Division by zero is never allowed. Thus, values of variables that make denominators in radicands zero are not allowed.

- The following general steps are used to solve radical equations. [10–8]

  1. Rewrite the equation with the radical isolated on one side.
  2. Square both sides of the equation to eliminate the radical.
  3. Solve the new equation.
  4. Check possible answers in the original equation.

- The Distance Formula [10–9]

  The distance d between two points (x_1, y_1) and (x_2, y_2) is
  $$d = \sqrt{(x_2 - x_1)^2 + (y_2 - y_1)^2}.$$

■ CHAPTER REVIEW

10–1   Objective: To simplify expressions containing square roots of perfect squares.

   1. Simplify. $-\sqrt{1.44}$ **2.** Find the square roots of 3600.   **3.** Solve. $x^2 = 81$

10–2   Objective: To use a table to find approximations for square roots of numbers.

   Use the table of square roots on page 641. Give your answers to the nearest hundredth.

   4. $\sqrt{24}$ **5.** $x^2 = 40$

   6. The area of a square is 50 cm². What is the length of each side of the square to the nearest hundredth of a centimeter?

10–3   Objective: To use the Pythagorean theorem to find a missing length of a side in a right triangle.

   Find the missing length.

   7.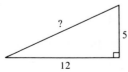

   8.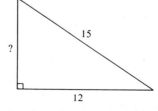

   9. A rectangle is 4 cm long and 2 cm wide. Use a table to find the length of the diagonal of the rectangle to the nearest tenth of a centimeter.

10–4 Objective: To simplify expressions involving sums, products, and differences of radicals.

Simplify.

10. $\sqrt{18}$ **11.** $\sqrt{2}(\sqrt{5} + \sqrt{2})$ **12.** $(3 + \sqrt{5})(3 - \sqrt{5})$

10–5 Objective: To simplify expressions involving quotients of radicals.

Simplify.

13. $\sqrt{\dfrac{5}{9}}$ **14.** $\dfrac{\sqrt{12}}{\sqrt{75}}$

15. The area of a rectangle is 20 cm² and its width is $\sqrt{5}$ cm. Write the length of the rectangle in simplest form.

10–6 Objective: To simplify sums and differences of square roots.

Simplify. Write "simplest form" if the expression cannot be simplified.

16. $3\sqrt{5} + \sqrt{5}$ **17.** $\sqrt{27} - \sqrt{18}$

18. Find the perimeter and area of the rectangle. Simplify the answer.

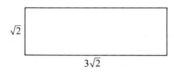

10–7 Objective: To simplify radicals containing variables.

Simplify. State restrictions, if any, on the values of the variables.

19. $\sqrt{4x^2}$ **20.** $\sqrt{\dfrac{x}{4}}$ **21.** $\sqrt{\dfrac{a^3}{b^2}}$

10–8 Objective: To solve equations containing radicals.

Solve.

22. $\sqrt{x} = 9$ **23.** $\sqrt{2x + 5} = 3$ **24.** $\sqrt{4x} + 3 = x$

10–9 Objective: To use the distance formula to find the distance between two points.

Find the distance between the two points. Simplify all radicals.

25. $(2, 4), (6, 7)$ **26.** $(4, -3), (4, 6)$

27. Find the value of x such that the distance between the points $(3, 0)$ and $(x, -2)$ is $2\sqrt{5}$.

28. Find the perimeter of triangle ABC with the following coordinates.
    $A\ (-2, 0)$ $B\ (-2, -5)$ $C\ (-6, -5)$

■ CHAPTER 10 SELF-TEST

10-1 Simplify, if possible. Write "NR" if not a real number.

1. $\sqrt{-64}$ **2.** $-\sqrt{0.36}$ **3.** $\sqrt{\dfrac{9}{4}}$

Solve.

4. $x^2 = 0.16$ **5.** $y = -\sqrt{81}$

10-2 **6.** State the consecutive integers that $\sqrt{59}$ is between.

Write the exact solution in simplest form.

7. $x^2 = 76$ **8.** $y = -\sqrt{180}$

10-3 Find the missing length.

9. 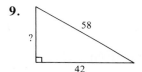 **10.**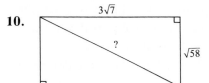

11. A flagpole 30 ft high has a rope attached to raise the flag. If the rope is 34 ft long, how far away from the base of the pole can the rope be stretched and still touch the ground?

10-4 Simplify.

12. $\sqrt{54}$ **13.** $\sqrt{7}(\sqrt{35} - 2\sqrt{7})$ **14.** $(\sqrt{2} + \sqrt{3})^2$

10-5

a	b	c
5	8	11

Substitute and simplify.

15. $\dfrac{\sqrt{a+c}}{\sqrt{b^2}}$ **16.** $\dfrac{\sqrt{b+2c}}{\sqrt{ab}}$

10-6 Write in simplest form.

17. $\sqrt{\dfrac{9}{5}}$ **18.** $\dfrac{\sqrt{35}}{\sqrt{15}}$

19. $\sqrt{98} - \sqrt{50}$ **20.** $\sqrt{18} + \sqrt{12} - \sqrt{3}$

10-7 Simplify if possible. State any restrictions on the values of the variables.

21. $\sqrt{\dfrac{5}{x^2}}$ **22.** $\sqrt{\dfrac{b^4}{a}}$

10–8 Solve.

23. $\sqrt{1 - 2a} = 7$ **24.** $x = \sqrt{15 - 2x}$

10–9   **25.** Find the distance between $(-3, 4)$ and $(2, -6)$.

       26. Find y such that the distance between $(2, y)$ and $(7, -5)$ is $\sqrt{29}$.

■ PRACTICE FOR COLLEGE ENTRANCE TESTS

Each question consists of two quantities, one in column A and one in column B. Compare the two quantities and select one of the following answers:

A   if the quantity in column A is greater

B   if the quantity in column B is greater

C   if the two quantities are equal

D   if the relationship cannot be determined from the information given

Comments:

- Letters such as a, b, x, and y are variables that can be replaced by real numbers.

- A symbol that appears in both columns in a question stands for the same thing in column A as in column B.

- In some questions information that applies to quantities in both columns is centered above the two columns.

	Column A	*Column B*
1.	The length of the hypotenuse of a right triangle with legs of lengths 1 and 3	The length of the hypotenuse of a right triangle with legs of lengths 2 and 2
2.	$\sqrt{a} = \sqrt{b}$	
	ab	b^2
3.	$a > 0,\ b > 0$	
	$\sqrt{a + b}$	$\sqrt{a} + \sqrt{b}$
4.	$3x \le 15$	
	$2x \ge 10$	
	x	5

	Column A	Column B

5.

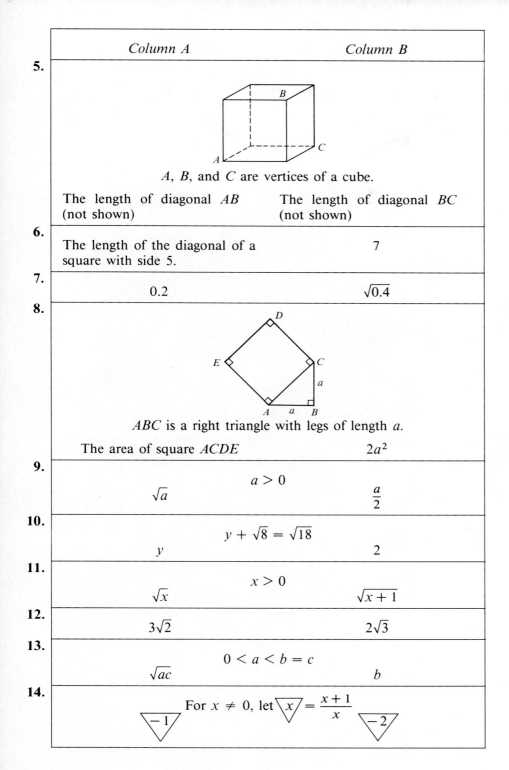

A, *B*, and *C* are vertices of a cube.

The length of diagonal *AB* (not shown)	The length of diagonal *BC* (not shown)

6.

The length of the diagonal of a square with side 5.	7

7.

0.2	$\sqrt{0.4}$

8.

ABC is a right triangle with legs of length *a*.

The area of square *ACDE*	$2a^2$

9.

$$a > 0$$

$\sqrt{a}$	$\dfrac{a}{2}$

10.

$$y + \sqrt{8} = \sqrt{18}$$

y	2

11.

$$x > 0$$

$\sqrt{x}$	$\sqrt{x+1}$

12.

$3\sqrt{2}$	$2\sqrt{3}$

13.

$$0 < a < b = c$$

$\sqrt{ac}$	*b*

14.

For $x \neq 0$, let $\boxed{x} = \dfrac{x+1}{x}$

$\boxed{-1}$	$\boxed{-2}$

11 Solving Quadratic Equations

Under the influence of gravity, a projectile follows a curved path called a parabola. The height (h) of a projectile at any time (t) is described by a second-degree (quadratic) equation.

For example, the equation $h = 80t - 16t^2$ describes the height of an object t seconds after it has been projected upward at a speed of 80 feet per second.

Imagine that a pitcher throws a baseball upward at a speed of 80 feet per second from the rim of the Grand Canyon. If the baseball falls to the base of the canyon one mile below, how long is the ball in the air?

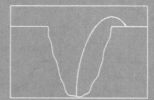

11–1 Quadratic Functions

Preview

The 25 members of a tennis club organized a round-robin tournament in which each player played against each of the other players. To find the number of required matches, one member looked at simpler problems and made a table.

Number of players	1	2	3	4	5	6	$\cdots$	25
Number of matches	0	1	3	6	10	?	$\cdots$	?

• How many matches would be played if there were only 6 players?
• Describe a way to find the number of matches to be played for all 25 players.

Another tennis club member found an equation for the number of matches (M) for any number of players (p).

$$M = \frac{1}{2}p^2 - \frac{1}{2}p$$

• Does the equation give the results shown in the table?
• Use the equation to find the number of matches required for 25 players.

Equations like the one above are called quadratic equations.

In this lesson you will learn about functions defined by quadratic equations.

■ LESSON

Polynomial equations of degree 2 are called **quadratic equations**. Here are examples of quadratic equations in two variables. In each equation, for a given value of x, there is a unique value of y. Therefore, each equation describes a function.

$$y = x^2 + 2x - 3 \qquad y = -2x^2 + 5x \qquad y = (x - 2)^2$$

$$y = x^2 + 6 \qquad y = \frac{1}{2}x^2 \qquad y - 5 = -3(x + 4)^2$$

The solution of each equation is the set of (x, y) number pairs that satisfies the equation. The set of solutions of each of the above equations is a **quadratic function**.

Definition: Quadratic Function

A *quadratic function* is a set of ordered pairs (x, y) described by an equation that can be written in the form

$$y = ax^2 + bx + c, \text{ where } a \neq 0.$$

Here is a table of some solutions of the equation $y = x^2 + 2x - 3$ and a graph of the function.

$$y = x^2 + 2x - 3$$

x	y
2	5
1	0
0	-3
-1	-4
-2	-3
-3	0
-4	5
-5	12

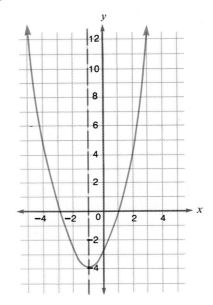

The graph of the equation is a **parabola**. Note these characteristics of the function and its graph.

- There is a "lowest" point, called the **minimum** of the function. The minimum of this quadratic function is $(-1, -4)$.
- The graph extends infinitely far up and to the right and left.
- If the graph is "folded" on the line $x = -1$, the two parts of the graph match each other. The line $x = -1$ is called the **axis of symmetry** of the function. The minimum is on the axis of symmetry.
- The graph crosses the y-axis at $(0, -3)$. The y-intercept is -3.
- The graph crosses the x-axis at $(-3, 0)$ and $(1, 0)$. The x-intercepts are -3 and 1.

By inspecting the y-values in the table, we can predict that $(-1, -4)$ is the minimum point and that the graph is symmetric about $x = -1$.

Example 1 Draw the graph of the quadratic function $y = -x^2 + 2x + 1$.

Solution To graph the function, first make a table of solutions. Then graph a point for each solution. Join the points with a smooth curve.

$$y = -x^2 + 2x + 1$$

x	y
3	-2
2	1
1	2
0	1
-1	-2
-2	-7

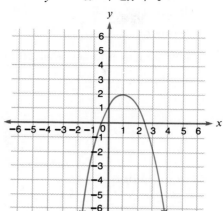

In Example 1 the function has a **maximum** point at (1, 2). In general, a quadratic function described by the equation $y = ax^2 + bx + c$ has a *maximum* point if a is negative and a *minimum* point if a is positive.

The axis of symmetry is the line $x = 1$. The maximum or minimum point is always on the axis of symmetry.

Example 2 Graph the quadratic function $y = 2x^2 - 2x - 1$, and give the maximum or minimum point and the axis of symmetry.

Solution

$$y = 2x^2 - 2x - 1$$

x	y
3	11
2	3
1	-1
0	-1
-1	3
-2	11

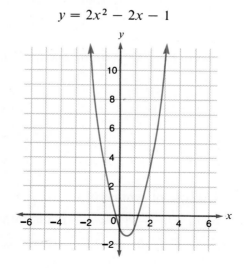

Example 2 (continued)

A minimum point does not appear in the table. However, we can see that the axis of symmetry passes midway between the points $(1, -1)$ and $(0, -1)$.

The average of the x-values for these points is $\frac{1}{2}$. This means that the line of symmetry is $x = \frac{1}{2}$.

Now we can find the y-coordinate of the minimum point by substituting $\frac{1}{2}$ for x in the equation $y = 2x^2 - 2x - 1$.

$$y = 2\left(\frac{1}{2}\right)^2 - 2\left(\frac{1}{2}\right) - 1$$

$$y = -\frac{3}{2}$$

Answer The minimum point is $\left(\frac{1}{2}, -\frac{3}{2}\right)$. The axis of symmetry is the line $x = \frac{1}{2}$.

In general, the equation for the axis of symmetry of the quadratic function $y = ax^2 + bx + c$ is $x = -\frac{b}{2a}$, where $a \neq 0$.

Example 3 Give the axis of symmetry and the minimum or maximum point of the quadratic function $y = 3x^2 - 2x - 4$.

Solution The axis of symmetry is

$$x = -\frac{b}{2a}$$

$$x = -\frac{-2}{2 \cdot 3}$$

$$x = \frac{1}{3}$$

Substitute $\frac{1}{3}$ for x in the quadratic equation.

$$y = 3x^2 - 2x - 4$$

$$y = 3\left(\frac{1}{3}\right)^2 - 2\left(\frac{1}{3}\right) - 4$$

$$y = \frac{1}{3} - \frac{2}{3} - 4$$

$$y = -\frac{13}{3}$$

Answer The axis of symmetry is $x = \frac{1}{3}$, and the minimum point is $\left(\frac{1}{3}, -\frac{13}{3}\right)$. The point $\left(\frac{1}{3}, -\frac{13}{3}\right)$ is a minimum, rather than a maximum, because the coefficient of x^2 in $y = 3x^2 - 2x - 4$ is positive.

CLASSROOM EXERCISES

The following quadratic equations can be written in the form $y = ax^2 + bx + c$. State the values of a, b, and c.

1. $y = 3x^2 + 2x - 4$

2. $y = -2x^2 - x + 5$

3. $y = -x^2 + x$

4. $y = 0.5x^2 - 6$

5. $y = (x + 5)(x + 6)$

6. $y = 2(x^2 - 7x - 1)$

Give the axis of symmetry and the minimum or maximum point.

7. $y = x^2 - 6x + 5$

8. $y = x^2 + 3x - 1$

9. $y = (x - 1)(x - 3)$

10. $y = (x + 4)^2$

11. What kind of function is the equation $y = 0x^2 + 2x + 4$?

WRITTEN EXERCISES

State whether the equation describes a quadratic function.

A
1. $y = 5x + 3$

2. $y = 3x - 2$

3. $y = \frac{1}{2}x^2 + 3x + 5$

4. $y = \frac{1}{3}x^2 + 2x + 1$

5. $y = (x + 1)(x - 1)$

6. $y = (x - 3)(x + 3)$

7. $y = x^2 - x$

8. $y = 2x^2 - 2x$

9. $y = x(x + 5)$

10. $y = x(x + 1)$

11. $y = x^2 + x - x^2$

12. $y = x^2 - 3x - x^2$

Write each equation in the form $y = ax^2 + bx + c$. Then give the values of a, b, and c.

13. $y = x^2 + 2x + 3$

14. $y = 3x^2 + x + 6$

15. $y = -3x^2 + 2x$

16. $y = -4x^2 + 3x$

17. $y = (x + 2)(x + 4)$

18. $y = (x + 3)(x + 6)$

Copy and complete each table of solutions. Then graph the function.

19. $y = x^2 - 2x - 3$

x	-1	0	1	2	3
y	?	?	?	?	?

20. $y = x^2 - 6x + 5$

x	1	2	3	4	5
y	?	?	?	?	?

21. $y = 2x^2 - 4x - 4$

x	-2	-1	0	1	2
y	?	?	?	?	?

22. $y = -2x^2 + 4x + 1$

x	-2	-1	0	1	2
y	?	?	?	?	?

Copy and complete each table of solutions. Then graph the function.

23. $y = \frac{1}{2}x^2$

x	-4	-2	0	2	4
y	?	?	?	?	?

24. $y = \frac{1}{4}x^2$

x	-4	-2	0	2	4
y	?	?	?	?	?

25. $y = \frac{1}{4}x^2 - x - 3$

x	-4	-2	0	2	4
y	?	?	?	?	?

26. $y = \frac{1}{2}x^2 - 2x - 1$

x	-4	-2	0	2	4
y	?	?	?	?	?

Give the axis of symmetry and the minimum or maximum point.

27. $y = x^2 - 2x - 3$

28. $y = x^2 - 6x + 5$

29. $y = 2x^2 + 3x - 2$

30. $y = 2x^2 - 3x - 2$

In each exercise, graph the four functions using the same coordinate axes. Label the graphs a, b, c, and d.

B **31.** **a.** $y = x^2$
    b. $y = x^2 + 2$
    c. $y = x^2 - 1$
    d. $y = x^2 - 3$

32. **a.** $y = x^2$
    b. $y = (x + 2)^2$
    c. $y = (x - 1)^2$
    d. $y = (x - 3)^2$

33. **a.** $y = x^2$
    b. $y = 3x^2$
    c. $y = \frac{1}{2}x^2$
    d. $y = 0.1x^2$

34. **a.** $y = -x^2$
    b. $y = -2x^2$
    c. $y = -\frac{1}{2}x^2$
    d. $y = -0.1x^2$

35. Describe how the value of c in the equation $y = x^2 + c$ affects the graph of the function. (See Exercise 31.)

36. Describe how the value of h in the equation $y = (x - h)^2$ affects the graph of the function. (See Exercise 32.)

37. Describe how the value of a in the equation $y = ax^2$ affects the graph of the function when a is positive. (See Exercise 33.)

38. Describe how the value of a in the equation $y = ax^2$ affects the graph of the function when a is negative. (See Exercise 34.)

For each exercise, copy and complete the table of solutions, graph the function, give the axis of symmetry, the minimum or maximum point, the y-intercept, and the x-intercepts.

39. $y = 2x^2 - 8x + 6$

x	-1	0	1	2	3	4	5
y	?	?	?	?	?	?	?

40. $y = 2x^2 - 4x - 6$

x	-2	-1	0	1	2	3	4
y	?	?	?	?	?	?	?

41. $y + \dfrac{9}{2} = \dfrac{1}{2}(x - 2)^2$

x	-1	0	1	2	3	4	5
y	?	?	?	?	?	?	?

42. $y - \dfrac{9}{2} = -\dfrac{1}{2}(x - 2)^2$

x	-1	0	1	2	3	4	5
y	?	?	?	?	?	?	?

43. $y = (x + 3)(x + 1)$

x	-5	-4	-3	-2	-1	0	1
y	?	?	?	?	?	?	?

44. $y = (x + 3)(x - 1)$

x	-4	-3	-2	-1	0	1	2
y	?	?	?	?	?	?	?

45. Is it possible for a parabola to lie entirely in one quadrant? in two quadrants? in three quadrants? in four quadrants?

46. Is it possible for the graph of a quadratic function to have two y-intercepts? no y-intercepts?

47. Is it possible for the graph of a quadratic function to have three x-intercepts? one x-intercept? no x-intercepts?

48. Graph the function $M = \dfrac{1}{2}p^2 - \dfrac{1}{2}p$ (described in the Preview) for this set of values of p: $\{0, 1, 2, 3, 4, 5\}$.

49. How does the graph in Exercise 48 differ from the graph of $y = \dfrac{1}{2}x^2 - \dfrac{1}{2}x$?

[*Hint:* The variable x can be any real number. The variable p is restricted to what kind of numbers?]

Answer these questions about the quadratic function $y = ax^2 + bx + c$, where $a \neq 0$.

50. What is the y-intercept?

51. What are the coordinates of the minimum (or maximum) point?

$\left[\textit{Hint: The point is on the line } x = -\dfrac{b}{2a}. \right]$

Answer these questions about the quadratic function $y = ax^2 + bx + c$, where $a \neq 0$.

52. Suppose that the minimum point is $(3, -4)$ and that the graph crosses the x-axis at $(0, 0)$. At what other point does the graph cross the x-axis? [*Hint:* Sketch the graph.]

53. Suppose that the minimum point is $(3, -4)$ and that the graph crosses the x-axis at $(1, 0)$. At what other point does the graph cross the x-axis?

54. Suppose that the minimum point is $(3, -4)$ and that the graph crosses the x-axis at two points. Is a positive or negative?

■ REVIEW EXERCISES

1. Simplify. $2x(x - 3) - x(2x + 4)$ [4–6]

2. Write the equation $2x - 3y = 6$ in slope–intercept form. [5–6]

3. Graph. $y = \frac{1}{2}x - 4$ [5–4]

4. Solve this system. $2x + 3y = 10$ [6–6]
$3x + 2y = 10$

5. Nuts that cost \$4 per lb are mixed with nuts that cost \$5 per lb. How many pounds of each kind must be used to obtain a 20-lb mixture worth \$4.75 per lb? [6–7]

Factor completely.

6. $2x^2 - 50$ [7–4]

7. $2x^2 + 4x - 30$ [7–10]

8. Solve. $x^2 - 7x = 18$ [7–9]

Mathematics and Your Future ——————

Your teacher knows both your mathematical background and the courses that are offered in your school. He or she has probably also received feedback from former students about how well prepared they were for college, vocational school, or employment. Students rarely return to thank teachers for letting them slide by; they often thank teachers for motivating them to apply themselves. Talk to your teacher about mathematics and your future. You will gain useful information.

11–2 Solving Quadratic Equations by Graphing

Preview

The height h (in meters) of a projectile with an initial vertical velocity of 40 meters per second is given by the formula

$$h = -5t^2 + 40t,$$

where t is the number of seconds after launch.

- How many seconds after launch does the projectile return to the ground?
- How many seconds after launch is the projectile 60 m above the ground?

The graph of the function can be used to answer these questions.

In this lesson you will learn how to use graphs to solve quadratic equations.

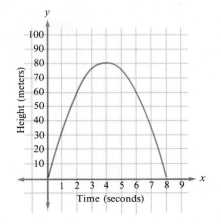

■ LESSON

In the study of linear functions, finding the y-intercept was helpful in determining the graph. Consider the equation $y = 2x - 4$, which is written in slope–intercept form $y = mx + b$. The y-intercept is the value of y when $x = 0$.

$$y = 2(0) - 4$$
$$y = -4$$

It is the constant term when the equation is in slope–intercept form.

Substitute other values for x to complete the table and graph the equation.

The x-intercept is the value of x when $y = 0$. The graph shows that the line crosses the x-axis at the point (2, 0). Thus the x-intercept is 2.

$$y = 2x - 4$$

x	0	-1	3
y	-4	-6	2

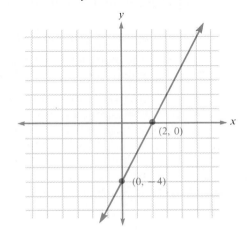

Consider the equation $y = x^2 - 2x - 3$.

The y-intercept is the value of y when $x = 0$.

$$y = (0)^2 - 2(0) - 3$$
$$y = -3$$

The y-intercept is the constant term c when the equation is in the form $y = ax^2 + bx + c$.

Substitute other values for x to complete the table and graph the equation.

x	y
-2	5
-1	0
0	-3
1	-4
2	-3
3	0

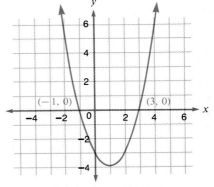

The x-intercepts are the values of x when $y = 0$. To find the x-intercepts we could let $y = 0$ and solve the equation $x^2 - 2x - 3 = 0$. However, since we have the graph of the equation $y = x^2 - 2x - 3$, we can determine the x-intercepts by noting where the graph crosses the x-axis. The graph crosses the x-axis at the points $(-1, 0)$ and $(3, 0)$, so the x-intercepts are -1 and 3.

Check by substituting these values in the equation $x^2 - 2x - 3 = 0$.

$$x^2 - 2x - 3 = 0 \qquad\qquad x^2 - 2x - 3 = 0$$
$$(-1)^2 - 2(-1) - 3 \stackrel{?}{=} 0 \qquad 3^2 - 2(3) - 3 \stackrel{?}{=} 0$$
$$1 + 2 - 3 \stackrel{?}{=} 0 \qquad\qquad 9 - 6 - 3 \stackrel{?}{=} 0$$
$$0 = 0 \quad \text{True.} \qquad\qquad 0 = 0 \quad \text{True.}$$

This example shows a method of solving quadratic equations in one variable: graph the related quadratic function and find the x-intercepts.

Example 1   Solve by graphing. $2x^2 - 7x + 3 = 0$

Solution   The x-intercepts of the function $y = 2x^2 - 7x + 3$ are the solutions of the equation $0 = 2x^2 - 7x + 3$.

Graph the quadratic function $y = 2x^2 - 7x + 3$.

x	y
0	3
1	-2
2	-3
3	0
4	7

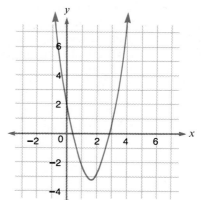

Example 1 (continued)

One of the x-intercepts of the function is 3. Another x-intercept is between 0 and 1, or approximately $\frac{1}{2}$. To check, substitute $\frac{1}{2}$ for x.

$$y = 2\left(\frac{1}{2}\right)^2 - 7\left(\frac{1}{2}\right) + 3$$

$$y = \frac{1}{2} - 3\frac{1}{2} + 3$$

$$y = 0$$

The other x-intercept of the function is $\frac{1}{2}$.

Answer $\left\{\frac{1}{2}, 3\right\}$

Example 2 Between which consecutive integers do the solutions of $x^2 - 4x - 1 = 0$ lie?

Solution Graph the quadratic function $y = x^2 - 4x - 1$.

x	y
-1	4
0	-1
1	-4
2	-5
3	-4
4	-1
5	4

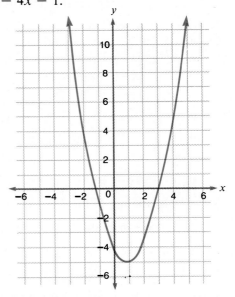

Answer The x-intercepts are between -1 and 0 and between 4 and 5. Therefore, one of the solutions is between -1 and 0, and another is between 4 and 5.

■ CLASSROOM EXERCISES

Solve by graphing.

1. $x^2 - 3x + 2 = 0$

2. $x^2 - 4 = 0$

3. $x^2 - 4x + 4 = 0$

In the following exercises, use the graph of the function $y = ax^2 + bx + c$ to determine the number of real-number solutions (0, 1, or 2) of the quadratic equation $ax^2 + bx + c = 0$.

4.

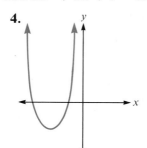

5.

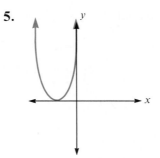

6.

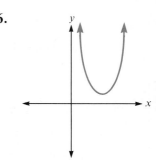

7. Use a graph of a quadratic function to find the consecutive integers between which the solutions of the equation $x^2 + 4x + 2 = 0$ lie.

■ WRITTEN EXERCISES

For each exercise, graph the function. Then solve the three equations using the graph.

A **1.** $y = x^2 + 2x - 3$
  a. $x^2 + 2x - 3 = 0$
  b. $x^2 + 2x - 3 = -3$
  c. $x^2 + 2x - 3 = -4$

2. $y = x^2 + 4x + 3$
  a. $x^2 + 4x + 3 = 0$
  b. $x^2 + 4x + 3 = 3$
  c. $x^2 + 4x + 3 = -1$

3. $y = x^2 - 6x + 8$
  a. $x^2 - 6x + 8 = 0$
  b. $x^2 - 6x + 8 = 3$
  c. $x^2 - 6x + 8 = -1$

4. $y = x^2 - 4x + 3$
  a. $x^2 - 4x + 3 = 0$
  b. $x^2 - 4x + 3 = 3$
  c. $x^2 - 4x + 3 = -1$

5. $y = -x^2 + 2x + 3$
  a. $-x^2 + 2x + 3 = 0$
  b. $-x^2 + 2x + 3 = 3$
  c. $-x^2 + 2x + 3 = 4$

6. $y = -x^2 - 2x + 3$
  a. $-x^2 - 2x + 3 = 0$
  b. $-x^2 - 2x + 3 = 3$
  c. $-x^2 - 2x + 3 = 4$

Solve by graphing. If there is no solution, write $\emptyset$.

7. $x^2 - 5x + 6 = 0$

8. $x^2 - 3x + 2 = 0$

9. $x^2 - 6x + 8 = 0$

10. $x^2 - 4x + 3 = 0$

11. $x^2 + 2x + 2 = 0$

12. $x^2 - 4x + 5 = 0$

Solve by graphing. State the consecutive integers between which the solutions lie.

13. $x^2 - 2x - 2 = 0$

14. $x^2 - 4x + 2 = 0$

15. $x^2 + 3x + 1 = 0$

16. $x^2 + x - 1 = 0$

17. $\frac{1}{2}x^2 - 2x + 1 = 0$

18. $\frac{1}{2}x^2 - 2x - 1 = 0$

Solve by graphing. Some solutions are not integers. Check your answers.

19. $2x^2 - 10x + 12 = 0$

20. $2x^2 - 13x + 20 = 0$

21. $3x^2 - x - 4 = 0$

22. $2x^2 - 6x - 20 = 0$

23. $4x^2 - 4x - 3 = 0$

24. $6x^2 + x - 2 = 0$

Use the graph in the Preview to answer Exercises 25 and 26. A projectile is launched with an initial vertical velocity of 40 m/s. After how many seconds will the projectile reach these heights?

25. 50 m   **26.** 70 m

Suppose that a projectile is launched with an initial velocity of 100 m/s. Its height h (in meters) after t seconds is given by the formula $h = -5t^2 + 100t$.

B **27.** Draw a graph of the function.

28. After how many seconds will the projectile return to the ground?

29. After how many seconds will the projectile reach its maximum height?

30. When will the projectile reach a height of 320 m?

31. When will the projectile reach a height of 480 m?

32. Estimate when the projectile will reach a height of 200 m.

33. Estimate when the projectile will reach a height of 300 m.

For each exercise, write an equation that fits the situation. Let x represent the number of meters in the width of the rectangle. Graph the related quadratic function. Use the graph to answer the question.

34. The length of a rectangle is 2 m more than the width. What is the width of the rectangle if its area is 11.25 m²?
[*Hint:* $x(x + 2) = 11.25$. Graph $y = x^2 + 2x - 11.25$.]

35. The length of a rectangle is 3 m more than the width. What is the width of the rectangle if its area is 8.64 m²?

36. A rectangle is twice as long as it is wide. Find the width of the rectangle if its area is 26 m².

Solve.

37. Bob wants to enlarge a 4-m by 1-m flower bed to twice its area by increasing the length and width by the same amount. By how much should each be increased?

38. The altitude of a triangle is 1.5 cm less than the base. Find the altitude and base if the area of the triangle is 26 cm².

The graphing method can be used to find the approximate solutions of any quadratic equation in one variable, but it does have some limitations. Describe the difficulties encountered in solving these equations by graphing.

C **39.** Equation: $8x^2 + 6x - 5 = 0$
     Graph: $y = 8x^2 + 6x - 5$

40. Equation: $x^2 + 14x - 240 = 0$
     Graph: $y = x^2 + 14x - 240$

41. Equation: $x^2 + 3x - 1 = 0$
     Graph: $y = x^2 + 3x - 1$

42. Equation: $\frac{1}{8}x^2 - \frac{1}{4}x - 1 = 0$

     Graph: $y = \frac{1}{8}x^2 - \frac{1}{4}x - 1$

Solve by graphing. Estimate the solutions to the nearest tenth. Do not check.

43. $x^2 - 2x - 1 = 0$ **44.** $x^2 - 2x - 4 = 0$

45. $x^2 - 4x + 2 = 0$ **46.** $x^2 + 6x + 6 = 0$

47. A farmer has 100 ft of fencing to make three sides of a rectangular pig pen beside the barn. The barn will serve as the fourth side. What should be the length and width of the pen in order to enclose the maximum area? [*Hint:* Let x be the width and $100 - 2x$ the length. If the area is y, then $y = x(100 - 2x)$. Graph this function.]

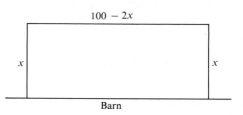

■ REVIEW EXERCISES

1. Which values of x are not admissible? $\dfrac{x - 2}{x(x + 1)}$ [8–1]

Simplify. Assume that no denominator is zero.

2. $\dfrac{2a}{5b} \cdot \dfrac{10b}{3c}$ **3.** $\dfrac{6(a + b)}{3} \div \dfrac{a^2 - b^2}{9}$ [8–2]

Solve.

4. $\dfrac{1}{5} = \dfrac{1}{R_1} + \dfrac{1}{10}$ **5.** $\dfrac{3}{5} = \dfrac{15}{x}$ [8–5]

6. $80\% = \dfrac{400}{x}$ **7.** $8 = x\%$ of 9

8. A circuit is made up of a 10-ohm resistance and a 50-ohm resistance connected in parallel. What is the total resistance of the circuit? $\left(\text{Use the equation } \dfrac{1}{R} = \dfrac{1}{r_1} + \dfrac{1}{r_2}.\right)$ [8–6]

9. One machine can complete a job in 8 h. Another machine can complete the job in 10 h. How long will it take both machines working together to complete the job?   [8–6]

10. If a coin is flipped 10 times and it comes up heads 7 times, what is the ratio of heads to tails for 10 flips?   [8–7]

Find the missing length.   [10–3]

11.
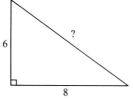

12.

Self-Quiz 1

11–1   **1.** Write the equation $y - 3 = (x - 2)(x + 5)$ in the form $y = ax^2 + bx + c$ and give the values of a, b, and c.

Given the quadratic function $y = x^2 + 4x + 1$.

2. Sketch its graph.

3. Give its axis of symmetry.

4. Give the maximum or minimum point, identifying which it is.

11–2   State the y-intercepts for these equations.

5. $y = x^2 - 5x + 6$ **6.** $y = -x^2 - x + 2$

7. $x^2 + 5x - 1$ **8.** $-x^2 - 3x + 1$

Solve by graphing. Write $\emptyset$ if there is no solution.

9. $2x^2 - 7x + 3 = 0$ **10.** $x^2 + x + 4 = 0$

Solve.

11. The height h in meters of a rocket after t seconds is given by the formula $h = -5t^2 + 60t$. In how many seconds after its launch will the rocket return to the ground?

12. A rectangle is three times as long as it is wide. Find the width of the rectangle if its area is 27 m².

Mathematics and Your Future

People used to think that mathematics was necessary only for those who planned careers in such areas as science, engineering, and mathematics. However, with today's increasing use of computers and greater analysis of numerical data, knowledge of mathematics is important to those working in virtually every field. Mathematical applications in such fields as agriculture, business, home economics, and psychology can be expected to continue to grow. As you continue to study mathematics, you will be preparing yourself for opportunities in a wide range of fields.

11–3 Solving Quadratic Equations Using Square Roots

Preview Historical note

In A.D. 825 the Arab mathematician al-Khowarizmi wrote the first textbook on algebra. Its title was

ilm al-jabr wa'l muqabalah

which means roughly "the art of bringing together unknowns to match a known quantity." The key word in the title is "al-jabr," or "bring together," which became our word "algebra." Curiously, in those times, the term "algebraist" could refer either to someone who "brought bones together" (a surgeon) or to a specialist in equations.

The most significant problem left unanswered by al-Khowarizmi and his predecessors was how to interpret negative numbers. It is believed that the Hindus were among the first to realize that quadratic equations can have negative solutions. For instance, the equation $x^2 = 9$ has the solution $x = -3$ as well as the obvious solution $x = 3$.

You already know two methods of solving quadratic equations: factoring and graphing. In this lesson you will learn how to use the square root property of equations in solving quadratic equations.

■ LESSON

We can solve quadratic equations by factoring as shown in the following examples.

$$x^2 = 9$$
$$x^2 - 9 = 0$$
$$(x + 3)(x - 3) = 0$$
$$x + 3 = 0 \quad \text{or} \quad x - 3 = 0$$
$$x = -3 \quad \text{or} \quad x = 3$$

$$x^2 = 10$$
$$x^2 - (\sqrt{10})^2 = 0$$
$$(x + \sqrt{10})(x - \sqrt{10}) = 0$$
$$x + \sqrt{10} = 0 \quad \text{or} \quad x - \sqrt{10} = 0$$
$$x = -\sqrt{10} \quad \text{or} \quad x = \sqrt{10}$$

The symbol $\pm$ can be used to shorten the solutions to

$$x = \pm 3 \qquad\qquad x = \pm \sqrt{10}$$

Since both of the quadratic equations above are of the form $x^2 = c$, where $c > 0$, we can use a shorter method, called the **square root property of equations**, to solve them.

The Square Root Property of Equations

For each real number c, $c > 0$,

$$\text{if } x^2 = c, \text{ then } x = \pm\sqrt{c}.$$

Applying this property gives the following.

$$\text{If } x^2 = 16, \text{ then } x = \pm 4.$$
$$\text{If } x^2 = 17, \text{ then } x = \pm \sqrt{17}.$$

There are many other quadratic equations that can be solved using this property.

Example 1 Solve. $(x + 2)^2 = 9$

Solution

Use the square root property.

$$(x + 2)^2 = 9$$
$$x + 2 = \pm 3$$
$$x + 2 = 3 \quad \text{or} \quad x + 2 = -3$$
$$x = 1 \quad \text{or} \qquad x = -5$$

Answer $\{1, -5\}$

Check

$(x + 2)^2 = 9$	$(x + 2)^2 = 9$
$(1 + 2)^2 \overset{?}{=} 9$	$(-5 + 2)^2 \overset{?}{=} 9$
$3^2 \overset{?}{=} 9$	$(-3)^2 \overset{?}{=} 9$
$9 = 9$ It checks.	$9 = 9$ It checks.

Example 2 Solve. $(x - 3)^2 = 7$

Solution

$$(x - 3)^2 = 7$$
$$x - 3 = \pm \sqrt{7}$$
$$x - 3 = \sqrt{7} \quad \text{or} \quad x - 3 = -\sqrt{7}$$
$$x = 3 + \sqrt{7} \quad \text{or} \qquad x = 3 - \sqrt{7}$$

Answer $\{3 + \sqrt{7}, 3 - \sqrt{7}\}$

Check

$(x - 3)^2 = 7$	$(x - 3)^2 = 7$
$[(3 + \sqrt{7}) - 3]^2 \overset{?}{=} 7$	$[(3 - \sqrt{7}) - 3]^2 \overset{?}{=} 7$
$(\sqrt{7})^2 \overset{?}{=} 7$	$(-\sqrt{7})^2 \overset{?}{=} 7$
$7 = 7$ It checks.	$7 = 7$ It checks.

Example 3 Solve. $x^2 - 2x + 1 = 4$

Solution

Factor the left side of the equation.
Use the square root property.

$$x^2 - 2x + 1 = 4$$
$$(x - 1)^2 = 4$$
$$x - 1 = \pm 2$$
$$x = 1 \pm 2$$
$$x = 3 \quad \text{or} \quad x = -1$$

Answer $\{3, -1\}$

Check The check is left to the student.

Example 4 Solve. $(x - 4)^2 = -5$

Solution The square root property of equations cannot be used because the right side of the equation is negative.

Answer $\emptyset$. No real number squared is negative.

Example 5 Solve. $(x - 5)^2 = 5$
Write the solution as a decimal correct to the nearest thousandth.

Solution
$$(x - 5)^2 = 5$$
$$x - 5 = \pm \sqrt{5}$$
$$x = 5 \pm \sqrt{5}$$

Use a table of square roots or a calculator to approximate $\sqrt{5}$.

$$\sqrt{5} \approx 2.236 \qquad x \approx 5 \pm 2.236$$

Answer $\{7.236, 2.764\}$

■ CLASSROOM EXERCISES

Solve.

1. $x^2 = 25$
2. $x^2 = 6$
3. $(x + 3)^2 = 16$

4. $(x + 1)^2 = 5$
5. $x^2 - 6x + 9 = 8$
6. $(x + 2)^2 = -2$

■ WRITTEN EXERCISES

Solve. If the equation has no solution, write $\emptyset$.

A
1. $x^2 = 49$
2. $x^2 = 81$
3. $(x - 2)^2 = 100$

4. $(x - 3)^2 = 64$
5. $x^2 = -81$
6. $x^2 = -16$

7. $(x + 5)^2 = 4$
8. $(x + 4)^2 = 1$
9. $(x + 3)^2 = -36$

10. $(x + 2)^2 = -25$
11. $(x - 9)^2 = 0$
12. $(x - 7)^2 = 0$

13. $3x^2 = 75$
14. $2x^2 = 72$
15. $x^2 + 22 = 122$

16. $x^2 + 26 = 170$
17. $2x^2 + 36 = 164$
18. $3x^2 + 42 = 150$

19. $4x^2 - 40 = 104$
20. $5x^2 - 30 = 50$
21. $4(x - 3)^2 = 36$

22. $5(x - 2)^2 = 20$
23. $4x^2 = 9$
24. $9x^2 = 25$

25. $x^2 + 6x + 9 = 100$
26. $x^2 + 10x + 25 = 36$
27. $x^2 - 12x + 36 = 4$

28. $x^2 - 8x + 16 = 1$
29. $x^2 - 20x + 100 = 0$
30. $x^2 - 16x + 64 = 0$

Solve. Write the solutions as simplified radicals.

31. $x^2 = 18$ **32.** $x^2 = 12$ **33.** $x^2 = 13$

34. $x^2 = 17$ **35.** $(x - 5)^2 = 7$ **36.** $(x - 7)^2 = 5$

37. $x^2 - 2x + 1 = 3$ **38.** $x^2 - 4x + 4 = 10$ **39.** $x^2 + 8x + 16 = 8$

40. $x^2 + 6x + 9 = 27$ **41.** $(2x - 1)^2 = 6$ **42.** $(3x - 2)^2 = 11$

Solve. Write the solutions as decimals to the nearest thousandth.

43. $(x - 2)^2 = 5$ **44.** $(x - 3)^2 = 7$ **45.** $(x + 3)^2 = 70$

46. $(x + 2)^2 = 60$ **47.** $(x - 1.1)^2 = 10$ **48.** $(x - 2.1)^2 = 90$

Write an equation that fits the situation. Solve the equation and answer the question. Write the solutions as simplified radicals.

B **49.** A lot in the shape of a square has an area of 400 m². What is the length of each side of the square?

50. Each side of a square is increased by 5 cm in order to make a square with an area of 80 cm². What is the length of each side of the original square?

51. A square has an area of 50 cm². By how much should each of its sides be increased in order to double its area?

52. A square has an area of 200 m². By how much should each of its sides be decreased in order to form a square with half the area of the original square?

53. A square garden has an area of 90 m². How long is the fence that completely encloses the garden?

54. A square room has a ceiling area of 14 m². How many meters of trim are needed to cover the edges where the ceiling joins the wall?

Solve. Write the solution as a decimal correct to the nearest thousandth.

55. The distance d in feet that an object falls in t seconds is given by the formula $d = 16t^2$. How long will it take an object to fall 1000 ft?

56. If the square of 3 more than a number equals 10, what is the number?

57. A number is subtracted from 50 and the difference is squared. If the result is 100, what is the number?

58. The area of a circle is 100π cm². If the area is doubled, how much is added to the circumference?

Solve. Write the solutions as simplified radicals.

59. $x^2 + 7x + 6 = 7x + 24$ **60.** $x^2 + 9x + 10 = 9x + 34$

61. $(x + 2)(x - 5) = 3(6 - x)$ **62.** $(x + 6)(x - 2) = 4(x + 9)$

Describe the values of k for which the equation has the given number of solutions.

C **63.** $x^2 = k$; 2 solutions

64. $x^2 = k$; 1 solution

65. $x^2 = k$; 0 solutions

66. $kx^2 = -2$; 2 solutions

67. $(x - k)^2 = 5$; 2 solutions

68. $(x - 5)^2 + k = 7$; 1 solution

■ REVIEW EXERCISES

Solve.

1. How much did Tom earn on $500 invested at 10% simple interest for two years?   [8–8]

2. Sarah answered 21 of 25 problems correctly on an algebra test. What percent of the problems did she answer correctly?   [8–9]

3. If Isabel can ride her bike 10 km in 40 min, how far will she ride at the same rate in 2 h?   [8–9]

4. Does y vary directly as x if $10y = 25x$?   [8–10]

5. If Henri solves 10 problems in 15 minutes, how long should it take him to complete a 30-problem assignment if he is solving at the same rate?   [8–10]

6. Suppose x and y vary inversely. If $y = 12$ when $x = 4$, what is the value of y when $x = 6$?   [8–11]

7. The electrical current I varies inversely as the resistance R. If the current is 5 amperes when the resistance is 20 ohms, what is the current when the resistance is 50 ohms?   [8–11]

8. Evaluate $x^2 + 2x - 3$ for these values of x.   [10–4]

a. $x = -1 + \sqrt{5}$

b. $x = -1 - \sqrt{5}$

9. Evaluate $x^2 - 3x + 1$ for these values of x.   [10–4]

a. $\dfrac{3 + \sqrt{33}}{2}$

b. $\dfrac{3 - \sqrt{33}}{2}$

Mathematics and Your Future ─────────

There tend to be more desirable job opportunities in positions that require a strong background in mathematics. Even in times of high unemployment, many higher-level jobs remain unfilled because of a lack of applicants with the necessary qualifications. Those qualifications often include a sound knowledge of mathematics. If you continue to take mathematics in high school you will be in a better position to compete for the best job opportunities.

11–4 Solving Quadratic Equations by Completing the Square

Preview Historical note

More than 2500 years ago the Greeks had already made great discoveries in geometry. Some of their discoveries were related to algebra, although they did not have the notation that we use today. For example, whereas we might write the algebraic expression $x^2 + 2xy$, the Greek mathematicians might represent the quantity geometrically as shown at the right.

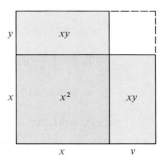

- What are the dimensions of the small square at the upper right that completes the square?

- What are the dimensions of the sides of the completed square?

- Write the area of the completed square as the square of a binomial.

How would you answer the questions above for the geometric representation of $a^2 + 6ab$?

In this lesson you will learn another method of solving quadratic equations: completing the square.

■ LESSON

Some quadratic equations can be solved by factoring, while some can be solved by graphing. But neither of these methods gives exact solutions for all quadratic equations.

In Section 11–3 we learned an interesting fact: It is possible to solve any quadratic equation that is in the form

$$(x + a)^2 = c, \qquad c \geq 0$$

Furthermore, every quadratic equation that has real-number solutions can be written in that form. For example, consider the equation:

$$x^2 + 6x = 7$$

We would like to factor the left side as the square of a binomial. Unfortunately, the expression is not a perfect square. If the left side were

$$x^2 + 6x + 9$$

we could factor it as we wish. By adding 9 to the left side (and to the right side), we will have what we need.

$$x^2 + 6x + 9 = 7 + 9$$

This process is called **completing the square**. Now we can use the square root property of equations to solve the equation.

$$x^2 + 6x + 9 = 16$$
$$(x + 3)^2 = 16$$
$$x + 3 = \pm 4$$
$$x = 1 \quad \text{or} \quad x = -7$$

We can use completing the square to solve any quadratic equation.

To complete the square of any equation in the form $x^2 + bx = c$:

1. find $\frac{1}{2}$ of the coefficient of x: $\frac{b}{2}$;

2. square the result of step 1: $\left(\frac{b}{2}\right)^2$; and

3. add the result of step 2 to both sides of the equation.

Example 1 Solve by completing the square. $x^2 - 4x = -3$

Solution
$$x^2 - 4x = -3$$

Complete the square: $\left(\frac{-4}{2}\right)^2 = 4$. $x^2 - 4x + 4 = -3 + 4$

Factor.
$$(x - 2)^2 = 1$$
$$x - 2 = \pm 1$$
$$x = 3 \quad \text{or} \quad x = 1$$

Answer $\{3, 1\}$

Check
$$x^2 - 4x = -3$$
$$3^2 - 4 \cdot 3 \stackrel{?}{=} -3$$
$$9 - 12 \stackrel{?}{=} -3$$
$$-3 = -3 \quad \text{It checks.}$$

$$x^2 - 4x = -3$$
$$1^2 - 4 \cdot 1 \stackrel{?}{=} -3$$
$$1 - 4 \stackrel{?}{=} -3$$
$$-3 = -3 \quad \text{It checks.}$$

Example 2 Solve by completing the square. $x^2 + 2x - 3 = 1$

Solution
$$x^2 + 2x - 3 = 1$$

Collect constants on the right side. $x^2 + 2x = 4$
Complete the square. $x^2 + 2x + 1 = 4 + 1$
Factor. $(x + 1)^2 = 5$
$$x + 1 = \pm \sqrt{5}$$
$$x = -1 \pm \sqrt{5}$$

Answer $\{-1 + \sqrt{5}, -1 - \sqrt{5}\}$

Check The check is left to the student.

Example 3   Solve. $x^2 - 3x + 1 = 7$

Solution

$$x^2 - 3x + 1 = 7$$

Collect constants on the right side. $x^2 - 3x = 6$

Complete the square. $x^2 - 3x + \dfrac{9}{4} = 6 + \dfrac{9}{4}$

Factor. $\left(x - \dfrac{3}{2}\right)^2 = \dfrac{33}{4}$

Simplify the radical. $x - \dfrac{3}{2} = \pm\sqrt{\dfrac{33}{4}}$

$$x - \dfrac{3}{2} = \pm\dfrac{\sqrt{33}}{2}$$

$$x = \dfrac{3}{2} \pm \dfrac{\sqrt{33}}{2}$$

$$x = \dfrac{3 \pm \sqrt{33}}{2}$$

Answer $\left\{\dfrac{3 + \sqrt{33}}{2}, \dfrac{3 - \sqrt{33}}{2}\right\}$

Check   The check is left to the student.

Example 4   Solve. $2x^2 + 8x = 9$

Solution

$$2x^2 + 8x = 9$$

Divide both sides by 2. $x^2 + 4x = \dfrac{9}{2}$

Complete the square. $x^2 + 4x + 4 = \dfrac{9}{2} + 4$

Factor. $(x + 2)^2 = \dfrac{17}{2}$

$$x + 2 = \pm\sqrt{\dfrac{17}{2}}$$

Simplify the radical. $x + 2 = \pm\sqrt{\dfrac{34}{4}}$

$$x + 2 = \pm\dfrac{\sqrt{34}}{2}$$

Solve for x. $x = -2 \pm \dfrac{\sqrt{34}}{2}$

Answer $\left\{-2 + \dfrac{\sqrt{34}}{2}, -2 - \dfrac{\sqrt{34}}{2}\right\}$

Check   The check is left to the student.

These steps are used to solve a quadratic equation by completing the square.

1. Divide both sides of the equation by the coefficient of x^2. The coefficient of x^2 will then be 1.
2. Isolate the x^2- and x-terms on one side of the equation.
3. Add the square of half the coefficient of x to both sides.
4. Factor.
5. Use the square root property of equations.
6. Simplify all radicals.
7. Write the solutions.

■ CLASSROOM EXERCISES

What number added to the expression will complete the square?

1. $x^2 + 6x$ **2.** $x^2 - 6x$ **3.** $x^2 + 12x$

What number must be added to both sides of the equation to complete the square?

4. $x^2 - 8x - 4 = 2$ **5.** $x^2 + 3x - 5 = 1$ **6.** $x^2 + 7x = 10$

Solve by completing the square.

7. $x^2 + 4x = 12$ **8.** $x^2 - 6x + 4 = 11$ **9.** $3x^2 + 12x = 8$

■ WRITTEN EXERCISES

What number added to the expression will complete the square?

A   **1.** $x^2 + 10x$ **2.** $x^2 + 8x$ **3.** $x^2 - 6x$

      4. $x^2 - 12x$ **5.** $x^2 + 5x$ **6.** $x^2 + 3x$

What number must be added to both sides of the equation to complete the square?

7. $x^2 - 4x = 3$ **8.** $x^2 - 14x = 40$ **9.** $x^2 + \frac{4}{5}x = 0$

10. $x^2 + \frac{2}{3}x = 0$ **11.** $x^2 - 8x = 6$ **12.** $x^2 - 10x = 5$

Solve by completing the square.

13. $x^2 + 10x = 11$ **14.** $x^2 + 8x = 20$ **15.** $x^2 - 6x = 16$

16. $a^2 - 12a = 28$ **17.** $x^2 - 6x + 12 = 4$ **18.** $b^2 - 8b + 18 = 3$

19. $t^2 + 5t + 10 = 4$ **20.** $p^2 + 3p + 1 = 5$ **21.** $n^2 + 4n - 20 = 12$

22. $x^2 + 4x - 8 = 4$ **23.** $q^2 - 6q - 3 = 4$ **24.** $x^2 - 8x - 20 = 13$

Solve by completing the square. Write the solutions as simplified radicals.

25. $x^2 + 4x = 14$ **26.** $x^2 + 6x = 15$

27. $x^2 - 10x = 15$ **28.** $x^2 - 8x = 29$

29. $x^2 - 20x + 25 = 0$ **30.** $x^2 - 12x + 26 = 40$

Solve by completing the square. Write the solutions as decimals correct to the nearest thousandth.

31. $x^2 - 6x = 17$ **32.** $x^2 - 4x = 7$

33. $x^2 + 8x - 3 = 66$ **34.** $x^2 + 10x - 4 = 36$

35. $x^2 - 7x + 2.25 = 16$ **36.** $x^2 - 5x + 1.25 = 85$

Write an equation that fits the problem. Then solve the equation by completing the square. Answer the question with a simplified radical.

B **37.** One positive number is 2 greater than another positive number. The product of the two numbers is 11. What is the smaller number?

38. A rectangle is 10 cm longer than it is wide. Its area is 38 cm². What are the length and width of the rectangle? [*Note:* Both must be positive.]

39. The length and width of a 5-cm by 7-cm rectangle are both increased by the same amount in order to form a rectangle with an area of 59 cm². By how much were the length and width increased?

Solve by completing the square.

40. $3x^2 - x = 2$ **41.** $2x^2 + x = 15$ **42.** $4x^2 + 8x + 1 = 4$

43. $9x^2 + 18x - 3 = 4$ **44.** $4x^2 - 12x - 2 = 25$ **45.** $5x^2 + 5x + 1 = 0$

46. $2x^2 - 6x + 1 = 0$ **47.** $3x^2 - 8x + 2 = 0$ **48.** $2x^2 - 8x - 3 = 0$

49. The base of a triangle is 4 cm more than its altitude. If the area of the triangle is 100 cm², find the base and altitude.

Solve by completing the square.

C **50.** $x^2 + 10x + c = 0$ Express the answer in terms of c.

51. $x^2 + bx + 2 = 0$ Express the answer in terms of b.

52. $ax^2 + 6x + 3 = 0$ Express the answer in terms of a.

53. $ax^2 + bx + c = 0$ Express the answer in terms of a, b, and c.

54. For Exercises 50–53, describe the restrictions on a, b, and c so that the equation has real-number solutions.

The formula $h = vt - 5t^2$ gives the height in meters of an object projected vertically at v meters per second after t seconds. Solve and give the answer to the following to the nearest tenth of a second.

55. How long will it take an object projected vertically at a speed of 100 m/s to reach a height of 200 m?

56. How long will it take an object projected vertically at a speed of 50 m/s to reach a height of 100 m?

■ REVIEW EXERCISES

1. Write the inequality that is graphed on this number line.   [9–1]

Graph on a number line.

2. $x \leq 3$ **3.** $x \neq 1$ [9–1]

4. $x \geq -3$ and $x \leq 0$ **5.** $x < 0$ or $x > 2$ [9–2]

Solve.

6. $x + 6 \leq 4$ **7.** $2x - 5 < x + 4$ [9–3]

Evaluate each of the expressions for the following values of a, b, and c.   [10–4]

i. $\dfrac{-b + \sqrt{b^2 - 4ac}}{2a}$ **ii.** $\dfrac{-b - \sqrt{b^2 - 4ac}}{2a}$

8. $a = 1$, $b = -2$, $c = -3$ **9.** $a = 2$, $b = -1$, $c = 5$ **10.** $a = 1$, $b = -6$, $c = 9$

Self-Quiz 2

11–3   Solve. Write $\emptyset$ if there is no solution.

    1. $x^2 = 125$ **2.** $(x + 6)^2 = 4$

    3. $y^2 + 49 = 0$ **4.** $y^2 - 14y + 49 = 100$

11–4   Solve by completing the square.

    5. $x^2 + 8x - 4 = 0$ **6.** $y^2 - 3y + 1 = 0$ **7.** $2x^2 + 8x - 5 = 0$

    8. One side of a square is increased by 3 cm. An adjacent side is decreased by 5 cm. The area of the rectangle formed is 32 cm². What is the length of a side of the original square?

Example 2   Solve using the quadratic formula. $2x^2 - x + 5 = 0$

Solution

$$x = \frac{-b \pm \sqrt{b^2 - 4ac}}{2a}$$

Substitute 2 for a, -1 for b, and 5 for c.

$$x = \frac{-(-1) \pm \sqrt{(-1)^2 - 4(2)(5)}}{2(2)}$$

$$x = \frac{1 \pm \sqrt{1 - 40}}{4}$$

$$x = \frac{1 \pm \sqrt{-39}}{4}$$

Since $\sqrt{-39}$ is not defined, there are no real-number solutions of the equation.

Answer $\emptyset$

Example 3   Solve using the quadratic formula. $x^2 + 4 = 6x - 5$

Solution

$$x^2 + 4 = 6x - 5$$

Collect all terms on the left side.

$$x^2 - 6x + 9 = 0$$

Use   the   quadratic formula.

$$x = \frac{-(-6) \pm \sqrt{(-6)^2 - 4(1)(9)}}{2(1)}$$

$$x = \frac{6 \pm 0}{2}$$

$$x = 3$$

Answer $\{3\}$

Check   The check is left to the student.

The three examples above illustrate that quadratic equations may have zero, one, or two solutions.

■ CLASSROOM EXERCISES

Write each equation in the form $ax^2 + bx + c = 0$.

1. $2x^2 - 3x = 7$ **2.** $3x^2 + 4 = x^2 + 6x$ **3.** $x(x - 1) = 5(x + 2)$

What are the values of a, b, and c?

4. $5x^2 - 3x + 4 = 0$ **5.** $6x^2 - 2x = 0$ **6.** $3x^2 - 4 = 2x - 3$

Solve using the quadratic formula.

7. $3x^2 + 2x - 4 = 0$ **8.** $x^2 - 3x + 6 = 0$ **9.** $2(2x^2 + 3) = 7$

■ WRITTEN EXERCISES

Write each equation in the form $ax^2 + bx + c = 0$.

A   **1.** $x^2 + 5x = 2$ **2.** $x^2 + 6x = 3$ **3.** $2x^2 = x - 6$

     4. $2x^2 = 3x - 5$ **5.** $3x^2 + 5x = 6 - x$ **6.** $4x^2 + x = x^2 + 2$

What are the values of a, b, and c?

     7. $5x^2 + 7x + 2 = 0$ **8.** $3x^2 + 6x + 2 = 0$ **9.** $x^2 + x - 1 = 0$

   10. $x^2 - x + 1 = 0$ **11.** $2x^2 - 11 = 0$ **12.** $3x^2 - 8x = 0$

   13. $2 + 3x + x^2 = 0$ **14.** $1 + 4x + 5x^2 = 0$ **15.** $\frac{1}{2}x^2 - x - \frac{1}{3} = 0$

   16. $\frac{1}{3}x^2 + x - \frac{1}{4} = 0$ **17.** $x^2 = 5x + 2$ **18.** $x^2 = 6x - 1$

Solve using the quadratic formula. Simplify the solutions.

   19. $x^2 + 2x - 15 = 0$ **20.** $x^2 - 2x - 15 = 0$ **21.** $5x^2 + 7x + 2 = 0$

   22. $4x^2 + 9x + 2 = 0$ **23.** $x^2 - 3x + 2 = 0$ **24.** $x^2 + 3x + 2 = 0$

   25. $6x^2 + x - 2 = 0$ **26.** $6x^2 - 5x - 6 = 0$ **27.** $4x^2 - 4x - 3 = 0$

   28. $5x^2 - x - 4 = 0$ **29.** $x^2 - 10x + 25 = 0$ **30.** $x^2 - 8x + 16 = 0$

   31. $x^2 - 5x = 0$ **32.** $x^2 - 7x = 0$

Solve. Give answers to the nearest thousandth.

   33. $x^2 + 5x + 2 = 0$ **34.** $x^2 + 6x + 2 = 0$

   35. $x^2 + 2x - 5 = 0$ **36.** $x^2 + 3x - 1 = 0$

Write the equation in the form $ax^2 + bx + c = 0$. Then solve the equation using the quadratic formula and simplify the solutions. Write $\emptyset$ if there are no solutions.

B **37.** $3x^2 + 5x + 6 = 2x^2 + 4x + 7$ **38.** $5x^2 + 2x + 3 = 4x^2 - x + 4$

   39. $(2x + 3)(x + 4) = (x - 4)(x - 5)$ **40.** $(2x - 3)(3x + 4) = 2x^2 - 12x + 11$

   41. $x^2 + 2x + 2 = 0$ **42.** $x^2 + 2x + 10 = 0$

   43. $-x^2 + 5x - 7 = 0$ **44.** $x^2 + 4x + 1 = 3x^2 + 2x + 3$

Write an equation that fits the problem. Solve the equation using the quadratic formula and answer the question.

   45. One number is 2 more than another number. The product of the two numbers is 3. What are the two numbers? (There are two possibilities.)

   46. The length of a rectangle is 10 cm greater than its width. What is the length of the rectangle if its area is 12 cm²?

   47. The length of a rectangle is 10 cm greater than its width. What is the length of the rectangle if its diagonal is 11 cm?

48. A rectangular garden borders a building along the garden's length. The length of the garden is 2 m more than its width. The area of the garden is 16 m². How long is the fence that is needed to enclose the garden? (No fence is needed on the building side of the garden.)

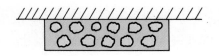

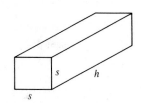

49. The surface area of a prism with a square base of side s and height h is given by the formula $S = 2s^2 + 4hs$. If the surface area is 200 square units and the height is 10, what is the length of the base?

Find the two solutions of each equation, the sum of the solutions, and the product of the solutions.

	Equation	Smaller solution	Larger solution	Sum of solutions	Product of solutions
	Sample: $2x^2 + 5x - 2 = 0$	$\dfrac{-5 - \sqrt{41}}{4}$	$\dfrac{-5 + \sqrt{41}}{4}$	$\dfrac{-5}{2}$	-1
C **50.**	$x^2 + 3x - 2 = 0$				
51.	$3x^2 + 6x + 2 = 0$				
52.	$6x^2 - 7x + 2 = 0$				
53.	$2x^2 - 11x + 12 = 0$				
54.	$ax^2 + bx + c = 0$				

■ REVIEW EXERCISES

Solve. Graph the solution on a number line.

1. $3x - 4 > 4x - 7$

3. $-2x \le -6$

5. $3(2 - x) > x$

7. $|x + 1| > 3$

2. $3x + 1 < 2x - 3$ [9–3]

4. $-4x > 12$ [9–4]

6. $5x - 12 \le 8$ [9–5]

8. $|x - 2| \le 1$ [9–6]

Graph the inequalities on the coordinate plane.

9. $x + 2y \le 4$ [9–7]

10. $x + y > 2$ and $y > 1$ [9–8]

11–6 The Discriminant

Preview

Recall that the solutions of these equations:

a. $x^2 + 6x + 9 = 0$
b. $x^2 - 6x + 7 = 0$
c. $x^2 - 6x + 12 = 0$

are the x-intercepts of their respective functions:

A. $y = x^2 + 6x + 9$
B. $y = x^2 - 6x + 7$
C. $y = x^2 - 6x + 12$

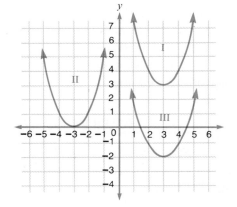

- Use the quadratic formula to solve equations a, b, and c.
- Use your solutions to the equations to match functions A, B, and C with the graphs.
- What part of the quadratic formula determines the number of x-intercepts on the graph of a quadratic function?

■ LESSON

The solutions of an equation are called the **roots** of the equation. The quadratic formula tells us that the roots of the equation $ax^2 + bx + c = 0$ are

$$\frac{-b + \sqrt{b^2 - 4ac}}{2a} \quad \text{and} \quad \frac{-b - \sqrt{b^2 - 4ac}}{2a}$$

The number and kind of roots of the equation depend on the value of the radicand, $b^2 - 4ac$. Four possibilities for the value of $b^2 - 4ac$ are shown in the following examples.

The value of $b^2 - 4ac$ is a positive perfect square, and there are two rational roots of the equation.

> **Example 1** Solve. $x^2 - 2x - 8 = 0$
>
> *Solution*
> $$x = \frac{-(-2) \pm \sqrt{(-2)^2 - 4(1)(-8)}}{2(1)}$$
>
> $$x = \frac{2 \pm \sqrt{36}}{2}$$
>
> $$x = -2 \quad \text{or} \quad x = 4$$
>
> *Answer* $\{-2, 4\}$

The value of $b^2 - 4ac$ is positive but not a perfect square, and there are two irrational roots.

> **Example 2**   Solve. $2x^2 - 3x - 1 = 0$
>
> *Solution*
> $$x = \frac{-(-3) \pm \sqrt{(-3)^2 - 4(2)(-1)}}{2(2)}$$
> $$x = \frac{3 \pm \sqrt{17}}{4}$$
>
> *Answer* $\left\{ \dfrac{3 + \sqrt{17}}{4}, \dfrac{3 - \sqrt{17}}{4} \right\}$

The value of $b^2 - 4ac$ is 0, and there is one rational root.

> **Example 3**   Solve. $x^2 - 4x + 4 = 0$
>
> *Solution*
> $$x = \frac{-(-4) \pm \sqrt{(-4)^2 - 4(1)(4)}}{2(1)}$$
> $$x = \frac{4 \pm \sqrt{0}}{2}$$
> $$x = 2$$
>
> *Answer* $\{2\}$

The value of $b^2 - 4ac$ is negative, and there are no real roots of the equation.

> **Example 4**   Solve. $x^2 + 3x + 5 = 0$
>
> *Solution*
> $$x = \frac{-3 \pm \sqrt{3^2 - 4(1)(5)}}{2(1)}$$
> $$x = \frac{-3 \pm \sqrt{-11}}{2}$$
>
> *Answer* $\emptyset$ There are no real roots.

Mathematics and Your Future

As we move from the industrial age into the information age, the number of jobs in the industrial sector are diminishing and the number in the information sector is increasing. Many of the new jobs in both the industrial and the information sectors require more education, including knowledge of mathematics. To prepare yourself for opportunities in the future, be sure that you take enough mathematics.

The expression $b^2 - 4ac$ is called the **discriminant**. Its value determines the number and kind of roots of the equation. The table below summarizes the relationship between the value of the discriminant and the number and kinds of real roots.

Value of discriminant $b^2 - 4ac$	Number of real roots of $ax^2 + bx + c = 0$	Kind of real roots (when a, b, and c are rational numbers.)
Positive and a perfect square	2	Rational
Positive but not a perfect square	2	Irrational
Zero	1	Rational
Negative	0	—

Example 5 Use the discriminant to find the number and kind of roots, and then solve. $3x^2 + 2x - 5 = 0$

Solution *Evaluate the discriminant.*
$$b^2 - 4ac = 2^2 - 4(3)(-5)$$
$$= 4 + 60$$
$$= 64$$

The value of the discriminant is a positive perfect square, so there are two rational roots.

Use the quadratic formula to solve.
$$x = \frac{-b \pm \sqrt{b^2 - 4ac}}{2a}$$

$$x = \frac{-2 \pm \sqrt{64}}{6}$$

$$x = \frac{-2 \pm 8}{6}$$

$$x = 1 \quad \text{or} \quad x = -\frac{5}{3}$$

Answer $\left\{1, -\frac{5}{3}\right\}$

■ CLASSROOM EXERCISES

1. What is the discriminant of the equation $ax^2 + bx + c = 0$?

In each of Exercises 2–7, the value of the discriminant of a quadratic equation is given. State the number and kind of roots.

2. Discriminant $= 25$ **3.** Discriminant $= 7$ **4.** Discriminant $= -9$

5. Discriminant $= 0$ **6.** Discriminant $= 12$ **7.** Discriminant $= 100$

State the number and kind of roots.

8. $x^2 - 6x + 8 = 0$ **9.** $x^2 - 6x + 7 = 0$ **10.** $x^2 + 6x + 10 = 0$

11. $9x^2 - 3x + 1 = 0$ **12.** $x^2 + x + 1 = 0$ **13.** $x^2 - 10x - 9 = 0$

14. $x^2 + 4x - 8 = 0$ **15.** $x^2 + 3x - 2 = 0$ **16.** $x^2 - 2x + 1 = 0$

■ WRITTEN EXERCISES

Write the discriminant of each equation in simplest form.

A   **1.** $x^2 - 2x + 3 = 0$ **2.** $x^2 - 3x + 4 = 0$

    3. $x^2 - 3x - 4 = 0$ **4.** $x^2 - 2x - 3 = 0$

    5. $x^2 - 4x + 4 = 0$ **6.** $x^2 + 2x + 1 = 0$

    7. $4x^2 + 8x + 3 = 0$ **8.** $4x^2 + 13x + 3 = 0$

State the number and kind of roots for a quadratic equation with the given discriminant.

9. 100   **10.** 50   **11.** 75   **12.** 25

13. -16 **14.** -36 **15.** 0   **16.** 3

17. 5   **18.** 8   **19.** 0.64   **20.** 0.16

State the number and kind of roots.

21. $x^2 - 20x + 100 = 0$ **22.** $x^2 + 20x + 100 = 0$

23. $x^2 + 20x - 100 = 0$ **24.** $x^2 - 20x - 100 = 0$

25. $x^2 - 20x + 99 = 0$ **26.** $x^2 + 20x + 99 = 0$

Solve using the quadratic formula. Simplify all solutions. If there is no solution, write ∅.

27. $x^2 + 8x + 6 = 0$ **28.** $x^2 + 6x + 4 = 0$

29. $x^2 + 6x - 4 = 0$ **30.** $x^2 + 8x - 6 = 0$

31. $x^2 + 6x + 10 = 0$ **32.** $x^2 + 5x + 8 = 0$

33. $x^2 - x - 6 = 0$ **34.** $x^2 - x - 12 = 0$

35. $3x^2 + x + 1 = 0$ **36.** $3x^2 - x + 2 = 0$

37. $4x^2 - 12x + 9 = 0$ **38.** $4x^2 - 4x + 1 = 0$

State the number of x-intercepts of each quadratic function.

B   **39.** $y = x^2 + 6x + 100$ **40.** $y = x^2 - 5x + 80$

    41. $y = 5x^2 - 4x - 100$ **42.** $y = 8x^2 - 5x - 80$

    43. $y = -2x^2 + 3x + 100$ **44.** $y = -5x^2 + 2x + 80$

For what value of k ($k \neq 0$) will the equation have exactly one root?

45. $3x^2 + 4x + k = 0$ **46.** $kx^2 + 2x + 5 = 0$

47. $kx^2 - 2x + 5 = 0$ **48.** $kx^2 + 2x - 5 = 0$

49. $3x^2 - 4x + k = 0$ **50.** $3x^2 + 4x - k = 0$

For what values of k (if any) will the equation have two roots?

51. $3x^2 + 5x + k = 0$ **52.** $3x^2 + 4x - k = 0$

53. $kx^2 + 4x + 3 = 0$ **54.** $kx^2 + 4x - 3 = 0$

55. $4x^2 + kx + 4 = 0$ **56.** $3x^2 + kx - 3 = 0$

Given $ax^2 + bx + c = 0$ and $a \neq 0$, is each of the following true or false?

C **57.** If $a > 0$ and $c < 0$, the equation has two roots.

58. If $a < 0$ and $c > 0$, the equation has two roots.

59. If $ac = b^2$, the equation has exactly one root.

60. If $ac = \frac{1}{4}b^2$, the equation has exactly one root.

Given $y = ax^2 + bx + c$ and $a \neq 0$, is each of the following true or false?

61. If $a > 0$ and $c < 0$, the graph crosses the x-axis twice.

62. If $a < 0$ and $c > 0$, the graph crosses the x-axis twice.

■ REVIEW EXERCISES

1. $\sqrt{\dfrac{9}{25}}$ **2.** $-\sqrt{64}$ **3.** $\sqrt{75}$ **4.** $\sqrt{2}(\sqrt{2} + \sqrt{6})$ [10–1, 10–4]

Self-Quiz 3

11–5 Solve using the quadratic formula. Write ∅ if there is no solution.

    1. $x^2 - 6x + 3 = 0$ **2.** $x(x + 5) = 1$

    3. $5x^2 + 10x = x^2 + x - 2$ **4.** $3x^2 + 7 = x^2 - 5x + 4$

11–6 Use the discriminant to give the number and kind of roots.

    5. $16x^2 + 6x - 1 = 0$ **6.** $x^2 - 2x + 7 = 0$

    For what values of k will the equation have one root?

    7. $x^2 + kx + 4 = 0$ **8.** $kx^2 - 6x + 1 = 0$

11–7 Choosing a Method of Solving Quadratic Equations

Preview

Five students in the Freedom High School algebra classes are specialists in solving quadratic equations.

Amy solves quadratic equations by using a graph.
Brad guesses the solutions and then checks his guesses.
Carla solves the equations by factoring.
David uses the method of completing the square.
Ellie applies the quadratic formula.

• Which students can solve any quadratic equation they are given?
• In a contest to determine who can solve quadratic equations fastest, who would probably finish first for each of these equations?

a. $x^2 + 6x + 8 = 0$
b. $x^2 + 4x = 6$

In this lesson you will study further the advantages of some of the methods of solving quadratic equations.

■ LESSON

We now know several methods of solving a quadratic equation including factoring, completing the square, graphing, and using the quadratic formula. Each time we wish to solve a quadratic equation we must choose one of the methods. We can, of course, use the quadratic formula for solving every quadratic equation. But in some cases one of the other methods may be easier to use.

When the coefficient of x^2 is 1 and the other coefficients are small, we can save time if the quadratic expression can be easily factored.

Example 1 Solve. $x^2 + x - 12 = 0$

Solution   *Factoring*

$$x^2 + x - 12 = 0$$
$$(x + 4)(x - 3) = 0$$
$$x + 4 = 0 \quad \text{or} \quad x - 3 = 0$$
$$x = -4 \quad \text{or} \quad x = 3$$

Quadratic formula

$$x^2 + x - 12 = 0$$

$$x = \frac{-b \pm \sqrt{b^2 - 4ac}}{2a}$$

$$x = \frac{-1 \pm \sqrt{1^2 - 4(1)(-12)}}{2(1)}$$

$$x = \frac{-1 \pm \sqrt{49}}{2}$$

$$x = \frac{-1 \pm 7}{2}$$

$$x = -4 \quad \text{or} \quad x = 3$$

Answer $\{-4, 3\}$

In Example 1 the factoring method is shorter and involves less computation than using the quadratic formula.

If the coefficient of x^2 is 1 and the coefficient of x is even, completing the square may be easier.

Example 2 Solve. $x^2 + 20x - 8 = 0$

Solution *Completing the square*   *Quadratic formula*

$$x^2 + 20x - 8 = 0 \qquad\qquad x^2 + 20x - 8 = 0$$
$$x^2 + 20x = 8$$
$$x^2 + 20x + 100 = 8 + 100 \qquad\qquad x = \frac{-20 \pm \sqrt{20^2 - 4(1)(-8)}}{2 \cdot 1}$$
$$(x + 10)^2 = 108$$
$$x + 10 = \pm \sqrt{108} \qquad\qquad x = \frac{-20 \pm \sqrt{432}}{2}$$
$$x + 10 = \pm 6\sqrt{3}$$
$$x = -10 \pm 6\sqrt{3} \qquad\qquad x = \frac{-20 \pm 12\sqrt{3}}{2}$$
$$x = -10 \pm 6\sqrt{3}$$

Answer $\{-10 + 6\sqrt{3},\ -10 - 6\sqrt{3}\}$

In Example 2 completing the square is easier than using the quadratic formula.

Remember that the quadratic formula can be used to solve any quadratic equation. If the coefficient of x^2 is not 1 or if the other constants are large, it may be the easiest method.

Example 3 Solve. $2x^2 - 3x - 6 = 0$

Solution $2x^2 - 3x - 6 = 0$

$$x = \frac{-b \pm \sqrt{b^2 - 4ac}}{2a}$$

$$x = \frac{-(-3) \pm \sqrt{(-3)^2 - 4(2)(-6)}}{2(2)}$$

$$x = \frac{3 \pm \sqrt{57}}{4}$$

Answer $\left\{\dfrac{3 + \sqrt{57}}{4},\ \dfrac{3 - \sqrt{57}}{4}\right\}$

Here is a suggested strategy for solving quadratic equations of the form

$$ax^2 + bx + c = 0, \qquad a \neq 0.$$

- If $a = 1$ and b and c are small integers, try factoring.
- If $a = 1$ and b is even, try completing the square.
- If the solution cannot be found quickly by either of the above methods, then use the quadratic formula.

■ CLASSROOM EXERCISES

1. Solve by factoring. $x^2 + 7x - 8 = 0$

2. Solve by completing the square. $x^2 + 24x + 9 = 0$

3. Solve by using the quadratic formula. $3x^2 + 4x - 3 = 0$

Solve.

4. $2x^2 + 5x - 4 = 0$

5. $x^2 + 8x + 2 = 0$

6. $x^2 + 8x - 16 = 0$

■ WRITTEN EXERCISES

Solve by factoring.

A

1. $x^2 - 3x - 28 = 0$

2. $x^2 - 5x - 24 = 0$

3. $x^2 - 9x + 18 = 0$

4. $x^2 - 11x + 18 = 0$

5. $x^2 + 15x + 14 = 0$

6. $x^2 + 16x + 15 = 0$

Solve by completing the square.

7. $x^2 + 4x = 6$

8. $x^2 + 6x = 4$

9. $x^2 - 8x = 5$

10. $x^2 - 2x = 6$

11. $x^2 - 12x + 12 = 0$

12. $x^2 - 10x + 7 = 0$

Solve by using the quadratic formula.

13. $x^2 + 5x - 4 = 0$

14. $x^2 + 7x - 4 = 0$

15. $3x^2 - 3x - 5 = 0$

16. $3x^2 - 5x - 3 = 0$

17. $5x^2 + 10x + 2 = 0$

18. $5x^2 + 10x + 4 = 0$

Solve by any method.

19. $x^2 - 3x = 0$

20. $x^2 - 5x = 0$

21. $x^2 - x - 1 = 0$

22. $x^2 - x - 3 = 0$

23. $x^2 + 6x + 8 = 0$

24. $x^2 + 4x + 3 = 0$

25. $4x^2 - 16x + 15 = 0$

26. $4x^2 - 19x + 15 = 0$

27. $2x^2 + 3x - 1 = 0$

28. $3x^2 + 2x - 2 = 0$

29. $x^2 - 8x = 33$

30. $x^2 - 6x = 27$

Write an equation that fits the problem. Solve the equation and answer the question.

31. The length of a rectangle is 3 m more than its width. What is the length of the rectangle if its area is 3 m²?

32. The length of a rectangle is 2 m more than its width. What is the length of the rectangle if its area is 2 m²?

33. The square of a number is 6 more than the number. What is the number?

34. The square of a number is 12 more than the number. What is the number?

35. Find the number such that its square is 3 more than the number.

36. Find the number such that its square is 1 more than the number.

B **37.** The sum S of the first n positive integers is given by the formula $S = \frac{1}{2}n^2 + \frac{1}{2}n$. How many consecutive integers, starting with 1, were added if their sum is 2556?

38. A vegetable garden is in the shape of a rectangle 6 m long and 4 m wide. It is surrounded by a border of flowers. What is the width of the border if the *total* area of the garden (including the flower border) is 30 m²?

39. A rocket shot in the air with a velocity of 24 m/s reaches a height of h meters in t seconds as given by the formula $h = -5t^2 + 24t$. After how many seconds is the rocket at each of the following heights?

   a. 10 m   **b.** 20 m   **c.** 25 m   **d.** 30 m

40. What is the perimeter of a right triangle if its hypotenuse is 15 cm and one of its legs is 3 cm longer than the other leg?

C **41.** Two cars leave a gas station at the same time. One travels west at 55 mph and the other travels south at 45 mph. After how many hours are the two cars 100 mi apart?

42. Two cars leave the same place at the same time. One travels east at a constant speed. The other travels north at a speed that is 5 km/h faster than the other car. When they have driven for 3 h, the two cars are 350 km apart. What was the average speed of the slower car?

43. A number is 4 more than its reciprocal. Find the number.

44. The height h in meters of a rocket t seconds after it is fired upward with an initial velocity of 130 m/s is given by the formula $h = -5t^2 + 130t$. Find the maximum height reached by the rocket? [Hint: First find the axis of symmetry.]

Solve by any method.

45. $3(x + 2)(x + 1) = x(2x + 3) + 7$

46. $8x^2 - 12x - 3 = (2x - 1)(x - 3)$

47. $(x^2 - 5x + 6)(x^2 - 5x + 5) = 0$

48. $(x^2 - 7)x - (x^2 - 7)5 = 0$

■ REVIEW EXERCISES

Simplify. List any restrictions on the variables.

1. $\sqrt{32} - \sqrt{18}$

2. $4\sqrt{2} - \sqrt{2}$ [10–6]

3. $\sqrt{x^2}$

4. $\sqrt{12a^2b^3}$ [10–7]

Solve.

5. $\sqrt{3x - 2} = 5$

6. $\sqrt{2x} + 4 = x$ [10–8]

Find the distance between the points. Simplify.

7. $(2, 5)$ and $(6, 8)$

8. $(-1, -1)$ and $(-2, 1)$ [10–9]

Factor completely.

9. $x^2 + 5x - 14$

10. $2x^2 + 5x + 2$ [7–9, 7–10]

■ CHAPTER SUMMARY

- **Vocabulary**

quadratic equation	[page 543]	x-intercept	[page 551]
quadratic function	[page 543]	completing the square	[page 564]
parabola	[page 544]	quadratic formula	[page 569]
minimum		roots of an equation	[page 574]
(of a quadratic function)	[page 544]	discriminant	[page 576]
axis of symmetry	[page 544]		
maximum			
(of a quadratic function)	[page 545]		

- The axis of symmetry of the function $y = ax^2 + bx + c$ $(a \neq 0)$ is the line $x = -\dfrac{b}{2a}$.

- The solutions of $ax^2 + bx + c = 0$ are the x-intercepts of the graph of $y = ax^2 + bx + c$.   [11–2]

- The Square Root Property of Equations   [11–3]

$$\text{For each real number } c, \, c > 0,$$
$$\text{if } x^2 = c, \text{ then } x = \pm\sqrt{c}.$$

- The Quadratic Formula   [11–5]

$$\text{If } ax^2 + bx + c = 0, \, a \neq 0, \text{ then}$$
$$x = \frac{-b \pm \sqrt{b^2 - 4ac}}{2a}.$$

- The following table lists the number and kind of roots of the quadratic equation $ax^2 + bx + c = 0$, $a \neq 0$.

Value of discriminant $b^2 - 4ac$	Number of real roots	Kind of real roots (when a, b, and c are rational numbers)
Positive and a perfect square	2	Rational
Positive but not a perfect square	2	Irrational
Zero	1	Rational
Negative	0	—

- A suggested strategy for solving $ax^2 + bx + c = 0$ [11–7]
  - If $a = 1$ and b and c are small integers, try factoring.
  - If $a = 1$ and b is even, try completing the square.
  - If the solution cannot be found quickly by either of the above methods, then use the quadratic formula.

■ CHAPTER REVIEW

11–1 Objective: To graph a quadratic function and find its axis of symmetry and minimum or maximum point.

1. Write the equation in the form $y = ax^2 + bx + c$ and give the values of a, b, and c.
$$y = (x - 2)(x + 4)$$

2. Give the axis of symmetry and the maximum or minimum point.
$$y = x^2 - 2x - 3$$

3. Graph. $y = 2x^2 + 5x - 3$

11–2 Objective: To solve quadratic equations by graphing.

Solve by graphing. If there is no solution, write $\emptyset$.

4. $2x^2 + 6x - 8 = 0$ **5.** $x^2 - 5x + 9 = 0$

6. The height h in meters of a football t seconds after it has been kicked upward is given by the formula $h = -5t^2 + 20t$. When will the football be at a height of 15 m?

11–3 Objective: To solve quadratic equations by using the square root property for equations.

Solve.

7. $(x - 3)^2 = 49$ **8.** $x^2 + 12x + 36 = 4$ **9.** $(x + 2)^2 = 5$

11–4 Objective: To solve quadratic equations by completing the square.

10. To complete the square, what number must be added to $x^2 - 8x$?

11. Solve $x^2 + 4x + 1 = 0$ by completing the square. Write the solutions as simplified radicals.

12. Solve $x^2 - 10x = 7$ by completing the square. Write the solutions as decimals to the nearest thousandth.

11–5 Objective: To use the quadratic formula to solve quadratic equations.

Solve using the quadratic formula.

13. $2x^2 - 7x + 6 = 0$ **14.** $x^2 + 1 = 4x$

15. The length of a rectangle is 6 cm more than the width. What are the length and width if the area is 30 cm²?

11–6 Objective: To determine the number and kind of roots of a quadratic equation from the value of the discriminant.

16. Write the discriminant of the equation $x^2 + 3x - 9 = 0$ in simplest form.

Give the number and kind of roots for the following.

17. An equation with discriminant -9

18. The equation $x^2 - 7x - 25 = 0$

11–7 Objective: To solve quadratic equations using the easiest method.

Solve.

19. $x^2 - 3x = 0$ **20.** $x^2 + 4x = 1$

21. The square of a number is 9 more than twice the number. What is the number?

■ CHAPTER 11 SELF-TEST

11–1   **1.** State whether the function described by this equation is a quadratic function.

$$y = (x + 4)(x - 1) - x^2$$

2. Write the equation $y = 4x(2 - x) + 3$ in the form $y = ax^2 + bx + c$.

3. Sketch the graph of the function defined by $y = x^2 + 6x + 5$ and identify its axis of symmetry and minimum point.

11–2   **4.** Graph the equation $y = 2x^2 - 9x - 5$ to determine the x- and y-intercepts.

5. Solve $x^2 + 3x - 15 = 0$ by graphing. Find the consecutive integers between which the roots lie.

6. What are the x-intercepts of the function $y = x^2 + x - 12$?

11–3   Solve. Write $\emptyset$ if there is no solution.

7. $x^2 = -144$ **8.** $x^2 + 20 = 120$

9. $9x^2 = 36$ **10.** $x^2 - 6x + 9 = 8$

11–4   Solve by completing the square.

11. $w^2 + 4w = 60$ **12.** $x^2 - 5x + \dfrac{1}{4} = 0$

13. $m^2 + 10m = 11$ **14.** $a^2 - 6a + 12 = 4$

11–5   Solve using the quadratic formula.

15. $2x^2 + 3x - 1 = 0$ **16.** $x^2 - 6x + 2 = 0$

17. $5a^2 - a - 4 = 0$ **18.** $a^2 - 8a + 16 = 0$

11–6   **19.** Complete this table by describing the discriminant of the quadratic equation $ax^2 + bx + c = 0$, $a \neq 0$.

Kind of roots	Number of solutions	Value of the discriminant $b^2 - 4ac$
Rational	2	?
Irrational	2	?
Rational	1	?
—	0	?

11–7   Solve using any method.

20. $x^2 + 20x - 44 = 0$ **21.** $4(x - 1)^2 = 9$

22. $5x^2 + x - 18 = 0$ **23.** $x^2 + 2x = 7$

11–7   Solve.

24. A rectangular pool is twice as long as it is wide. There is a 2-m wide concrete walk around the outside of the pool. The pool and walk cover a combined area of 160 m². Find the dimensions of the pool.

25. The height h in meters of a rocket t seconds after it has been fired upward with an initial velocity of 120 m/s is given by the formula $h = -5t^2 + 120t$. After how many seconds is the rocket 640 meters high?

■ PRACTICE FOR COLLEGE ENTRANCE TESTS

Each question consists of two quantities, one in column A and one in column B. Compare the two quantities and select one of the following answers:

A if the quantity in column A is greater;
B if the quantity in column B is greater;
C if the two quantities are equal;
D if the relationship cannot be determined from the information given.

Comments:

- Letters such as a, b, x, and y are variables that can be replaced by real numbers.
- A symbol that appears in both columns in a question stands for the same thing in column A as in column B.
- In some questions information that applies to quantities in both columns is centered above the two columns.

	Column A	Column B
1.	y is the least possible value of $x^2 + 4$.	
	y	4
2.	The surface area of a cube is $24s^2$.	
	The volume of the cube	$64s^3$
3.	$x > 0$	
	$x^3 - 8$	x^2
4.	$y = x^2 - 4x + 4$	
	The value of y when $x = 0$	The value of y when $x = 2$
5.	$2^n = 2^5 - 2^4$	
	3	n
6.	a and b are integers and $a^b = 16$.	
	a	b
7.	$x < 0$	
	x^2	$(x - 1)^2$
8.	$x^2 + 1 > y^2$	
	x	y
9.	a is a positive integer.	
	$2^{(a + 1)}$	$2 \cdot 2^a$

■ CUMULATIVE REVIEW (*Chapters* 7–11)

7–1   **1.** Write the prime factorization of 84.

7–2   **2.** State the greatest common factor of $4a^2b$ and $8ab$.

7–3   **3.** Solve. $(a + 4)(2a - 1) = 0$

7–5   Expand and simplify.

      4. $(x + 7)(x - 2)$ **5.** $(c - 5)^2$

7–4,   Factor.
7–6,
7–7,   **6.** $6x + 12y$ **7.** $y^2 - 64$
7–8,   **8.** $a^2 + 6a + 9$ **9.** $x^2 - 8x + 15$
7–9,
7–10   **10.** $y^2 + 2y - 8$ **11.** $2x^2 + 7x - 4$

7–8,   Solve.
7–9
      12. $w^2 - 5w + 6 = 0$ **13.** $t^2 + 2t = 15$

8–1   **14.** List the value(s) of x for which the fraction is undefined.

$$\frac{x + 4}{x - 1}$$

8–1,   Simplify. Assume that no denominator is zero.
8–2,
8–3,   **15.** $\frac{25ab}{5b}$ **16.** $\frac{3x}{y} \cdot \frac{5y}{9}$
8–4

      17. $\frac{18xy}{7} \div \frac{3x}{14}$ **18.** $\frac{7y}{3} + \frac{14y}{3}$

      19. $\frac{x}{4} - \frac{x}{6}$ **20.** $\frac{8}{x - 1} + \frac{2}{x + 1}$

8–5,   Solve.
8–7,
8–8   **21.** $\frac{1}{x} + \frac{7}{2x} = 9$ **22.** $\frac{x}{16} = \frac{15}{64}$ **23.** 30% of $x = 18$

8–7   **24.** Write the measurements in the same units and then simplify the ratio.
      24 in. to 2 yd

8–9   **25.** A 15-oz box of cereal costs 87¢. What does a 25-oz box cost if the price
      per ounce is the same?

8–10 **26.** Suppose y varies directly as x. If $x = 12$ when $y = 16$, find y when $x = 21$.

8–11 **27.** Suppose y varies inversely as x. If $y = 6$ when $x = 36$, find y when $x = 18$.

9–1 **28.** Write an inequality for this statement.

The number n is not more than 17.

9–1,
9–2 Graph on a number line.

29. $x < -3$ **30.** $y < 2$ and $y > 0$ **31.** $x \geq 5$ or $x = 1$

32. Write a compound inequality for this graph.

9–3 Solve.

33. $b + 3 > -2$ **34.** $5x + 1 < 6x - 2$

9–3 Write an inequality for the situation and solve.

35. If Glenda gained 15 lb, she would be at least 128 lb. What is Glenda's present weight (w)?

9–4 **36.** The price of $\frac{1}{3}$ lb of imported cheese is slightly more than $1.50. What is the price (p) per pound?

9–4,
9–5 Solve and graph on a number line.

37. $-3a > 12$ **38.** $\frac{1}{2} d \geq 2$ **39.** $-2y + 1 < 5$

9–5 Solve.

40. The length of a rectangle is 4 cm more than its width. What is the width if the perimeter is less than 48 cm?

9–6 Write as a compound inequality without absolute value.

41. $|x| > 11$ **42.** $|b - 2| < 4$

9–7 **43.** State whether the point $(2, -3)$ is on the graph of $y > 2x - 6$.

44. Graph. $x + y < 3$

9–8 **45.** Graph the compound inequality
$y > 5$ and $x < 2$.

46. Which region of the plane satisfies the graph of the compound ineqality
$y < x$ and $y < -3$?

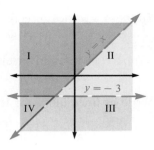

10–1 Simplify.

47. $\sqrt{81}$ **48.** $-\sqrt{36}$

49. The area of a square is 49 cm². What is the length of a side of the square?

10–2 **50.** Between what two consecutive integers does $\sqrt{31}$ lie?

51. The square root of an integer is 8.3 (accurate to the nearest tenth). What is the integer?

10–3 **52.** Could 10 cm, 24 cm, and 26 cm be the lengths of the sides of a right triangle?

53. Find the missing length in the right triangle.

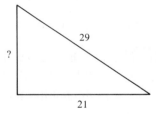

10–4, Simplify.
10–5,
10–6 **54.** $\sqrt{72}$ **55.** $-\sqrt{48}$ **56.** $\sqrt{2}(\sqrt{6} + \sqrt{2})$

57. $\sqrt{\dfrac{9}{16}}$ **58.** $\dfrac{\sqrt{25}}{\sqrt{27}}$ **59.** $8\sqrt{7} - \sqrt{7}$

60. $\sqrt{75} + \sqrt{12}$

10–6 **61.** Find the perimeter of the triangle.

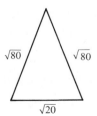

10–7 Simplify. Assume that all variables are positive.

    62. $\sqrt{9xy^2}$ **63.** $\sqrt{32x^5}$

10–8 Solve.

    64. $\sqrt{y-2} = 6$ **65.** $\sqrt{12 + c} = c$

10–9 **66.** Find the distance between the points $(0, 7)$ and $(8, 1)$.

    67. Find the value(s) of y that makes the distance between the points $(5, 2)$ and $(3, y)$ equal to $2\sqrt{5}$.

11–1 **68.** Write the equation $y = (x - 4)(2x + 1)$ in the form $y = ax^2 + bx + c$ and state the values of a, b, and c.

    69. Graph the equation $y = x^2 + 2$.

11–2 **70.** Graph the function $y = x^2 - x - 6$ to solve the equation $x^2 - x - 6 = 0$.

11–3 Solve. If there is no solution, write $\emptyset$.

    71. $x^2 = -25$ **72.** $(x - 2)^2 = 100$

11–4 **73.** What number will complete the square when added to the expression $x^2 - 8x$?

    74. Solve by completing the square. $x^2 - 2x = 24$

11–5 **75.** In applying the quadratic formula to solve the equation $6x^2 - x + 1 = 0$, what values should be substituted for a, b, and c?

Solve using the quadratic formula.

    76. $x^2 - 14x + 48 = 0$ **77.** $2x^2 - x - 10 = 0$

11–6 **78.** Use the discriminant to state the number and kind of roots of the equation $4x^2 - 20x + 25 = 0$.

    79. Write the discriminant in simplest form for the equation $x^2 - 8x + 4 = 0$.

11–7 Solve.

    80. A picture is 4 in. wide and 6 in. long. It has a white border around it. What is the width of the border if the total area of the picture, including the border, is 29 square inches?

    81. A rocket shot in the air with a velocity of 40 m/s reaches a height of h meters in t seconds, as given by the formula $h = -5t^2 + 40t$. After how many seconds will the rocket be at a height of 80 m?

Supplementary Topic
Probability and Statistics

Probability

Consider the experiment of tossing a "fair" coin into the air. The coin is just as likely to land heads up as tails up. Since there are 2 equally likely outcomes, we say that the **probability** of obtaining heads is $\frac{1}{2}$. We write $P(\text{heads}) = \frac{1}{2}$. Similarly, $P(\text{tails}) = \frac{1}{2}$.

Now consider a more complex situation. Suppose a jar contains 4 red marbles and 2 white marbles. The probability of selecting any given marble without looking is $\frac{1}{6}$, but what is the probability of selecting a red marble without looking?

$$P(\text{red}) = \frac{\text{Number of red marbles}}{\text{Total number of marbles}} = \frac{4}{6} = \frac{2}{3}$$

An event is a collection of one or more outcomes of an experiment. If all outcomes are equally likely, the probability of an event is the quotient of the number of outcomes in the event (called successes) divided by the total number of outcomes.

Definition: Probability of an Event

$$P(\text{event}) = \frac{\text{Number of successes}}{\text{Number of outcomes}}$$

Example 1 In the throw of a die, what is the probability that the number of dots on the top face will be fewer than 3?

Solution There are 6 different faces that can be up.

Two of those faces have fewer than 3 dots.

Answer $P(\text{fewer than 3 dots}) = \frac{2}{6} = \frac{1}{3}$

Example 2 A deck of playing cards has 52 cards. There are 13 cards in each of four suits. Two of the suits (hearts and diamonds) are red and two of the suits (spades and clubs) are black. What is the probability that one card drawn at random from a deck will be a red ace?

Solution There are 2 red aces in the 52 cards so there are 2 ways of drawing a red ace. There are 52 possible ways of drawing one card.

$$P(\text{red ace}) = \frac{2}{52} = \frac{1}{26}$$

Answer The probability of drawing a red ace is $\frac{1}{26}$.

Consider rolling a die. What is the probability of getting a 7? What is the probability of getting a number less than 7? In the first case, there is no face with 7 dots. Therefore,

$$P(7) = \frac{0}{6} = 0.$$

In the second case, all faces have fewer than 7 dots. Therefore,

$$P(\text{less than } 7) = \frac{6}{6} = 1.$$

The first case is an example of an **impossible event**. The second is an example of a **certain event**.

Definitions: Impossible and Certain Events

The probability of an impossible event is 0.
The probability of a certain event is 1.

Note that the probability of an event occurring plus the probability of the event not occurring is 1.

Example 3 A jar has 5 red marbles, 4 green marbles, and 3 white marbles. A marble is selected without looking. Find $P(\text{red})$ and $P(\text{not red})$.

Solution $P(\text{red}) = \dfrac{\text{Number of red marbles}}{\text{Number of marbles}} = \dfrac{5}{12}$

$P(\text{not red}) = \dfrac{\text{Number of not red marbles}}{\text{Number of marbles}} = \dfrac{7}{12}$

Check $P(\text{red}) + P(\text{not red}) = \dfrac{5}{12} + \dfrac{7}{12} = 1$ It checks.

> **Example 4** What is the probability of this spinner landing on yellow?
>
> **Solution** The yellow section is a 90-degree section of the circle.
>
> $$P(\text{yellow}) = \frac{90}{360} = \frac{1}{4}$$

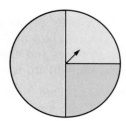

■ CLASSROOM EXERCISES

In a jar of 15 marbles, 5 of the marbles are red and 6 are black. Find the probability.

1. $P(\text{red})$ **2.** $P(\text{not red})$ **3.** $P(\text{neither red nor black})$

4. A deck of playing cards has 52 cards, 13 in each of 4 suits (hearts, spades, diamonds, and clubs). What is the probability of selecting a spade?

■ WRITTEN EXERCISES

Twenty identical slips of paper are marked with the numbers 1 to 20, put in a box, and thoroughly mixed. A slip of paper is randomly drawn. Find each probability, where n represents the number that was randomly drawn.

A **1.** $P(n = 7)$ **2.** $P(n > 12)$ **3.** $P(n \text{ is a multiple of } 3)$

4. $P(n \text{ is odd})$ **5.** $P(n \text{ is prime})$ **6.** $P(n \text{ is a 2-digit number})$

A die is rolled and the number on the top face is observed. Find each probability.

7. $P(2)$ **8.** $P(\text{less than } 10)$ **9.** $P(\pi)$

10. $P(\text{multiple of } 3)$ **11.** $P(\text{not prime})$ **12.** $P(\text{factor of } 60)$

A card is selected at random from a deck of 52 thoroughly shuffled playing cards. Find each probability.

13. $P(\text{club})$ **14.** $P(\text{not club})$ **15.** $P(\text{king or queen})$

The possible outcomes of tossing two coins are HH, HT, TH, TT. State the probability of the following outcomes.

B **16.** both heads **17.** heads on at least one coin **18.** no heads

List the possible outcomes of tossing three coins. State the probabilities of the following outcomes.

19. three tails **20.** no tails **21.** two heads and one tail

22. more tails than heads **23.** at least one tail **24.** half tails and half heads

C **25.** What is the probability of getting more than two heads when you toss four coins?

Compound Events

Consider tossing a "fair" coin three times. What is the probability of getting heads on each toss? To answer the question we can list all possible outcomes.

HHH   HHT   HTH   HTT   THH   THT   TTH   TTT

Since the eight outcomes are equally likely, the probability of each outcome is $\frac{1}{8}$. Therefore, $P(\text{HHH}) = \frac{1}{8}$. Note that

$$P(\text{HHH}) = \frac{1}{8} = \frac{1}{2} \cdot \frac{1}{2} \cdot \frac{1}{2} = P(\text{H}) \cdot P(\text{H}) \cdot P(\text{H})$$

This fact illustrates the following property.

Probability—Independent Events

For two independent events A and B, if $P(A)$ is the probability of A occurring and $P(B)$ is the probability of B occurring, then

$$P(A \text{ and } B) = P(A) \cdot P(B)$$

Example 1 What is the probability of rolling a die twice and getting 6 on the first roll and 3 on the second roll?

Solution $P(6 \text{ on the first roll}) = \frac{1}{6}$

$P(3 \text{ on the second roll}) = \frac{1}{6}$

$$P(6, 3) = P(6) \cdot P(3) = \frac{1}{6} \cdot \frac{1}{6} = \frac{1}{36}$$

Answer $P(6, 3) = \frac{1}{36}$

Check List all the possible outcomes:

(1, 1)	(1, 2)	(1, 3)	(1, 4)	(1, 5)	(1, 6)
(2, 1)	(2, 2)	(2, 3)	(2, 4)	(2, 5)	(2, 6)
(3, 1)	(3, 2)	(3, 3)	(3, 4)	(3, 5)	(3, 6)
(4, 1)	(4, 2)	(4, 3)	(4, 4)	(4, 5)	(4, 6)
(5, 1)	(5, 2)	(5, 3)	(5, 4)	(5, 5)	(5, 6)
(6, 1)	(6, 2)	(6, 3)	(6, 4)	(6, 5)	(6, 6)

One of the 36 outcomes is (6, 3). The answer checks.

In rolling dice, an outcome of one roll does not affect an outcome of another roll. Therefore, the probability of getting a given number is the same on every roll. However, there are situations in which the outcome of a first event may change the probability of a second event.

Example 2 A deck of playing cards has 52 cards. If two cards are selected by chance, what is the probability of getting the ace of hearts followed by the ten of clubs? The first card is not replaced before the second selection.

Solution On the first draw there are 52 cards in the deck. Therefore, the probability of getting any specific card is $\frac{1}{52}$. On the second draw there are only 51 cards. The probability of getting any specific one of the remaining cards is $\frac{1}{51}$.

$$P(\text{ace of hearts, ten of clubs}) = \frac{1}{52} \cdot \frac{1}{51}$$

$$= \frac{1}{2652}$$

Answer $P(\text{ace of hearts, ten of clubs}) = \frac{1}{2652}$

■ CLASSROOM EXERCISES

There are 5 red beads, 6 green beads, and 9 white beads in a jar. Two beads are drawn by chance.

1. What is the probability of getting red on the first draw and green on the second draw if the first bead is replaced before the second draw?

2. What is the probability of getting white on the first draw and white on the second draw if the first bead is not replaced before the second draw?

■ WRITTEN EXERCISES

Two coins are tossed. Find each probability.

A 1. $P(2 \text{ heads})$ 　　　　　　　　　　　　　 2. $P(\text{same outcome})$

　　 3. $P(\text{different outcomes})$ 　　　　　　　　 4. $P(\text{heads on first, tails on second})$

A card is randomly selected from a deck of 52 playing cards, observed, and returned to the deck. Then another card is randomly selected and observed. Find each probability.

　　 5. $P(\text{both red})$ 　　　　　　 6. $P(\text{red, club})$ 　　　　　　 7. $P(\text{diamond, heart})$

　　 8. $P(\text{first card is a 2, 3, or 4 and second card is a 2 or 3})$

　　 9. $P(\text{first card is a 5, 6, 7, or 8 and second card is a 5 or 6})$

A card is randomly selected from a deck of 52 playing cards and not replaced. Then a second card is randomly selected. Find each probability.

10. P(both red)

11. P(first card is black and second card is red)

12. P(first card is an ace and second card is a king)

13. P(first card is the ace of spades and second card is the ace of spades)

The 36 equally likely outcomes of rolling two dice are given at the right. Find each probability.						
	(1, 1)	(1, 2)	(1, 3)	(1, 4)	(1, 5)	(1, 6)
	(2, 1)	(2, 2)	(2, 3)	(2, 4)	(2, 5)	(2, 6)
	(3, 1)	(3, 2)	(3, 3)	(3, 4)	(3, 5)	(3, 6)
	(4, 1)	(4, 2)	(4, 3)	(4, 4)	(4, 5)	(4, 6)
	(5, 1)	(5, 2)	(5, 3)	(5, 4)	(5, 5)	(5, 6)
	(6, 1)	(6, 2)	(6, 3)	(6, 4)	(6, 5)	(6, 6)

14. P(sum $= 6$)

15. P(sum > 3)

16. P(first die < 3 and second die < 4)

17. P(first die < 3 or second die < 4)

18. P(first die > 2 and sum $= 7$)

19. P(first die < 3 or sum $= 7$)

20. P(sum > 1)

21. P(sum > 12)

A jar contains 6 blue beads, 3 green beads, and 1 brown bead. Two beads are drawn at random without replacement. Find each probability.

22. P(both blue)

23. P(1 blue and 1 brown)

24. P(neither blue)

25. P(neither green)

B In an experiment a card is drawn at random from a deck of 52 playing cards and a die is rolled. Find each probability.

26. P(the card is a heart and the die shows one)

27. P(red card and number less than 5)

28. P(ace and one)

29. P(not ace and not one)

30. P(club and odd number)

31. P(not club and even number)

C The 200 students in a small school may be enrolled in mathematics (M), in social studies (S), in both, or in neither. The diagram at the right shows that 130 students are enrolled in mathematics and 120 are enrolled in social studies. A student is selected at random. Find each probability.

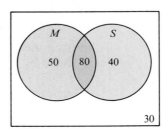

32. P(enrolled in mathematics or in social studies)

33. P(not enrolled in mathematics)

34. P(enrolled in mathematics but not in social studies)

35. P(enrolled neither in mathematics nor in social studies)

Statistics

The **frequency** table at the right shows student scores on an algebra test. Each **tally mark** represents one student score. The most common score is 86, achieved by 4 students. The most common score is called the **mode.**

There are 25 scores in all. The middle score is 85. The middle score is called the **median.** The median is found by arranging the scores in order from largest to smallest and locating the middle score. If there is an even number of scores, the median is midway between the two middle scores. For example, the median of these four scores is 6.

$$1, \quad 5, \quad 7, \quad 18$$
$$\text{Median} = 6$$

The **arithmetic mean** of a set of scores is the quotient of the sum of the scores divided by the number of scores.

$$\text{Mean} = \frac{\text{Total of scores}}{\text{Number of scores}} = \frac{2121}{25} \approx 84.8$$

Score	Tallies	Frequency
95	/	1
92	//	2
90	/	1
89	/	1
87	//	2
86	////	4
85	///	3
84	/	1
83	///	3
82	/	1
81	//	2
80	//	2
78	/	1
75	/	1
Total		25
Total of scores	=	2121

Note that the test scores are correct to the nearest whole-number integer and the mean is given to the nearest tenth. Means are usually given to one more place of accuracy than the score themselves.

The mean, median, and mode are called **averages,** or **measures of central tendency.**

Example 1   Find the mean, median, and mode of this set of kilogram weights.

$$81, \quad 84, \quad 75, \quad 58, \quad 75, \quad 82, \quad 84, \quad 75$$

Solution   Arrange the numbers in order.

$$58, \quad 75, \quad 75, \quad 75, \quad 81, \quad 82, \quad 84, \quad 84$$

To find the mean, first find the sum of the scores: 614.
Then divide by the number of scores: 8.

$$\text{Mean} = \frac{614}{8} = 76.8$$

The median, midway between 75 and 81, is 78.
The mode is the most frequent score, 75.

Answer   Mean = 76.8 kg   Median = 78 kg   Mode = 75 kg

Example 2 Carrie scores 87, 84, and 93 on her first three algebra tests. What does she have to score on the next test in order to have a mean of at least 90?

Solution Let x = the next test score.

$$\frac{87 + 84 + 93 + x}{4} \geq 90$$

$$\frac{264 + x}{4} \geq 90$$

$$264 + x \geq 360$$
$$x \geq 96$$

Answer Carrie must score 96 or more in order to have a mean of at least 90.

The fact that the mean, median, and mode are called averages can create confusion. For example, these are the yearly salaries of the seven employees of a small company.

$200,000 $20,000 $18,000 $15,000 $12,000 $12,000 $12,000

The mean salary is $41,286.
The median salary is $15,000.
The mode salary is $12,000.

During salary negotiations a union representative might claim that the average salary is $12,000. The owner might claim that the average salary is over $41,000. Both claims are correct, but communication will be clear only if the words "mode" and "mean" are used in place of the word "average."

■ CLASSROOM EXERCISES

1. In the example of the yearly salaries given above, which best describes the "average" salary: the mean salary, the median salary, or the mode salary? Why?

2. Compute the mean, median, and mode of the frequency distribution shown here.

Score	Frequency
135	1
145	2
155	4
165	5
175	8
185	7
195	6
205	4

■ WRITTEN EXERCISES

Compute the mean, median, and mode of each set of numbers.

A **1.** 3, 6, 2, 3, 8, 2, 5, 5, 2

2. 7, 9, 6, 37, 5, 9, 6, 9

3. 23, 26, 27, 27, 28, 30

4. 2, 2, 2, 6, 7, 10, 11, 20

Compute the mean, median, and mode of each frequency distribution.

5.

Score	Frequency
10	1
13	2
15	1
18	2
20	3
23	1

6.

Score	Frequency
50	3
51	2
55	2
59	1
65	1

7.

Score	Frequency
65	2
70	4
72	2
73	1
78	1

Solve.

8. Eva had scores of 127 and 139 in her first two bowling games. What score must she get in her third game in order to have a mean score of 150?

9. On his fifth test, Zach's score was 40, which made the mean of his scores 50. What was the mean of Zach's scores for the first four tests?

10. The mean score on a recent test was 56 when one student was absent. When the absent student returned and took the test, the class mean was 57. What was the absent student's score if there are 25 students in the class?

11. Three of the most popular shows in the winter series of the Nurnberg Music Theater were "Die Fledermaus" (297 performances; 182,000 attending), "The Marriage of Figaro" (216 performances; 183,000 attending), and "La Boheme" (216 performances; 172,000 attending). Which of the three shows was most popular, on the basis of average attendance? Explain.

Compute the mean, median, and mode of each frequency distribution. Consider the height for each frequency to be at the midpoint of the interval.

B **12.**

Height (in cm)	Frequency
136–140	4
141–145	3
146–150	13

13.

Height (in cm)	Frequency
150–154.9	5
155–159.9	5
160–164.9	10

Find the frequency distribution that satisfies each set of conditions.

C **14.** Ten players.
Points scored are 1, 3, and 6.
Mean is 4 points.
Mode is 3 points.

15. Ten players.
Points scored are 1, 3, and 6.
Mean is 4 points.
Median is 3 points.

Variation

These two sets of data have the same mean, 60, but are very different from each other.

$$59, \quad 60, \quad 61 \qquad\qquad 1, \quad 2, \quad 177$$

In the first set, the data are nearly the same whereas in the second, they vary widely. It is often useful to measure the variation of the data.

One measure of variation is the **range**, the difference between the least number and the greatest number in the set. The range of the first set of data given above is $61 - 59$, or 2. The range of the second set is $177 - 1$, or 176. The range is the difference between the extremes of a set of data, but it tells nothing about how the data are distributed between the extreme values.

The following two sets of data have the same mean (60) and the same range (40). But note that in Set 1 the data "cluster" near the mean, whereas in Set 2 the data are near the extreme values.

	Set 1					Set 2			
40	59	60	61	80	40	41	60	79	80
	59	60	61		40	41		79	80
	59	60	61		40				80

This example shows another component of variation: the amount of dispersion from the mean of numbers in the set.

A commonly used statistic that describes this characteristic is called the **standard deviation**. The standard deviation of the data in Set 1 is computed as follows.

1. Compute the sum of the squares of the differences from the mean.

$$(40 - 60)^2 + 3(59 - 60)^2 + 3(60 - 60)^2 + 3(61 - 60)^2 + (80 - 60)^2 = 806$$

2. Compute the mean of the squares. $806 \div 11 \approx 73.27$

3. Compute the square root of the mean of the squares. $\sqrt{73.27} \approx 8.6$

The standard deviation of Set 1 is about 8.6. Using the same method, we can compute the standard deviation of Set 2. The standard deviation of Set 2 is about 18.7. The larger standard deviation of Set 2 indicates that its scores vary more from the mean than those in Set 1.

Example   Compute the mean, range, and standard deviation of this set of scores from a sharpshooter's competition.

$$60, \quad 65, \quad 68, \quad 68, \quad 68, \quad 70, \quad 70, \quad 71, \quad 75$$

Solution $\text{Mean} = \dfrac{60 + 65 + 68 + 68 + 68 + 70 + 70 + 71 + 75}{9} = \dfrac{615}{9} \approx 68.3$

$\text{Range} = 75 - 60 = 15$

Example (continued)

Compute the standard deviation.
1. Square the differences from the mean. (Use a calculator.)

$$(68.3 - 60)^2 = 68.89$$
$$(68.3 - 65)^2 = 10.89$$
$$(68.3 - 68)^2 = \ \ 0.09 \qquad 3(0.09) = 0.27$$
$$(68.3 - 70)^2 = \ \ 2.89 \qquad 2(2.89) = 5.78$$
$$(68.3 - 71)^2 = \ \ 7.29$$
$$(68.3 - 75)^2 = 44.89$$

2. Compute the sum of the squares.

$$138.01$$

3. Compute the mean of the sum of the squares.

$$\frac{138.01}{9} \approx 15.33$$

4. Compute the square root of the mean of the sum of the squares. (Use a calculator.)

$$\sqrt{15.33} \approx 3.9$$

Answer   The mean is 68.3, the range is 15, and the standard deviation is 3.9.

■ CLASSROOM EXERCISES

Match each standard deviation with the correct set of data.

Standard deviations: 1.6   1.1

1. Set A

Score	20	19	18	17	16	15
Number	3	3	4	4	3	3

2. Set B

Score	20	19	18	17	16	15
Number	1	2	6	8	2	1

3. What is the range of data in Set A?

■ WRITTEN EXERCISES

Three sets of data are given. Without computing, state which set has the greatest standard deviation and which set has the least standard deviation.

A   **1.** Set A: {91, 94, 97, 100, 103, 106, 109}
  Set B: {94, 96, 98, 100, 102, 104, 106}
  Set C: {97, 98, 99, 100, 101, 102, 103}

Find the range and standard deviation of each set of data. (Use a calculator.)

2. {10, 16, 17, 20, 21, 28, 28}

3. {−15, −17, −17, −18, −24, −24, −25}

4. {35, 35, 45, 45}

5. {36, 39, 39, 40, 40, 40, 41, 41, 44}

6. {34, 34, 40, 40, 40, 40, 40, 46, 46}

7. {63, 72, 72, 75, 75, 75, 78, 78, 87}

Find the mean, range, and standard deviation of each frequency distribution. (Use a calculator.)

8.

Score	9	22	23	24	26	27
Frequency	1	2	1	2	2	2

9.

Score	−9	−22	−23	−24	−26	−27
Frequency	10	20	10	20	20	20

True or false?

B **10.** If 10 is added to each score in a frequency distribution, the mean, median, and mode are each increased by 10.

11. If 10 is added to each score in a frequency distribution, the range and standard deviation are increased by 10.

12. If each score in a frequency distribution is doubled, the mean, median, and mode are doubled.

13. If each score in a frequency distributuion is doubled, the range and standard deviation are doubled.

14. The standard deviation of a set of numbers is never negative.

15. There are as many numbers larger than the mean as there are smaller than the mean.

In many frequency distributions, approximately 68% of the scores are within one standard deviation of the mean and approximately 95% are within two standard deviations of the mean.

The distribution of heights for a class of students is given.

Height of students (in.) (midpoint of interval)	54	57	60	63	66	69	72	75	78
Number of students	2	8	22	32	72	32	23	6	3

C **16.** Describe the heights using the following statistics.

    a. The mean.
    b. The standard deviation (rounded to the nearest tenth).
    c. The percent of heights within one standard deviation of the mean.

Supplementary Topic
Trigonometry

Similar Triangles

Triangles exist in a variety of shapes and sizes. However, the sum of the degree measures of the angles of a triangle is constant.

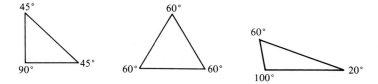

Angle-Sum Property

The sum of the measures of the angles of a triangle is 180°.

The two triangles below have the same shape. They are called **similar** triangles. Angle *A* and angle *D* are **congruent** angles. (The angles are the same size.) Angle *B* is congruent to angle *E*, and angle *C* is congruent to angle *F*.

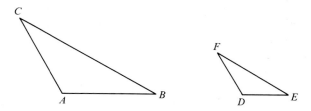

Definition: Similar Triangles

Two triangles are similar if and only if their corresponding angles are congruent.

Example 1 Determine whether triangle *MNO* and triangle *RST* are similar.

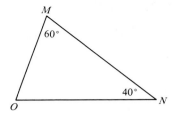

Solution Since each triangle has a 40°-angle and a 60°-angle, the third angles are [180 − (60 + 40)]° angles or 80°-angles. Corresponding angles of the two triangles are congruent.

Answer The triangles are similar.

The two triangles at the right are similar. Note that the ratios of the lengths of corresponding sides are equal.

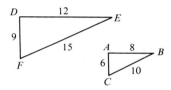

$$\frac{AB}{DE} = \frac{8}{12} = \frac{2}{3} \qquad \frac{BC}{EF} = \frac{10}{15} = \frac{2}{3} \qquad \frac{CA}{FD} = \frac{6}{9} = \frac{2}{3}$$

Property of Corresponding Sides of Similar Triangles

Corresponding sides of similar triangles are proportional.

The fact that corresponding sides of similar triangles are proportional gives us a method of computing missing lengths of triangles.

Example 2 Find the length of side *GH*.

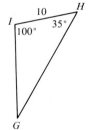

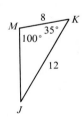

Solution Since two angles of one triangle are congruent to two angles of the other triangle, the triangles are similar and corresponding sides are proportional.

$$\frac{GH}{JK} = \frac{IH}{MK}$$

Substitute.
$$\frac{GH}{12} = \frac{10}{8}$$

$$GH = \frac{10}{8} \cdot 12$$

$$GH = 15$$

Answer The length of side *GH* is 15.

Similar triangles can be used to solve some practical problems.

Example 3 A person 2 m tall casts a 3-m shadow at the same time that a tree casts a 10-m shadow. How tall is the tree?

Solution Sketch a picture. The two triangles are similar.

$$\frac{2}{x} = \frac{3}{10}$$

$$3x = 20$$

$$x = 6\frac{2}{3}$$

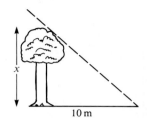

Answer The tree is about 6.7 m tall.

■ CLASSROOM EXERCISES

1. The measures of two angles of a triangle are 56° and 39°. What is the measure of the third angle?

Find the missing lengths in these similar triangles.

2.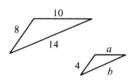

3.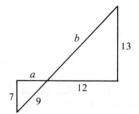

■ WRITTEN EXERCISES

The measures of two angles of a triangle are given. Find the measure of the third angle.

A **1.** 25°, 100°   **2.** 90°, 12°   **3.** 16°, 56°   **4.** 62.6°, 72°

Find the missing lengths in these similar triangles.

5.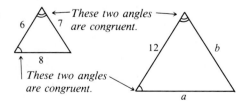

6.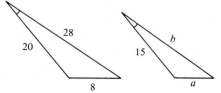

606   Supplementary Topic   Trigonometry

7.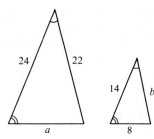

8.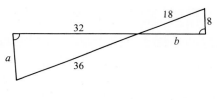

Solve.

9. At the time a 2-m pole casts a 3-m shadow, the shadow of a building is 15 m. What is the height of the building?

10. A scale model of a building is 7 cm tall and 15 cm long. If the actual building is 6 m long, what is the height of the building?

Triangles *ABC* and *DEF* are similar. Find the missing measures of the angles and sides.

B **11.**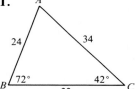

12.

A **regular polygon** has all angles congruent and all sides congruent. An equilateral triangle and a square are examples of regular polygons. The sum S of the degree measures of the angles of an n-sided regular polygon is given by the formula $S = (n - 2) \cdot 180$. Find the measure of each angle of these regular polygons.

C **13.** Regular pentagon   **14.** Regular hexagon

15. Regular octagon   **16.** Regular decagon

17. Regular 20-sided polygon   **18.** Regular 100-sided polygon

Mathematics and Your Future

Many entry-level jobs require only a limited knowledge of mathematics. Opportunities for promotion are often tied to education or training with a strong mathematics component. If you acquire a good background in mathematics you will increase your opportunities for employment and advancement.

The Tangent Ratio

Recall that a right triangle has a right (or 90°) angle and two acute angles. The side opposite the right angle is called the **hypotenuse**. Side BC is called the **side opposite angle A** and the **side adjacent to angle B**. Similarly, side AC is called the **side opposite angle B** and the **side adjacent to angle A**.

Right triangles can be thought of as belonging to "families." For example, all right triangles with a 40° angle are similar, and all right triangles with a 30° angle are similar. For the triangles in any "family" of similar right triangles, the corresponding sides are proportional. For each right triangle "family," this ratio is constant:

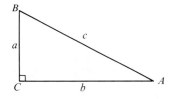

$$\frac{\text{Length of side opposite an acute angle}}{\text{Length of side adjacent to that acute angle}}$$

The ratio is called the **tangent** (or **tan**) of the acute angle. In triangle ABC:

$$\text{Tangent of angle } A = \frac{\text{Length of side opposite angle } A}{\text{Length of side adjacent to angle } A} \qquad \tan A = \frac{a}{b}$$

$$\text{Tangent of angle } B = \frac{\text{Length of side opposite angle } B}{\text{Length of side adjacent to angle } B} \qquad \tan B = \frac{b}{a}$$

Example 1 For triangle XYZ compute tan X and tan Y correct to the nearest hundredth.

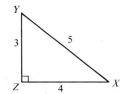

Solution $\tan X = \dfrac{3}{4} = 0.75 \qquad \tan Y = \dfrac{4}{3} = 1.33$

The tangent of an angle depends on the size of the angle. The following table lists the tangents of some acute angles. A complete table of tangents is on page 642. These ratios can be used to compute the missing measures of a right triangle.

Angle	Tangent	Angle	Tangent
5°	0.0875	50°	1.1918
10°	0.1763	55°	1.4281
15°	0.2679	60°	1.7321
20°	0.3640	65°	2.1445
25°	0.4663	70°	2.7475
30°	0.5774	75°	3.7321
35°	0.7002	80°	5.6713
40°	0.8391	85°	11.4301
45°	1.0000		

Example 2   Find the length of side *MN* to the nearest tenth.

Solution

$$\tan 35° = \frac{l}{12}$$

Find tan 35° in the table and substitute.

$$0.7002 \approx \frac{l}{12}$$

$$l \approx 12(0.7002)$$
$$\approx 8.4$$

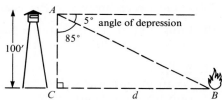

Answer   Side *MN* is about 8.4 m long.

Example 3   A fire ranger in a 100-ft tower spotted smoke near the ground. The angle of depression was 5°. About how far from the base of the tower was the fire?

Solution   Make a sketch. If the angle of depression is 5°, then angle *A* of the triangle is 85°.

$$\tan 85° = \frac{d}{100}$$

$$11.4301 \approx \frac{d}{100}$$

$$d \approx 1143$$

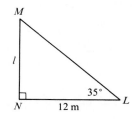

Answer   The fire is about 1143 ft from the base of the tower.

■ CLASSROOM EXERCISES

1. Compute tan *A* and tan *B* to the nearest hundredth.

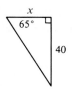

In each triangle, find the length *x* to the nearest tenth.

2.

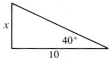

3.

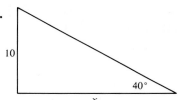

4.

■ WRITTEN EXERCISES

Compute tan *A* and tan *B* to the nearest hundredth.

A **1.**

2.

3.

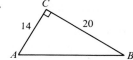

Find the length x to the nearest tenth.

4.

5.

6.

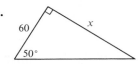

Find the measures of angle A and angle B to the nearest degree.

7.

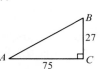

8.

9.

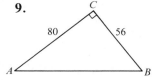

Solve. Give each answer to the nearest tenth.

10. From a point 20 m from the base of a tree, the angle of elevation to the top of the tree is 30°. What is the height of the tree?

11. A lighthouse has a telescope mounted 120 m above the surface of the ocean. A ship was sighted at an angle of depression of 5°. How far was the ship from the lighthouse?

Find the lengths x and y to the nearest tenth.

B 12.

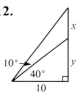

13.

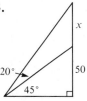

14.

Find α and β in each figure to the nearest degree. The letters α (alpha) and β (beta) are the Greek letter equivalents of a and b.

C 15.

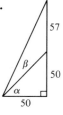

16.

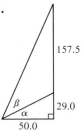

17.

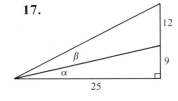

Solve. Give each answer to the nearest tenth.

18. Looking due north from the observation deck of a lighthouse 60 m above the sea, a lighthouse keeper sees two ships. The angles of depression to the ships are 5° and 10°. How far apart are the ships?

Sine and Cosine

The tangent ratio is an example of a **trigonometric ratio**. Two other trigonometric ratios are the **sine ratio (sin)** and **cosine ratio (cos)**.

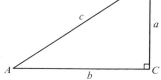

Sine of angle $A = \dfrac{\text{Length of side opposite angle } A}{\text{Length of hypotenuse}}$ $\sin A = \dfrac{a}{c}$

Cosine of angle $A = \dfrac{\text{Length of side adjacent to angle } A}{\text{Length of hypotenuse}}$ $\cos A = \dfrac{b}{c}$

Example 1 Compute sin R, cos R, sin T, and cos T to the nearest hundredth.

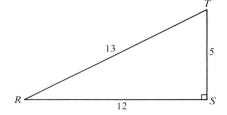

Solution $\sin R = \dfrac{5}{13}$ $\cos R = \dfrac{12}{13}$

≈ 0.38 ≈ 0.92

$\sin T = \dfrac{12}{13}$ $\cos T = \dfrac{5}{13}$

≈ 0.92 ≈ 0.38

Example 1 illustrates the facts that sin $A =$ cos $(90 - A)$ and cos $A =$ sin $(90 - A)$. A table of trigonometric ratios for angles 0° to 90° is on page 642.

The sine and cosine ratios can be used to solve problems involving right triangles.

Example 2 Find the lengths of sides ST and RS to the nearest tenth.

Solution Solve for r.

$$\sin 35° = \frac{r}{18}$$

$$0.5736 \approx \frac{r}{18}$$

$$r \approx 10.3$$

Solve for t.

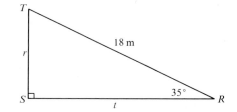

$$\cos 35° = \frac{t}{18}$$

$$0.8192 \approx \frac{t}{18}$$

$$t \approx 14.7$$

Answer The length of side ST is about 10.3 m.
The length of side RS is about 14.7 m.

Example 3   A ladder is needed to reach a window that is 15 ft above the ground. How long must the ladder be if the angle that it makes with the ground is 60°?

Solution $\sin 60° = \dfrac{15}{x}$

$0.8660 \approx \dfrac{15}{x}$

$0.8660x \approx 15$

$x \approx 17.3$

Answer   The ladder must be at least 17.3 ft long.

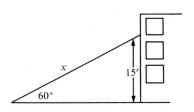

■ CLASSROOM EXERCISES

1. Compute sin A, cos A, sin B, and cos B to the nearest hundredth.

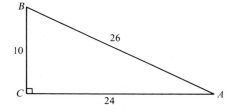

Find the length x to the nearest tenth.

2.

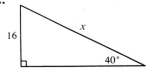

3.

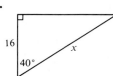

4.

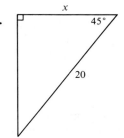

■ WRITTEN EXERCISES

Compute sin A, cos A, sin B, and cos B to the nearest hundredth.

A   **1.**

2.

3.

Find the length x to the nearest tenth.

4.

5.

6.

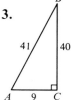

Find the measures of angle A and angle B to the nearest degree.

7.

8.

9.

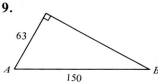

Solve. Give each answer to the nearest tenth.

10. A 15-ft ladder leaning against a building forms a 60° angle with the ground. How high above the ground does the ladder touch the building?

11. A 12-ft ladder leaning against a wall forms a 65° angle with the ground. How far from the wall is the foot of the ladder?

12. The diagonal of a rectangle forms a 35° angle with the longer side. How long is the diagonal if the longer side of the rectangle is 150 cm long?

Find the lengths x and y to the nearest tenth.

B **13.**

14.

15.

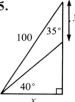

Solve. Give each answer to the nearest tenth.

16. The diagonal of a rectangle forms a 25° angle with the longer side. If the rectangle is 100 cm long, what is the perimeter of the rectangle?

17. There are two routes from A to B, as shown in the figure. The direct route is 100 mi long at an angle of 35° south of east. The other route is east to point C, then south to B. How much longer is the route by way of C?

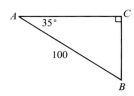

Find the measures of angles α and β to the nearest degree.

C **18.**

19.

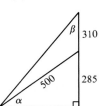

20.

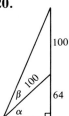

Using Trigonometric Ratios

If the measures of an acute angle and one side of a right triangle are known, the other measures can be computed.

Example 1 Find the length of side *AC*.

Solution Let x = the length of side *AC* in meters.

$$\tan B = \frac{x}{5}$$

$$\tan 30° = \frac{x}{5}$$

$$0.5774 \approx \frac{x}{5}$$

$$x \approx 2.9$$

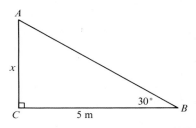

Answer The length of side *AC* is about 2.9 m.

Example 2 Find the missing measures of the angle and sides in the triangle below.

Solution *Find angle A.* Measure of angle $A = 90° - 35° = 55°$

Find side BC.

$$\tan A = \frac{a}{18}$$

$$\tan 55° = \frac{a}{18}$$

$$1.4281 \approx \frac{a}{18}$$

$$a \approx 25.7$$

Find side AB.

$$\sin B = \frac{18}{c}$$

$$\sin 35° = \frac{18}{c}$$

$$0.5736 \approx \frac{18}{c}$$

$$0.5736c \approx 18$$
$$c \approx 31.4$$

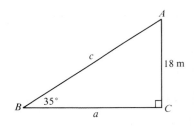

Answer The measure of angle *A* is 55°. Side *BC* is about 25.7 m long. Side *AB* is about 31.4 m long.

Example 3   Find the area of a right triangle in which one acute angle measures 40° and the side opposite that angle is 30 m long.

Solution   The area of a right triangle is equal to one half the product of the lengths of the perpendicular sides.

Find x.

$$\tan 40° = \frac{30}{x}$$

$$0.8391 \approx \frac{30}{x}$$

$$x \approx 36$$

Find the area.

$$A \approx \frac{1}{2} \cdot 30 \cdot 36$$

$$\approx 540$$

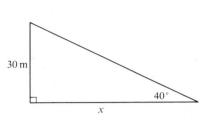

Answer   The area of the triangle is about 540 m².

■ CLASSROOM EXERCISES

1. Find the missing measures to the nearest tenth.

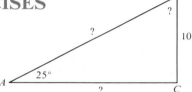

2. A flagpole casts a shadow 30 m long when the angle of elevation of the sun is 60°. How tall is the flagpole?

■ WRITTEN EXERCISES

Find sin *A*, cos *A*, and tan *A* to the nearest hundredth.

A   **1.**

2.

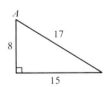

3.

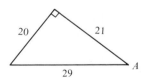

Find α and compute *a* and *b* to the nearest tenth.

4.

5.

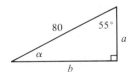

6.

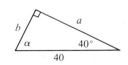

Find α and compute b and c to the nearest tenth.

7.

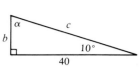

8.

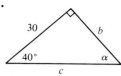

9.

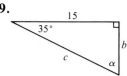

Solve. Give each answer to the nearest whole number.

10. From a point 20 mi from the base of a mountain, the angle of elevation to the summit of the mountain is 5°. How many miles is the summit above the base of the mountain?

11. A scout found an angle of elevation to the tip of a tree to be 40° at a point 50 ft from the tree. What is the height of the tree?

12. A 50-ft supporting cable attached to the top of a 41-ft pole is anchored to the ground. What angle does the cable make with the ground?

13. The base of a 25-ft ladder leaning against a building is 16 ft from the building. What angle does the ladder make with the ground?

14. In a right triangle, the side opposite a 35° angle is 28 cm. What is the area of the triangle?

15. A side 18 cm long is opposite a 20° angle in a right triangle. What is the area of the triangle?

True or false? Refer to triangle ABC.

B **16.** $\sin A = \dfrac{a}{c}$

17. $\sin A = \cos B$

18. $(\sin A)^2 + (\cos A)^2 = 1$

19. Area $= \dfrac{ac \cdot \sin B}{2}$

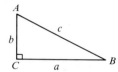

Find the missing length in each right triangle. Then compute $\sin A$ to the nearest hundredth.

20.

21.

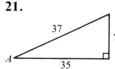

22.

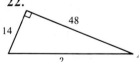

Find the measure of α to the nearest degree.

C **23.**

24.

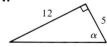

25.

■ EXTRA PRACTICE

Chapter 1

1–1

j	k	K	r	R
$\frac{1}{3}$	$\frac{1}{2}$	2	0	3

Substitute and simplify.

1. $k \times K$ **2.** $r \times R$ **3.** $R - K$ **4.** $k - r$ **5.** $j \times k$

6. $\frac{1}{4} \times j$ **7.** $\frac{1}{4} + k$ **8.** $R - k$ **9.** $\frac{r}{k}$ **10.** $\frac{R}{j}$

w	y	t	s	r
4	16	0.25	12	0.15

Substitute and simplify.

11. $\frac{w}{y}$ **12.** $\frac{y}{w}$ **13.** $r - t$ **14.** $w - r$ **15.** $\frac{s}{w}$

16. $\frac{y}{t}$ **17.** $1.3 - r$ **18.** $\frac{s}{0.6}$ **19.** $1.5 \times w$ **20.** $y \times y$

1–2   Write $+$, $-$, $\times$, or $\div$ to tell which operation is done first.

1. $7 \times 9 + 10$ **2.** $7 + 9 \times 10$ **3.** $\frac{5 + 10}{3}$ **4.** $6 + 4 \div 2$

5. $(7 + 3) \times 9$ **6.** $9 \times (10 - 2)$ **7.** $12 \div (6 - 2)$ **8.** $12 \div 6 - 2$

Simplify.

9. $(7 + 5) \times 3$ **10.** $30 \div (6 - 1)$ **11.** $5 + 3 \times 10$ **12.** $(48 - 8) \div 5$

13. $(18 - 3) \times 2$ **14.** $36 + 6 \div 3$ **15.** $(6.8 - 1.8) \times 3$ **16.** $4.2 + 2.4 \div 6$

17. $(4.3 + 2.1) \times 5$ **18.** $\frac{12 + 18}{6 + 4}$ **19.** $\left(\frac{7}{8} - \frac{1}{4}\right) \times \frac{1}{2}$ **20.** $8 + 12 \div \frac{1}{2}$

1–3   Evaluate for $a = 12$.

1. $5(a + 3)$ **2.** $7(a - 4)$ **3.** a^2 **4.** $3a$ **5.** $\frac{1}{4}a$

6. $2a - 7$ **7.** $3a + 4$ **8.** 1^a **9.** $(a - 2)^3$ **10.** $0.4a$

x	y	a	b	c
$\frac{1}{3}$	$\frac{1}{2}$	0	2	6

Substitute and simplify.

11. bc **12.** $(c - b)y$ **13.** $xc + 5$

14. $c\left(x + \frac{2}{3}\right)$ **15.** $(10 - c)y$ **16.** $(b + c)^2$

17. $b^4 + c$ **18.** $xyabc$ **19.** $c(y - x)$

For each pair of expressions, substitute numbers for the variable and simplify the expressions. State whether you think the expressions in each pair are equivalent.

1. $2x + x$
$3x$

2. $(2a)^2$
$2a^2$

3. $7 + r$
$r + 7$

4. $7(8y)$
$(7 \cdot 8)y$

5. $9(b - 3)$
$9b + 6$

6. $20 - (s - 4)$
$(20 - s) - 4$

7. $50 - (z - 10)$
$(50 - z) + 10$

8. $(c + 1)^2$
$c^2 + 2c + 1$

9. $t \cdot 6$
$6t$

10. $12(8 + k)$
$96 + 12k$

11. $1 \cdot m$
m

12. d^3
$d^2 + d$

13. $(j + 2)(j + 3)$
$j^2 + 6$

1–5 Name the property illustrated in each equation.

1. $(5 + 3) + 0 = 5 + 3$

2. $(1 + 5) + 2 = 2 + (1 + 5)$

3. $4 + (1 + 8) = (4 + 1) + 8$

4. $7(9 + 6) = 63 + 42$

5. $24(5a) = (24 \cdot 5)a$

6. $13(7r) = (7r) \cdot 13$

7. $10(1.3 + 0.5d) = 13 + 5d$

8. $1q = q$

9. $(0 + m) + 6 = m + 6$

10. $12r + 36 = 12(r + 3)$

11. $20(5n) = (20 \cdot 5)n$

12. $(z + 36) + 64 = z + (36 + 64)$

1–6 Simplify each expression.

1. $3p + 7p$

2. $4r - r$

3. $2(m + 3) - 1$

4. $6(3z) - 3z$

5. $3(c + 9) - 4$

6. $12y - 10y$

7. $5(n + 2) - 2n$

8. $2(r + s) - 2r$

9. $7 - g - 5$

10. $2.5z - 2 - 0.5z$

11. $2.3w - 2w - x$

12. $\frac{1}{6}c + \frac{5}{6}c$

13. $2(5a) + 3(4a)$

14. $7h + 8h + 9$

15. $7(10b) - 2(3b)$

16. $5 + 6p + 3p$

17. $w + 3(w - 1)$

18. $12j + 6j + j$

19. $k + 10k + 12k$

20. $4(x + y) - 4y$

1–7 Write an expression for each.

1. Mr. Adams is 3 years older than Mrs. Adams.

   a. Let x be Mr. Adams' age in years. What is Mrs. Adams' age?
   b. Let y be Mrs. Adams' age in years. What is Mr. Adams' age?

2. A rectangle is 4 times as long as it is wide.

   a. Let w be the width. What is the length?
   b. Let l be the length. What is the width?

3. There are c coins in a case.

   a. If the average weight of a coin is 4.3 g, what is the total weight of the coins?
   b. If the total weight of the coins is 128 g, what is the average weight of one coin?

2–1 Replace ⑦ with $<$, $>$, or $=$ to make a true statement.

1. $+4.7$ ⑦ -5.1 **2.** $-2\frac{1}{2}$ ⑦ $-\frac{5}{2}$ **3.** -1.7 ⑦ -1.8

4. $-\frac{3}{8}$ ⑦ $-\frac{5}{7}$ **5.** $-\frac{5}{4}$ ⑦ $-\frac{4}{5}$ **6.** $\frac{3}{4}$ ⑦ $\frac{2}{3}$

7. -2.5 ⑦ -1.9 **8.** -4.9 ⑦ 3.2 **9.** $\frac{5}{4}$ ⑦ $-1\frac{1}{4}$

Simplify.

10. $-(-4.7)$ **11.** $|9.2|$ **12.** $|-2.5|$ **13.** $-|0.5|$

14. $-|-6.2|$ **15.** $-(-5.3)$ **16.** $|0.6|$ **17.** $|-19.3|$

18. $-|7|$ **19.** $-\left|-\frac{7}{2}\right|$ **20.** $-\left|\frac{4}{3}\right|$ **21.** $|-12.8|$

2–2 Simplify.

1. $5 + (-2)$ **2.** $-9 + (-9)$ **3.** $7 + (-10)$ **4.** $-12 + (-6)$

5. $-16 + (-23)$ **6.** $12 + (-4)$ **7.** $-28 + (-75)$ **8.** $15 + (-45)$

9. $-4.93 + (-2.85)$ **10.** $4.75 + (-5.5)$ **11.** $-9.53 + 2.2$

12. $-6.4 + 8.2$ **13.** $1.7 + (-1.2)$ **14.** $-5.4 + 1.18$

15. $\frac{1}{2} + \left(-\frac{2}{3}\right)$ **16.** $\frac{5}{3} + \left(-\frac{3}{4}\right)$ **17.** $-\frac{1}{3} + \frac{1}{2}$

18. $1\frac{1}{2} + \left(-3\frac{1}{4}\right)$ **19.** $2\frac{5}{8} + \left(-1\frac{5}{6}\right)$ **20.** $-5\frac{3}{4} + 1\frac{2}{3}$

21. $-10 + (-4) + 3$ **22.** $-1.3 + 5.4 + (-1.2)$

23. $-\frac{2}{3} + \left(-\frac{1}{2}\right) + \left(-\frac{1}{3}\right)$ **24.** $-\frac{5}{4} + \left(-\frac{2}{3}\right) + \frac{1}{4}$

25. $|1.2 + (-3.6)| + (-2)$ **26.** $|-2.8 + 5.3| + 7$

2–3 Simplify.

1. $-3 - (-10)$ **2.** $43 - 27$ **3.** $26 - 30$

4. $-43 - 20$ **5.** $-1.5 - (-2.5)$ **6.** $3.7 - 4.2$

7. $6.4 - (-4.2)$ **8.** $-17.2 - 12.8$ **9.** $\frac{1}{2} - \left(-\frac{1}{2}\right)$

10. $-\frac{2}{3} - \frac{2}{3}$ **11.** $\frac{3}{4} - \left(-\frac{1}{2}\right)$ **12.** $-\frac{2}{3} - \frac{1}{2}$

13. $-5 - 2 - 6$ **14.** $-7 + 10 - 2$ **15.** $8 - 10 - 5$

16. $-7 - 6 + 2$ **17.** $12 - (8 - 3)$ **18.** $2 - (10 - 7)$

2–4   Simplify.

1. $-4 \cdot 7$ 2. $-9 \cdot -3$ 3. $10 \cdot -4$

4. $-1.5 \cdot -4$ 5. $\frac{1}{2} \cdot -\frac{1}{3}$ 6. $-\frac{3}{4} \cdot -\frac{2}{3}$

w	x	y	z
$-\frac{5}{4}$	$\frac{2}{3}$	-12	-1

Substitute and simplify.

7. wx 8. wy 9. $wxyz$ 10. $(-x)y$

11. $w(9 + z)$ 12. $w(9 - z)$ 13. $z(y - z)$ 14. $(yz)(yz)$

2–5   Write each division as a multiplication and simplify.

1. $2 \div 3$ 2. $\frac{3}{4} \div \frac{2}{3}$ 3. $-\frac{1}{2} \div \frac{1}{3}$ 4. $0 \div 4$

5. $-6 \div -3$ 6. $-\frac{1}{2} \div 5$ 7. $2\frac{1}{3} \div -2$ 8. $-1\frac{1}{4} \div -1\frac{1}{8}$

A	r	R	c
-8	-12	$\frac{3}{2}$	$-\frac{2}{3}$

Substitute and simplify.

9. $\frac{r}{A}$ 10. $\frac{A}{r}$ 11. $\frac{R}{r}$ 12. $\frac{r}{R}$

13. $\frac{A + r}{R}$ 14. $\frac{A - r}{c}$ 15. $\frac{Ar}{Rc}$ 16. $\frac{r}{R + c}$

2–6   Simplify.

1. $6 + (-a)$ 2. $5 - (-2b)$ 3. $-c + (-2c)$ 4. $-2c - (-c)$

5. $t + (-r) + s$ 6. $7 - (-x) - y$

7. $5 + (-p) + (-q)$ 8. $4 - (-a) - (-b)$

2–7   State the property illustrated in each equation.

1. $p + q = q + p$ 2. $(2k)r = 2(kr)$ 3. $5(m + n) = 5m + 5n$

4. $3a + 0 = 3a$ 5. $t + (-t) = 0$ 6. $(2 + r) + t = 2 + (r + t)$

Simplify.

7. $3t + t$ 8. $4r - r$ 9. $2a + 3 + 7a$

10. $3z - 7 - 7 - 3z$ 11. $-5t + v + t + v$ 12. $14w + 6 + 6w - w$

13. $b + 3a + 5 + a$ 14. $b + 2a + 2b - 4$ 15. $4 + 7v - 16w - 6v - w$

2–8 Write an equivalent expression without parentheses.

1. $3(c + 10)$ **2.** $-4(t + 8)$ **3.** $7(t - 9)$ **4.** $-8(b - 3)$

5. $2(3r - 7)$ **6.** $6(a + 5)$ **7.** $-1(4c + 2)$ **8.** $-4(b + 1)$

Simplify.

9. $2(q + 3) + 4$ **10.** $3(h - 6) + 9$ **11.** $-2(k - 3) + 10$

12. $-4(2 - j) - 3$ **13.** $5(x + 2) + x$ **14.** $3(y - 4) + 2y$

15. $-8a + 3(2a + 3)$ **16.** $3a - 5(a - 2) + 7$ **17.** $-4 + 6(b + 2) - b$

18. $2(c + 3) + 4(c + 5)$ **19.** $6(p - 2) + 2(2p + 3)$

20. $4(q + 3) - 3(q - 1)$ **21.** $5(r + 2) - (r - 10)$

2–9 Make a table to solve each problem.

1. Andrew Rose has 20 coins consisting of nickels and dimes. How many of the coins are nickels if the total value of the coins is $1.30?

2. Machine A produces 8 units per hour and machine B produces 10 units per hour. Machine A starts producing at 9:00 A.M. and machine B starts producing 2 h later. At what time will they have produced a total of 115 units?

3. A painter can paint $\frac{1}{30}$ of a house in 1 h. The painter's assistant can paint $\frac{1}{60}$ of the house in 1 h. Working together, how long would it take them to paint the house?

4. What number added to the numerator and to the denominator of $\frac{2}{11}$ makes a new fraction equal to $\frac{1}{2}$?

5. Russ Fullbright's light truck goes 25 mi on 1 gal of gasoline. He drives at a rate of 55 mph. How many hours can he drive using 11 gal of gasoline?

6. Mama's Deli makes a profit of $225 per day on Fridays and Saturdays. It makes a $175 profit on Mondays and Wednesdays. On Tuesdays and Thursdays it makes a $50 profit. What is the profit for a 50-week period?

Chapter 3

3–1 Which of the numbers 0, 1, 2, 3, and 4 are solutions of these equations?

1. $2x + 5 = 13$ **2.** $-x + 3 = 1$ **3.** $3x + 1 = 1$

4. $3(x + 1) = 6$ **5.** $x^2 + 2 = 3x$ **6.** $x^2 + 12 = 7x$

7. $2 - (x - 1) = 0$ **8.** $4 - (3 - x) = 5$ **9.** $x^3 + 4x = 5x^2$

Solve these equations for the replacement set $\{-2, -1, 0, 1, 2\}$.
Write $\emptyset$ if there is no solution.

10. $5 - x = 6$ **11.** $2x - 1 = 5$ **12.** $10x + 5 = 25$

13. $x - 10 = -8$ **14.** $3x - x = 2$ **15.** $7 - x = 7$

16. $5x = 0$ **17.** $x^2 + x = 2$ **18.** $x^2 = x + 2$

What equations result from following these directions in order?

1. $-3x + 2 = 10$

   a. Subtract 2 from both sides.

   b. Multiply both sides by $-\frac{1}{3}$.

3. $2x + 3x = 8$

   a. Combine the x-terms.

   b. Divide both sides by 5.

2. $\frac{3}{4}x - 3 = -15$

   a. Multiply both sides by 4.

   b. Add 12 to both sides.

   c. Divide both sides by 3.

4. $4x + 3 + 2x = 7$

   a. Add -3 to both sides.

   b. Combine the x-terms.

   c. Multiply both sides by $\frac{1}{6}$.

5. $\frac{1}{2}x + 2 + \frac{1}{3}x = 2\frac{1}{6}$

   a. Combine the x-terms.

   b. Subtract 2 from both sides.

   c. Multiply both sides by $\frac{6}{5}$.

6. $x - 5 - 4x = -7$

   a. Add 5 to both sides.

   b. Combine the x-terms.

   c. Multiply both sides by $-\frac{1}{3}$.

3–3 Solve each equation by writing a series of equivalent equations.

1. $x + 8 = -3$ **2.** $y - 4 = 10$ **3.** $2z = z + 3$

4. $6a = 5a - 4$ **5.** $7 + b = 2b$ **6.** $6 + 3p = 4p$

7. $5r + 2 = 4r + 12$ **8.** $7s - 5 = 6s - 1$ **9.** $-7 + s = 2s + 3$

10. $x + \frac{1}{3} = \frac{5}{3}$ **11.** $x - \frac{1}{2} = 5$ **12.** $4v + \frac{1}{2} = 3v + 1$

13. $-\frac{1}{2}w + \frac{4}{5} = \frac{1}{2}w + \frac{1}{5}$ **14.** $\frac{1}{3}y + 1 = \frac{4}{3}y + \frac{1}{6}$

3–4 Solve.

1. $4c = 7$ **2.** $-3k = 16$ **3.** $\frac{1}{4}p = -3$ **4.** $-\frac{1}{3}r = 9$

5. $\frac{v}{4} = 1.3$ **6.** $0.3s = 30$ **7.** $\frac{5}{8}w = \frac{3}{4}$ **8.** $5a = -15$

9. $\frac{y}{2.5} = -20$ **10.** $-7b = -21$ **11.** $0.4z = 5$ **12.** $-\frac{1}{8}m = -8$

13. $1.5t = 21$ **14.** $\frac{4}{3}d = 12$ **15.** $-\frac{1}{3}c = -\frac{1}{2}$ **16.** $\frac{k}{0.3} = 4$

3–5 Solve.

1. $4x - 3 = 2$ **2.** $\frac{2x + 5}{3} = 4$ **3.** $\frac{1}{2}x - 2 = 3$ **4.** $6x + 2 = 2x + 10$

5. $3 + \frac{1}{3}x = 4$ **6.** $3x - 5 = 9$ **7.** $4x + 7 = 2$ **8.** $\frac{x - 1}{2} = 12$

3–6 Solve.

1. $4 - 3x = 13$ **2.** $2 - 4x = -10$ **3.** $-5 - 6x = 7$ **4.** $7 - \frac{1}{5}x = 9$

5. $-2 - \frac{1}{4}x = 2$ **6.** $-10 - \frac{2x}{3} = -4$ **7.** $\frac{x}{4} - 2 = 3$ **8.** $4x - 5 = 20$

9. $4(2 - x) = 3$ **10.** $3(x - 4) = 1$ **11.** $-2(x - 5) = 3$ **12.** $-6(3 - x) = 5$

3–7 Solve.

1. $3(t - 4) = 15$ **2.** $6(v + 2) = -12$ **3.** $\frac{1}{3}(w + 4) = 4$

4. $\frac{1}{5}(y - 10) = 3$ **5.** $-\frac{1}{2}(z + 4) = -3$ **6.** $-\frac{1}{4}(a - 6) = 1$

7. $2c + 3(c + 2) = -4$ **8.** $7s + 2(6 - s) = 7$

9. $5x - 3(x - 2) = 18$ **10.** $3(b + 1) + 2(b + 3) = 34$

3–8 Solve the literal equation for the variable indicated.

1. $pq = r$, for p **2.** $3 + a = b$, for a

3. $p - 4 = q$, for p **4.** $\frac{2}{3}t = d$, for t

5. $7jk = 28$, for k **6.** $2x + 3y = 12$, for y

7. $2x - 3y = 12$, for x **8.** $a(2 + b) = 7$, for b

9. $(3 + r)s = 2$, for r **10.** $(7 - t)2 = v$, for t

11. $2(t + 3) = 3(r + 5)$, for r **12.** $4(w - z) = z$, for w

13. $2j + 3k = 3j - 3k$, for j **14.** $\frac{7 - 2y}{3} = x - 1$, for y

3–9 For each problem, write (a) what the variable represents, (b) an equation that fits the problem, (c) the solution to the equation, and (d) the answer to the question.

1. The length of a rectangle is 5 cm more than its width. What is the width if the length is 6.8 cm?

2. Lance has 4 times as many dimes as quarters. How many quarters does he have if he has 44 dimes?

3. Amy is 4 cm taller than Christa. How tall is Amy if Christa is 165 cm tall?

4. Mark saved $\frac{1}{5}$ the price of a camera. He saved $24. What is the cost of the camera?

5. Sabrina scored 8 points more on the second test than she scored on the first test. What was her score on the first test if she scored 98 points on the second test?

6. Matt is 6 lb heavier than Scott. How much does Scott weigh if Matt weighs 150 lb?

3–10 For each problem, write (a) what the varible represents, (b) a figure that helps organize the information, (c) an equation that fits the problem, (d) the solution to the equation, and (e) the answer to the question.

1. A wire 100 cm long is bent to form a triangle. The middle-sized side of the triangle is 2 cm longer than one side and 6 cm shorter than the other side. What are the lengths of the three sides?

2. A 12-ft board is cut so that one piece is 5 ft longer than the other piece. How long is each piece?

3. The body of a fish is 4 times the length of its head. Its tail is 2 cm longer than the head. The total length of the fish is 20 cm. How long are the head, body, and tail of the fish?

4. The perimeter of a triangle is 50 cm. The longest side is twice the length of the shortest side. The other side is 5 cm shorter than the longest side. Find the length of each side.

Chapter 4

4–1 Add these polynomials.

1. $5a + 6b - 2c$
   $2a - 8b - 4c$

2. $3r + 7s$
   $\phantom{3r + {}}5s + 4t$

3. $x^2 - 2x - 9$
   $x^2 - 5x + 4$

Subtract these polynomials.

4. $5x^2 + 6x + 7$
   $x^2 + 2x + 9$

5. $a + b - c$
   $a - b + c$

6. $3a^2 - 6a - 9$
   $a^2 + 4a + 1$

Simplify each sum or difference.

7. $(x^2 - x - 10) - (x^2 + x - 20)$

8. $(p - q - r) - (2p - 2q - 2r)$

9. $(2a - 3b + 4c) + (a + b - c)$

10. $(7x + 10y + z) + (3x - 10y + z)$

11. $(a^2 + 5a + 8) - (5a - 2)$

12. $(t^2 - 16t) - (5t + 4)$

4–2 State whether each polynomial is a monomial, binomial, or trinomial.

1. $2a + 3b$

2. $7rst$

3. $2 + c$

4. $a^2 + a + 1$

5. $p + q - r$

6. $0.5x^2$

7. $a^4 - a^3$

8. $5x^2 - 5xy + y^2$

State the degree of each polynomial.

9. $5 + 4x$

10. $2t^3 + 4t$

11. $-2x^2 + 3x + 7$

12. 7

13. $2x^2y^3$

14. $xy + x - y$

15. $r^2 + rs^2$

16. $2a^2 - 3ab + 4b^2$

Write each polynomial in descending order.

17. $2 + 3a + 4a^2$

18. $7a - 6a^2 - 4$

19. $r + 2 + r^2$

20. $7x^2 - 2 - x$

21. $4p + 5p^2 + 1$

22. $7 - a - 2a^2$

23. $x^2 + x^3 + 4x$

24. $r^2 + 2r^3 - r^4$

4-3 Complete.

1. $5^7 \cdot 5^6 = 5^?$

2. $(0.7)(0.7)^2 = (0.7)^?$

3. $12 \cdot 12 = 12^?$

4. $6^? \cdot 6^3 = 6^{12}$

5. $(-4)^2(-4)^3 = (-4)^?$

6. $\left(\frac{1}{2}\right)^{10}\left(\frac{1}{2}\right)^{20} = \left(\frac{1}{2}\right)^?$

Simplify.

7. $p^3 \cdot p^2$

8. $q^5 \cdot q$

9. $(2r)(3r)$

10. $(6s^2)(5s^4)$

11. $(0.5t)(7t)$

12. $\left(\frac{1}{2}v^2\right)\left(\frac{1}{2}v^3\right)$

13. $(3w^2)(w^{10})$

14. $xy \cdot x^2y^3$

15. $(2yz^2)(5z^3)$

4-4 Write each product in scientific notation.

1. $(3 \times 10^2)(3.1 \times 10^4)$

2. $(2 \times 10^4)(8 \times 10^5)$

3. $(2.5 \times 10)(6 \times 10^8)$

4. $(7.3 \times 10^4)(4 \times 10^5)$

5. $(8.2)(2 \times 10^6)$

6. $(4 \times 10^9)(5)$

7. $(2 \times 10^4)^3$

8. $(3 \times 10^5)^2$

9. $(4 \times 10^6)^3$

10. $2,500 \times 50,000$

11. $73,000,000 \times 2,000$

12. $8,500,000 \times 4,000,000$

13. $340 \times 2,000,000$

4-5 Simplify.

1. $(c^4)^3$

2. $(2k)^4$

3. $(3m)^2$

4. $(z^4)^5$

5. $(-2x^3)^2$

6. $(-2x^2)^3$

7. $\left(\frac{1}{2}a^4\right)^4$

8. $(r^3s^4)^5$

9. $(2a^2b^3)^4$

10. $(0.2jk^2)^2$

11. $(5m^2n)^4$

12. $(10rs^2t^3)^3$

State which number is larger.

13. $(3^3)^4$ or $3^3 \cdot 3^4$

14. 6^{10} or $(6^7)^3$

15. 10^{10} or 100^1

16. $\left(\frac{1}{3} \cdot 3\right)^4$ or $\frac{1}{3} \cdot 3^4$

17. $(-2)^4$ or $(-2)^5$

18. $\left(\frac{1}{2}\right)^{10}$ or $\left(\frac{1}{2}\right)^{11}$

4-6 Simplify.

1. $8(a + 3)$

2. $-7(b - 4)$

3. $\frac{1}{2}(c + 6)$

4. $5(2d + 3)$

5. $10(0.1m - 2.4)$

6. $-1(-p - 2)$

7. $t(t + 3)$

8. $v(v^2 - 3)$

9. $w(6w - 3)$

10. $2x(x - 4)$

11. $3y(x + y)$

12. $-z(2 - z)$

13. $2a(a - 3b)$

4–7 The distances for A and B are equal. Write an equation for each problem, and solve for x.

   1. A: Rate is x km/h, and time is 17 h.
      B: Rate is $(x + 20)$ km/h, and time is 13 h.

   2. A: Rate is 800 km/h, and time is x h.
      B: Rate is 750 km/h, and time is $(x + 1)$ h.

The sum of the distances for A and B is 400 km. Write an equation for each problem, and solve for x.

   3. A: Rate is 50 km/h, and time is x h.
      B: Rate is 40 km/h, and time is $(x - 3.5)$ h.

   4. A: Rate is 600 km/h, and time is x h.
      B: Rate is 500 km/h, and time is $2x$ h.

4–8 Simplify. Assume that no variable is equal to zero.

1. $\dfrac{p^{12}}{p^4}$ **2.** $\dfrac{6r^8}{3r^6}$ **3.** $\dfrac{24t^4}{-18t^6}$ **4.** $\dfrac{-14jk^2}{21j^2k}$

Solve.

5. $\dfrac{7^6}{7^2} = 7^x$ **6.** $\dfrac{3^{10}}{3^2} = 3^x$ **7.** $\dfrac{2^{20}}{2^{17}} = x$ **8.** $\dfrac{5^6}{5^8} = x$

Write the quotient in scientific notation.

9. $\dfrac{9 \times 10^8}{2 \times 10^2}$ **10.** $\dfrac{4.8 \times 10^6}{3.0 \times 10^5}$ **11.** $\dfrac{9.6 \times 10^{12}}{4.8 \times 10}$ **12.** $\dfrac{2 \times 10^{15}}{5 \times 10^{10}}$

4–9 Solve.

   1. The cost of 5 soccer balls including a $3.75 delivery charge is $128.70. What is the cost of 1 soccer ball not including a delivery charge?

   2. Ranson's Rapid Repair charges for parts plus $25 per hour for labor. A vacuum-cleaner repair bill was $16.25 including $3.75 for parts. How many hours of labor were needed?

   3. The four members of the Peterson family all ordered the Breakfast Special at the Blue Diner. They paid $2.20 for beverages not included in the Breakfast Special. Their bill, including the beverages, was $11.20. What was the cost of one Breakfast Special?

4–10 Let n represent an integer. Express each of these numbers.

   1. The next integer

   2. The preceding integer

   3. The preceding odd number if n is odd

   4. The next three consecutive integers

Solve.

   5. The sum of three consecutive integers is 369. What is the smallest integer?

   6. The sum of four consecutive even integers is 348. What are the integers?

EXTRA PRACTICE

Chapter 5

5-1 Identify the quadrant that each point is in or the axis that it is on. Then state the coordinates of each point.

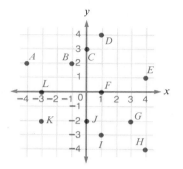

1. A	**2.** B	**3.** C	**4.** D
5. E	**6.** F	**7.** G	**8.** H
9. I	**10.** J	**11.** K	**12.** L

Graph the points P, Q, R, and S. Draw line segments from P to Q, Q to R, R to S, and S to P.

13. $P(1, 2)$, $Q(-1, 1)$, $R(0, -1)$, $S(2, 0)$

14. $P(2, 1)$, $Q(6, 2)$, $R(5, 4)$, $S(1, 3)$

15. $P(-1, -1)$, $Q(-2, 3)$, $R(2, 3)$, $S(1, -1)$

16. $P(4, 0)$, $Q(-2, -3)$, $R(-3, -1)$, $S(3, 2)$

5-2 Each exercise describes a relation.

a. List the ordered pairs. **c.** List the range.
b. List the domain. **d.** Draw the graph.

1. The set of ordered pairs in which the components are integers and the product of the components is 6.

2. The set of ordered pairs in which the components are positive integers having a product of 6 or less.

3. The set of ordered pairs in which the components are negative integers having a sum greater than -6.

4. The set of ordered pairs in which the first component is a positive integer less than 6, and the second component is 1 less than the first component.

5. The set of ordered pairs in which the first component is a positive even integer less than 10, and the second component is half the first component.

5-3 Copy and complete each table.

1. $y = 4 - x$

x	0	2	4	6
y	?	?	?	?

2. $y = -2x + 4$

x	-1	0	1	2
y	?	?	?	?

3. $2x + 3y = 24$

x	?	?	0	12
y	4	6	?	?

4. $2x - 3y = 24$

x	6	9	?	?
y	?	?	0	1

5. $4x - 3y = 5$

x	1	?	0	?
y	?	0	?	1

6. $6x + 2y = -3$

x	0	?	2	?
y	?	0	?	2

5–4 Graph each equation.

1. $x + y = 4$ 2. $x + y = -4$ 3. $x - y = -2$

4. $xy = 8$ 5. $xy = -6$ 6. $2x - 4y = 5$

7. $3y - 4x = 24$ 8. $y = 0.1x^3$ 9. $y = -x$

10. $y = 0x + 2$ 11. $y = -x^2$ 12. $y = 2(x + 1)$

5–5 Graph the point. Then draw lines with the stated slopes through the point. Label the lines a, b, and c.

1. Point $(4, 2)$ a. Slope 1   b. Slope 2   c. Slope $\frac{1}{4}$

2. Point $(1, -3)$ a. Slope 0   b. Slope -3 c. Slope $\frac{1}{2}$

3. Point $(-3, 3)$ a. Slope $-\frac{1}{3}$ b. Slope -1 c. Slope -2

5–6 Write an equation in slope–intercept form. State the slope and y-intercept of the line.

1. $2x - 3y = 12$ 2. $\frac{1}{2}y = x + 4$ 3. $4y - 6x = 9$

4. $10y + x = 20$ 5. $y - \frac{3}{4}x = 7$ 6. $\frac{1}{3}y = \frac{1}{2}x - 1$

7. $5x + 2y = 10$ 8. $-2x - 3y = 6$

State whether the graphs of these pairs of equations are parallel, perpendicular, or neither.

9. $x + y = 6$
$y = -x + 3$

10. $2x - 4y = 13$
$4x - 8y = 5$

11. $2y - 3x = 4$
$3y + 2x = 6$

12. $5x + 6y = 30$
$4x + 5y = 20$

5–7 Write an equation in the form $y = mx + b$ for the line with the given characteristics.

1. Has slope 2 and passes through $(3, 1)$

2. Has slope 0 and passes through $(1, -5)$

3. Has no slope and passes through $(-2, -6)$

4. Has slope $-\frac{1}{2}$ and passes through $(8, 3)$

5. Has slope -10 and passes through $(3, 3)$

Write an equation in the form $y = mx + b$ for the line that passes through the two points.

6. $(6, 6), (12, 10)$ 7. $(6, -6), (3, -5)$ 8. $(4, 1), (-4, -1)$

9. $(-2, 10), (-1, 6)$ 10. $(4, 1), (-2, 4)$ 11. $(8, 12), (-4, 3)$

5–8 State whether the equation defines a function.

    1. $y = x^2$ **2.** $x = y^2$ **3.** $x = |y|$ **4.** $y = |x|$

    5. $y = 2x$ **6.** $x = 2y$ **7.** $y = (x - 1)^2$ **8.** $xy = 1$

State whether the equation defines a *linear* function.

    9. $y = \frac{1}{2}x^2$ **10.** $y = 2x$ **11.** $y = x + 2$ **12.** $y = \frac{2}{x}$

    13. $y = x^3$ **14.** $y = \frac{1}{3}x$ **15.** $y = 3 - x$ **16.** $y = -\frac{3}{x}$

5–9 For the linear functions, determine the slope, y-intercept, and equation in slope–intercept form.

1.

x	-1	0	1	2
y	7	4	1	-2

2.

x	-1	1	3	5
y	-9	-1	7	15

3.

x	-10	-9	-8	-7
y	4	3	2	1

4.

x	4	8	12	16
y	6	8	10	12

Each table shows the number of spectators at a game and the revenue at the concession stands. State whether the number pairs belong to a *linear* function.

5.

Spectators	500	750	1000
Revenue (in dollars)	750	1125	1500

6.

Spectators	1000	2000	3000
Revenue (in dollars)	2000	3500	4500

Chapter 6

6–1 Solve each system of equations by graphing.

    1. $x + 2y = 4$ **2.** $2x - y = 0$ **3.** $2x + y = 1$ **4.** $y = x + 6$
        $x - y = -5$ $x - y = 2$ $3x + y = 3$

                                                                         $y = \frac{1}{2}x + 4$

    5. $y = \frac{1}{3}x$ **6.** $x + 2y = -4$ **7.** $y = -x - 2$ **8.** $-x + y = 4$
                           $x - y = -4$ $y = 2x - 2$ $-2x - y = 5$

       $y = x + 2$

6–2 Solve by substitution.

    1. $y = 6x$ **2.** $x = -3y$ **3.** $y = 2x - 4$ **4.** $x = y + 3$
        $4x + 3y = 11$ $5x - 6y = 7$ $5x - 2y = 4$ $2x + 2y = -1$

    5. $3x = y - 10$ **6.** $y = 1 - 3x$ **7.** $2x = y - 2$ **8.** $3x + y = 3$
        $6x + 5y = -6$ $4x + y = 4$ $2x - 3y = -9$ $5x + y = 1$

6–3 If possible, solve the system by substituting from the first equation into the second equation. State the number of solutions that each system has.

1. $y = 2x + 3$
   $y = 2x + 5$

2. $y = x + 4$
   $2x = 2x + 8$

3. $x = 2y + 3$
   $2x + 3 = 4y$

4. $x = -2y + 1$
   $3x + 6y = 3$

5. $2y = 3x - 7$
   $2y + 3x = 11$

6. $3x = 2y - 2$
   $y + 6x = -4$

7. $y = -2x$
   $x = -2y$

8. $y = -x + 4$
   $x + 2y = 8$

6–4 For each problem, let t represent the tens digit and u represent the ones digit of a two-digit number. Write two equations. Solve the system and answer the question.

1. The tens digit is 1 more than the ones digit. The tens digit plus twice the ones digit is 13. What is the number?

2. The ones digit is 1 more than the tens digit. If the digits are reversed and the new number is doubled, the result is 41 more than the original number. What is the original number?

Solve.

3. One number is 3 less than another number. Their sum is 53. What are the numbers?

4. One number is 12 more than another number. If twice the smaller number is subtracted from the larger number, the result is 16. What are the numbers?

6–5 Solve by addition.

1. $x + y = 6$
   $x - y = 9$

2. $x - 2y = 6$
   $3x + 2y = -14$

3. $3x + 5y = -1$
   $-5x - 5y = 0$

4. $3x + y = 1$
   $-3x - 2y = 4$

5. $3x + 2y = 2$
   $-3x + 2y = 6$

6. $x = 4y - 2$
   $12y = x + 8$

7. $-2x + 5y = 12$
   $2x + 3y = -12$

8. $x + 8y = 1$
   $-x + 4y = 2$

6–6 Solve each system by multiplication and addition.

1. $2x - 3y = 18$
   $4x - 2y = 4$

2. $x + 3y = 2$
   $2x + 5y = 2$

3. $4x - y = 4$
   $6x - 3y = 3$

4. $3x - y = 7$
   $5x - y = 5$

5. $x - 3y = 5$
   $2x - 9y = 9$

6. $3x + 4y = 9$
   $5x + 8y = 14$

7. $3x + 4y = -15$
   $-2x + 3y = 10$

8. $5x - 4y = 6$
   $10x - 5y = 5$

9. $9x - 8y = 1$
   $3x - 4y = 11$

10. $7x - 4y = 3$
    $8x - 5y = 4$

11. $6x - 2y = 12$
    $-4x - 3y = 18$

12. $9x + 4y = 8$
    $3x + 5y = 43$

6–7 Write two equations. Solve the system, and answer the question.

1. Some $1.50-per-oz spice and some $2.50-per-oz spice are to be mixed to produce 20 oz of a mixture worth $1.90 per oz. How many ounces of each should be used?

2. Bird feed worth 50¢ per lb is mixed with bird feed worth 75¢ per lb to produce 200 lb of a mixture worth 55¢ per lb. How many pounds of each type of bird feed are used?

3. Cheeses costing $2.00 per lb and $2.75 per lb are mixed to produce a 10-lb package worth $21.05. How many pounds of each type of cheese are used?

4. Milk that is 5% butterfat is mixed with milk that is 1% butterfat to produce 50 gal of milk that is 2% butterfat. How much milk of each type is used?

EXTRA PRACTICE

Chapter 7

7–1 List all positive integer factors of the given number.

1. 28	**2.** 45	**3.** 57	**4.** 46
5. 75	**6.** 81	**7.** 64	**8.** 60

Write the prime factorization of each number. List the factors in increasing order.

9. 92	**10.** 48	**11.** 52	**12.** 84
13. 80	**14.** 72	**15.** 68	**16.** 96

7–2 List the common factors of each pair of monomials.

1. $10a$, $15a^2$
2. b^2c, $2bc^2$
3. $3st^2$, $6s^2$
4. ab^3, b^2c
5. $7a$ $8b$
6. $18m^2$, $27mn$

List the greatest common factor of each pair of monomials.

7. $2a^2$, $6a$
8. $4bc$, $10cd$
9. $6e^2g^2$, $9eg^2$
10. $3j^2k$, jk^2
11. $9a$, $4b^2$
12. $24pq^2$, $36q$

7–3 Solve.

1. $(n - 2)(n - 7) = 0$
2. $(m + 4)(m + 9) = 0$
3. $p(p - 10) = 0$
4. $q(q + 8) = 0$
5. $(s - 1)(s + 1) = 0$
6. $(t + 12)(t + 12) = 0$
7. $(2a - 5)(3a - 2) = 0$
8. $(b - 7)(2b - 1) = 0$
9. $(5d + 3)(3d + 5) = 0$
10. $4(c + 7)(2c - 13) = 0$
11. $(s + 4)(s + 4) = 0$
12. $3 \cdot 4 \cdot 5 \cdot x = 0$

7–4 Factor. If the expression cannot be factored, write "not factorable."

1. $5x + 5y$ 2. $a^2 + 3a$ 3. $x^2 + x$ 4. $a^2 + 4$

5. $14x^2 - 21x$ 6. $3a - 4b$ 7. $x^2y^2 - xy^3$ 8. $16 - 16t$

Solve.

9. $t^2 - 4t = 0$ 10. $w^2 + 3w = 0$ 11. $2p^2 - 3p = 0$ 12. $3q^2 - 6q = 0$

13. $z^2 = 10z$ 14. $2c^2 = -3c$ 15. $b = b^2$ 16. $-a = a^2$

7–5 Expand and simplify.

1. $(n + 2)(n + 6)$ 2. $(p + 7)(p - 1)$ 3. $(a - 3)(a - 4)$

4. $(m - 8)(m + 3)$ 5. $(q - 10)(q + 10)$ 6. $(b + 3)^2$

7. $(r - 5)^2$ 8. $(2c + 1)(c + 2)$ 9. $(3v - 4)(3v + 4)$

10. $(d - 4)(3d - 2)$ 11. $(2s + 1)(3s + 2)$ 12. $(2h - 3)^2$

7–6 Factor.

1. $a^2 - 16$ 2. $x^2 - 100$ 3. $c^2 - 1$ 4. $2n^2 - 50$

5. $3p^2 - 12$ 6. $24 - 6t^2$ 7. $25k^2 - 36$ 8. $d^2 - 144$

9. $49m^2 - 4$ 10. $9j^2 - 1$ 11. $p^2q^2 - 81$ 12. $5r^2s^2 - 5$

Solve.

13. $c^2 - 25 = 0$ 14. $j^2 - 64 = 0$ 15. $g^2 = 400$ 16. $m^2 - 4 = 0$

17. $a^2 - 121 = 0$ 18. $4n^2 - 36 = 0$ 19. $2b^2 - 200 = 0$ 20. $3h^2 - 3 = 0$

21. $5k^2 = 45$ 22. $4p^2 = 9$ 23. $4d^2 = 100$ 24. $25q^2 - 36 = 0$

7–7 Factor.

1. $a^2 - 10a + 25$ 2. $d^2 + 18d + 81$ 3. $p^2 + 2pq + q^2$

4. $r^2 - 4rs + 4s^2$ 5. $b^2 - 14b + 49$ 6. $c^2 + 16c + 64$

7. $g^2 + 22g + 121$ 8. $4t^2 + 16t + 16$ 9. $w^2 - 100$

Solve.

10. $t^2 - 2t + 1 = 0$ 11. $s^2 - 18s + 81 = 0$

12. $v^2 - 16 = 0$ 13. $r^2 + 10r + 25 = 0$

14. $b^2 - 100 = 0$ 15. $c^2 - \frac{2}{3}c + \frac{1}{9} = 0$

16. $j^2 + j + \frac{1}{4} = 0$ 17. $k^2 - 49 = 0$

7-8 Factor.

1. $m^2 - 5m + 6$ 2. $h^2 - 3h + 2$ 3. $c^2 + 9c + 8$

4. $a^2 + 9a + 20$ 5. $k^2 - 13k + 42$ 6. $t^2 - 12t + 20$

7. $w^2 - 7w + 10$ 8. $d^2 - 14d + 24$ 9. $t^2 + 11t + 18$

Solve.

10. $k^2 + 13k + 30 = 0$ 11. $a^2 + 5a + 4 = 0$

12. $s^2 - 13s + 12 = 0$ 13. $t^2 - 10t + 9 = 0$

14. $j^2 - 11j + 24 = 0$ 15. $c^2 - 16c + 64 = 0$

16. $w^2 + 15w + 54 = 0$ 17. $h^2 - 10h + 24 = 0$

7-9 Factor.

1. $g^2 + 4g - 12$ 2. $z^2 - 16$ 3. $x^2 + 3x - 10$

4. $m^2 - 3m - 28$ 5. $c^2 + c - 30$ 6. $y^2 - 11y - 12$

7. $a^2 + 5a - 36$ 8. $h^2 + 6h - 16$ 9. $k^2 - 5k - 24$

Solve.

10. $c^2 + 6c - 27 = 0$ 11. $j^2 + 10j - 11 = 0$ 12. $p^2 + p - 56 = 0$

13. $s^2 - 2s - 15 = 0$ 14. $u^2 - 3u - 18 = 0$ 15. $d^2 - 10d - 24 = 0$

16. $m^2 + 2m - 24 = 0$ 17. $k^2 - 2k - 8 = 0$ 18. $v^2 - 2v - 35 = 0$

19. $x^2 - 36 = 0$ 20. $z^2 + 3z - 54 = 0$ 21. $g^2 + 8g - 48 = 0$

22. $w^2 + 6w - 40 = 0$ 23. $h^2 - 7h - 60 = 0$ 24. $k^2 + 8k - 20 = 0$

7-10 Factor.

1. $2a^2 + 5a + 2$ 2. $3m^2 - 20m - 7$ 3. $3b^2 - 11b + 6$

4. $2g^2 + 7g - 4$ 5. $7t^2 - 10t + 3$ 6. $5n^2 - 13n - 6$

7. $4c^2 + 11c - 3$ 8. $6q^2 + q - 12$ 9. $4d^2 - 4d - 3$

10. $8p^2 - 2p - 15$ 11. $3h^2 - 7h + 2$ 12. $4k^2 - 4k - 15$

Solve.

13. $5z^2 - 13z - 6 = 0$ 14. $7a^2 + 19a - 6 = 0$

15. $4m^2 + 15m - 4 = 0$ 16. $4h^2 - 12h + 9 = 0$

17. $4y^2 + 23y - 6 = 0$ 18. $6n^2 - 19n + 15 = 0$

19. $8b^2 + 37b - 15 = 0$ 20. $8c^2 - 2c - 15 = 0$

21. $8d^2 + 14d - 15 = 0$ 22. $9j^2 - 16 = 0$

8–1 Simplify. List any necessary restrictions on the variables.

1. $\dfrac{12x}{18x}$

2. $\dfrac{20a^2}{15a^3}$

3. $\dfrac{16b^4}{24b}$

4. $\dfrac{3x - 6}{3x + 12}$

5. $\dfrac{2t}{t^2 - 4t}$

6. $\dfrac{c^2 - c}{c^2}$

7. $\dfrac{h^2 + 3h + 2}{h^2 + 6h + 8}$

8. $\dfrac{k^2 - 25}{k^2 + 10k + 25}$

9. $\dfrac{4t + 20}{t^2 + t - 20}$

8–2 Write each product in simplest form. Assume that no denominator is zero.

1. $\dfrac{2}{r} \cdot \dfrac{r^2}{8}$

2. $\dfrac{15h}{7} \cdot \dfrac{2}{5h}$

3. $12\left(\dfrac{x}{3} + \dfrac{1}{4}\right)$

4. $6a\left(\dfrac{4}{3a} + \dfrac{1}{2a^2}\right)$

5. $\dfrac{r(s - 2)}{s} \cdot \dfrac{s}{r(r + 2)}$

6. $\dfrac{2(w + 3)}{3} \cdot \dfrac{9}{4(w + 3)}$

Write each quotient in simplest form. Assume that no denominator is zero.

7. $\dfrac{r^3}{s^2} \div \dfrac{r}{s}$

8. $\dfrac{6mn^4}{5} \div \dfrac{4n^3}{15m}$

9. $\dfrac{a^2b^2}{c} \div \dfrac{ab^3}{c^2}$

10. $\dfrac{3x^3y}{6x} \div \dfrac{2x^3y^2}{5x}$

11. $\dfrac{b(c + 2)}{c + 3} \div \dfrac{c}{b(c + 3)}$

12. $\dfrac{h}{k + 1} \div \dfrac{h^3}{k + 1}$

13. $\dfrac{(r - 3)^2}{r} \div \dfrac{r - 3}{r^3}$

14. $\dfrac{5(s - 3)}{s} \div \dfrac{s + 3}{s^2}$

15. $\dfrac{7(t - 4)}{t} \div \dfrac{5(t - 4)}{t}$

8–3 Simplify. Assume that no denominator is zero.

1. $\dfrac{3t}{2} + \dfrac{5t}{2}$

2. $\dfrac{b}{a} - \dfrac{5b}{a}$

3. $\dfrac{7}{3c} + \dfrac{5}{3c}$

4. $\dfrac{h}{h - 4} - \dfrac{2}{h - 4}$

5. $\dfrac{4j + 3}{2j + 1} - \dfrac{2j + 3}{2j + 1}$

6. $\dfrac{2r - 4}{5r - 20} + \dfrac{-r}{5r - 20}$

7. $\dfrac{t}{t^2 - 9} - \dfrac{3}{t^2 - 9}$

8. $\dfrac{z}{(z + 2)^2} + \dfrac{2}{(z + 2)^2}$

9. $\dfrac{3c}{(c + 2)(c - 3)} + \dfrac{6}{(c + 2)(c - 3)}$

10. $\dfrac{5r}{(r - 4)(r - 5)} - \dfrac{4r + 4}{(r - 4)(r - 5)}$

8–4 Simplify. Assume that no denominator is zero.

1. $\dfrac{2}{5r} + \dfrac{3}{r}$

2. $\dfrac{4}{w} - \dfrac{5}{6}$

3. $\dfrac{7}{v^2} + \dfrac{5}{v}$

4. $\dfrac{2}{3b} + \dfrac{5}{3b^2}$

5. $\dfrac{7}{2a} - \dfrac{10}{3a^2}$

6. $\dfrac{a}{4} - \dfrac{b}{a}$

7. $\dfrac{2}{x} + \dfrac{2}{x + 3}$

8. $\dfrac{1}{x + 2} + \dfrac{2}{x + 3}$

9. $\dfrac{4}{x^2 + 7x + 12} + \dfrac{4}{x + 4}$

8–5 Solve. Write $\emptyset$ if there is no solution.

1. $\dfrac{3}{x} = \dfrac{9}{12}$

2. $\dfrac{21}{35} = \dfrac{a}{5}$

3. $\dfrac{c}{2} + \dfrac{1}{2} = 4$

4. $\dfrac{x}{3} - \dfrac{x}{5} = \dfrac{2}{3}$

5. $\dfrac{2}{y} + \dfrac{3}{4} = \dfrac{11}{4}$

6. $\dfrac{3}{2} - \dfrac{7}{k} = 5$

7. $\dfrac{1}{8}n - \dfrac{7}{8}n = 3$

8. $\dfrac{1}{3}x + \dfrac{1}{2}x = 5$

9. $\dfrac{1}{b} + \dfrac{3}{2b} = 5$

10. $\dfrac{7}{x} - \dfrac{9}{x} = 1$

11. $\dfrac{3}{4}r + \dfrac{9}{4}r = 2$

12. $\dfrac{2}{3}p - \dfrac{6}{p} = 16$

8–6 Solve these work problems. Write an equation that fits the problem. Solve the equation and answer the question.

1. Working together, Chrystal and Gayle can clean their house in 3 h. Working alone, Chrystal takes twice as long to clean the house as Gayle takes when she cleans it by herself. How long does Gayle take to clean the house by herself?

2. Steve can deliver the newspapers in his neighborhood in 2 h. Alexander takes 3 h to do the same job. How long does it take Steve and Alexander to deliver the papers if they work together?

3. Libby takes 2 h to mow the lawn. When her sister helps her, the job takes $\dfrac{5}{4}$ h. How long does it take Libby's sister to mow the lawn by herself?

4. Machine A and machine B produce the same item. Machine A produces 3 times as many items as machine B in the same time. When both machines work, they produce the day's quota in 4 h. How long would it take machine B alone to produce the quota? How long would it take machine A alone to produce the quota?

5. A small pipe, a large pipe, or both pipes can be used to fill a pool. The large pipe takes 20 h to fill the pool. When both pipes are used, the pool is filled in 18 h. How long does it take the small pipe alone to fill the pool?

8–7 Solve these proportions.

1. $\dfrac{8}{11} = \dfrac{x}{5.5}$

2. $\dfrac{10}{15} = \dfrac{12}{x}$

3. $\dfrac{35}{x} = \dfrac{40}{32}$

4. $\dfrac{x}{24} = \dfrac{15}{40}$

5. $\dfrac{2.5}{x} = \dfrac{15}{24}$

6. $\dfrac{3.5}{6} = \dfrac{1.4}{x}$

7. $\dfrac{2}{3} = \dfrac{x+4}{x+9}$

8. $\dfrac{7}{4} = \dfrac{x+9}{x}$

Solve.

9. A 12-ft board is cut into two pieces whose lengths are in the ratio of 1 to 5. How long are the pieces?

10. A prize of $6000 is divided between two people in the ratio of 3 to 5. How much does each person get?

11. The ratio of student tickets to adult tickets for a drama production is 4 to 3. How many of each kind were sold if 840 tickets were sold in all?

8–8 Solve.

1. 15% of 260 = x
2. 10% of x = 320
3. x% of 450 = 99
4. 3% of 20 = x
5. 2% of x = 500
6. x% of 540 = 27
7. 125% of 16 = x
8. 150% of x = 375
9. x% of 56 = 42

Solve.

10. Owen Lott borrowed $7000 for 3 years at an annual rate of 15%. How much interest did he pay on his loan?

11. A television station reported that 45% of the registered voters questioned favored candidate Lincoln. If 81 of the voters questioned favored Lincoln, how many voters were questioned in all?

12. A convenience store reported that 156 out of 240 customers spent more than $10. What percent of the customers spent more than $10?

8–9 Write a proportion for each problem. Solve the proportion, and answer the question.

1. How much will 10 oz of cheese cost if it sells for $3.60 per lb?

2. Mr. VanBuren drove at a constant rate for $1\frac{1}{2}$ h and traveled 81 mi. What was his rate in miles per hour?

3. Catherine Harrison received 52% of the votes in an election in which 2575 persons voted. How many votes did she get?

4. Robert Stewart got 44% of the votes in an election. If he got 1463 votes, how many people voted?

5. On a map of Idaho, 3.2 cm represents 40 km. How far is Payette from Gooding if the distance is 8.4 cm on the map?

6. Mrs. Spruce drove 391 mi and used 11.5 gal of gasoline. What was her average fuel consumption (in miles per gallon)?

8–10 Suppose that y varies directly as x in the following exercises.

1. If y = 2.5 when x = 4, what is y when x = 20?

2. If y = 6 when x = 3.5, what is y when x = 28?

3. If y = 200 when x = 15, what is x when y = 50?

4. If y = 3 when x = 125, what is x when y = 4.5?

Suppose that the quantities in the following exercises vary directly. Write an equation that fits the problem. Solve the equation, and answer the question.

5. A 50-ft length of wire weighs 20 oz. How long is a wire that weighs 5 lb?

6. A 10.75-oz can of cream of chicken soup contains 2.75 servings. Each can has 302.5 calories. How many calories are in each serving?

Chapter 9

9–1 Graph on a number line.

1. $x > 6$ **2.** $x < 10$ **3.** $x \geq 2$ **4.** $x \leq 4$ **5.** $x \neq 3$

6. $x \neq -2$ **7.** $x > -2$ **8.** $x < -5$ **9.** $x \geq -10$ **10.** $x \leq -7$

9–2 Graph on a number line.

1. $a < 1$ or $a > 5$ **2.** $b > 3$ and $b < 8$ **3.** $c \leq -2$ or $c \geq 2$

4. $d \geq -4$ and $d \leq 4$ **5.** $h \neq 3$ and $h \leq 5$ **6.** $j \geq -5$ and $j \neq 0$

7. $k = 0$ or $k > 4$ **8.** $q < -3$ and $q < -6$ **9.** $r < -3$ or $r < -6$

9–3 Solve.

1. $a - 2 \geq 3$ **2.** $d - 1 \leq -4$ **3.** $h - 5 > -3$

4. $2m + 3 \geq 3m$ **5.** $4p + 8 < 3p + 10$ **6.** $7r - 1 > 8r + 5$

Solve the compound sentence. Graph the solutions on a number line.

7. $s + 1 \geq 4$ or $s + 4 \leq 4$ **8.** $t - 2 \leq 6$ and $t + 2 \geq 6$

9. $2x - 3 \leq 3x + 4$ and $3x + 4 \leq 2x + 2$

9–4 Solve and graph on a number line.

1. $3a \leq 15$ **2.** $4b \geq 12$ **3.** $-2c \leq 8$

4. $-5d \geq -30$ **5.** $\frac{1}{2}h > -4$ **6.** $-\frac{1}{3}j < 6$

9–5 Solve and graph on a number line.

1. $3v + 2 \leq 14$ **2.** $2w - 3 \geq 7$ **3.** $\frac{1}{2}x - 1 > 5$

4. $3c + 5 \leq 5c - 7$ **5.** $-4d + 2 \geq d - 8$ **6.** $2(j + 3) \leq 4(j - 1)$

9–6 Write a compound sentence without absolute value.

1. $|a| \geq 5$ **2.** $|b| \leq 3$ **3.** $|c| < 4$ **4.** $|d| > 6$

5. $|h + 3| \leq 2$ **6.** $|j - 4| \geq 7$ **7.** $|k - 2| > 1$ **8.** $|m + 4| < 4$

Graph on a number line.

9. $|r + 3| < 8$ **10.** $|2t| > 7$ **11.** $\left|-\frac{y}{4}\right| \geq 2$ **12.** $|-u + 2| \geq 5$

9–7 Graph.

1. $y \geq 2x + 3$ **2.** $y > -\frac{1}{2}x - 2$ **3.** $x + y \geq -2$ **4.** $y \geq \frac{1}{3}x^2$

9–8 Graph.

1. $y \geq 2x$ or $y \leq -2x$ **2.** $y \geq 2x$ and $y \leq -2x$ **3.** $y \leq 2$ and $x \geq 3$

10-1 Simplify.

1. $\sqrt{64}$

2. $-\sqrt{81}$

3. $\sqrt{0.36}$

4. $-\sqrt{1.96}$

5. $\sqrt{\dfrac{16}{25}}$

6. $-\sqrt{\dfrac{9}{100}}$

7. $\dfrac{-\sqrt{4}}{-\sqrt{81}}$

8. $\dfrac{\sqrt{25}}{-\sqrt{36}}$

Solve. The equations may have two solutions, one solution, or no solutions. Write $\emptyset$ if the equation has no solution.

9. $x = \sqrt{100}$

10. $x^2 = 100$

11. $x = \sqrt{-100}$

12. $x^2 = -100$

13. $\sqrt{x} = 16$

14. $x = \sqrt{-9}$

15. $x^2 = 81$

16. $x = -\sqrt{81}$

10-2 State the consecutive integers between which the given number lies.

1. $\sqrt{70}$

2. $\sqrt{25.4}$

3. $\sqrt{3.61}$

4. $\sqrt{40.32}$

5. $\sqrt{90.67}$

Use the table on page 641 to find a decimal approximation to the nearest hundredth.

6. $\sqrt{5}$

7. $-\sqrt{30}$

8. $-\sqrt{87}$

9. $\sqrt{75}$

10. $\sqrt{40}$

Solve. List approximations for the solutions to the nearest hundredth.

11. $x^2 = 45$

12. $x^2 = 68$

13. $x = -\sqrt{83}$

14. $x^2 = 90$

15. $x^2 = 7$

16. $x^2 = 200$

10-3 Find the length of the diagonal of a rectangle with the given length and width. Use the table on page 641 to approximate to the nearest tenth.

1. length: 4 cm
   width: 2 cm

2. length: 9 cm
   width: 4 cm

3. length: 6 m
   width: 6 m

4. length: 8 m
   width: 5 m

5. length: 23 km
   width: 12 km

6. length: 40 km
   width: 25 km

10-4 Simplify.

1. $\sqrt{98}$

2. $\sqrt{48}$

3. $-\sqrt{108}$

4. $\sqrt{162}$

5. $-\sqrt{500}$

6. $\sqrt{288}$

7. $\sqrt{5}(3 + \sqrt{5})$

8. $\sqrt{2}(\sqrt{8} + \sqrt{3})$

9. $\sqrt{3}(\sqrt{27} - \sqrt{3})$

10. $(3 + \sqrt{5})(3 - \sqrt{5})$

11. $(4 - \sqrt{3})^2$

12. $(3 + \sqrt{2})^2$

10-5 Simplify.

1. $\sqrt{\dfrac{5}{36}}$

2. $\sqrt{\dfrac{17}{64}}$

3. $\sqrt{\dfrac{5}{3}}$

4. $\sqrt{\dfrac{9}{2}}$

5. $\sqrt{\dfrac{7}{8}}$

6. $\dfrac{\sqrt{10}}{\sqrt{2}}$

7. $\dfrac{\sqrt{12}}{\sqrt{8}}$

8. $\dfrac{\sqrt{20}}{\sqrt{24}}$

9. $\dfrac{\sqrt{27}}{\sqrt{18}}$

10. $\dfrac{\sqrt{54}}{\sqrt{12}}$

Write a decimal approximation to the nearest hundredth.

11. $\dfrac{1}{\sqrt{5}}$

12. $\dfrac{\sqrt{2}}{\sqrt{3}}$

13. $\sqrt{\dfrac{7}{2}}$

14. $\sqrt{4\dfrac{4}{5}}$

15. $\sqrt{2\dfrac{1}{7}}$

10–6 Simplify. Write "simplest form" if the expression cannot be simplified.

1. $\sqrt{8} + \sqrt{2}$ 2. $2\sqrt{5} - 5\sqrt{5}$ 3. $\sqrt{18} - \sqrt{3}$ 4. $\sqrt{18} - \sqrt{2}$

5. $\sqrt{72} - \sqrt{2}$ 6. $\sqrt{20} + \sqrt{5}$ 7. $\sqrt{10} + \sqrt{6}$ 8. $\sqrt{108} + \sqrt{48}$

10–7 Simplify. Assume that the domain of each variable is the set of positive numbers.

1. $\sqrt{b^3}$ 2. $\sqrt{\dfrac{c}{d}}$ 3. $\sqrt{25h}$ 4. $\sqrt{\dfrac{3}{j^2}}$ 5. $\sqrt{\dfrac{m^3}{2}}$

Simplify, stating restrictions where necessary.

6. $\sqrt{x^2}$ 7. $\sqrt{z^5}$ 8. $\sqrt{\dfrac{n}{2}}$ 9. $\sqrt{\dfrac{8}{p}}$ 10. $\sqrt{\dfrac{u^4}{v^5}}$

10–8 Solve. If there is no solution, write $\emptyset$.

1. $\sqrt{q + 11} = 3$ 2. $\sqrt{r} + 5 = 2$ 3. $\sqrt{3s - 5} = s - 1$

4. $\sqrt{2t + 24} = t + 8$ 5. $\sqrt{4v - 1} = 2v$ 6. $\sqrt{6w + 6} + 3 = w + 4$

7. $6 = \sqrt{y + 1}$ 8. $\sqrt{2 - z} + 3 = 7$ 9. $a - 1 = \sqrt{4a + 1}$

10–9 Find the distance between the two points. Simplify the distances.

1. $(-12, 3), (12, 10)$ 2. $(3, 2), (8, 14)$ 3. $(10, -5), (-5, 3)$

4. $(6, 10), (8, 6)$ 5. $(-2, 3), (1, 6)$ 6. $(6, 5), (2, -1)$

Chapter 11

11–1 State the axis of symmetry and the minimum or maximum point.

1. $y = x^2 - 8x + 10$ 2. $y = x^2 - 3x - 2$

3. $y = x^2 + 5x + 2$ 4. $y = -x^2 + 6x$

5. $y = -x^2 - 2x$ 6. $y = 2x^2 + 11x + 14$

11–2 Graph the function. Then solve the three equations using the graph. If there is no solution, write $\emptyset$.

1. $y = x^2 - 5x + 6$ 2. $y = x^2 + x - 12$

a. $x^2 - 5x + 6 = 0$ a. $x^2 + x - 12 = 0$
b. $x^2 - 5x + 6 = -1$ b. $x^2 + x - 12 = -6$
c. $x^2 - 5x + 6 = 2$ c. $x^2 + x - 12 = 8$

3. $y = x^2 - 2x - 8$ 4. $y = -x^2 + 6x - 5$

a. $x^2 - 2x - 8 = 0$ a. $-x^2 + 6x - 5 = 0$
b. $x^2 - 2x - 8 = 7$ b. $-x^2 + 6x - 5 = 4$
c. $x^2 - 2x - 8 = -9$ c. $-x^2 + 6x - 5 = 6$

11–3 Solve. If the equation has no solution, write ∅.

    1. $x^2 = 64$ **2.** $(x - 3)^2 = 4$ **3.** $(x + 5)^2 = -1$

    4. $(x + 4)^2 = 36$ **5.** $(x - 1)^2 = 0$ **6.** $(x + 6)^2 = 2$

    7. $3(x + 2)^2 = 12$ **8.** $2(x - 5)^2 = 36$ **9.** $x^2 + 7 = 23$

    10. $(2x - 7)^2 = 36$ **11.** $(x - 4)^2 = 20$ **12.** $x^2 + 4x + 4 = 9$

    13. $x^2 - 6x + 9 = 25$ **14.** $x^2 - 4x + 4 = 18$ **15.** $x^2 - 8x + 16 = 11$

11–4 Solve by completing the square. Write the solutions as simplified radicals.

    1. $x^2 + 8x = 9$ **2.** $x^2 + 7x = 8$ **3.** $x^2 - 6x = -8$

    4. $x^2 - 4x = 5$ **5.** $x^2 - 2x = 4$ **6.** $x^2 + 6x + 2 = 5$

    7. $x^2 + 2x + 5 = 4$ **8.** $x^2 - 9x = 2$ **9.** $x^2 - 5x = 3$

11–5 Solve using the quadratic formula. Simplify the solutions. If there is no solution, write ∅.

    1. $x^2 - 6x + 4 = 0$ **2.** $x^2 + 3x + 1 = 0$

    3. $x^2 + 5x + 3 = 0$ **4.** $2x^2 + 5x + 5 = 0$

    5. $3x^2 + x - 2 = 0$ **6.** $x^2 + 6x - 2 = 0$

    7. $x^2 - 4x - 3 = 0$ **8.** $3x^2 + x - 1 = 0$

    9. $x^2 + 6x = 0$ **10.** $x^2 - 2 = 0$

11–6 Solve using the quadratic formula. Simplify all solutions. If there is no solution, write ∅.

    1. $x^2 + 6x + 10 = 0$ **2.** $x^2 - 4x + 4 = 0$

    3. $4x^2 - 5x - 6 = 0$ **4.** $3x^2 - 4x + 2 = 0$

    5. $4x^2 - 4x + 1 = 0$ **6.** $3x^2 - 5x + 1 = 0$

    7. $2x^2 + 6x - 3 = 0$ **8.** $-2x^2 + 3x + 1 = 0$

11–7 Solve by any method. If there is no solution, write ∅.

    1. $x^2 - 2x - 1 = 0$ **2.** $x^2 + 8x - 15 = 0$

    3. $x^2 - 3x = 0$ **4.** $x^2 - 16 = 0$

    5. $x^2 - 4x + 2 = 0$ **6.** $x^2 - x - 12 = 0$

    7. $x^2 + 5x + 2 = 0$ **8.** $x^2 - 3x + 8 = 0$

    9. $x^2 + 7x = 0$ **10.** $x^2 - 3x - 18 = 0$

    11. $x^2 + 7x + 5 = 0$ **12.** $2x^2 - 5x - 2 = 0$

    13. $2x^2 - 5x + 2 = 0$ **14.** $2x^2 - 9x + 7 = 0$

    15. $3x^2 + 2x = 0$ **16.** $6x^2 + 5x - 21 = 0$

Table of Squares and Approximate Square Roots

n	n^2	$\sqrt{n}$	$\sqrt{10n}$	n	n^2	$\sqrt{n}$	$\sqrt{10n}$
1	1	1.000	3.162	51	2601	7.141	22.583
2	4	1.414	4.472	52	2704	7.211	22.804
3	9	1.732	5.477	53	2809	7.280	23.022
4	16	2.000	6.325	54	2916	7.348	23.238
5	25	2.236	7.071	55	3025	7.416	23.452
6	36	2.449	7.746	56	3136	7.483	23.664
7	49	2.646	8.367	57	3249	7.550	23.875
8	64	2.828	8.944	58	3364	7.616	24.083
9	81	3.000	9.487	59	3481	7.681	24.290
10	100	3.162	10.000	60	3600	7.746	24.495
11	121	3.317	10.488	61	3721	7.810	24.698
12	144	3.464	10.954	62	3844	7.874	24.900
13	169	3.606	11.402	63	3969	7.937	25.100
14	196	3.742	11.832	64	4096	8.000	25.298
15	225	3.873	12.247	65	4225	8.062	25.495
16	256	4.000	12.649	66	4356	8.124	25.690
17	289	4.123	13.038	67	4489	8.185	25.884
18	324	4.243	13.416	68	4624	8.246	26.077
19	361	4.359	13.784	69	4761	8.307	26.268
20	400	4.472	14.142	70	4900	8.367	26.458
21	441	4.583	14.491	71	5041	8.426	26.646
22	484	4.690	14.832	72	5184	8.485	26.833
23	529	4.796	15.166	73	5329	8.544	27.019
24	576	4.899	15.492	74	5476	8.602	27.203
25	625	5.000	15.811	75	5625	8.660	27.386
26	676	5.099	16.125	76	5776	8.718	27.568
27	729	5.196	16.432	77	5929	8.775	27.749
28	784	5.292	16.733	78	6084	8.832	27.928
29	841	5.385	17.029	79	6241	8.888	28.107
30	900	5.477	17.321	80	6400	8.944	28.284
31	961	5.568	17.607	81	6561	9.000	28.460
32	1024	5.657	17.889	82	6724	9.055	28.636
33	1089	5.745	18.166	83	6889	9.110	28.810
34	1156	5.831	18.439	84	7056	9.165	28.983
35	1225	5.916	18.708	85	7225	9.220	29.155
36	1296	6.000	18.974	86	7396	9.274	29.326
37	1369	6.083	19.235	87	7569	9.327	29.496
38	1444	6.164	19.494	88	7744	9.381	29.665
39	1521	6.245	19.748	89	7921	9.434	29.833
40	1600	6.325	20.000	90	8100	9.487	30.000
41	1681	6.403	20.248	91	8281	9.539	30.166
42	1764	6.481	20.494	92	8464	9.592	30.332
43	1849	6.557	20.736	93	8649	9.644	30.496
44	1936	6.633	20.976	94	8836	9.695	30.659
45	2025	6.708	21.213	95	9025	9.747	30.822
46	2116	6.782	21.448	96	9216	9.798	30.984
47	2209	6.856	21.679	97	9409	9.849	31.145
48	2304	6.928	21.909	98	9604	9.899	31.305
49	2401	7.000	22.136	99	9801	9.950	31.464
50	2500	7.071	22.361	100	10000	10.000	31.623

Table of Trigonometric Ratios

Angle	Sine	Cosine	Tangent	Angle	Sine	Cosine	Tangent
1°	0.0175	0.9998	0.0175	46°	0.7193	0.6947	1.0355
2°	0.0349	0.9994	0.0349	47°	0.7314	0.6820	1.0724
3°	0.0523	0.9986	0.0524	48°	0.7431	0.6691	1.1106
4°	0.0698	0.9976	0.0699	49°	0.7547	0.6561	1.1504
5°	0.0872	0.9962	0.0875	50°	0.7660	0.6428	1.1918
6°	0.1045	0.9945	0.1051	51°	0.7771	0.6293	1.2349
7°	0.1219	0.9925	0.1228	52°	0.7880	0.6157	1.2799
8°	0.1392	0.9903	0.1405	53°	0.7986	0.6018	1.3270
9°	0.1564	0.9877	0.1584	54°	0.8090	0.5878	1.3764
10°	0.1736	0.9848	0.1763	55°	0.8192	0.5736	1.4281
11°	0.1908	0.9816	0.1944	56°	0.8290	0.5592	1.4826
12°	0.2079	0.9781	0.2126	57°	0.8387	0.5446	1.5399
13°	0.2250	0.9744	0.2309	58°	0.8480	0.5299	1.6003
14°	0.2419	0.9703	0.2493	59°	0.8572	0.5150	1.6643
15°	0.2588	0.9659	0.2679	60°	0.8660	0.5000	1.7321
16°	0.2756	0.9613	0.2867	61°	0.8746	0.4848	1.8040
17°	0.2924	0.9563	0.3057	62°	0.8829	0.4695	1.8807
18°	0.3090	0.9511	0.3249	63°	0.8910	0.4540	1.9626
19°	0.3256	0.9455	0.3443	64°	0.8988	0.4384	2.0503
20°	0.3420	0.9397	0.3640	65°	0.9063	0.4226	2.1445
21°	0.3584	0.9336	0.3839	66°	0.9135	0.4067	2.2460
22°	0.3746	0.9272	0.4040	67°	0.9205	0.3907	2.3559
23°	0.3907	0.9205	0.4245	68°	0.9272	0.3746	2.4751
24°	0.4067	0.9135	0.4452	69°	0.9336	0.3584	2.6051
25°	0.4226	0.9063	0.4663	70°	0.9397	0.3420	2.7475
26°	0.4384	0.8988	0.4877	71°	0.9455	0.3256	2.9042
27°	0.4540	0.8910	0.5095	72°	0.9511	0.3090	3.0777
28°	0.4695	0.8829	0.5317	73°	0.9563	0.2924	3.2709
29°	0.4848	0.8746	0.5543	74°	0.9613	0.2756	3.4874
30°	0.5000	0.8660	0.5774	75°	0.9659	0.2588	3.7321
31°	0.5150	0.8572	0.6009	76°	0.9703	0.2419	4.0108
32°	0.5299	0.8480	0.6249	77°	0.9744	0.2250	4.3315
33°	0.5446	0.8387	0.6494	78°	0.9781	0.2079	4.7046
34°	0.5592	0.8290	0.6745	79°	0.9816	0.1908	5.1446
35°	0.5736	0.8192	0.7002	80°	0.9848	0.1736	5.6713
36°	0.5878	0.8090	0.7265	81°	0.9877	0.1564	6.3138
37°	0.6018	0.7986	0.7536	82°	0.9903	0.1392	7.1154
38°	0.6157	0.7880	0.7813	83°	0.9925	0.1219	8.1443
39°	0.6293	0.7771	0.8098	84°	0.9945	0.1045	9.5144
40°	0.6428	0.7660	0.8391	85°	0.9962	0.0872	11.4301
41°	0.6561	0.7547	0.8693	86°	0.9976	0.0698	14.3007
42°	0.6691	0.7431	0.9004	87°	0.9986	0.0523	19.0811
43°	0.6820	0.7314	0.9325	88°	0.9994	0.0349	28.6363
44°	0.6947	0.7193	0.9657	89°	0.9998	0.0175	57.2900
45°	0.7071	0.7071	1.0000	90°	1.0000	0.0000	.....

■ GLOSSARY

abscissa (p. 198) The first number in the ordered pair for a point in the coordinate plane.

absolute value (p. 40) On the number line, the distance of a number from zero is called the absolute value of that number. The absolute value of any number (except zero) is a positive number, and the absolute value of zero is zero. Absolute value is indicated by the symbol, $|\ |$.

absolute value property of inequalities (p. 464) For all real numbers x and a, if a is 0 or positive, $|x| < a$ is equivalent to $-a < x$ and $x < a$, which is equivalent to $-a < x < a$. For all real numbers x and a, if a is 0 or positive, $|x| > a$ is equivalent to $x < -a$ or $x > a$.

addition property of equality (p. 279) For all numbers a, b, c, and d, if $a = b$ and $c = d$, then $a + c = b + d$.

addition property of equations (p. 98) An equivalent equation is obtained when the same number is added to (or subtracted from) both sides of an equation. For all real numbers a, b, and c, if $a = b$, then $a + c = b + c$.

addition property of exponents (p. 153) For all real numbers a, and all positive numbers m and n, $a^m \cdot a^n = a^{m+n}$.

additional property of inequalities (p. 447) For all numbers a, b, and c, if $a < b$, then $a + c < b + c$ and if $a > b$, then $a + c > b + c$.

admissible values (p. 359) The values of the variables for which a fraction is defined.

algebraic fractions (p. 359) Fractions that contain variables.

angle-sum property (p. 604) The sum of the measures of the angles of a triangle is $180°$.

arithmetic mean (p. 598) For a set of numbers, the arithmetic mean is the quotient of the sum of the numbers divided by the number of numbers.

ascending order (p. 149) The order of the terms of a polynomial when written in order of their degrees from lowest to highest.

associative property of addition (p. 20) For all numbers a, b, and c, $(a + b) + c = a + (b + c)$.

associative property of multiplication (p. 20) For all numbers a, b, and c, $(ab)c = a(bc)$.

averages (p. 598) The mean, median, and mode.

axes (p. 197) Two perpendicular lines selected in a plane.

axis of symmetry (p. 544) The line at which, if the graph of a quadratic equation is folded, the two parts of the graph will match each other.

base (of an exponent) (p. 11) The number that is used as a factor.

binomial (p. 148) A polynomial with exactly two terms.

braces (p. 1) Symbols, $\{\ \}$, used to indicate a set of numbers.

brackets (p. 6) Grouping symbols, $[\ ]$, similar to parentheses.

certain event (p. 593) An event that will always occur. The probability of a certain event is 1.

coefficient (p. 10) The number in a product that includes a number and variables.

common factors (p. 310) Numbers that are factors of more than one number.

commutative property of addition (p. 19) For all real numbers a and b, $a + b = b + a$.

commutative property of multiplication (p. 19) For all real numbers a and b, $ab = ba$.

completing the square (p. 564) A process used to solve any quadratic equation.

composite number (p. 305) A positive number with more than two factors.

compound inequality (p. 435) An inequality with $\geq$ or $\leq$.

congruent angles (p. 604) Angles that are the same size.

conjugates (p. 514) If $\sqrt{y}$ is an irrational number, then the two binomials of the form $x + \sqrt{y}$ and $x - \sqrt{y}$ are conjugates of each other.

consecutive integers (p. 185) Integers that "follow" one after another.

constant (p. 148) A monomial that is a number.

constant of variation (p. 411) The constant k in the relationships of variation.

converse of the Pythagorean theorem (p. 501) If sides a, b, and c of a triangle are such that $a^2 + b^2 = c^2$, then the triangle is a right triangle. The right angle is opposite the longest side.

coordinate (of a point) (p. 197) The number associated with a particular point on a number line.

corresponding sides of similar triangles property (p. 605) Corresponding sides of similar triangles are proportional.

cosine of an angle (p. 611) In a right triangle, the ratio of the length of the side adjacent to an acute angle to the length of the hypotenuse.

cross-multiplication property (p. 395) For all numbers a, b, c, and d ($b \neq 0$ and $d \neq 0$), if $\dfrac{a}{b} = \dfrac{a}{d}$, then $ad = bc$.

degree of monomial (p. 148) The sum of the exponents of a monomial's variables.

degree of a polynomial (p. 149)   The highest degree of any of a polynomial's terms *after* the polynomial has been simplified.

descending order (p. 149)   The order of the terms of a polynomial when written in order of their degrees from highest to lowest.

difference of squares property (p. 327)   For all numbers a and b, $(a+b)(a-b) = a^2 - b^2$.

direct variation (p. 411)   A function such that

$y = kx$ or $\dfrac{y}{x} = k$ is called direct variation, where

y varies directly with x and k is called the consonant of variation.

discriminant (p. 576)   The expression $b^2 - 4ac$.

distance formula (p. 531)   The distance d between two points (x_1, y_1) and (x_2, y_2) is
$d = \sqrt{(x_2 - x_1)^2 + (y_2 - y_1)^2}$.

distributive property of exponents over multiplication (p. 163)   For all real numbers a and b, and all positive integers n, $(ab)^n = a^n b^n$.

distributive property of multiplication over addition (p. 20)   For all numbers a, b, and c, $ab + ac = a(b + c)$ and $ba + ca = (b + c)a$.

distributive property of multiplication over subtraction (p. 24)   For all numbers a, b, and c, $ca - cb = c(a - b)$ and $ac - bc = (a - b)c$.

distributive property of opposites (p. 67)   For all real numbers a and b, $-(a + b) = -a + (-b)$.

dividing opposites property (p. 67)   For all real numbers a and b, $b \neq 0$, $\dfrac{-a}{b} = -\dfrac{a}{b} = \dfrac{a}{-b}$.

division property for equivalent fractions (p. 360)   For all admissible values of a, b, and c,
$\dfrac{a}{b} = \dfrac{a \div c}{b \div c}$.

division property for radicals (p. 510)   For all nonnegative numbers a and b ($b \neq 0$),
$\sqrt{\dfrac{a}{b}} = \dfrac{\sqrt{a}}{\sqrt{b}}$.

domain (of a relation) (p. 201)   The set of first components of a relation.

domain (of a variable) (p. 1)   The set of numbers to be substituted for a variable.

empty set (p. 89)   The solution set of equations that have no solutions given in the replacement set.

equal ordered pairs (p. 198)   Pairs that have the same first coordinates and the same second coordinates. For all real numbers a, b, c, and d, $(a, b) = (c, d)$ if and only if $a = c$ and $b = d$.

equivalent equations (p. 92)   Equations that have the same solution set.

equivalent expressions (p. 15)   Two expressions that have the same values for every possible substitution.

equivalent fractions (p. 360)   Two fractions are equivalent if one of them can be transformed into the other by multiplication or division by a number or expression equal to 1.

equivalent inequalities (p. 447)   Inequalities with the same set of solutions.

evaluate (p. 2)   To substitute for a variable and then simplify the resulting numerical expression.

expanding (p. 318)   The process of changing from factored form to polynomial form.

exponent (p. 11)   The number that indicates how many times the base number is used as a factor.

factor (*noun*) (p. 10)   In a multiplication expression, a number that is multiplied.

factor (*verb*) (p. 306)   To write a number or an expression as the product of two or more of its factors.

factoring (p. 318)   The process of changing from polynomial form to factored form.

first component (p. 198)   The x-coordinate of an ordered pair.

frequency table (p. 598)   A list of the number of outcomes of an experiment that shows the frequency with which each outcome occurs.

function (p. 234)   A relation that has no two ordered pairs with the same first component. A relation is a function if and only if each first component in the relation is paired with exactly one second component.

graph of the equation (p. 211)   The graph of the set of all solutions of an equation.

graph (of an ordered number pair) (p. 197)   A point indicating the location of an ordered pair in the plane.

greatest common factor of two or more integers (GCF) (p. 310)   The greatest integer that is a factor of each given integer.

greatest common factor of two or more monomials (p. 311)   A monomial whose coefficient is the greatest common factor of the coefficients of the given monomials and whose variables have the greatest common degree of each variable in the given monomials.

horizontal line (p. 217)   A line parallel to the x-axis.

hyperbola (p. 419)   The graph of a second-degree equation formed by inverse variation.

hypotenuse (p. 500)   The longest side of a right triangle.

identity property for addition (p. 21)   For all numbers a, $a + 0 = a$ and $0 + a = a$.

identity property for multiplication (p. 21)   For all numbers a, $1a = a$ and $a \cdot 1 = a$.

impossible event (p. 593)   An event that will never occur. The probability of an impossible event is 0.

independent events (p. 595) Two events where either event can occur without affecting the other.

inequality (p. 435) A mathematical sentence stating that two quantities are not equal.

interest (p. 402) The amount of money earned (or paid) from (for) an investment (loan).

inverse operation (p. 108) The operation that will "undo" an operation.

inverse property of addition (p. 66) For all real members a, $a + (-a) = 0$.

inverse property of multiplication (p. 66) For all real numbers a, $a\left(\dfrac{1}{a}\right) = 1$ $(a \neq 0)$.

inverse variation (p. 419) A function defined by an equation of the form $xy = k$ or $y = \dfrac{k}{x}$, where k is a constant greater than 0, y is said to vary inversely as x, and k is called the constant of variation.

irrational number (p. 494) A real number that cannot be expressed as the quotient of two integers.

least common denominator (p. 376) The least common multiple of the denominators of two or more fractions.

legs (p. 500) The two shorter sides of a right triangle.

like terms (p. 24) Terms that contain the same variable factors.

linear equation in two variables (p. 235) Equations that define linear functions.

linear function (p. 235) A function whose graph is a straight line.

literal equation (p. 123) Formula or other equation containing more than one variable.

maximum (of a quadratic function) (p. 545) The highest point on the graph of a quadratic function.

measures of central tendency (p. 598) The mean, median, and mode.

median (p. 598) For a set of numbers, the median is found by arranging the numbers in order from greatest to least and locating the middle number. If there is an even number of numbers, the median is midway between the two middle numbers.

minimum (of a quadratic function) (p. 544) The lowest point on the graph of a quadratic function.

mode (p. 598) For a set of numbers, the most frequently occurring number.

monomial (p. 143) A number, a variable, or the product of numbers and/or variables.

multiplying opposites property (p. 67) For all real numbers a and b, $(-a)b = a(-b) = -(ab)$ and $(-a)(-b) = ab$.

multiplication property of equations (p. 103) An equivalent equation is obtained when both sides of an equation are multiplied (or divided) by the same number. For all real numbers a, b, and c, if $a = b$, then $ac = bc$.

multiplication property for equivalent fractions (p. 360) For all admissible values of a, b, and c, $\dfrac{a}{b} = \dfrac{ac}{bc}$.

multiplication property of exponents (p. 162) For all numbers a, and all positive integers m and n, $(a^m)^n = a^{mn}$.

multiplication property of inequalities (p. 452) For all numbers a, b, and c, $c > 0$, if $a > b$, then $ca > cb$, and if $a < b$, then $ca < cb$. For all numbers a, b, and c, $c < 0$, if $a > b$, then $ca < cb$, and if $a < b$, then $ca > cb$.

multiplication property for radicals (p. 506) For all numbers a and b, $\sqrt{ab} = \sqrt{a}\sqrt{b}$.

negative numbers (p. 39) Numbers that are to the left of zero on the number line.

negative one (-1) **property of multiplication** (p. 67) For all real numbers a, $(-1)a = -a$.

number line (p. 39) A model in which real numbers are represented as points.

numerical coefficient (p. 10) The number in a product that includes a number and variables.

opposite directions (p. 39) Numbers involved in measurements that can be described by positive and negative numbers.

opposite of an opposite property (p. 67) For all real numbers a, $-(-a) = a$.

opposites (p. 40) Numbers that have the same absolute value and opposite signs. Zero is the opposite of zero.

order of operations (p. 5) Rules used in simplifying numerical expressions. Mathematicians have agreed on the following rules for the order of operations:
1. Simplify inside parentheses and other grouping symbols first.
2. Do multiplications and divisions next, in order from left to right.
3. Do additions and subtractions next, in order from left to right.

order of real numbers (p. 39) The way in which the real numbers are set up on the number line.

ordered pair (p. 197) A pair of numbers that has a graph in the plane.

ordinate (p. 198) The second number in the ordered pair for a point in the coordinate plane.

origin (p. 197) The point of intersection of the two axes.

parabola (p. 544) The graph of a quadratic equation.

percent (p. 400) A term meaning "hundredth," usually denoted by the symbol %.

perfect square property (page 332) For all numbers a and b, $(a + b)^2 = a^2 + 2ab + b^2$ and $(a - b)^2 = a^2 - 2ab + b^2$.

polynomial (p. 143) The sum (or difference) of monomials. Monomials are also polynomials.

positive numbers (p. 39) Numbers that are to the right of zero on the number line.

prime factorization (p. 306) A factorization of a number such that all of its factors are prime numbers.

prime number (p. 305) A number that has exactly two factors.

principal (p. 402) An amount of money borrowed or invested in dollars.

principal square root (p. 489) The positive square root of a number, and indicated by the radical sign.

probability (p. 592) The quotient of the number of outcomes in an event divided by the total number of outcomes.

product (p. 10) The result obtained when factors are multiplied.

proportion (p. 394) An equation that states that two ratios are equal.

Pythagorean theorem (p. 500) For a right triangle, the square of the hypotenuse is equal to the sum of the squares of the legs.

quadrants (p. 197) The four regions in the plane formed by the axes.

quadratic equation (p. 543) A polynomial equation of degree two.

quadratic formula (p. 569) If $ax^2 + bx + c = 0$, $a \neq 0$, then $x = \dfrac{-b \pm \sqrt{b^2 - 4ac}}{2a}$.

quadratic function (p. 543) A set of solutions of a polynomial function of degree two. A quadratic function is a set of ordered pairs (x, y) described by an equation that can be written in the form $y = ax^2 + bx + c$, $a \neq 0$.

radical (p. 489) An expression that includes a radical sign ($\sqrt{}$) and the expression under the radical sign called the radicand.

radical equation (p. 526) An equation that contains a radical with a variable in the radicand.

radical sign ($\sqrt{}$) (p. 489) The symbol used for taking a square root.

radicand (p. 489) The expression under the radical sign.

raise to a power (p. 11) To evaluate numbers with exponents.

range (measure of central tendency) (p. 601) The difference between the least number and the greatest number in a set of numbers.

range (of a relation) (p. 201) The set of second components of the relation.

ratio (p. 394) A comparison of two numbers by division.

rate (p. 402) A comparison of two quantities by division; for example, 5 miles per hour, 6% per year.

rationalizing the denominator (p. 511) The process of eliminating a radical from the denominator of a fraction.

real numbers (p. 39) The set of numbers consisting of the positive numbers, the negative numbers, and zero.

reciprocals (p. 56) The numbers whose product is 1.

regular polygon (p. 607) A polygon that has all angles congruent and all sides congruent.

relation (p. 201) A set of ordered pairs.

replacement set (p. 89) The set of numbers that can be substituted for the variable. Also called the domain of the variable.

roots of an equation (p. 574) The solutions of an equation.

scientific notation (p. 157) A method of writing a number as the product of two factors. One factor is a number greater than or equal to 1 and less than 10, and the other factor is a power of 10 written in exponential form.

second component (p. 198) The y-coordinate of an ordered pair.

sense (p. 452) The inequalities $a > b$ and $c > d$ are said to be inequalities in the *same sense,* whereas $a < b$ and $c > d$ are said to be inequalities of the *opposite sense.*

similar triangles (p. 604) Two triangles are similar if and only if their corresponding angles are congruent.

simplest expression (p. 2) A numerical or variable expression that most closely represents a number.

simplest form (of a radical) (p. 507) When a radicand does not contain a fraction, has no perfect-square factors other than 1, and has no radicals in the denominator, we say that it is in simplest form.

simplify (p. 2) To write the simplest expression that represents the number.

sine of an angle (p. 611) In a right triangle, the ratio of the length of the side adjacent to an acute angle to the length of the hypotenuse.

slope–intercept form (of a linear equation) (p. 223) The equation $y = mx + b$. The graph of the equation is a straight line with slope m and y–intercept b.

slope of a line (p. 216) The ratio of the vertical change to the horizontal change. The slope of a line is the ratio of the change in y to the corresponding change in x between two points on the line (x_1, y_1) and (x_2, y_2).

$$m = \frac{y_2 - y_1}{x_2 - x_1} = \frac{y_1 - y_2}{x_1 - x_2}$$

slopes of parallel lines property (p. 225) Two lines are parallel if and only if they have the same slope or they have no slope.

slopes of perpendicular lines property (p. 226) Two lines are perpendicular if and only if the product of their slopes is -1, or if one slope is zero and the other is undefined.

solution (p. 89) The number that changes an equation into a true statement when it is substituted for the variable.

solution set (p. 89) The set of numbers from the replacement set that are solutions of the equation.

solution of a system of equations (p. 258) An ordered pair that is a solution of each of the two equations.

solutions of inequalities (p. 436) Those numbers that make the inequality true when they are substituted for the variable.

solve (p. 89) To find all the solutions of an equation.

solve an inequality (p. 447) To change to simpler equivalent inequalities until the variable is alone on one side.

square (p. 489) The square of a number is the product of the number multiplied by itself.

square root (p. 489) If $x^2 = y$, then x is a square root of y.

square root property of equations (p. 558) For each real number c, $c > 0$, if $x^2 = c$, then $x = \pm \sqrt{x}$.

standard deviation (p. 601) A commonly used statistic that describes the amount of dispersion from the mean numbers in the set.

subscript (p. 123) A number that indicates that variables are related but different.

substitution (p. 1) The process by which variables are replaced by numbers.

subtraction property of exponents (p. 175) For all numbers a, $a \neq 0$, and for all positive integers m and n, if $m > n$, then $\frac{a^m}{a^n} = a^{m-n}$ and if $m < n$, then $\frac{a^m}{a^n} = \frac{1}{a^{n-m}}$.

symmetric property of equations (p. 93) For all real numbers a and b, if $a = b$, then $b = a$.

system of equations (p. 258) A pair of equations.

tangent of an angle (p. 608) In a right triangle, the ratio of the length of the side opposite an acute angle to the length of the side adjacent to the acute angle.

terms (p. 24) Parts of an expression "connected" by addition or subtraction signs.

trigonometric ratios (p. 608) Ratios related to triangles, such as the sine, cosine, and tangent ratios.

trinomial (p. 148) A polynomial with exactly three terms.

uniform motion (p. 169) When motion is uniform, the relationship among distance (d), rate of speed traveled (r), and time traveled (t) is given by the formula $d = rt$.

unlike terms (p. 143) Terms that have different variable expressions.

variable (p. 1) A letter such as x, y, A, and g that is used in an expression in place of a number.

vertical line (p. 217) A line parallel to the y-axis.

vertical-line test (p. 235) A test used to determine whether a relation is a function. A vertical line can never cross the graph of a function more than once since there are no two ordered pairs with the same first coordinate.

x-axis (p. 197) The horizontal axis.

x-coordinate (p. 198) The first coordinate in an ordered pair.

x-intercept (p. 229) The x-coordinate of the point at which a line crosses the x-axis.

y-axis (p. 197) The vertical axis.

y-coordinate (p. 198) The second coordinate in an ordered pair.

y-intercept (p. 223) The y-value of the point at which a graph line crosses the y-axis.

zero exponent (p. 176) For all numbers a ($a \neq 0$), $a^0 = 1$.

zero (0) property of multiplication (p. 67) For all real numbers a, $0a = 0$.

zero-product property (p. 314) For all numbers a and b, if $ab = 0$, then $a = 0$ or $b = 0$.

■ SYMBOLS

{ }	Set
∅	Empty set
=	Is equal to
≠	Is not equal to
>	Is greater than
<	Is less than
≥	Is greater than or equal to
≤	Is less than or equal to
≈	Is approximately equal to
5 · 4	5 times 4
$^-5$	Negative 5
$^+5$	Positive 5
5^4	5 to the fourth power (5 · 5 · 5 · 5)
$-n$	Opposite or additive inverse of a number n
$\|n\|$	Absolute value of a number n
$x \overset{?}{=} y$	Does $x = y$?
π	Pi

°	Degree
%	Percent
(x, y)	Ordered pair x, y
$A(x, y)$	Point A with coordinates x, y
$x : y$	Ratio of x to y
$\sqrt{x}$	Principal or positive square root of x
$\pm\sqrt{5}$	Positive or negative $\sqrt{5}$
$3 \pm \sqrt{5}$	3 plus or minus $\sqrt{5}$
$\overrightarrow{AB}$	Line AB
$\overline{AB}$	Segment AB
AB	Measure of segment AB
$\angle A$	Angle A
$P(A)$	The probability of the outcome A
$\tan A$	Tangent of angle A
$\sin A$	Sine of angle A
$\cos A$	Cosine of angle A

■ FORMULAS

$P = a + b + c$	Perimeter of a triangle
$P = 4s$	Perimeter of a square
$P = 2(l + w)$	Perimeter of a rectangle
$C = \pi d$	Circumference of a circle
$A = lw$	Area of a rectangle
$A = s^2$	Area of a square
$A = bh$	Area of a parallelogram
$A = \frac{1}{2}bh$	Area of a triangle
$A = \frac{1}{2}h(b_1 + b_2)$	Area of a trapezoid
$A = \pi r^2$	Area of a circle
$S = 4\pi r^2$	Surface area of a sphere
$S = \pi rs$	Area of the slant surface of a cone

$V = lwh$	Volume of a rectangular prism
$V = Bh$	Volume of a prism
$V = \frac{1}{3}Bh$	Volume of a pyramid
$V = \frac{1}{3}\pi r^2 h$	Volume of a cone
$V = \frac{4}{3}\pi r^3$	Volume of a sphere
$V = \pi r^2 h$	Volume of a cylinder
$i = prt$	Interest
$d = rt$	Distance
$F = \frac{9}{5}(C + 32)$	Temperature conversion to Fahrenheit
$C = \frac{5}{9}(F - 32)$	Temperature conversion to Celsius
$a^2 + b^2 = c^2$	Pythagorean theorem

■ METRIC SYSTEM

1 kilometer (km) = 1000 meters (m)
1 centimeter (cm) = 0.01 meter
1 millimeter (mm) = 0.001 meter

1 kilogram (kg) = 1000 grams (g)
1 milligram (mg) = 0.001 gram

1 kiloliter (kL) = 1000 liters (L)
1 milliliter (mL) = 0.001 liter

km/h	Kilometers per hour
m/s	Meters per second
°C	Degrees Celsius
m^2	Square meter
cm^2	Square centimeter
m^3	Cubic meter
cm^3	Cubic centimeter

■ SELECTED ANSWERS

Chapter 1

Page 2, Classroom Exercises
1. t **2.** b **3.** A **4.** x **5.** 12 **6.** 14 **7.** 6 **8.** $\frac{1}{4}$
9. 3.45 **10.** 9 **11.** 10 **12.** 5 **13.** 1 **14.** 7
15. 9 **16.** My dog is 5 years old. **17.** 3; 5; 7
18. 10; 8; 6 **19.** 0; 6; 12 **20.** 0; 1; 2

Pages 3–4, Written Exercises
1. $17 - 1$; $17 - 4$; $17 - 3.7$ **3.** $\frac{2}{1}$; $\frac{2}{4}$; $\frac{2}{3.7}$ **5.** $1 + 1$;
$1 + 4$; $1 + 3.7$ **7.** 6×1; 6×4; 6×3.7 **9.** 10
11. 7 **13.** 12 **15.** 1 **17.** 30 **19.** 12 **21.** 12
23. $\frac{2}{3}$ **25.** 4 **27.** $3\frac{2}{3}$ **29.** 0 **31a.** 10×5
31b. $10 + 5 + 10 + 5$ **33.** Answers will vary.
35. Answers will vary. **37a.** 1024 **37b.** 64 **37c.** 576
37d. 1600 **39.** 3.5; 3.5 **41.** 3; 4 **43.** 8; 2; 4

Page 4, Review Exercises
1. $\frac{5}{8}$ **2.** $\frac{11}{15}$ **3.** $\frac{1}{3}$ **4.** $\frac{19}{40}$ **5.** $\frac{1}{6}$ **6.** $\frac{1}{10}$ **7.** $\frac{2}{9}$ **8.** $2\frac{2}{5}$

Pages 7–8, Classroom Exercises
1. See page 5. **2.** 2nd; 1st; 23 **3.** 1st; 2nd; 35
4. 2nd; 1st; 3rd; 48 **5.** 3rd; 1st; 2nd; 14 **6.** 1st or
2nd: $+$, $\times$; 3rd: $\div$; 4 **7.** 1st; 2nd; 3rd; 4th; 6

Pages 8–9, Written Exercises
1. $+$ **3.** $\times$ **5.** $\div$ **7.** $\div$ **9.** $\times$ **11.** $\div$ **13.** 34
15. 18 **17.** 26 **19.** 3 **21.** 36 **23.** 36 **25.** 10
27. 22 **29.** 28 **31.** 180 **33.** 210 **35.** 155
37. 60 **39.** 9 **41.** 56 **43.** 27 **45.** 1 **47.** 9
49. T **51.** F **53.** T **55.** F **57.** $7 + 3 - 4$ (Answers
may vary.) **59.** $15 + 1 - 2 \times 3$ (Answers may vary.)
61. Yes

Page 9, Review Exercises
1. $1\frac{1}{2}$ **2.** $\frac{13}{20}$ **3.** $\frac{1}{2}$ **4.** $\frac{3}{10}$ **5.** $\frac{1}{16}$ **6.** 1 **7.** $\frac{7}{24}$
8. $\frac{5}{6}$

Page 12, Classroom Exercises
1. 2; x; y; $2x$; $2y$; xy; $2xy$; 1 **2.** 5; a; $5a$; a^2; $5a^2$; 1
3. 4; $(a + 1)$; $4(a + 1)$; 1 **4.** 7 **5.** 3 **6.** m squared
7. 6 to the fourth power **8.** y cubed **9.** 4 plus
a squared **10.** the quantity (four plus a) squared
11. 16 **12.** 125 **13.** 100,000 **14.** 11 **15.** 25
16. 49

Pages 12–13, Written Exercises
1. 9 **3.** 1,000,000 **5.** 90 **7.** 14 **9.** 25 **11.** 12.25
13. 6 **15.** 9 **17.** 8 **19.** 45 **21.** 60

23. 1,000,000 **25.** 18 **27.** 144 **29.** 108 **31.** 1
33. 4 **35.** 9 **37.** 125 cm^3 **39.** 1000 m^3
41. 3.375 cm^3 **43.** $\frac{1}{8}$ in.^{3} **45.** 4 ft **47.** 400 ft
49. \$1210 **51.** \$1464.10 **53.** 0.013 mm
55. 0.4 mm **57.** 0 **59.** 152 **61.** 16 **63.** 8
65. 64 **67.** 3

Page 14, Review Exercises
1. 90 cm^2; 42 cm **2.** 33 cm^2; 28 cm **3.** 50 cm^2; 40 cm

Page 14, Self-Quiz 1
1. 4 **2.** 64 **3.** 10 **4.** 18 **5.** 23 **6.** 35 **7.** 40
8. 5 **9.** 8 **10.** 16 **11.** 5 **12.** 38

Page 17, Classroom Exercises
1. Two expressions are not equivalent if they have
different values for at least one substitution. **2a.** 2; 2
2b. 8; 6 **2c.** 11; 8 **2d.** No **3a.** 36; 36 **3b.** 16; 16
3c. 81; 81 **3d.** Yes

Pages 17–18, Written Exercises
1a. 0; 0 **1b.** 4; 4 **1c.** 40; 40; yes **3.** 15, 21, 30;
5, 11, 20; no **5.** 10, 14, 35; 0, 14, 35; no **7.** Yes
9. No, 0 **11.** No, 1 **13.** No, 1 **15.** $x = 0$ gives
different values. **17.** No **19.** 100, 100; 2, 2; 0, 0; yes
21. Yes **23.** Yes **25.** No, 1 and 1 **27.** No, 1 and 1
29. $5x + 10y$; $5(x + 2y)$ **31.** $10 - (a - b)$;
$(10 - a) + b$ I: $2x + 2y$, $2y + 2x$, $y + y + x + x$,
$x + y + x + y$, $2(x + y)$; II: $x + 2y$, $2y + x$, $x + y + y$;
III: $x + y + x$, $y + x + x$

Page 18, Review Exercises
1. 23 **2.** 3 **3.** 15 **4.** 15 **5.** 18 **6.** 10 **7.** 25
8. 13

Page 21, Classroom Exercises
Examples for exercises 1–5 will vary.
1. $5(3 \cdot 2) = (5 \cdot 3)2$ **2.** $3 + 4 = 4 + 3$
3. $8(4 + 5b) = 8 \cdot 4 + 8 \cdot 5b$ **4.** $6b + 0 = 6b$
5. $6 \cdot 1 = 6$ **6.** Associative property of addition
7. Distributive property **8.** Commutative property of
multiplication **9.** Commutative property of
multiplication

Pages 21–23, Written Exercises
1. Associative property of addition **3.** Commutative
(addition) **5.** Commutative (addition) **7.** Associative
(addition) **9.** addition **11.** addition **13.** multiplication **15.** multiplication **17.** Commutative property of
multiplication **19.** Associative property of multiplication **21.** Distributive property of multiplication over

addition   **23.** Associative property of addition
25. $2x + 5x$ **27.** $5x + 5 \cdot 2$ **29.** $(x + 2)5$ **31.** Commutative property of addition   **33.** Distributive property of multiplication over addition

35. $6 \cdot 3 + 6 \cdot \frac{1}{3}$ **37.** $43 \cdot 100 + 43 \cdot 1$

39. $29 \cdot 1000 + 29 \cdot 1$ **41.** 123,123   **43.** 123,123
45. 100; 100; yes   **47.** 12; 6; no

Page 23, Review Exercises

1. 10.93   **2.** 3.61   **3.** 0.365   **4.** 71.4   **5.** $\frac{11}{12}$ **6.** $\frac{1}{6}$

7a. For $a = 2$, $b = 3$: 13; 25; 5; 1. For $a = 1$, $b = 4$: 17; 25; 15; 9. For $a = 5$, $b = 5$: 50; 100; 0; 0.   **7b.** No
7c. No

Page 26, Classroom Exercises

1. $2x$; $2y$; 7, $4y$ **2.** No   **3.** No   **4.** Yes   **5.** $3a$ and $5a$; $2b$ and $3b$; 9 and 7   **6.** $12x$ **7.** $4c$ **8.** $5x$
9. $9y$ **10.** $7ab$ **11.** $5a + 6$ **12.** $3x + 6y$
13. $2x + 13$ **14.** $12m + 12$ **15.** $2a + 5$

Pages 26–27, Written Exercises

1. Two   **3.** Three   **5.** Two   **7.** Three   **9.** $2x$; $3y$; $4x$; $5xy$ **11.** $7a^2$; $8b^2$; 6; $3ab$ **13.** Like   **15.** Unlike
17. Unlike   **19.** Unlike   **21.** $5m$ **23.** $4x$ **25.** $3a$
27. $4b$ **29.** $8a + 4b$ **31.** $4a + 8$ **33.** $14a + 2b + 3$
35. $11x$ **37.** $6x + 10$ **39.** $7x + 16$ **41a.** Distributive property   **41b.** Associative property of multiplication
41c. Commutative property of addition   **41d.** Identity property for multiplication   **41e.** Distributive property
43. $x = 0$ gives different values.   **45.** $x = 1$ gives different values.   **47.** $x = 3$ gives different values.
49. $8x + 13$ **51.** $x + 4$ **53.** $x^2 + 15$
55. $5x + 15 + 10x$ (Distributive); $5x + 10x + 15$ (Commutative property of addition); $15x + 15$ (Distributive)

Page 27, Review Exercises

1. 10   **2.** 32   **3.** $\frac{11}{12}$ **4.** $\frac{13}{30}$ **5.** $a = 0$ gives different values   **6.** $a = 2$ and $b = 1$ gives different values
7. $x = 0$ gives different values   **8.** Associative property of addition   **9.** Commutative property of addition
10. Identity property for multiplication   **11.** Distributive property

Page 28, Self-Quiz 2

1a. 15; 15   **1b.** 0; 0   **1c.** 27; 45   **1d.** No   **2.** Distributive property   **3.** Identity property for multiplication   **4.** Associative property of addition   **5.** Commutative property of multiplication   **6.** $7x$ **7.** $6a$
8. $2b + 11$ **9.** $7a + 23$ **10.** $12xyz$ **11.** $7m + 26$

Page 28, Extension

1. 10   **2.** 1   **3.** 48   **4.** 36   **5.** PRINT $7 \uparrow 2$

6. PRINT $(4 + 2) \uparrow 3$ **7.** PRINT $(5 \uparrow 2) * 3 \uparrow 2$
8. PRINT $5 * 6 \uparrow 2$ **9.** PRINT $(4 \uparrow 2) \uparrow 3$
10. PRINT $6 * 4 - 2 \uparrow 3$

Page 30, Classroom Exercises

1a. $(t + 3)$ **1b.** $(t - 5)$ **1c.** $4t$ **1d.** $\frac{1}{2}t$ **2a.** $2r$
2b. $\frac{1}{2}b$ **3a.** $(n + 7)$ **3b.** $(n - 10)$ **3c.** $3n$ **3d.** $\frac{1}{2}n$

Pages 31–33, Written Exercises

1a. 20   **1b.** $(A + 3)$ **3a.** $(d - 3)$ **3b.** $(t + 3)$ **5a.** 7
5b. $(s - 5)$ **7a.** 13   **7b.** $(m + 5)$ **7c.** $(m - 4)$
9a. $(w + 3)$ **9b.** $(l - 3)$ **11a.** $\frac{1}{154}x$ **11b.** $154y$
13a. r; $(r - 8)$ **13b.** p; $(p + 8)$ **15a.** a; $(a + 7.2)$
15b. w; $(w - 7.2)$ **17a.** d; $\frac{1}{3}d$ **17b.** c; $3c$
19. Morris: m; Ken: $3m$ or Ken: k; Morris: $\frac{k}{3}$
21. Frank: f; Chris: $(f + 3)$ or Chris: c; Frank $(c - 3)$
23. Chris: c; Frank: $(c + 3)$ or Frank: f; Chris: $(f - 3)$
25. Millie: m; Janell: $(48 - m)$ or Janell: j; Millie: $(48 - j)$
27. Phil: p; Sandy: $(47 - p)$ or Sandy: s; Phil: $(47 - s)$
29a. Bill: b; Jack: $\frac{1}{2}b$ or Jack: j; Bill: $2j$
29b. Howard $\left(\frac{1}{2}b + 1\right)$ or $(j + 1)$

Page 33, Review Exercises

1. $\frac{8}{9}$ **2.** $\frac{8}{9}$ **3.** Let $x = 0$.   **4.** Let $x = 0$.   **5.** Let $x = 0$ and $y = 1$.   **6.** Let $a = 1$ and $b = 1$.   **7.** Commutative property of multiplication   **8.** Distributive property

Pages 34–35, Chapter Review

1. 13   **3.** 6   **5.** 23   **7.** 40   **9.** 81   **11.** 100
13a. $x = 0$: 6, 2; $x = 1$: 9, 5; $x = 5$: 21, 17   **13b.** No
15. Distributive property of multiplication over addition   **17.** Associative property of multiplication
19. $7x + 5$ **21.** $7x + 15$ **23.** $3x + 20$ **25.** $6b + 3$
27. $3w$; $\frac{1}{3}l$

Page 36, Chapter 1 Self-Test

1. 12   **2.** 0   **3.** 24   **4.** 24   **5.** 57   **6.** 40   **7.** 24
8. 195   **9.** 6   **10.** 23   **11.** 38   **12.** 1   **13.** 9
14. 125   **15.** 125   **16.** 625   **17a.** 6; 0   **17b.** 12; 12
17c. 18; 24   **17d.** No   **18a.** 28; 7   **18b.** 32; 11
18c. 36; 15   **18d.** No   **19.** Distributive property
20. Associative property of multiplication   **21.** Commutative property of addition   **22.** Identity property for addition   **23.** $11x$ **24.** $6a + 3b$ **25.** $7m + 10n$
26. $2y + 21$ **27.** $6m + 9$ **28.** $9m + 7n$ **29.** $2n + 1$
30. $3k - 16$

Page 37, Practice for College Entrance Tests
1. D   2. D   3. C   4. E   5. E   6. E   7. E   8. B
9. B   10. D   11. E

Chapter 2

Pages 40–41, Classroom Exercises
1. Positive eight   2. Negative sixteen   3. Is less than
4. Is greater than   5. The absolute value of negative
three   6. T   7. T   8. T   9. F   10. T   11. F
12. T   13. F   14. T   15. F   16. T   17. $|^-3|$
18. $-^+6$ 19. $-^-3 = {}^+3$ 20. $|^-2| = {}^+2$

Pages 41–42, Written Exercises
1. $^-30$ 3. $^+23$ 5. $^-380$ 7. $^+18$ 9. $^+5 > {}^-3$
11. $\frac{1}{4} > {}^-1$ 13. $<$ 15. $>$ 17. $>$ 19. $>$ 21. 8
23. 4   25. $3\frac{2}{3}$ 27. 13.67   29. T   31. T   33. F
35. F   37. F   39. T   41. T   43. $-2, 0, 1$ 45. $-1,$
$-\frac{1}{2}, 0, 1.5$ 47. 2   49. 11   51. 6   53. 3   55. $\frac{1}{2}$
57. $-\frac{1}{2}$ 59. $\{6\}$ 61. $\{-2\}$ 63. $\{0\}$

Page 42, Review Exercises
1. 9.84   2. 25.24   3. $\frac{5}{6}$ 4. $7\frac{1}{4}$ 5. 4.73   6. 16.01
7. $\frac{1}{6}$ 8. $1\frac{3}{4}$

Page 42, Extension
$613.46

Pages 45–46, Classroom Exercises
1. $-3 + (-2) = -5$ 2. $5 + (-3) = 2$
3. $-4 + 5 = 1$
4. -5
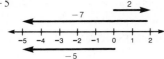
5. -2
6. 0
7. 0

8. -1

9. 3   10. 2   11. 1   12. 0   13. -1 14. 0
15. 6, 7, and 12   16. -2 17. 2   18. 4   19. 8

Pages 46–47, Written Exercises
1. -4 3. -14 5. 7   7. 0   9. 0   11. 0   13. 17.3
15. -2.09 17. 0   19. 8.05   21. -12 23. -12
25. -6 27. -6 29. 7, 9, 11, and 17   31. 6
33. 6   35. $\frac{3}{4}$ 37. $5\frac{3}{4}$ 39. 43.07   41. 125.63
43. -2.2 45. -2.2 47. 5.4   49. 12.6   51a. 10
51b. 1-yard gain   53. 2 under par

Page 47, Review Exercises
1. $\frac{1}{2}$ 2. 30   3. 26   4. 40   5. 15   6. 0

Page 47, Classroom Exercises
1. $6 + 3$ 2. $-4 + (-5)$ 3. $-2 + 7$ 4. $3 + (-5)$
5. -7 6. 10   7. -3 8. 6   9. 5   10. -1
11. 17   12. 2

Pages 50–51, Written Exercises
1. $11 + (-3)$ 3. $7 + 2$ 5. $6 + (-9)$ 7. $10 + 13$
9. 21   11. -42 13. 72   15. 9   17. 22   19. 10
21. 13   23. -22 25. $-1\frac{1}{4}$ 27. $3\frac{1}{4}$ 29. 1.5
31. 1.5   33. 55 yards   35. 19   37. 9   39. 26.4
41. -23.6 43. $19

Page 51, Review Exercises
1. 25   2. 49   3. 16   4. 4   5. 2.52   6. $\frac{8}{15}$
7. 3040   8. $1\frac{1}{4}$ 9. 0.0807   10. $\frac{5}{6}$ 11. 1.93
12. 7.65   13. $\frac{1}{6}$ 14. $1\frac{3}{4}$ 15. $\frac{5}{12}$ 16. 7.3

Page 51, Self-Quiz 1
1. 5   2. -8 3. x 4. -3 5. 4   6. $3\frac{1}{2}$ 7. $<$
8. $<$ 9. $>$ 10. -3 11. 8   12. -9 13. 14
14. -23 15. -1 16. 11   17. -31 18. 2
19. 33   20. 22   21. -2

Page 54, Classroom Exercises
1. 24   2. 15   3. -32 4. -54 5. -50 6. 40
7. 36   8. 0   9. -2

Pages 54–55, Written Exercises
1. 24   3. -99 5. 0   7. 0   9. -48 11. 54
13. -48 15. 54   17. 126   19. -72 21. 126
23. -72 25. 16   27. 16   29. 135   31. 135
33. 3   35. 3   37. $-\frac{1}{6}$ 39. $\frac{2}{3}$ 41. $\frac{1}{6}$ 43. $1\frac{1}{6}$

45. $1\frac{1}{2}$ **47.** -3 **49.** $-2 \cdot 3 = n;\ -6°$
51. $-5 \cdot -7 = m;\ \$35$ **53.** $2 \cdot -10 = w;\ -20$ liters
55. Positive **57.** Positive **59.** Negative **61.** Positive
63. Negative **65.** Positive

Page 55, Review Exercises

1. 0.6 **2.** 2.3 **3.** $\frac{3}{4}$ **4.** 3 **5.** 1 **6.** 0 **7.** Equivalent **8.** Not equivalent **9.** Not equivalent **10.** Not
equivalent

Page 55, Extension

$16\frac{7}{8};\ 36\frac{3}{8}$

Page 58, Classroom Exercises

1. $\frac{1}{8}$ **2.** $-\frac{1}{8}$ **3.** 2 **4.** -2 **5.** none **6.** 1
7. $-\frac{4}{3}$ **8.** 2.5 **9.** reciprocal **10.** $\frac{1}{6}$ **11.** $-\frac{3}{2}$
12. -1 **13.** 2 **14.** -4 **15.** -5 **16.** $-\frac{1}{3}$ **17.** 0
18. -16 **19.** -64 **20.** $-\frac{1}{3}$

Pages 58–59, Written Exercises

1. -6 **3.** $\frac{3}{2}$ **5.** $-\frac{1}{6}$ **7.** $1\frac{1}{5}$ **9.** $-\frac{8}{3}$ **11.** -5
13. $\frac{2}{3} \cdot \frac{8}{3}$ **15.** $-6 \cdot 4$ **17.** $\frac{3}{8} \cdot -\frac{5}{3}$ **19.** $-\frac{3}{4} \cdot -\frac{3}{2}$
21. $-\frac{5}{12}$ **23.** $-\frac{5}{3}$ **25.** $-\frac{9}{4}$ **27.** $\frac{3}{5}$ **29.** -4
31. $-\frac{1}{4}$ **33.** -12 **35.** -9 **37.** -1 **39.** -16
41. -100 **43.** -6.9 **45.** $w = -10 \div 4;\ -2.5$ pounds
47. $p = \frac{-840}{120}$ **49.** Positive **51.** Positive **53.** Negative **55.** Positive **57.** Positive **59.** Positive **61.** As
the value of n increases, the value of $\frac{12}{n}$ decreases.

Page 60, Review Exercises

1. Associative property of addition **2.** Distributive
property of multiplication over addition **3.** Identity
property for addition **4.** Commutative property of
addition **5.** $4x + 5$ **6.** $15a + 4b$ **7.** $14x + 22$
8. $3a + 9b$ **9.** $(w + 5)$ **10.** $(l - 5)$

Page 60, Extension

1. 6 **2.** -8 **3.** 20 **4.** 6 **5.** 2 **6.** 2
7. PRINT $-9--2$ **8.** PRINT $-2 \cdot -3 \cdot -4$
9. PRINT $8 + -12$ **10.** PRINT $(-3) \uparrow 2$
11. PRINT ABS(-12) **12.** PRINT ABS$(-7 + 4)$

Page 62, Classroom Exercises

1. T **2.** F; $a = -1$ **3.** F; $b = 1$ **4.** F; $x = -1$
5. $a + (-7)$ **6.** $2x + 8$ **7.** $9 + (-y)$ **8.** $-a + y$
9. $b - 3$ **10.** $2a - y$ **11.** $2a - (-y)$ **12.** $-x - 2$
13. $a + 4$ **14.** $b - 2$ **15.** $2x + 4$ **16.** $3a - 2b$

17. $c - 5$ **18.** $3x - y$ **19.** $2n + 3m$ **20.** $4x - 3$

Page 63, Written Exercises

1. $-a$ **3.** $-(-4)$ **5.** T **7.** T **9.** F; $a = -1$
11. F; $c = -1$ **13.** $h + (-5)$ **15.** $k + 3$
17. $x + (-y)$ **19.** $-2 + (-p)$ **21.** $r - 9$ **23.** $t - u$
25. $k - (-n)$ **27.** $2x - x$ **29.** 11 **31.** -15
33. $x + 2$ **35.** $2x - 4$ **37.** $2x + 8 + 2y$
39. $-3w + 3$ **41.** $5.6 + (-4.9) - (-3.6)$
43. $|17 + (-20) - (-1)|$ **45.** 0 **47.** $-5\frac{1}{2}$ **49.** -10
51. -25

Page 64, Review Exercises

1. Associative property of multiplication **2.** Commutative property of addition **3.** Identity property for
multiplication **4.** Distributive property **5.** $4a + 4b$
6. $7x + 8$ **7.** $3x + 18$ **8.** $10a + 17b$ **9.** $(R - 10)$
dollars **10.** $(M + 10)$ dollars

Page 64, Self-Quiz 2

1. 54 **2.** -6 **3.** -25 **4.** -12 **5.** $\frac{1}{2}$ **6.** -6
7. -18 **8.** $\frac{1}{3}$ **9.** $\frac{5}{3}$ **10.** $-\frac{1}{2}$ **11.** $-\frac{1}{7}$
12. $-\left(-\frac{1}{2}\right)$ **13.** $x + (-5)$ **14.** $b - 3$
15. $21 + (-m)$ **16.** $-6p - 1$

Page 65, Extension

1. 4 **2.** -13.5 **3.** -6 **4.** 6.8 **5.** -5 **6.** -25
7. -1 **8.** -2

Page 68, Classroom Exercises

Answers for exercises 1–4 will vary. **1.** $(xy)z = x(yz)$
2. $x(y + z) = xy + xz$ **3.** $x + y = y + x$
4. add: $x + 0 = x$; mult: $x \cdot 1 = x$ **5.** $8x$ **6.** $8x$
7. $4a$ **8.** b **9.** $-3m$ **10.** $6n$ **11.** $-6n$ **12.** $a + 5$
13. $-4c + 3a$

Pages 69–70, Written Exercises

1. $5a + (-a)$ **3.** $5 + 10s + (-8s)$
5. $-5a + (-4a) + b$ **7.** $-3 + 2a$ **9.** $15a$ **11.** $-6c$
13. $9q$ **15.** $-11s$ **17.** $6a + 3b$ **19.** $10r + 5s$
21. $-4w + 5z$ **23.** Distributive property **25.** Commutative property of addition **27.** Identity property for
multiplication **29.** Associative property of multiplication **31.** $25n$ **33.** $25n + 18m$ **35.** $40a + 170b$
37. $2x^2 + x$ **39.** $xy + 3x$ **41.** $-0.5x$ **43.** $2x - x^2$
45. $-5 + 5x^2$ **47.** $-16x$ **49.** $-4a^2 + 12a - 8$
51a. Commutative property of addition **51b.** Subtracting is equivalent to adding the opposite. **51c.** Associative property of addition **51d.** Computation
51e. Subtracting is equivalent to adding the opposite.
53. $(a - b) + b + b + (a - b) + (a - b) + b + b + (a - b) = 4a$

Page 70, Review Exercises

1. $(s - 5)$ 2. $2S$ 3. $2w + 10$ 4. 6   5. x 6. 8
7. -9 8. $-\dfrac{2}{3}$

Page 72, Classroom Exercises

1. $6a + 3$ 2. $7a - 35$ 3. $-8y - 16$ 4. $-14 + 6m$
5. $-4 + 3b$ 6. $6 + 2n$ 7. $-8g - 2$ 8. $-7a + 5$
9. $6x - 5$ 10. $12a - 3$ 11. $10y - 3x$ 12. $10n + 3$
13. $-y - 6x$ 14. $3c + 13$ 15. $15b - 3a$
16. $4g - 2$ 17. $12x - 2$ 18. $30x - 1$ 19. $5x - 8y$
20. $6x + 2y$ 21. $2x + 1$ 22. $4x$ 23. $-3x - 2$
24. $-2x$

Page 73, Written Exercises

1. $3x + 21$ 3. $7a - 28$ 5. $-6y - 12$
7. $-10b + 80$ 9. $15x - 30$ 11. $-3a + 5$
13. $-3a + 5$ 15. $6m + 12n$ 17. $8b + 20$
19. $2x + 16$ 21. $6x - 10$ 23. $-2x + 5$
25. $-x + 8$ 27. $7x - 10$ 29. $-2x - 35$
31. $x - 10$ 33. $2 + 3h$ 35. $20 + 0.2r$
37. $5h + 8t + 8n$; yes   39. $5a + 16b$ 41. $-a + 16b$
43. $-10x - 5y$ 45. $-11x + 8y$ 47. $2x - y$ 49. 0
51. $-a^2 - a + 1$ 53. $(b - 2a) + a + a + (b - 2a) +$
$a + a + (b - 2a) + a + a + (b - 2a) + a + a = 4a$
55. $f + (f + 1) + (f + 1) + (f + 1) + f + (g + 4) + 2g$
$+ 2g + (g + 4) = 5f + 6g + 11$

Page 75, Review Exercises

1a. $(w + 3)$ 1b. $(l - 3)$ 2a. $\left(j - 1\dfrac{1}{2}\right)$ 2b. $\left(b + 1\dfrac{1}{2}\right)$
3. -5 4. -13 5. 2   6. 8   7. -29 8. 10
9. 32   10. 7

Page 75, Self-Quiz 3

1. c   2. e   3. f   4. d   5. b   6. a   7. g   8. $3x$
9. $2y$ 10. $6x + 8$ 11. $15x - 32$ 12. $-2x + 3$
13. $a + 4b$ 14. $8x - 1$ 15. $-2x + 1$ 16. $6x + 3$
17. $7x + 15$ 18. $-2x - 2$ 19. $5x + 3y$

Page 78, Classroom Exercises

1. Mrs., 50 mi; Mr., 27.5 mi; 22.5 mi   2. Mrs., 50 mi;
Mr., 55 mi; 17.5 mi

3.

Time	Total miles (Mrs.)	Total miles (Mr.)
8:00	0	0
8:30	50	0
9:00	100	55
9:30	150	110
10:00	200	165
10:30	250	220
11:00	300	275
11:30	350	330
12:00	400	385
12:30	450	440
1:00	500	495
1:30	550	550

4. 1:30 P.M.

Pages 78–79, Written Exercises

1. 7   3. 4 days   5. 4
7. 4

Hours	Km driven	Liters used
1	84	7
2	168	14
3	252	21
4	336	28

9. Oysters   11a. $1400 profit
11b. $35,000

Sun.	$400
Mon.	−$300
Tues.	$400
Wed.	−$300
Thurs.	$400
Fri.	−$300
Sat.	$400
Total	$700

13. $8\dfrac{1}{2}$ or 9

	Position (m)
1 (A.M.)	−8
1 (P.M.)	−9
2 (A.M.)	−7
2 (P.M.)	−8
3 (A.M.)	−6
3 (P.M.)	−7
4 (A.M.)	−5
4 (P.M.)	−6
5 (A.M.)	−4
5 (P.M.)	−5
6 (A.M.)	−3
6 (P.M.)	−4
7 (A.M.)	−2
7 (P.M.)	−3
8 (A.M.)	−1
8 (P.M.)	−2
9 (A.M.)	0

15. She will be ahead by 171 points and stop on the 9th play.

Play	Points scored	Total points
1	+1	+1
2	−2	−1
3	+4	+3
4	−8	−5
5	+16	+11
6	−32	−21
7	+64	+43
8	−128	−85
9	+256	171

Page 81, Review Exercises

1. 148.33   2. 2.675   3. 0.0441   4. 44.97   5. $4\dfrac{1}{6}$
6. $2\dfrac{3}{4}$ 7. $1\dfrac{1}{8}$ 8. $2\dfrac{1}{4}$

Pages 84–85, Chapter Review

1. -10 3. 2.4   5. T   7. -4 9. -2.25 11. -10
13. -20 15. 18   17. $-\dfrac{1}{5}$ 19. $-\dfrac{3}{4}$ 21. -8
23. $a - (-b)$ 25. $1 + x$ 27. $r + 10$ 29. $4x - 6$
31. $8x + 4y$

33. Shorter length, 25 ft; longer length, 50 ft.

Length (ft)	Width (ft)	Area (sq. ft)
5	90	450
10	80	800
15	70	1050
20	60	1200
25	50	1250
30	40	1200

Pages 85–86, Chapter 2 Self-Test

1. 5 **2.** -35 **3.** -31 **4.** 7 **5.** -8 **6.** 3
7. -24 **8.** -7 **9.** 60 **10.** 25 **11.** $-\frac{1}{2}$ **12.** 12
13. 12 **14.** $x - 3$ **15.** $2y + 2$ **16.** x **17.** $7r$
18. $-w + 3t$ **19.** $5a + 2$ **20.** $2y - 3$ **21.** $2a$
22. $14b - 7$ **23.** $7a + 5$ **24.** $y + xy - x$
25. Distributive property **26.** Inverse property for
addition **27.** Subtracting is equivalent to adding the
opposite. **28.** Associative property of addition **29.** 0
property of multiplication **30.** Commutative property
of addition
31.–32.

Width	Length	Area
5	10	50
6	12	72
7	14	98
8	16	128
9	18	162
10	20	200

33. 7 cm by 14 cm

Pages 86–87 Practice for College Entrance Tests

1. B **2.** D **3.** C **4.** A **5.** E **6.** A **7.** D **8.** A
9. E **10.** E **11.** B **12.** E

Chapter 3

Page 90, Classroom Exercises

1. The set of numbers from the replacement set that
are solutions of the equation **2.** Empty set **3.** No
4. No **5.** Yes **6.** Yes **7.** -2.2 **8.** 2 **9** -1
10. 0, 1, 2

Pages 90–91, Written Exercises

1. No **3.** Yes **5.** No **7.** Yes **9.** Yes **11.** No
13. 3 **15.** 4 **17.** 1, 2, 3, 4, 5 **19.** {2} **21.** $\{-2\}$
23. $\{-1\}$ **25.** {2} **27.** $\{-1\}$ **29.** 40 **31.** $\emptyset$
33. 11 **35.** 14 **37.** $\{-4, 2\}$ **39.** $\{-4\}$ **41.** $\emptyset$
43. $\{-4, -2, 0, 2, 4\}$ **45.** {2} **47.** {3} **49.** {1, 3}
51. $\emptyset$ **53.** {2}

Page 91, Review Exercises

1. 75 **2.** $4\frac{1}{2}$ **3.** 0.15 **4.** 23 **5.** 99 **6.** 9 **7.** 25
8. 49 **9.** Yes **10.** No

Page 95, Classroom Exercises

1. Equations that have the same solution sets **2.** Add
(or subtract) same number to (from) each side;
interchange the expressions on each side; multiply (or
divide) each side by same number; rearrange (by
properties) terms of each side; simplify (by order of
operations). **3.** $2x + 3 = x + 5$ **4.** $2y = 20$
5. $6a = 27$ **6.** $t = 15$ **7.–8.** Answers will vary.
9. Yes; both have {3} as the solution set. **10.** No; each
has a different solution set. **11a.** $2 = 13x + 7$
11b. $-5 = 13x$ **11c.** $x = -\frac{5}{13}$ **12a.** $6x + 4 = 7$
12b. $6x = 3$ **12c.** $x = \frac{1}{2}$

Pages 96–97, Written Exercises

1. $6x = 18$ **3.** $5x + 10 = 8$ **5.** $4n = 3$
7. $2 = 3n + 1$ **9.** $n = 18$ **11.** $n = -12$ **13.** $x = -3$
15. $2x + 3 = 5$ **17.** Yes **19.** Yes **21.** Yes
23. Yes **25a.** $4x = 20$ **25b.** $x = 5$ **27a.** $\frac{1}{2}x = 30$
27b. $x = 60$ **29a.** $8x + 2 = 20$ **29b.** $8x = 18$
29c. $x = \frac{9}{4}$ **31a.** $2x + 6 = 25$ **31b.** $2x = 19$
31c. $x = \frac{19}{2}$ **33a.** $\frac{1}{4}x - 3 = 100$ **33b.** $\frac{1}{4}x = 103$
33c. $x = 412$ **35a.** $5x = 35$ **35b.** $x = 7$
37a. $x + 3 = 10$ **37b.** $x = 7$ **39.** F **41a.** Add $2x$ to
both sides. **41b.** Subtract 5 from both sides.
41c. Divide by 5 and simplify.

Page 97, Review Exercises

1. 21 **2.** $\frac{1}{9}$ **3.** 75 **4.** $\frac{4}{3}$ **5.** 108 **6.** 324

Page 100, Classroom Exercises

1. For all real numbers a, b, and c, if $a = b$, then
$a + c = b + c$. **2.** Subtraction is equivalent to addition
of the opposite. **3.** $-x$ **4.** -3 **5.** 9 **6.** {6}
7. {20} **8.** {0} **9.** $\{-3\}$ **10.** {13} **11.** {9}

Pages 100–101, Written Exercises

1. $-7a$ **3.** -3 **5.** -8 **7.** {3} **9.** $\{-2\}$ **11.** {16}
13. $\left\{1\frac{1}{2}\right\}$ **15.** {2} **17.** {1} **19.** {8} **21.** {2}
23. {0.4} **25.** {6} **27.** {9.09} **29.** {3} **31.** $\{-5\}$
33. {4} **35.** $\{-4.5\}$ **37.** Add -13; then multiply
by -1. Multiply by -1; then add 13. **39.** $\{-7x + 7\}$
41. {3} **43a.** $x = 3$: 2, 7; $x = 4$: 4, 6; $x = 5$: 6, 5;
$x = 6$: 8, 4; $x = 7$: 10, 3; $x = 8$: 12, 2
43b. increases, 2 **43c.** decreases, 1 **43d.** 4 and 5

Page 102, Review Exercises

1. Commutative property of addition   2. Associative property of addition   3. Distributive property   4. $4x$
5. a 6. $11x + 19$ 7. $f + 25$ 8. $j - 25$ 9. $3w$
10. $\frac{1}{3}l$

Page 102, Self-Quiz 1

1. 3   2. -1 3. $-3, 3$ 4. 0, 3   5. $4x = 14$
6. $9x - 6 = 39$ 7. $x = -12$ 8. $4x + 4 - 2x = -6$
9. $\{9\}$ 10. $\left\{1\frac{1}{4}\right\}$ 11. $\{-4\}$ 12. $\{-3\}$

Page 105, Classroom Exercises

1. For all real numbers a, b, and c, if $a = b$, then $ac = bc$.   2. Dividing is multiplying by the reciprocal.
3. $\times \frac{1}{2}$ or $\div 2$ 4. $\times -4$ 5. $\times \frac{5}{2}$ 6. $\div 0.2$ or $\times 5$
7. $\left\{-\frac{6}{5}\right\}$ 8. $\left\{\frac{9}{4}\right\}$ 9. $\{16\}$ 10. $\{50\}$ 11. $\{14\}$
12. $\{20\}$

Pages 105–106, Written Exercise

1. $-\frac{1}{3}$ 3. $\frac{8}{3}$ 5. 7   7. -1 9. 7   11. -3
13. -4 15. $\frac{1}{2}$ 17. Wrong   19. Wrong   21. $\left\{\frac{19}{2}\right\}$
23. $\left\{-\frac{31}{5}\right\}$ 25. $\{-48\}$ 27. $\{-72\}$ 29. $\{48\}$
31. $\left\{\frac{18}{5}\right\}$ 33. $\left\{\frac{1}{2}\right\}$ 35. $\left\{-\frac{6}{7}\right\}$ 37. $\left\{\frac{8}{5}\right\}$ 39. $\left\{\frac{8}{5}\right\}$
41. $\{2.73\}$ 43. $\{-1.1\}$ 45. 0; there is no division by 0.
47. $\frac{a}{b}\left(\frac{bc}{a}\right) - c = c - c = 0$

Page 106, Review Exercises

1. $7x + 9$ 2. $5x + 10$ 3. $w + 4$ 4. $\frac{1}{2}x$ 5. $17 - n$
6. $\frac{24}{w}$

Page 110, Classroom Exercises

1. Subtracting -5 2. Multiplying by 3   3. Dividing by -2 4. Adding 5   5. Dividing by $-\frac{1}{3}$ 6. 1, 2
7a. Subtract 8   7b. Divide by 6   8a. Multiply by 2
8b. Subtract 1   8c. Divide by 3   9. $\{8\}$ 10. $\{3\}$
11. $\{20\}$ 12. $\{-18\}$

Pages 111–112, Written Exercises

1. 1, 2   3. 1, 2   5. Add 2   7. Subtract 7   9. Subtract 7
11. Multiply by 3   13. Subtract 8   15. Subtract 8
17. $\left\{\frac{4}{3}\right\}$ 19. $\left\{\frac{18}{5}\right\}$ 21. $\{8\}$ 23. $\{48\}$ 25. $\{37\}$
27. $\{-4\}$ 29. $\{4\}$ 31. $\{2\}$ 33. $\{-3\}$ 35. $\left\{-\frac{7}{2}\right\}$
37. $\{6\}$ 39. $\{-2\}$ 41. $\{4\}$ 43. $\{5\}$ 45. $\left\{\frac{2}{3}\right\}$
47. 3, 1, 2   49. 2, 1, 3   51. $\{4\}$ 53. $\{-3\}$

55. $\left\{\frac{38}{3}\right\}$ 57. 5   59. 6   61. $\frac{3n + 2}{2} - 1 = 6$

Page 112, Review Exercises

1. $x + (-4) + y$ 2. $a + b + (-c) + (-d)$
3. $m + (-n) + 2$ 4. $3x + (-5y) + (-x)$
5. $4a - 11b$ 6. $16 - 2x$ 7. $-x - 13$ 8. $6m - 10$
9. $21x - 33$ 10. $4x - 28$ 11. No   12. No
13. Identity property for addition   14. Associative property of addition

Page 115, Classroom Exercises

1. $2 + (-x) = 3$ 2. $5 + (-2a) = 4$
3. $8 + \left(-\frac{w}{4}\right) = 12$ 4. $7 + \left(-\frac{b}{5}\right) = 8$ 5. Add -3 to both sides.   6. Add 1 to both sides.   7. Add $-\frac{2}{3}$ to both sides.   8. Add -0.4 to both sides.   9. $\{-3\}$
10. $\left\{-\frac{8}{5}\right\}$ 11. $\{-2\}$ 12. $\{6\}$

Pages 115–116, Written Exercises

1. $\{-2\}$ 3. $\{1\}$ 5. $\left\{\frac{1}{5}\right\}$ 7. $\{2\}$ 9. $\{2\}$ 11. $\{12\}$
13. $\{40\}$ 15. $\{-26\}$ 17. $\left\{\frac{11}{2}\right\}$ 19. $\{-3\}$ 21. $\{1\}$
23. $\{2\}$ 25. $\left\{-\frac{4}{3}\right\}$ 27. $\{-6\}$ 29. $\left\{\frac{5}{2}\right\}$ 31. $\{2\}$
33. $\left\{-\frac{15}{2}\right\}$ 35. $\{1\}$ 37. $\{7\}$ 39. $\left\{-\frac{1}{9}\right\}$
41. $\{-3, 3\}$ 43. $\{-4, 4\}$ 45. $\left\{-\frac{5}{2}, \frac{5}{2}\right\}$ 47. $\{-5, 5\}$

Page 116, Review Exercises

1. Commutative property of addition   2. Identity property for multiplication   3. Identity property for addition   4. Distributive property
5. 15 days

Day	Blue Team	Red Team
1	2600	1200
2	2700	1400
3	2800	1600
4	2900	1800
5	3000	2000
6	3100	2200
7	3200	2400
8	3300	2600

Day	Blue Team	Red Team
9	3400	2800
10	3500	3000
11	3600	3200
12	3700	3400
13	3800	3600
14	3900	3800
15	4000	4000

Page 116, Self-Quiz 2

1. $\{7\}$ 2. $\{-9\}$ 3. $\{85\}$ 4. $\{99\}$ 5. $\{7\}$ 6. $\{9\}$
7. $\{9\}$ 8. $\{80\}$ 9. $\left\{\frac{4}{3}\right\}$ 10. $\{-6\}$ 11. $\{-9\}$
12. $\{11\}$

Page 117, Extension

3. $\{84\}$ 4. $\{40\}$ 6. $\{48\}$ (All other answers are correct.)

Page 120, Classroom Exercises

1. Multiply both sides by $\frac{1}{5}$.　**2.** Expand the left side.
3. Simplify each side.　**4.** {11}　**5.** $\left\{\frac{59}{5}\right\}$　**6.** $\left\{\frac{55}{6}\right\}$
7. $\{-16\}$　**8.** {16}　**9.** {5}

Pages 120–121, Written Exercises

1. {7}　**3.** {6}　**5.** $\left\{\frac{26}{3}\right\}$　**7.** $\left\{\frac{8}{3}\right\}$　**9.** $\left\{\frac{5}{3}\right\}$　**11.** {8}
13. $\left\{\frac{7}{2}\right\}$　**15.** $\{-6\}$　**17.** $\left\{\frac{5}{3}\right\}$　**19.** $\{-6\}$　**21.** $\left\{\frac{10}{3}\right\}$
23. $\left\{\frac{22}{17}\right\}$　**25.** $\emptyset$　**27.** {3}　**29.** {6}　**31.** $\left\{-\frac{1}{2}\right\}$;
answers will vary.　**33.** $\left\{\frac{10}{3}\right\}$　**35.** $\{-1\}$　**37.** $\left\{\frac{11}{14}\right\}$
39. {1}　**41.** {2}　**43.** {7}

Page 121, Review Exercises

1. $-(-5)=|-5|$　**2.** $|-3|=|3|$　**3.** -8　**4.** 1.6
5. -2　**6.** -18　**7.** -125　**8.** 9　**9.** -9　**10.** 27

Page 124, Classroom Exercises

1. The four formulas are equivalent.　**2.** $h=\frac{A}{b}$
3. $b=\frac{2A}{h}$　**4.** $n=\frac{C-5}{2}$

Page 124, Written Exercises

1. $\frac{7}{a}$　**3.** $0.2r$　**5.** $15-a$　**7.** $\frac{j+10}{2}$　**9.** $a=\frac{c-2}{b}$
11. $p=\frac{r+3}{q}$　**13.** $r=3q-s$　**15.** $y=\frac{12-3x_1}{4}$
17. $x=2y+2$　**19.** $x=2y-6$　**21.** $y=\frac{2x-18}{3}$
23. $x_2=\frac{24-3x}{4}$　**25.** $r=\frac{w-t}{s}$　**27.** $x=\frac{z-by}{a}$
29a. $h=\frac{2A}{b_1+b_2}$　**29b.** $h=15$　**31.** $x_1=\frac{10-x_2}{x_3}$,
$x_2=10-x_1x_3$　**33.** $x_1=\frac{10x_2}{x_3}$, $x_2=\frac{x_1x_3}{10}$
35. $x_1=10+x_2x_3$, $x_2=\frac{x_1-10}{x_3}$

Page 124, Review Exercises

1. $8x-10$　**2.** $2a-4b$　**3.** $x+4y$　**4.** $-2r+7s$

Pages 127–128, Classroom Exercises

1a. $m=$ Marcia's age　**1b.** $m-4=$ Sidney's age
1c. $m-4=17$　**1d.** 21　**2a.** $k=$ Kevin's points
2b. $2k=$ Tracy's points　**2c.** $2k=30$　**2d.** 15
3a. $w=$ width　**3b.** $4w=$ length
3c. $4w+w+4w+w=50$　**3d.** Width is 5 cm; length
is 20 cm.　**4a.** $c=$ Carrie's points　**4b.** $c+6=$ Joan's
points　**4c.** $c+c+6=26$　**4d.** Carrie: 10; Joan: 16

Pages 128–129, Written Exercises

1. $\frac{1}{2}w=30$　**3.** $w+5=12.3$　**5.** $1.3w+w=10.35$
7. $(w+5)+w+(w+5)+w=78$

---

9. $x+(x+15)=237$　**11.** $0.10x=11.20$
13. $x+12=200$　**15a.** $t=$ Tom's height
15b. $t+8=150$　**15c.** {142}　**15d.** 142 cm
17a. $c=$ Carla's weight　**17b.** $c-32=111$
17c. {143}　**17d.** 143 pounds　**19a.** $c=$ cost of car
19b. $\frac{1}{3}c=2715$　**19c.** {8145}　**19d.** \$8145
21a. $g=$ amount Gerry paid　**21b.** $g+(g+10)=110$
21c. {50}　**21d.** Gerry: \$50; Sandy: \$60
23a. $w=$ width　**23b.** $w+3w+w+3w=48$
23c. {6}　**23d.** Width: 6 m; length: 18
25a. $c=$ Connie's records　**25b.** $c+\frac{1}{2}c=54$
25c. {36}　**25d.** 36 records　**27a.** $w=$ width
27b. $w(1.4w)=315$　**27c.** {15}　**27d.** 15 cm

Page 129, Review Exercises

1. $23-4a$　**2.** $-x-25$　**3.** $7x-2$　**4.** $\left\{-\frac{16}{3}\right\}$
5. {4}　**6.** {1}

Page 130, Self-Quiz 3

1. {4}　**2.** {5}　**3.** {3}　**4.** $\left\{\frac{19}{2}\right\}$　**5.** $\left\{\frac{7}{2}\right\}$　**6.** {16}
7. $l=\frac{E}{R}$　**8.** $t=\frac{A-7}{8}$　**9.** $4l=84$; 21 cm
10. $m+(m+4)=24$; Mike: 10, Drew: 14

Page 130, Extension

1. $A=L*W$　**2.** $X=(D-B)/(A-C)$　**3.** No

Page 132, Classroom Exercises

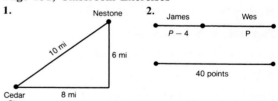

Pages 133–134, Written Exercises

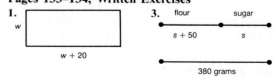

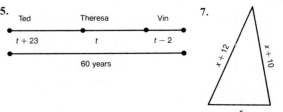

9a. w = width of board

9b.

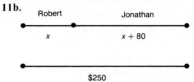

w + 7

w

9c. $w + (w + 7) + w + (w + 7) = 94$ **9d.** {20}   **9e.** 20 cm

11a. x = amount Robert paid

11b.

Robert   Jonathan

x $x + 80$

$250

11c. $x + (x + 80) = 250$ **11d.** {85}   **11e.** $85

13c. $x + (x - 7) = 27$ **13e.** 17

15c. $x + (x + 4) + (x + 5) + (x + 10) = 139$ **15e.** 30 cm

17c. $16 + x = 35 + 7$ **17e.** 26 miles

19c. $x + \left(x + \frac{1}{2}\right) + (x + 1) = 12$ **19e.** $3\frac{1}{2}$ ft, 4 ft, $4\frac{1}{2}$ ft

21c. $x + (x + 1) + (x + 2) = 96$ **21e.** 31 cm

Page 134, Review Exercises

1. 1000   **2.** 1000   **3.** 1500   **4.** 0.273   **5.** 3   **6.** 3

Pages 135–137, Chapter Review

1. -1 **3.** 1, 2   **5.** Not equivalent   **7.** {5}

9. $\left\{-\frac{10}{3}\right\}$ **11.** $\left\{-\frac{13}{4}\right\}$ **13.** {72}   **15.** $\left\{\frac{7}{2}\right\}$ **17.** {−9}

19. {6}   **21.** {3}   **23.** $r = \frac{d}{t}$ **25.** $x = \frac{y + 7}{3}$

27. $j + 23 = 151$; 128 pounds

Pages 137–138, Chapter 3 Self-Test

1. None   **2.** 0, 1   **3.** $x = -\frac{3}{4}$ **4.** $y = -15$

5. {−11}   **6.** {4}   **7.** $\left\{-\frac{4}{3}\right\}$ **8.** {32}   **9.** {2}

10. {8}   **11.** {2.6}   **12.** {7}   **13a.** Multiply by 3.

13b. Add -6.   **13c.** Multiply by $-\frac{1}{2}$.   **14.** {−28}

15. $\left\{-\frac{17}{2}\right\}$ **16.** {−2}   **17.** {2}   **18.** $r = \frac{A}{2\pi h}$

19. $x = \frac{y - 3}{5}$ **20.** $x + (x + 3) + (x + 8) = 59$; 16 cm

21. $x + (x + 5) = 5.50$; Kim: $5.25, Karen: $.25

Pages 138–139, Practice for College Entrance Tests

1. C   **2.** D   **3.** A   **4.** C   **5.** B   **6.** D   **7.** E   **8.** C

9. C   **10.** A   **11.** A   **12.** E

Pages 139–141, Cumulative Review (Chapters 1–3)

1. 9   **3.** 0   **5.** 44   **7.** 54   **9.** 30   **11.** 1   **13.** Yes

15. Commutative (add)   **17.** Distributive   **19.** $4x$

21. $2B$ **23.** $B - 4$ **25.** T   **27.** F (0 is)   **29.** 35

31. 2.5   **33.** -64 **35.** -24 **37.** -12 **39.** -7

41. $4.5 - x$ **43.** $t + 3$ **45.** $5a - 16$ **47.** $26x + y$

49. 11:30 A.M.

Time	Taxi (mi)	Bus (mi)	Difference
8:30	30		30
9:00	60	25	35
9:30	90	50	40
10:00	120	75	45
10:30	150	100	50
11:00	180	125	55
11:30	210	150	60

51. {−2}   **53.** $7y = 18$ **55.** No   **57.** {−35}

59. {10}   **61.** {13}   **63.** {−3}   **65.** {−18}   **67.** {6}

69. $n = \frac{t + 1}{6}$ **71a.** w = width

71b. $2(w + 8) + 2w = 40$ **71c.** {6}   **71d.** 6 cm

Chapter 4

Page 145, Classroom Exercises

1.–5. $x^2 y^2$, $\frac{1}{7}$ **6.–10.** All except $\frac{1}{x} + y$ **11.** $2x + 2y$

12. $5x + 3$ **13.** $-2y$ **14.** $-2c - 7$

Pages 145–147, Written Exercises

1. $2a$, $5b^2$, $\frac{8}{9}d$ **3.** 2   **5.** 3   **7.** 4   **9.** $7x^2 - 4x + 11$

11. $8a + 3b - 5c$ **13.** $4x^2 + 7x + 1$

15. $2a + 7b + c$ **17.** $8x + 12$ **19.** $14n + 6m + 6$

21. $2a^2 + a + 5$ **23.** $5x - 9y + 6z$

25. $4x^2 + 7x + 16$ **27.** $3a + 3b - 2c$ **29.** $6x^2 - 6$

31. $-8a - 2$ **33.** $13t^3 + 8t^2 + 7t + 6$

35. $5t^3 + t^2 + t$ **37.** $8t^3 + 4t^2 + 9t + 6$

39. $n + (n + 1) + (n + 2) + (n + 3) + (n + 4) = 5n + 10$

41. $n + (n + 2) + (n + 4) + (n + 6) + (n + 8) +$

$(n + 10) = 6n + 30$ **43.** $2wl + 2wh + 2lh$

45. $2\pi r^2 + 2\pi rh$

Page 147, Review Exercises

1. 11   **2.** 7   **3.** 12   **4.** 50   **5.** 49   **6.** 42   **7.** 32

8. 110   **9.** 100   **10.** 400   **11.** 3; 1; 2; 4

Pages 149–150, Classroom Exercises

1. $6a$, $x^2 y$ **2.** $5a + 2b$ **3.** $5 + 3x + 2x^2$, $7x - 3y - z$

4. 3   **5.** 0   **6.** 1   **7.** 2   **8.** 5   **9.** 2   **10.** 4   **11.** 0

12. $3 + 5x - 2x^2$ **13.** $2a^3 - 3a^2 + 5a - 5$

14. $3x^3 + 4x^2 y - 2xy + y$

Pages 150–151, Written Exercises

1. $5x$, xy, -6, 6^2 **3.** All but $3xy^3$ **5.** 2   **7.** 0

9. 5   **11.** 3   **13.** 3   **15.** 2   **17.** 1   **19.** 0   **21.** 3

23. 4   **25.** $4 + 5x + 9x^2$ **27.** $5 - 3b + 9b^3$

29. $8x^2 + 2x + 5$ **31.** $5y^3 - 7y^2 + 4y$ **33.** 1   **35.** 2

37. $-5y^2 + 3xy + 7x^2$ **39.** $8 - 7xy + 6x^2y^2$
41. $2x^3 - 3xy^2 + 5y^3 + 6x^2y$ **43.** $6x^2y^2 - 7xy + 8$
45a. 2 **45b.** 2 **45c.** 2 **47a.** 2 **47b.** 2 **47c.** 1

Page 151, Review Exercises
1. Yes **2.** No **3.** No **4.** No **5.** No **6.** No
7. Identity property for multiplication **8.** Distributive property **9.** Commutative property of addition
10. Commutative property of multiplication **11.** Associative property of addition **12.** Commutative property of multiplication

Page 152, Extension
1. -300; 1700; 1700; -3; -4; 1 **2.** 1.7
3. 50 PRINT X ↑ 2 + 7 * X * Y − Y ↑ 2
4. 50 PRINT X ↑ 4 + 3 * X ↑ 2 * Y − Y ↑ 2

Page 154, Classroom Exercises
1. 6 **2.** 3^{12} **3.** 4^7 **4.** 8 **5.** 9 **6.** r^{a+b} **7.** y^{11}
8. $12x^5$ **9.** $-10x^3y^3$

Pages 154–156, Written Exercises
1. 7 **3.** 8 **5.** 5 **7.** 6 **9.** 10 **11.** 4 **13.** a^{20}
15. $6b^5$ **17.** $12q^{20}$ **19.** $100r^{101}$ **21.** $\frac{1}{8}t^6$
23. $15a^3b^5$ **25.** 10 **27.** 23 **29.** 5 **31.** 7 **33.** 6
35. \$25,600 **37.** 2^{63} **39.** $-360\,x^3y^3$ **41.** $2a^3$
43. $-a^6$ **45.** x^8 **47.** Yes **49.** Yes **51.** $6ab$
53. $6k^2$ **55.** $\frac{1}{2}abc$

Page 156, Review Exercises
1. $5x$ **2.** $5x - 5$ **3.** $2x + 7$ **4.** $7x + 12$

Page 156, Self-Quiz 1
1. Binomial **2.** Monomial **3.** Trinomial
4. $11x^2 + 3x - 5$ **5.** $4a^2 + ab$ **6.** $3x^2 + 2x + 10$
7. $5x^2 + x + 2$ **8.** x^9 **9.** $-2y^5$ **10.** $15a^3b^4$
11. $-6x^6y^5$

Page 159, Classroom Exercises
1. Yes **2.** No **3.** No **4.** No **5.** 6×10^3
6. 4.68×10^5 **7.** 5.675×10^7 **8.** 2.6×10^7
9. 30,000 **10.** 890,000 **11.** 70.7 **12.** 4,505,000,000
13. 8×10^8 **14.** 2.673×10^8

Pages 159–160, Written Exercises
1. No **3.** Yes **5.** No **7.** No **9.** 5.3×10^5
11. 3.28×10^8 **13.** 2.5×10^7 **15.** 2.17×10^{14}
17. 5,000,000 **19.** 650.3 **21.** 81.23
23. 4,000,000,000 **25.** 1.52×10^8 km
27. 4.071×10^{13} km **29.** 1.66×10^8 km
31. 2×10^{10} **33.** 1.21×10^{50} **35.** 7.29×10^{20}
37. 8×10^{12} **39.** 2×10^{12} **41.** 1.922×10^{13}
43. 3.75×10^{15} cubic meters **45.** 1.073×10^{13} cubic meters

Page 160, Review Exercises
1. $3\frac{1}{2}$ **2.** 3 **3.** -3 **4.** -6 **5.** 10 **6.** -1
7. -13.75 **8.** 14 **9.** 7 **10.** -11 **11.** 16

Page 161, Extension
1. 52,180,000 **2.** 3,946,281,000 **3.** 460,000
4. 957 **5.** 48.7 **6.** 300,100 Answers for exercises 7–12 will vary according to the number of digits in the calculator display. **7.** $1.446 \cdot 10^9$ **8.** $1.754245 \cdot 10^{13}$
9. $7.223988895 \cdot 10^{11}$ **10.** $1.37 \cdot 10$ **11.** $8.2 \cdot 10^3$
12. $4.12 \cdot 10^4$

Page 163, Classroom Exercises
1. a^6 **2.** a^8 **3.** m^9 **4.** m^6 **5.** x^{15} **6.** $9x^2$
7. $8x^3$ **8.** $16x^8y^4$

Page 164, Written Exercises
1. x^{20} **3.** a^{21} **5.** $16c^4$ **7.** $64y^6$ **9.** $\frac{1}{8}b^9$
11. $0.01r^8$ **13.** $4x^6$ **15.** $-w^{15}$ **17.** x^8y^4
19. $32c^{15}d^5$ **21.** $(5 \cdot 7)^3$; 5^2 times as large **23.** 4^{50}; 4^{35} times as large **25.** $18x^5$ **27.** $-a^5b^3$ **29.** $8x^5y^3$
31. $16a^{12}$ **33.** x^{10} **35.** $-8t^5$ **37.** $a^7b^5c^6$
39. $-108x^7y^8$ **41.** 10^4 **43.** 10^5

Page 164, Review Exercises
1. 30 **2.** 34 **3.** 0 **4.** 3 **5.** -7 **6.** $-\frac{8}{9}$
7. $-5 + 4x$ **8.** $4x + 7$ **9.** 24 **10.** 1 **11.** 5

Page 166, Classroom Exercises
1. $12x + 6y$ **2.** $6a - 10b$ **3.** $-15x + 20$
4. $3x^2 + 4x$ **5.** $6a^2 + 3a$ **6.** $-27c^3 - 12c^4$
7. $-3a^3b - 3ab^3$ **8.** $-7y + 3y^2$ **9.** $8a^3 - 12a^2$
10. $42a + 14b - 7c$ **11.** $10ab - 6ac - 4ad$

Pages 166–167, Written Exercises
1. $4a + 12$ **3.** $10d + 20$ **5.** $-5n + 10m$
7. $-15d^2 + 10d$ **9.** 396 **11.** 1495 **13.** 143
15. 0.52 **17.** $-15r^3 + 10r^2 + 40r$
19. $30t^4 + 60t^3 + 50t^2$ **21.** $a^3 + a^2b$ **23.** $4xy^3 + 2y^4$
25. $-a^2b - ab^2$ **27.** $-2s^2t + st^2$ **29.** $x^3y + x^2y^2$
31. $x^3y + x^3y^2 + x^2y^2$ **33.** $-a^2b^2 + 2a^2b^3 - ab^3$
35. $2x(3x + 7) = 6x^2 + 14x$
37. $\frac{1}{2}a(b - 10) = \frac{1}{2}ab - 5a$
39. $\frac{1}{2}ab(4a + 10) = 2a^2b + 5ab$

Page 168, Review Exercises
1. $4x$ **2.** $6x - 7$ **3.** $x + 6y - 6$ **4.** $x + y$
5. $7x - 14$ **6.** $2x + 21$ **7.** 8 **8.** $3a - 8b$
9. $13a + 3$ **10.** $a - b$ **11.** $2a - b + 3$ **12.** $4a + 4b$

13. Tenth

Day	Total $	
1	1	$\}+2$
2	3	$\}+4$
3	7	$\}+8$
4	15	$\}+16$
5	31	$\}+32$
6	63	$\}+64$
7	127	$\}+128$
8	255	$\}+156$
9	511	$\}+512$
10	1023	

Page 168, Self-Quiz 2
1. 2.89×10^6 **2.** 57,000 **3.** 4.56×10^8 **4.** y^{12}
5. $16a^8$ **6.** $a^{10}b^{15}$ **7.** $-27b^9$ **8.** $8x + 20y$
9. $24a - 18b$ **10.** $-6r^2 + 16r$ **11.** $6ac^2 - 9bc^2 + 3c^3$

Page 172, Classroom Exercises
1. $d = rt$ **2.** 30 km **3.** $3.5x$ mi **4.** 50 km/h
5. $\frac{y}{5}$ km/min **6.** $\frac{1}{8}$ h **7.** $(55x + 27.5)$ mi **8.** 2.5

Pages 172–173, Written Exercises
1. 5 **3.** 5 **5.** 150 mi **7.** 45 m **9.** $600x$ mi
11. 4 h **13.** 50 m/s **15.** $10x$ mi
17. $10x = 8(x + 10)$; 40 **19.** $60x = 80(x - 1)$; 4
21. $40x + 60x = 1000$; 10
23. $40x + 60(x + 5) = 1000$; 7 **25.** 1300 **27.** 3 h
29. 9 min **31.** 1 **33a.** Mrs. Douglas: 3, Mr. Douglas: 7
33b. 285 **33c.** 630

Page 174, Review Exercises
1. $-1, 1$ **2.** 0, 1 **3.** $-2, 1$ **4.** $x = 5$ **5.** $x = 50$
6. $x = 30$ **7.** {7} **8.** {1} **9.** {−8.5}

Page 177, Classroom Exercises
1. x^2 **2.** $\frac{1}{x^2}$ **3.** $\frac{y^2}{4}$ **4.** $3y^4$ **5.** $\frac{2}{3a^2}$ **6.** $\frac{2a^2}{3b}$
7. $\frac{-5y^2}{2x}$ **8.** $\frac{4m}{3n^2}$ **9.** 1 **10.** 1 **11.** 2×10^3
12. 4×10^2

Pages 177–178, Written Exercises
1. c^6 **3.** $\frac{1}{b^9}$ **5.** $\frac{3a^2}{2}$ **7.** $\frac{4}{3c^6}$ **9.** j **11.** 5
13. $\frac{-3rs^2}{2}$ **15.** $\frac{1}{y}$ **17.** $\frac{-1}{b}$ **19.** 1 **21.** {8}
23. {1} **25.** $\left\{\frac{1}{4}\right\}$ **27.** {8} **29.** {2} **31.** $\left\{\frac{3}{2}\right\}$
33. $\left\{\frac{1}{3}\right\}$ **35.** {7} **37.** 3×10^5 **39.** $\approx 8.1 \times 10^2$
41. $2a^2b$; $3a^3b^3$; $3a^2b^2$ **43.** $4x^2y^4$; $-\frac{4}{3}x^2y^4$; $-3x^2y$
45. 1×10^2 **47.** 7×10^3

Page 178, Review Exercises
1. $\left\{-\frac{7}{4}\right\}$ **2.** {45} **3.** {32} **4.** $\left\{\frac{7}{2}\right\}$ **5.** {26}
6. {25} **7.** {3} **8.** $\left\{\frac{25}{2}\right\}$ **9.** {12}

Page 181, Classroom Exercises
1. $25x$ **2.** $0.1y$ **3.** $39g$ **4.** $5c$ **5.** 16 cents
6. $\frac{68}{b}$ cents **7.** $\frac{d}{c}$ cents **8a.** $2q$ **8b.** $20q$ **8c.** $25q$
9. $d =$ the number of lines; $25 \cdot 2d + 10d = 420$;
7 dimes, 14 quarters **10.** $d =$ the number of dimes;
$5(d + 4) + 10d = 155$; 9 dimes, 13 nickels

Pages 182–183, Written Exercises
1. $50h$ **3.** $0.1d$ **5.** $30s$ **7.** $\frac{1}{6}$ dollars **9a.** $d + 4$
9b. $5(d + 4)$ **9c.** $10d$ **11.** $2x$ **13.** $\frac{1}{2}x$
15. $x + 4.20$ **17.** $2.05 **19.** $0.85 **21.** 2.5 **23.** 6
25. 10 **27.** 19 **29.** 12 **31.** 20 quarters, 26 dimes
Only the answer to the problem and the check are the
same.

Page 183, Review Exercises
1. {10} **2.** $\left\{\frac{1}{2}\right\}$ **3.** {12} **4.** {5} **5.** $h = \frac{2A}{b}$
6. $w = \frac{V}{lh}$ **7.** $h = \frac{2A}{b_1 + b_2}$ **8.** $b_1 = \frac{2A}{h} - b_2$

Page 184, Self-Quiz 3
1. 63 mi **2.** 9s **3.** 84 km/h **4.** 3 P.M. **5.** $\frac{1}{x}$
6. $\frac{z^8}{3}$ **7.** $2q^2$ **8.** $\frac{-b}{a}$ **9.** $25(n + 1)$ **10.** 7 quarters,
11 dimes

Page 184, Extension
$\frac{121}{100}$

Page 187, Classroom Exercises
1. Answers will vary. **2.** Answers will vary.
3. $-2, -1, 0$ **4.** $-7, -5, -3$ **5.** 16, 18, 20
6. $x - 1, x + 1$ **7.** $y + 2, y + 4$ **8.** $z - 2, z - 4$
9. 15 and 16 **10.** 86 and 88

Pages 187–189, Written Exercises
1. 17, 18, 19, 20 **3.** 30, 32, 34 **5.** $-17, -15, -13$
7. $n + 1, n + 2$ **9.** $n + 2, n + 4$ **11.** 37
13. $n =$ the smallest integer; $n + (n + 1) + (n + 2) = 171$;
56, 57, 58 **15.** $n =$ the smallest integer; $n + (n + 1)$
$+ (n + 2) = 0$; -1 **17.** $n =$ the smallest integer;
$n + (n + 2) + (n + 4) = 390$; 128, 130, 132 **19.** 31,
32, 33, 34, 35 **21.** 84, 86, 88, 90, 92 **23.** Odd
25. 91 **27.** 49, 50, 51 **29.** 0 **31.** $\frac{1}{2}$ and $1\frac{1}{2}$
33. Ken; sum should be even.

Page 189, Review Exercises

1. $r = \dfrac{d}{t}$ 2. $p = \dfrac{i}{rt}$ 3. $x = \dfrac{-b}{a}$ 4. $l = \dfrac{p}{2} - w$

5. $-2a$ 6. $-4 - 2x$ 7. $x + 7$

Pages 191–192, Chapter Review

1. $4x^2 - x + 5$ 3. $7x^3 - 5x^2 - 3x - 1$ 5. 2   7. 2

9. $-6x^2y^3$ 11. 36,400,000   13. x^{12} 15. $25x^2b^6$

17. $10y^2 - 5xy$ 19. 2 P.M., 110 miles   21. $\dfrac{1}{3x^2}$

23. $7 \cdot 10^2$ 25. 2.5   27. 93 and 95

Page 193, Chapter 4 Self-Test

1. $2x^2 + 2xy$ 2. $10a^2 + bc$ 3. $5x^2 - x - 19$ 4. A

5. B   6. 3   7. $6a^5$ 8. $-15x^7$ 9. $-24x^8y^5$

10. 8.2×10^8 11. $8.4 \cdot 10^2$ 12. $2.38 \cdot 10^8$ 13. x^{10}

14. $16y^{12}$ 15. $\dfrac{1}{4}a^4b^6$ 16. $6x^2 - 10x$ 17. $6y^2 - 8x^2$

18. $3a^3 + 6a^2b$ 19. 480 km   20. $\dfrac{x}{2y}$ 21. $3x^2y^2$

22. $\dfrac{3a}{b}$ 23. $\approx 2.4 \cdot 10^3$ 24. 16   25. 58, 60, 62

Pages 194–195, Practice for College Entrance Tests

1. C   2. A   3. E   4. A   5. B   6. E   7. A   8. C
9. A   10. D   11. B   12. E   13. A

Chapter 5

Page 199, Classroom Exercises

1. (2, 4)   2. (−1, 4)   3. (−2, 0)   4. (0, 1)
5. (−2, −3)   6. (0, −4)   7. (2, −5)   8. (4, 0)
9. *A* and *G, C* and *E, D* and *F*   10. *A* and *B, C* and
H   11. (0, 0)   12. I   13. IV   14. II   15. *y*-axis
16. III

Pages 199–200, Written Exercises

1. (−5, 3)   3. (0, 3)   5. (4, 1)   7. (2, −1)
9. (0, −5)   11. (−5, −2)   13. II   15. *y*-axis   17. I
19. IV   21. *y*-axis   23. III   25. *Q*   27. *M*   29. *R*
31. *W*

33. 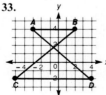   35.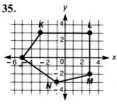

37. IV   39. I   41. I   43. (4, 4)   45. (5, 1)

---

47. Answers will vary.   49. Answers will vary.

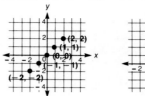 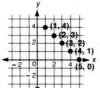

Pages 200, Review Exercises

1. 60   2. 30   3. 25   4. 1   5. −5   6. $6\dfrac{2}{3}$ 7. $5\dfrac{1}{2}$
8. −6

Page 203, Classroom Exercises

1. A set of ordered pairs   2. No   3. Yes   4. {3}
5. (−1, 4), (−2, 2), (−2, 3), (2, 4); domain: {−1, −2, 2};
range: {4, 2, 3}   6. (2, 3), (2, 1), (2, −2), (−2, 0),
(−2, 1); domain: {2, −2}; range: {3, 1, −2, 0, 1}

7.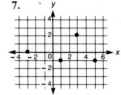

8a.–8b. Answers will
vary.
8c. A line
8e. Real numbers

Pages 203–205, Written Exercises

1. Yes   3. No   5. No   7. (2, −2), (−3, 3),
(0.5, −0.5), (−100, 100)   9. (6, 6), (−6, 6), (8, 8),
(−8, 8)   11. Domain: {1, 3, 5}; range: {2, 4, 6}

13. Domain: $\left\{\dfrac{1}{2}\right\}$; range: $\left\{\dfrac{1}{3}\right\}$ 15. Domain: {5, −1, −4};
range: {6, 0, −3}   17. Domain and range: {1, 2, 3}
19a. {(1, 1), (−1, 3), (−3, 2), (3, −2)}
19b. {1, −1, −3, 3}   19c. {1, 3, 2, −2}

21. 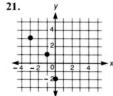   23.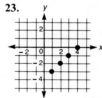

25a. {(1, 10), (10, 1), (−1, −10), (−10, −1), (2, 5),
(5, 2), (−2, −5), (−5, −2)}   27a. {(−5, 5), (−4, 4),
(−3, 3), (−2, 2), (−1, 1)}   29a. {(−4,16), (−3, 9),
(−2, 4), (−1, 1)}   31a. {(5, 4), (5, 3), (5, 2), (5, 1),
(4, 3), (4, 2), (4, 1), (3, 2), (3, 1), (2, 1)}   33a. {(3, 3),
(−3, 3), (3, −3), (−3, −3)}   35a. {(3, 4), (−3, 4),
(3, −4), (−3, −4), (0, 5), (0, −5), (4, 3), (4, −3),
(−4, 3), (−4, −3), (5, 0), (−5, 0)}

Page 205, Review Exercises

1. 36   2. 38   3. −52   4. 8   5. $\dfrac{1}{4}$ 6. 1   7. $1\dfrac{2}{5}$

8. $-\dfrac{1}{24}$ **9.** 10   **10.** 7

Page 208, Classroom Exercises

1a. Yes   **1b.** Yes   **1c.** Yes   **2a.** No   **2b.** Yes
2c. Yes   **3a.** Yes   **3b.** No   **3c.** No   **4.** 1
5. Infinitely many

6.

x	4	10	-3	3
y	-2	4	-9	-3

7.

p	7	10	-2	1
q	4	7	-5	-2

Pages 208–210, Written Exercises

1a. No   **1b.** Yes   **1c.** Yes   **3a.** Yes   **3b.** No
3c. Yes   **5.** {(−2, −2), (0, 0), (2, 2)}   **7.** {(−2, −4),
(0, 0), (2, 4)}   **9.** {(−4, −16), (0, 0), (4, 16)}

11. 3; 0; 8; 16   **13.** 4; 0; 5; 9   **15.** 4; 6; $3\dfrac{1}{3}$; $4\dfrac{1}{2}$

17. -4; 6; $-3\dfrac{1}{3}$; $7\dfrac{1}{2}$ **19.** 5; 0; -5; 10

21. 2; 3; 1; $2\dfrac{1}{3}$ **23.** 0; 48; 48; $\dfrac{1}{3}$ **25.** 2; 1; -1; $\dfrac{1}{2}$

27. {(0, 3), (1, 4), (2, 5)}; {(0, −3), (1, −2),
(2, −1)}; no   **29.** {(−3,3), (0, 0), (3, 3)}; {(−3, 3),
(0, 0), (3, 3)}; yes   **31.** {(0, 0), (3, 36), (5, 100)}; {(0, 0),
(3, 36), (5, 100)}; yes   **33.** No   **35.** Yes

Page 210, Review Exercises

1. $a + 5$ **2.** $5x + 8 - y$ **3.** $a - 3 + 3b$ **4.** $-x - 7y$
5. $a + 3b$ **6.** $7a + 5b$

Page 210, Self-Quiz 1

1.
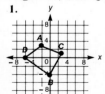

2a. III
2b. y-axis
2c. II

3a.
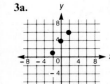

3b. Domain: {1, −2, 0}
3c. Range: {5, 1, 4}

4a.
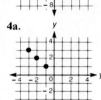

4b. Domain:
{−3, −2, −1}
4c. Range: {1, 2, 3}

5.

x	-2	1	0.5	3
y	-5	1	0	5

6.

a	0	$1\dfrac{1}{2}$	3	-3
b	4	2	0	8

Page 214, Classroom Exercises

1.
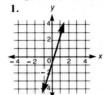

2.
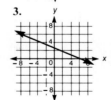

3.
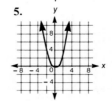

4.
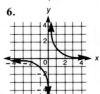

5.

6.
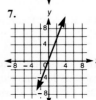

Pages 214–215, Written Exercises

1. -1; 3; 7
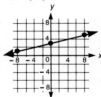

3. 1; 3; 5
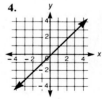

5. 2; 0; -2

7.
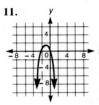

9.
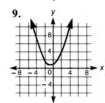

11.
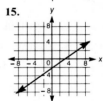

13.
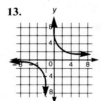

15.

17.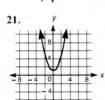

19.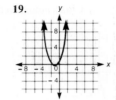

21.

23.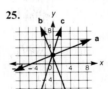

21.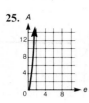

23.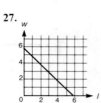

25.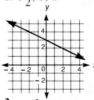

27. $\frac{2}{3}$ **29.** 3

31. $\frac{1}{2}$ **33.** None

25.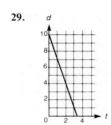

27.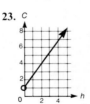

35.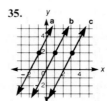

37a. $-1; -1; -1; -1$
37b. $-1; -1; -1; -1$
37c. Product is -1.

29.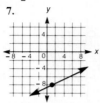

Page 215, Review Exercises

1. $\{-3.25\}$ **2.** $\left\{-7\frac{3}{4}\right\}$ **3.** $\{-27\}$ **4.** $\{120\}$

Page 218, Classroom Exercises

1. Vertical lines have no slope, horizontal lines have 0 slope.   **2a.** Falls   **2b.** Horizontal   **2c.** Rises   **3.** 2

4. $\frac{1}{2}$ **5.** -1 **6.** -3

7.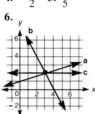

8.

Pages 219–221, Written Exercises

1. Positive   **3.** Negative   **5.** Negative   **7.** $\frac{1}{3}$ **9.** 3

11. 0   **13.** -2 **15.** -3 **17a.** 5   **17b.** $-\frac{1}{5}$ **17c.** 5

17d. $-\frac{1}{5}$ **19a.** $-\frac{2}{3}$ **19b.** 0   **19c.** None

Page 221, Review Exercises

1. $\{-4\}$ **2.** $\{8\}$ **3.** $\{12\}$ **4.** $\{-3\}$ **5.** $\{10\}$ **6.** $\{48\}$

Page 221, Self-Quiz 2

1. $3\frac{1}{2}$; 3; 2

2. $-8; -1; 0; 1; 8$

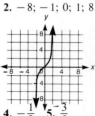

3.

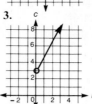

4. $-\frac{1}{2}$ **5.** $-\frac{3}{5}$

6.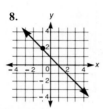

Page 222, Extension

1. 0.5   **2.** -4 **3.** 5   **4.** 0.5   **5.** -7.5 **6.** 0.4
7. $-0.\overline{3}$ **8.** -0.44

Page 227, Classroom Exercises

1. $y = -\frac{5}{2}x + 3$ **2.** Slope: -0.5, y-intercept: 7

3. Slope: $\frac{5}{3}$, y-intercept: $\frac{2}{3}$ **4.** $y = -5x$ **5.** Yes
6. No **7.** No

Pages 227–229, Written Exercises

1. 5; 3 **3.** $-\frac{1}{2}$; -6 **5.** $\frac{1}{2}$; 0 **7.** 2; 2 **9.** 1; 0

11. $\frac{1}{2}$; -1 **13.** 0; -2 **15.** $y = 3x + 10$

17. $y = -6x + 0.75$ **19.** $y = 0x - 3$ **21.** $y = \frac{2}{3}x - 2$

23. Neither **25.** Neither **27.** Perpendicular

29. $y = -\frac{2}{3}x + 2$; $-\frac{2}{3}$; 2 **31.** $y = \frac{3}{2}x + 3$; $\frac{3}{2}$; 3

33. $y = 2x - \frac{3}{2}$; 2; $-\frac{3}{2}$ **35.** Perpendicular

37. Parallel **39.** Neither **41.** $y = \frac{5}{3}x - 3$; $\frac{5}{3}$; -3

43. $y = -\frac{1}{3}x + \frac{7}{6}$ **45.** $y = -6x + 8$; -6; 8

47. $y = -\frac{1}{2}x + \frac{4}{3}$; $-\frac{1}{2}$; $\frac{4}{3}$ **49.** $y = -\frac{3}{4}x + \frac{3}{2}$; $-\frac{3}{4}$; $\frac{3}{2}$

51. q **53.** s **55.** t **57.** $y = -2x$ **59.** $y = \frac{1}{2}x$

61. 3 **63.** 2 **65.** $y = \frac{3}{5}x + 2$

Page 229, Review Exercises

1. $1\frac{7}{12}$ **2.** $-\frac{1}{12}$ **3.** $\frac{5}{8}$ **4.** $\frac{9}{10}$ **5.** $b + (b + 10) +$
$(b + 10) = 125$; 45 cm; 45 cm; 35 cm

Page 229, Extension

2. ≈ 5.2, ≈ 5.4
3. The graph with the smaller
slope produces the better buy.

Page 232, Classroom Exercises

1. $y = 4x - 11$ **2.** $y = -2x$

Pages 232–233, Written Exercises

1. $y = x - 2$ **3.** $y = \frac{1}{4}x + 1$ **5.** $y = 2x + 10$

7. $y = -\frac{1}{2}x + 4$ **9.** $y = 2x + 2$ **11.** $y = 2x + 4$

13. $y = 2x + 8$ **15.** $y = 2x$ **17.** $y = 3x - 6$

19. $y = \frac{1}{2}x + 4$ **21.** $y = -\frac{2}{3}x + 4$ **23.** $y = -2x + 4$

25. $y = -3x + 21$ **27.** 3 **29.** 2 **31.** -2 **33.** $-\frac{1}{2}$

35. $y = -2x + 7$ **37.** $y = 2x - 10$ **39.** $y = \frac{1}{8}x + 5$

41. $y = \frac{3}{2}x + \frac{19}{2}$ **43.** $y = 25x + 20$

45. $y = \frac{5}{9}x - \frac{160}{9}$ **47.** $\frac{y_2 - y_1}{x_2 - x^1}$ **49.** $y_1 - mx_1$

Page 233, Review Exercises

1. 2 **2.** $x - 3$, $a^2 + b^2$ **3.** $-3x^2 + 5x + 4$ **4.** {6}
5. a^6 **6.** $-12a^2b^3$ **7.** 180 **8.** 1590

Page 236, Classroom Exercises

1. See page 234. **2.** Yes **3.** No **4.** Yes **5.** Yes

6. Yes **7.** No **8.** Yes **9.** Yes **10.** No **11.** Yes
12. Yes **13.** Yes **14.** Yes **15.** No

Pages 237–238, Written Exercises

1. Yes **3.** Yes **5.** No **7.** Yes **9.** Yes **11.** No
13. Yes **15.** Yes **17.** Yes **19.** Yes **21.** Yes
23. No **25.** Yes **27.** Yes **29.** No

31. No 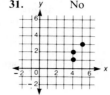 **33.** Yes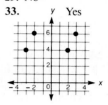

35. Yes; no 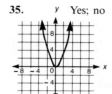 **37.** Yes; yes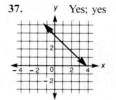

39. Yes; yes 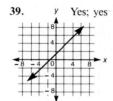 **41.** Yes; yes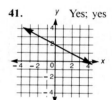

43. Yes; yes **45.** Yes; no **47.** Yes; no
49.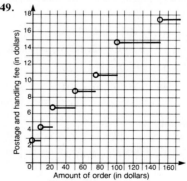

51.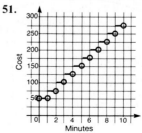

Page 238, Review Exercises
1. $3 \cdot 10^{10}$ 2. $345{,}000$ 3. $2.4 \cdot 10^8$ 4. a^8 5. $4x^6$
6. $10{,}000\ a^8 b^{12}$ 7. $25q$ 8. 18 dimes, 6 quarters

Page 239, Self-Quiz 3
1. $\frac{2}{3}$; 4 2. $y = 2x + 5$ 3. $\frac{5}{2} \cdot -\frac{2}{5} = -1$
4. $y = \frac{2}{3}x + 4$ 5. $y = 2x - 3$ 6. $y = -2x + 4$
7. No 8. No 9. Yes

Page 240, Extension
1. 7; 7.3 2. 5; 4.6 3. 1; 0.5 4. 30 Y = INT
$(100 * X + 0.05)/100$ 40 PRINT X; " ROUNDED
TO THE NEAREST HUNDREDTH IS "; Y

Page 244, Classroom Exercises
1. Yes 2. No 3. Yes 4. Exercise 1: slope 2;
y-intercept 1; $y = 2x + 1$ Exercise 2: slope $\frac{1}{2}$;
y-intercept -4; $y = \frac{1}{2}x - 4$

Pages 244–247, Written Exercises
1. Yes 3. Yes 5. No 7. Yes 9. Slope: 2;
y-intercept: 3; $y = 2x + 3$ 11. Slope: 5; y-intercept: -2;
$y = 5x - 2$ 13. Slope: 3; y-intercept: -1; $y = 3x - 1$
15. Slope: -2; y-intercept: 3; $y = -2x + 3$ 17. Yes
19. Yes 21. No 23. Yes 25. Yes
27. $V = \frac{5}{3}T + 455$ 29. $w = 20l - 1000$
31. $t = \frac{3}{2}b + 100$ 33. Yes 35. 21 ohms 37. For a
linear function, the second differences are all zero.
39. For a second-degree function, the third differences
are all zero.

Page 247, Review Exercises
1. $6x^2 - 18x$ 2. $2x^3 - 8x^2 + 6x$ 3. $6a^2 - 3ab + 3ac$
4. $4a^4$ 5. $\frac{-4y}{x}$ 6. $8 \cdot 10^3$ 7. 18, 19, 20, 21, 22
8. $-26, -24, -22$ 9. $x + (x + 10) = 100$
10. $w + 3w + w + 3w = 40$; length: 15 cm, width: 5 cm

Page 248, Extension
1. $Y = -2X + 10$ 2. $Y = -1X - 1$ 3. $Y = 4X + 9$
4. $Y = 1.5X - 9$ 5. $Y = 0X + 2$ 6. Division by zero
error in 50.

Pages 250–252, Chapter Review
1a. $(2, -3)$ 1b. $(-2, -1)$ 1c. $(-3, 0)$ 1d. $(2, 2)$
3a. $\{1, 2, 3\}$ 3b. $\{2, 6\}$
3c.
5a. Yes
5b. No
5c. Yes

7.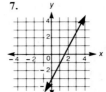
9a. $\frac{1}{2}$ 9b. -3
9c. 0 9d. No slope
11. $\frac{1}{2}$; -3

13.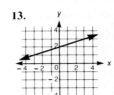
15. Yes 17. $y = \frac{5}{2}x - 2$
19. No 21. No
23. No

Pages 252–253, Chapter 5 Self-Test
1. $(-3, 2)$ 2. $-\frac{5}{4}$ 3. II 4. $\{(4, 1), (1, -2), (0, 3),$
$(-2, 1)\}$ 5. Domain: $\{4, 1, 0, -2\}$; range: $\{1, -2, 3\}$
6. $\{(1, 12), (2, 6), (3, 4), (4, 3), (6, 2), (12, 1)\}$
7.

x	6	-3	2
y	-8	10	0

8. No 9. Yes

10. 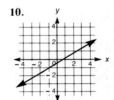   11.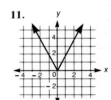

12. $\frac{1}{3}$

13.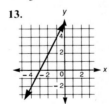

14. $\frac{3}{2}$ 15. $y = -\frac{1}{4}x + 3$ 16. $y = 3x - 1$
17. $y = -4x + 8$ 18. $y = -\frac{1}{2}x + 4$ 19a. No
19b. Yes 19c. No 19d. Yes 19e. No 20. No
21. $C = \frac{5}{9}F - \frac{160}{9}$

**Pages 254–255, Practice for College Entrance
Tests**
1. C 2. D 3. E 4. C 5. D 6. E 7. C 8. D
9. A 10. C 11. A 12. B

Chapter 6

Page 259, Classroom Exercises

1. Yes 2. No 3. $\{(3, 1)\}$ 4. $\{(2, 2)\}$ 5. $\{(0, 0)\}$
6. $\{(3, -6)\}$ 7. $\{(0, 0), (1, 1)\}$

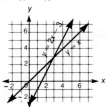

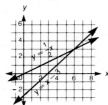

Pages 260–262, Written Exercises

1. $\{(-2, 3)\}$ 3. $\{(2, 5)\}$ 5. $\{(-2, -2)\}$ 7. $\{(0, 2)\}$
9. $\emptyset$ 11. Yes 13. Yes 15. No 17. No
19. $\{(3, 3)\}$ 21. $\{(6, 3)\}$

23. $\{(4, 1)\}$ 25. $\{(4, -1)\}$

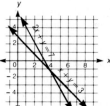

27. $\{(2, 6)\}$; width: 2 cm, length: 6 cm

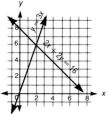

29. $\{(2, 6)\}$; 2 nickels, 6 dimes

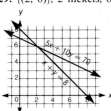

31. $y = 2x + 3$; $(2, 7)$; 2   33. $\{(2, 2), (-6, 6)\}$

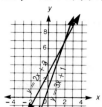

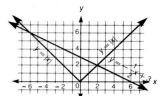

35. $\{(3, 9), (-1, 1)\}$ 37. $\left\{\left(1\frac{1}{2}, 1\frac{2}{3}\right)\right\}$;

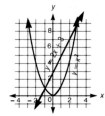

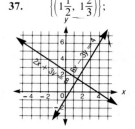

answer contains fractions.

39. $\{(12, 1)\}$;

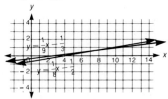

y-intercepts are fractions.

Page 262, Review Exercises

1. 7 2. 1 3. $13\frac{1}{3}$ 4. 1 5. $2\frac{1}{7}$ 6. 4
7. $x = \dfrac{24 - 4y}{3}$ 8. $y = \dfrac{12 - 3x}{-4}$

Page 266, Classroom Exercises

1. $2x + 3(3x + 1) = 4$ 2. $\left(\frac{1}{2}x - 2\right) - 3 = x + 2$
3. $y + 7 = 2(y + 6)$ 4. $2\left(\frac{x}{2}\right) - x = 0$ 5. $y = 5 - x$
6. $y = x + 3$ 7. $y = \frac{3}{2}x - 3$ 8. $y = \frac{1}{3} - \frac{1}{3}x$
9. $x = 3 - \frac{5}{3}y$ 10. $x = \frac{1}{2}y + 3$ 11. $x = \frac{2}{3}y - 2$
12. $x = \frac{1}{2} - \frac{1}{2}y$ 13. $\left\{\left(\frac{7}{4}, \frac{7}{2}\right)\right\}$ 14. $\{(4, 2)\}$
15. $\left\{\left(-\frac{1}{3}, -2\right)\right\}$ 16. $\{(0, 4)\}$

Pages 267–268, Written Exercises

1. $\{(4, 12)\}$ 3. $\{(4, 11)\}$ 5. $\{(2, 3)\}$ 7. $\left\{\left(-\frac{1}{3}, 4\right)\right\}$
9. $y = 2 - \frac{1}{2}x$ 11. $y = -5 + 2x$ 13. $x = 4 - 2y$

15. $x = -\frac{5}{2} + \frac{1}{2}y$ **17.** $\{(8, -2)\}$ **19.** $\{(2, -1)\}$

21. $\{(3, 9)\}$ **23.** $\{(4, 7)\}$ **25.** $\{(9, -1)\}$ **27.** $\left\{\left(3, \frac{4}{3}\right)\right\}$

29. $\left\{\left(-\frac{5}{3}, -\frac{7}{3}\right)\right\}$ **31.** $\{(-350, 100)\}$ **33.** $\{(-1, 3)\}$

35. $\left\{\left(3, \frac{3}{2}\right)\right\}$ **37.** $l = 3w;\ 2l + 2w = 20;\ \{(2.5, 7.5)\};$
width: 2.5 cm, length: 7.5 cm   **39.** $n = d + 4;$
$5n + 10d = 410;\ \{(26, 30)\};$ 26 dimes, 30 nickels
41. $x - 10 = y;\ \{(45, 35)\};$ first number: 45, second
number: 35   **43.** $\{(0, 12, 12)\}$ **45.** $\left\{\left(6, \frac{15}{2}, \frac{3}{2}\right)\right\}$
47. $d = 50\left(t + \frac{1}{2}\right);\ d = 55t;\ \{(5,275)\};$ 5 hours

Page 268, Review Exercises
1. 0.37   **2.** 35%   **3.** 40%   **4.** 32   **5.** $\{0, 1\}$ **6.** $\{7\}$
7. $\{96\}$ **8.** $\{4\}$ **9.** $\left\{-\frac{4}{3}\right\}$

Page 271, Classroom Exercises
1. 1   **2.** Many   **3.** None   **4.** Many   **5.** 1   **6.** None
7. Many   **8.** 1   **9.** None

Pages 272–273, Written Exercises
1. 1   **3.** None   **5.** Many   **7.** 1
9. None   **11.** 1

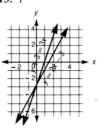

13. 1   **15.** Many

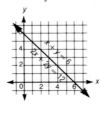

17. Many   **19.** Many   **21.** None   **23.** $\{(2, 5)\};$ 1
25. $\left\{\left(\frac{1}{2}, \frac{5}{4}\right)\right\};$ 1   **27.** Many   **29.** None   **31.** 1
33. 1   **35.** None or many   **37.** None or 1
39. None or 1   **41.** 1

Page 273, Review Exercises
1. 21   **2.** 74   **3.** $x + 6y$ **4.** $-a - 10$ **5.** 1

Page 273, Self-Quiz 1
1. $\{(4, 3)\}$ **2.** $\{(1, -2)\}$

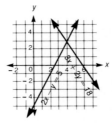

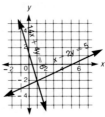

3. $\{(3, 1)\}$ **4.** $\{(3, 1)\}$ **5.** 1   **6.** 1   **7.** Many
8. None

Page 276, Classroom Exercises
1. $10t + u$ **2.** $10u + t$ **3.** $t + u = 7$ **4.** $t = u + 3$
5. $(10t + u) + (10u + t) = 85$ **6.** $y = x + 15;$
$x + y = 123;$ 54 and 69   **7.** $n = d + 8;$ 1032 nickels, 24 dimes

Pages 276–278, Written Exercises
1. $t = 3u$ **3.** $u = t + 4$ **5.** $10t + u = 58$
7. $10u + t = 31$ **9.** $10u + t = (10t + u) + 9$ **11.** 41
13. 93   **15.** 53   **17.** 21 and 63   **19.** Width: 50 cm,
length: 75 cm   **21.** 83   **23.** 325   **25.** Carl's: 10 km/h,
Barbara's: 6 km/h   **27.** Carl's: 77 km/h,
Barbara's: 88 km/h   **29.** Pat's: \$4.50, Carol's: \$4.25

Page 278, Review Exercises
1. 1.25   **2.** 3%   **3.** 8%   **4.** 20.8   **5.** x^5 **6.** $-6x^4y^2$
7. x^8

Page 281, Classroom Exercises
1. $\{(4, 1)\}$ **2.** $\{(8, -1)\}$ **3.** $\{(4, -3)\}$

Pages 281–282, Written Exercises
1. $\{(5, 3)\}$ **3.** $\{(2, 6)\}$ **5.** $\{(7, 1)\}$ **7.** $\{(5, 2)\}$
9. $\{(4, -3)\}$ **11.** $\{(-1, -2)\}$ **13.** $\{(-3, -5)\}$
15. $\{(2, 3)\}$ **17.** $\{(3, 4)\}$ **19.** $\{(7, 5)\}$ **21.** $\{(-2, 3)\}$
23. $\{(0, 5)\}$ **25.** 9 and 2   **27.** -2 and 5   **29.** $\frac{2}{3}$ and $\frac{1}{2}$
31. John is 25, Mary is 20.   **33.** Sally: \$60, John: \$40
35.

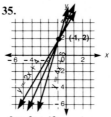

37a. $y = \left(\frac{m_1 + m_2}{2}\right)x + \frac{b_1 + b_2}{2}$ **37b.** $\frac{3}{2}$ **37c.** -1

Page 282, Review Exercises
1. $2.56 \cdot 10^5$ **2.** 2,500,000   **3.** 1.2×10^7

4. $3a^3 - 6a^2b + 3ab^2$ **5.** $-2x^3 + 8x^2 - 10x$ **6.** $\dfrac{3}{4x}$
7. $\dfrac{b^2}{a^3}$ **8.** 16 minutes

Pages 285–286, Classroom Exercises
1. Multiply either by -1 and add.   **2.** Multiply 2nd by -3 and add.

3.

Multiply 1st by	-3	3
Multiply 2nd by	2	-2

and add

4. Multiply 2nd by 2 and add.   **5.** Multiply either by -1 and add.

6.

Multiply 1st by	-3	3
Multiply 2nd by	5	-5

and add

7. $\{(5, 1)\}$ **8.** $\{(-2, 1)\}$

Pages 286–287, Written Exercises
1. Multiply 1st by -3 and add.   **3.** Multiply 2nd by 4 and add.   **5.** Multiply 1st by 2 and add.

7.

Multiply 1st by	3	-3
Multiply 2nd by	-5	5

and add

9. $\{(3, 2)\}$ **11.** $\{(8, -7)\}$ **13.** $\{(-2, -6)\}$
15. $\left\{\left(\frac{1}{2}, -2\right)\right\}$ **17.** $\left\{\left(1, \frac{1}{4}\right)\right\}$ **19.** $\{(5, 6)\}$
21. $\{(-5, 2)\}$ **23.** $\{(2, 5)\}$ **25.** $\{(7, -1)\}$
27. $\{(4, 5)\}$ **29.** $\left\{\left(5, \frac{1}{2}\right)\right\}$ **31.** $\left\{\left(-10\frac{1}{2}, 9\right)\right\}$
33. None   **35.** None   **37.** 12 and 21   **39.** 160 adult
and 340 student   **41.** $x = \dfrac{b_2c_1 - b_1c_2}{b_2a_1 - b_1a_2}$ or
$y = \dfrac{a_2c_1 - a_1c_2}{a_2b_1 - a_1b_2}$ or $\dfrac{a_1c_2 - a_2c_1}{a_1b_2 - a_2b_1}$

Pages 287–288, Review Exercises
1a. II   **1b.** y-axis   **1c.** IV   **2a.** Domain:
$\{-2, -1, 0, 1, 2\}$; range: $\{0, 1, 2\}$
2b.

3.

c	0	2.5	2	3
d	-10	0	-2	2

4. 4; 3; 2; 1; 0; 1; 2   **5.** -8; -1; $-\frac{1}{8}$; 0; $\frac{1}{8}$; 1; 8

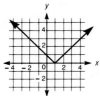

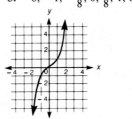

1. $u = t + 6$ **2.** $10t + u = 47$ **3.** 39   **4.** $\{(2, -1)\}$
5. $\left\{\left(-2, \frac{1}{2}\right)\right\}$ **6.** $\{(2, 5)\}$ **7.** $\left\{\left(-\frac{3}{2}, 3\right)\right\}$

Page 291, Classroom Exercises
1a. 55   **1b.** 28   **2a.** 64   **2b.** 64%   **3.** $0.55x$
4. $0.7y$ **5a.** $0.55x + 0.7y$ **5b.** $x + y$ **6.** $x + y = 10$;
$0.55x + 0.7y = 0.6(10)$

Pages 291–294, Written Exercises
1a. $20x$ cents   **1b.** $(20x + 40y)$ cents   **1c.** $x + y$
1d. $x + y = 10$ **1e.** $20x + 40y = 350$ **3.** 2 lb of $4/lb,
3 lb of $1/lb   **5.** 7 lb of $2/lb, 3 lb of $3/lb   **7.** 54 lb
of $1.75/lb, 46 lb of $2.25/lb   **9.** 44 lb of $3.25/lb,
56 lb of $2.25/lb   **11.** 6.25 lb of $3/lb, 18.75 lb of
$2.20/lb   **13.** 16 kg of 40%, 64 kg of 60%   **15.** $2\frac{1}{2}$L
of 80%, $7\frac{1}{2}$L of 40%   **17.** 10 mL of 25%, 40 mL of
40%   **19.** 6.25 L   **21.** 21   **23.** 2

Page 294, Review Exercises
1a. 1   **2.**
1b. $-\dfrac{1}{3}$
1c. None

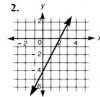

3. $\dfrac{2}{3}$, -6 **4.** $y = 2x - 1$ **5.** Yes   **6.** No   **7.** 4 P.M.,
200 mi

Page 295, Extension
8. 2 lb

Pages 297–298, Chapter Review
1. $\{(0, 2)\}$
3. $\{(3, -1)\}$

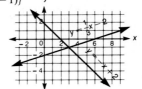

5. $\{(-1, -3)\}$ **7.** Many
9. 75   **11.** Length: 27 cm,
width: 7 cm   **13.** $\left\{\left(\frac{2}{3}, 6\right)\right\}$
15. $\{(2, -1)\}$
17. 6 lb of $3.75/lb, 4 lb of $5.00/lb

Pages 298–299, Chapter 6 Self-Test
1. No   **2.** $\{(3, 5)\}$ **3.** $\{(5, 0)\}$ **4.** $y = -2x + 6$

5. $\{(3, -2)\}$ **6.** 0 **7.** 0 **8.** $\left\{\left(-\frac{6}{5}, \frac{4}{5}\right)\right\}$
9. $\{(75, -175)\}$ **10.** $0.45x$ **11.** 61 and 23 **12.** 73
13. 60 lb of 60¢/lb, 40 lb of 35¢/lb

Pages 299–300, Practice for College Entrance Tests
1. B **2.** C **3.** D **4.** E **5.** C **6.** A **7.** E **8.** E
9. B

Pages 300–303, Cumulative Review (Chapters 1–6)
1. 21 **3.** 36 **5.** 16 **7.** d **9.** T **11.** 2 **13.** 8
15. $a + 2$

17. 4 days

	Start	1	2	3	4
Robert	40	43	46	49	52
Chico	32	37	42	47	52

Day

19. $\{-6\}$ **21.** $\{5\}$ **23.** $13x^2 + 2x = 5$ **25.** d
27. $9x^3 + 12x^2 - x + 6$ **29.** $-12y^3$ **31.** $5.73 \cdot 10^5$
33. a^{10} **35.** x^3y^{12} **37.** $b^2 - 5b^4$ **39.** x^5 **41.** $-\frac{1}{b}$
43. 54 mph

45.
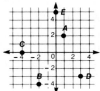

47. D **49.** No

51.
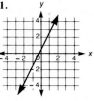

53. -1

55.
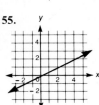

57. $y = 2x - \frac{5}{2}$
59. Yes **61.** $y = 3x$

63. $\{(2, 3)\}$
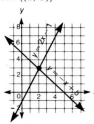

65. $\{(3, 5)\}$
67. Infinitely many
69. $\{(1, 2)\}$
71. $\{(2, 2)\}$
73. 39

Chapter 7

Pages 306–307, Classroom Exercises
1. T **2.** F **3.** T **4.** T **5.** T **6.** F **7.** F **8.** F
9. $2^2 \cdot 3$ **10.** $3 \cdot 5$ **11.** 2 **12.** 3^3 **13.** $1 \cdot 4; -1 \cdot -4;$
$-2 \cdot -2; 2 \cdot 2$ **14.** $1 \cdot 8; -1 \cdot 8; -2 \cdot 4; -4 \cdot 2;$
$-2 \cdot 2 \cdot 2; -2 \cdot -2 \cdot -2$ **15.** $-3 \cdot 2; -2 \cdot 3; -1 \cdot 6;$
$6 \cdot -1$ **16.** $5 \cdot 2; -5 \cdot -2; -10 \cdot -1; 1 \cdot 10$

Pages 307–308, Written Exercises
1. $\{1, 3, 5\}$ **3.** $\{4, 13\}$ **5.** $\{7\}$ **7.** $\{1, 2, 3, 6, 9, 18\}$
9. $\{1, 7, 49\}$ **11.** $\{1, 5, 7, 35\}$ **13.** $\{1, 2, 3, 6, 7, 14,$
$21, 42\}$ **15.** $\{1, 2, 3, 4, 6, 7, 12, 14, 21, 28, 42, 84\}$
17. $\{1, 47\}$ **19.** composite **21.** prime **23.** composite **25.** prime **27.** prime **29.** composite
31. prime **33.** prime **35.** $1 \cdot 44, 2 \cdot 22, 4 \cdot 11$
37. $1 \cdot 63, 3 \cdot 21, 7 \cdot 9$ **39.** $1 \cdot 75, 3 \cdot 25, 5 \cdot 15$
41. $1 \cdot 76, 2 \cdot 38, 4 \cdot 19$ **43.** $2 \cdot 3 \cdot 5$ **45.** $2^3 \cdot 3$
47. $2 \cdot 3^3$ **49.** $7 \cdot 13$ **51.** $2 \cdot 5^2$ **53.** $2^3 \cdot 3^2$
55. $2^3 \cdot 7$ **57.** 3^4 **59.** $1 \cdot -15; -1 \cdot 15; -3 \cdot 5;$
$-5 \cdot 3$ **61.** $10; -1 \cdot -10; 2 \cdot 5; -2 \cdot -5$ **63.** $1 \cdot -77;$
$-1 \cdot 77; -7 \cdot 11; -11 \cdot 7$ **65.** $1 \cdot 69; -1 \cdot -69; 3 \cdot 23;$
$-3 \cdot -23$ **67.** $2^2 \cdot 3 \cdot 5$ **69.** $2^2 \cdot 3 \cdot 7$ **71.** 3^4
73. $2 \cdot 3^2 \cdot 5$ **75a.** 2 **75b.** 3 **75c.** 4 **75d.** 5
75e. 6 **77a.** 2 **77b.** 3 **77c.** 4 **77d.** 5 **77e.** 6
79. $n + 1$ **81.** $(m + 1)(n + 1)$ or $1 + m + n + mn$

Page 309, Review Exercises
1. Associative property of addition **2.** Distributive property **3.** $-1, 0, 1$ **4.** $-2, 2$ **5.** $-4a^3b$
6. $9x^2y^2$ **7.** $3xy^2$ **8.** $-3a$

Page 309, Extension
1. Yes **2.** No **3.** No **4.** Yes **5.** Prime **6.** 3, 43
7. Prime **8.** 7, 19

Page 312, Classroom Exercises
1. 1, 2, 7, 14 **2.** 1, 17 **3.** 1, 2, 3, 6 **4.** 1 **5.** 3
6. 8 **7.** 1, 5, 25, x, $5x$, $25x$ **8.** 1, 7, x, x^2, $7x$, $7x^2$
9. x, y, xy **10.** 2, x, $2x$ **11.** $3a^2b$ **12.** $4mn^3$

Pages 312–313, Written Exercises
1. 1, 2, 29, 58 **3.** 1, 31 **5.** 1, 2, 3, 4, 6, 12 **7.** 1, 13 **9.** 1 **11.** 1, 2, 5, 10 **13.** 4 **15.** 1 **17.** 50
19. 2 **21.** 1, 2, 3, 4, 6, 12 **23.** 1, 2, x, x^2, $2x$, $2x^2$
25. 1, 3, 9, a, $3a$, $9a$ **27.** 1, a, b, b^2, ab, ab^2
29. 1, x **31.** 1, 3, a, b, $3a$, $3b$, ab, $3ab$ **33.** x^3
35. $2xy$ **37.** $5a$ **39.** $9abc^2$ **41.** 2 **43.** 6 **45.** xy
47. $2xy^2$ **49.** $4xy$ **51.** T **53.** T **55.** T
57. $x^{37}y^{50}$ **59.** 3003 **61.** 6! **63.** 4!

Page 313, Review Exercises
1. $\{-3\}$ **2.** $\left\{2\frac{1}{2}\right\}$ **3.** $\{9\}$ **4.** $\{32\}$ **5.** $I = \frac{E}{R}$
6. $y = \frac{c - ax}{b}$

Page 315, Classroom Exercises

1. {0, 2}   **2.** {0, −3}   **3.** $\left\{0, \frac{1}{2}\right\}$ **4.** $\left\{0, -\frac{1}{3}\right\}$

5. {1, 2}   **6.** {5, 2}   **7.** $\left\{-\frac{1}{2}, 3\right\}$ **8.** $\left\{-\frac{2}{3}, -\frac{1}{2}\right\}$

Pages 315–316, Written Exercises

1. {3, 0}   **3.** {0, −4}   **5.** {3, 8}   **7.** {−5}

9. {5, −5}   **11.** $\left\{-\frac{1}{3}, \frac{1}{2}\right\}$ **13.** $\left\{-\frac{5}{3}, \frac{3}{4}\right\}$

15. $\left\{\frac{3}{2}, -\frac{2}{3}\right\}$ **17.** {0, 3, 4}   **19.** $(x − 5)(x + 7) = 0$;

5 or − 7   **21.** $(x − 7)(x + 3) = 0$; 7 or − 3

23. $2x(x + 2) = 0$; 0 or − 2   **25.** $(x − 3)(2x − 5) = 0$;

3 or $2\frac{1}{2}$ **27.** Chances are good the Cubs won't score any runs in at least 1 game. So, product will be 0.

29. 1   **31.** 1   **33.** 2   **35.** 2   **37.** 3   **39.** 1

Page 317, Review Exercises

1. 14   **2.** 14   **3.** 40   **4.** 36   **5.** $2x^2 + 4x$

6. $3abc − 2a + 3ab$ **7.** $12m^2n − 18mn^2$ **8.** $20x − 10$

9. $A = \pi 10^2$ **10.** $A = \frac{1}{2}\pi x^2$

Page 317, Self-Quiz 1

1. 1, 3, 5, 9, 15, 45   **2.** 1, 2, 4, 13, 26, 52

3. $2^3 \cdot 3 \cdot 5$ **4.** $2 \cdot 3^2 \cdot 7$ **5.** 15   **6.** 35   **7.** $2xy$

8. {0, 4}   **9.** {7, −2}   **10.** $\left\{\frac{3}{2}, -\frac{2}{5}\right\}$

Page 320, Classroom Exercises

1. $6x^2 + 8x$ **2.** $3a^2 − 6a^2b + 3ab$ **3.** $3(x + b)$

4. $3(2x + y)$ **5.** $x(4x + 1)$ **6.** $a(7a − 3)$

7. $7bc(3b − 2)$ **8.** $2ab(3a^2 + 2ac − 4c)$ **9.** {0, −4}

10. {0, 2}   **11.** {0, 2}

Pages 320–321, Written Exercises

1. 5   **3.** x **5.** $6x$ **7.** $2ab$ **9.** $2(x + 5b)$

11. $7a(3a − 1)$ **13.** Not factorable   **15.** $a(b + 1)$

17. $2xy(1 + 2x + 3y)$ **19.** Not factorable   **21.** {0, 3}

23. {0, −4}   **25.** {0, 6}   **27.** $\left\{0, \frac{3}{2}\right\}$ **29.** $x^2 = 10x$;

0 or 10   **31.** $x^2 + 2x = 0$; 0 or − 2   **33.** 700

35. 80   **37.** 67   **39a.** $\frac{1}{2}\pi r^2 + 4r^2$ **39b.** $r^2\left(\frac{\pi}{2} + 4\right)$

41a. $4r^2 − \pi r^2$ **41b.** $r^2(4 − \pi)$ **43a.** $w_1 l − w_2 l$

43b. $l(w_1 − w_2)$ **45a.** $\frac{1}{2}\pi r^2 + r^2$ **45b.** $r^2\left(\frac{\pi}{2} + 1\right)$

47a. $2xy + 2x^2$ **47b.** $2x(y + x)$

49a. $lw − 2xw − 2xl + 4x^2$ **49b.** $(l − 2x)(w − 2x)$

Page 321, Review Exercises

1. $4x$ **2.** a **3.** $3a − 3b$ **4.** $x^2 + x − 12$ **5.** $x^2 − 25$

6. $3x^2 + 3x + 3$

Page 324, Classroom Exercises

1. $mp + mq + np + nq$ **2.** $m^2 + 5m + 6$

3. $x^2 − 2xy + y^2$ **4.** $a^2 − a − 2$ **5.** $s^2 − 9$

6. $2t^2 − 5t − 3$

Pages 324–325, Written Exercises

1. $6x^2, 4xy, 9xy, 6y^2$ **3.** $3a^2, −6ab, 4ab, −8b^2$

5. $p^2, −6pq, −pq, 6q^2$ **7.** $ac + ad + bc + bd$

9. $ab + ad − bc − cd$ **11.** $ab − ac − bd + cd$

13. $a^2 + 10a + 21$ **15.** $b^2 − 5b + 6$ **17.** $c^2 − 64$

19. $d^2 − 9$ **21.** $x^2 + 6x + 9$ **23.** $y^2 − 10y + 25$

25. $x^2 + 3x + 2x + 6 = x^2 + 5x + 6$

27. $x^2 + xy + xy + y^2 = x^2 + 2xy + y^2$ **29.** e   **31.** f

33. b   **35.** $6n^2 + 19n + 10$ **37.** $3n^2 + 2n − 21$

39. $5a^2 + 24ab − 5b^2$ **41.** $4x^2 − 12x + 9$

43. $x^2 − 0.25$ **45.** 1764   **47.** 63.99   **49.** $\left\{-\frac{5}{2}\right\}$

51. {−9}

Page 325, Review Exercises

1. 14   **2.** 10   **3.** 3,140,000   **4.** $63 \cdot 10^5$ **5.** $1\frac{1}{4}$ h

6. 2 h

Page 328, Classroom Exercises

1. $(3x)^2$ **2.** $(4ab)^2$ **3.** $\left(\frac{2}{3}x^2\right)^2$ **4.** $(4ab)^2$ **5.** $(5m^2n)^2$

6. $\left(\frac{1}{3}a\right)^2$ **7.** $x^2 − 16$ **8.** $4a^2 − 25$ **9.** $4x^2 − 1$

10. $4a^2 − 36$ **11.** $16a^2 − 9$ **12.** $a^2 − 1$

13. $(3a + 5)(3a − 5)$ **14.** $(4x + 1)(4x − 1)$

15. $(2y + z)(2y − z)$ **16.** {−3, 3}   **17.** {−4, 4}

18. {−5, 5}

Pages 329–330, Written Exercises

1. $(5a)^2$ **3.** $(3cd)^2$ **5.** $(p^2)^2$ **7.** $\left(\frac{3}{5}s^6 t^4\right)^2$ **9.** $x^2 − 25$

11. $x^2 − 36$ **13.** $36 − x^2$ **15.** $4x^2 − 25$

17. $x^2 − a^2$ **19.** $x^2 − 25y^2$ **21.** $(x + 2)(x − 2)$

23. $(3x + 1)(3x − 1)$ **25.** $(x + 10)(x − 10)$

27. $(3x + 4)(3x − 4)$ **29.** $2(x + 3)(x − 3)$

31. $(cd + 7)(cd − 7)$ **33.** {7, −7}   **35.** {8, −8}

37. $\left\{\frac{4}{3}, -\frac{4}{3}\right\}$ **39.** $\left\{\frac{5}{4}, -\frac{5}{4}\right\}$ **41.** {5, −5}

43. {3, −3}   **45.** 4 or −4   **47.** $(a + b)(a − b)$

49. $\pi(R + r)(R − r)$ **51.** $(x + 0.5)(x − 0.5)$

53. $\left(x + \frac{2}{3}\right)\left(x - \frac{2}{3}\right)$ **55.** $(x^2 + 1)(x + 1)(x − 1)$

57. $(4x^2 + 9)(2x + 3)(2x − 3)$ **59.** 10

61. $(x + 3)(x − 3)$ **63.** $(x + 5)(x − 5)$

Page 330, Review Exercises

1. a^6 **2.** $\frac{4}{9}x^4y^2$ **3.** $8x^6$ **4.** $27x^3y^6$ **5.** $x^2 + 6x + 9$

6. $y^2 − 10y + 25$ **7.** $a^2 + 2ab + b^2$

8. $a^2 − 2ab + b^2$ **9.** Length: 14 in., width: 10 in.

Page 330, Self-Quiz 2

1. $5x(5x + 1$ **2.** $7b(3 − 2a)$ **3.** Not factorable

4. $x(x + y)$ **5.** $c^2 + 10c + 21$ **6.** $2r^2 − 11r − 40$

7. $(x + 7)(x - 7)$ **8.** $(2y + 9)(2y - 9)$
9. $2(b + 5)(b - 5)$ **10.** $\{0, -7\}$ **11.** $\{4, -4\}$
12. $\left\{\dfrac{10}{3}, -\dfrac{10}{3}\right\}$

Page 334, Classroom Exercises

1. $(2x)^2$ **2.** $(3ab)^2$ **3.** $(x^2)^2$ **4.** $\left(\dfrac{3}{4}ab^2\right)^2$
5. $x^2 - 6x + 9$ **6.** $2x^2 + 20x + 50$ **7.** No   **8.** Yes
9. Yes   **10.** $(x + y)^2$ **11.** $(m - n)^2$ **12.** $(3a + 2b)^2$
13. $(x^2 - 2y)^2$ **14.** $(3a + 2b)(3a - 2b)$ **15.** $3(x + 1)^2$
16. $\{-7\}$ **17.** $\{9, -9\}$

Pages 334–336, Written Exercises

1. $(3a)^2$ **3.** $(4cd)^2$ **5.** $(p^3)^2$ **7.** $\left(\dfrac{4}{9}s^4t^5\right)^2$ **9.** No
11. No   **13.** Yes   **15.** $x^2 + 10x + 25$
17. $a^2 - 12a + 36$ **19.** $9y^2 + 12y + 4$
21. $9b^2 - 24b + 16$ **23.** $25c^2 - 10cd + d^2$
25. $2x^2 + 40x + 200$ **27.** $(x + 3)^2$ **29.** $(x - 4z)^2$
31. $2(x + 2y)^2$ **33.** $(x + 4)(x - 4)$
35. $5(\blacksquare + 3)(x - 3)$ **37.** $(2x - 3)^2$ **39.** $\{5\}$
41. $\{-6\}$ **43.** $\{0.5\}$ **45.** $\{1\}$ **47.** $\{-1.5\}$
49. $\{10, -10\}$ **51.** $x^2 + 6x + 9$ **53.** $4a^2 + 4ab + b^2$
55. $x(x - 3)^2$ **57.** $x(x + 3)(x - 3)$
59. $(x + 2)^2(x - 2)^2$ **61.** $(x + y)^2 = x^2 + 2xy + y^2$
63. 1225; 2025; 3025; 4225   **65.** $100t^2 + 100t + 25$
67. $(10t + 5)^2 = 100(t^2 + t) + 25$
69. $100t^2 + 40t + 4 = 4(25t^2 + 10t + 1)$

Page 336, Review Exercises

1. $x^2 + 8x + 12$ **2.** $x^2 - 8x + 15$ **3a.** III   **3b.** II
3c. y-axis   **3d.** IV   **4.** Domain: $\{1, 2, 3, 4\}$; range:
$\{4, 5\}$
5.

6a. Yes
6b. No
6c. No

Page 339, Classroom Exercises

1. $(x + 3)(x + 2)$ **2.** $(x - 3)(x - 2)$ **3.** $(a + 6)(a + 1)$
4. $(b - 6)(b - 1)$ **5.** $(c + 2)^2$ **6.** $(a + b)(a - b)$
7. $(x + 3y)^2$ **8.** $3(a - 5)(a - 1)$ **9.** $(y + 7z)(y + z)$
10. $\{-4, -3\}$ **11.** $\{1, 5\}$ **12.** $\{1, 4\}$

Pages 340–341, Written Exercises

1. $(a + 3)(a + 1)$ **3.** $(x + 4)(x + 2)$ **5.** $(a - 6)(a - 2)$
7. $(x + 5)(x + 4)$ **9.** $(a - 6)(a - 3)$
11. $(c + 18)(c + 1)$ **13.** $(b - 7)(b - 3)$ **15.** $(a - 4)^2$
17. $(a + 25)(a + 1)$ **19.** $(x + 4)(x + 1)$
21. $(x + 3y)(x + 2y)$ **23.** $(x - 8y)(x - y)$ **25.** $\left\{\dfrac{3}{2}, 4\right\}$

27. $\{3, 1\}$ **29.** $\{-6, -1\}$ **31.** $\{-3\}$ **33.** $\{7, 2\}$
35. $\{8, 3\}$ **37.** $\{4, 2\}$ **39.** $\{3, 2\}$ **41.** $\{-4, -1\}$
43. $\{7, 1\}$ **45.** $\{3, 2\}$ **47.** 4, -4 **49.** 10, -10, 6,
-6 **51.** 19, -19, 11, -11, 9, -9 **53.** 4, 3
55. 6, 4   **57.** 12, 10, 6   **59.** $-b$ **61.** 4, 3
63. Length: 18 m, width: 10 m   **65.** $(x + 6)(x + 2)$
67. $(x - 3)(x - 1)$ **69.** $(x - 6)(x - 2)$ **71.** $\{2\}$
73. $\{4, 3\}$

Page 341, Review Exercises

1. $x^2 + 2x - 15$ **2.** $x^2 - 3xy - 4y^2$
3.

4. 7; 2; -1; $-\dfrac{7}{4}$;
-2; $-\dfrac{7}{4}$; -1; 2; 7

5a. $\dfrac{2}{3}$ **5b.** $-\dfrac{3}{2}$
5c. $-\dfrac{1}{5}$ **5d.** $\dfrac{2}{5}$

Page 345, Classroom Exercises

1. $(x + 4)(x - 2)$ **2.** $(y - 3)(y + 2)$ **3.** $(b + 3)(b - 2)$
4. $(a - 9b)(a + 2b)$ **5.** $\{4, -1\}$ **6.** $\{5, -3\}$
7. $\{3, -4\}$ **8.** $\{-7, -4\}$

Pages 345–347, Written Exercises

1. $(a - 6)(a + 1)$ **3.** $(b + 3)(b - 2)$ **5.** $(c + 4)(c - 3)$
7. $(a - 12)(a + 1)$ **9.** $(b - 9)(b + 1)$
11. $(c + 4)(c - 4)$ **13.** $2(x + 7)(x - 1)$
15. $(y - 9)(y + 4)$ **17.** $(z + 18)(z - 2)$
19. $(x + 8)(x - 3)$ **21.** $(x + 3y)(x - y)$
23. $(x + 2y)(x - y)$ **25.** $\{4, -1\}$ **27.** $\{7, -1\}$
29. $\{2, -5\}$ **31.** $\{2, -6\}$ **33.** $\{12, -1\}$
35. $\{-6, -3\}$ **37.** 4 or -5 **39.** 7 or -5
41. 6, -6 **43.** 3, -3, 0   **45.** 11, -11, 4, -4, 1,
-1 **47.** r, $r + s$ **49.** 6 m   **51.** -17, -16 **53.** 10
and 13, or -13 and -10 **55.** 6, 9   **57.** $\{6, -2\}$
59. $\{7, -2\}$ **61.** $\{2, -5\}$

Page 347, Review Exercises

1. $6x^2 + 11x + 3$ **2.** $6a^2 - 10ab - 4b^2$ **3.** $\dfrac{2}{3}$, 4

4.

5. No **6.** No

Page 347, Self-Quiz 3

1. $(x + 4)^2$ **2.** $(y - 7)^2$ **3.** $3(w - 1)^2$
4. $(a + 3)(a + 2)$ **5.** $(t - 7)(t - 4)$ **6.** $(r - 8)(r - 2)$
7. $(x - 6)(x + 1)$ **8.** $(b + 14)(b - 2)$
9. $(y + 16)(y - 1)$ **10.** $\{7, 9\}$ **11.** $\{4, -5\}$

Page 350, Classroom Exercises

1. $4x$ and x, $2x$ and $2x$ **2.** -3 and 1, 3 and -1
3. One is positive, one is negative **4.** $(2x - 3)(2x + 1)$
5. $(3a + 2)(2a + 1)$ **6.** $(2x - 3)(x - 1)$
7. $(5m - n)(2m + 3n)$ **8.** $\left\{2, -\dfrac{1}{2}\right\}$ **9.** $\left\{\dfrac{3}{2}, -1\right\}$
10. $\left\{4, -\dfrac{1}{3}\right\}$

Pages 350–352, Written Exercises

1. $(2x + 3)(x + 1)$ **3.** $(3x - 1)(x - 2)$
5. $(5a + 2)(a - 3)$ **7.** $(5a - 3)(a + 2)$
9. $(5n + 1)(n - 6)$ **11.** $(3n + 1)(2n - 5)$
13. $(7y + 2)(y + 4)$ **15.** $(7y - 1)(y - 8)$
17. $2(7a - 3)(a - 1)$ **19.** $(8x - y)(x + 3y)$
21. $(4a + 5b)(a - b)$ **23.** $(3x + 2y)(2x - 5y)$
25. $\left\{1, \dfrac{5}{2}\right\}$ **27.** $\left\{\dfrac{4}{3}, -2\right\}$ **29.** $\left\{\dfrac{5}{2}, -\dfrac{3}{2}\right\}$
31. $\left\{-\dfrac{1}{3}, -\dfrac{7}{2}\right\}$ **33.** $\left\{\dfrac{11}{6}, \dfrac{1}{2}\right\}$ **35.** $\left\{\dfrac{5}{4}, -\dfrac{4}{3}\right\}$
37. $1\dfrac{1}{2}$ or -5 **39.** $1\dfrac{1}{2}$ or -1 **41.** 5, -5, 7, -7
43. 9, -9, 11, -11, 19, -19 **45.** 1, -1, 4, -4, 11, -11 **47.** 1.5 m **49.** 2 m

51.

1st	2nd	3rd	4th
4	-1	3	2
3	2	4	-1

53. 16 **55.** $\left\{1, -\dfrac{9}{2}\right\}$
57. 5 m

Page 352, Review Exercises

1. $\{(1, -1)\}$ **2.** $\{(-3, -1)\}$ **3.** $\{(7, -8)\}$ **4.** Adult: 150, student: 50

Page 352, Extension

1a. 1300 **1b.** 1338.23 **2.** b **3.** $38.23

Pages 354–355, Chapter Review

1. 2, 3, 5, 7 **3.** 1, 2, 3, 5, 6, 10, 15, 30 **5.** 1, 2, 5, 10
7. 1, 3, a, $3a$ **9.** $\{0, 4\}$ **11.** $\left\{\dfrac{1}{2}, -\dfrac{3}{2}\right\}$ **13.** $2x(x - 5)$
15. 0 or 3 **17.** $a^2 - 100$ **19.** $2x^2 - 5x - 12$

21. $(3x + 1)(3x - 1)$ **23.** $\{4, -4\}$ **25.** $2(x - 3)^2$
27. $(x + 3)(x + 2)$ **29.** $\{10, 2\}$ **31.** $(a + 8)(a - 1)$
33. $\{4, -5\}$ **35.** $(3x - 2y)(x + y)$ **37.** $\left\{\dfrac{5}{2}, -1\right\}$

Page 356, Chapter 7 Self-Test

1. $2^2 \cdot 3^2 \cdot 7$ **2.** $2^3 \cdot 3 \cdot 5^2$ **3.** 12 **4.** $12ab^2c$
5. $\{0, -3\}$ **6.** $\left\{\dfrac{1}{2}, -5\right\}$ **7.** $15x^2 - 27x$
8. $4x^2 - 49y^2$ **9.** $6a^2 + 5a - 99$ **10.** $t^2 - 18t + 81$
11. $5x(x - 7)$ **12.** $(3a + 2)(3a - 2)$ **13.** $(y + 5)^2$
14. $2(w - 3)^2$ **15.** $(x + 5)(x + 4)$ **16.** $(c + 8)(c - 3)$
17. $(a - 3b)(a + 2b)$ **18.** $(r - 3)(r - 2)$
19. $(2x + 9)(x - 6)$ **20.** $\{5, -5\}$ **21.** $\{7\}$ **22.** $\{2, 5\}$
23. $\{6, -13\}$ **24.** 11 or -13 **25.** $2\dfrac{1}{2}$ in.

Page 357, Practice for College Entrance Tests

1. E **2.** C **3.** A **4.** D **5.** E **6.** B **7.** A **8.** B
9. B **10.** A **11.** C

Chapter 8

Page 362, Classroom Exercises

1. $x \neq 0$ **2.** $x \neq -1$ **3.** $x \neq 0$, 1 **4.** $x \neq 2$, -2
5. $a \neq b$ **6.** $a \neq 0$, -6 **7.** None **8.** None
9. No **10.** Yes **11.** Yes **12.** Yes **13.** $\dfrac{a + b}{a - b}$, $a \neq b$
14. $\dfrac{2}{5}$, $c \neq -d$ **15.** 1, $a \neq b$ **16.** $\dfrac{x + 1}{2}$, $x \neq -1$

Pages 362–364, Written Exercises

1. 3 **3.** $\dfrac{3}{2}$ **5.** 2, 4 **7.** 2, 3 **9.** $a \neq 5$ **11.** $b \neq -3$
13. $a \neq 0$, $b \neq 2$ **15.** $a \neq 0$, -3 **17.** Yes **19.** No
21. Yes **23.** No **25.** $\dfrac{1}{3}$, $x \neq 0$ **27.** $\dfrac{4}{15x}$, $x \neq 0$
29. $\dfrac{b + 2}{b + 4}$, $b \neq -3$, -4 **31.** $\dfrac{2}{5}$, $x \neq -2$
33. $\dfrac{x + 5}{2x - 7}$, $x \neq 0$, $\dfrac{7}{2}$ **35.** $\dfrac{5a}{(a + 1)^2}$, $a \neq 0$, -1
37. $\dfrac{1}{c + 2}$, $c \neq -3$, -2 **39.** $\dfrac{3}{t + 5}$, $t \neq -5$, 4
41. $\dfrac{a - 1}{a + 1}$, $a \neq -2$, -1 **43.** $\dfrac{x}{x + 1}$, $x \neq -\dfrac{3}{2}$, -1
45. $x + 3$ **47.** $a + 2$ **49.** $a + 6$ **51.** $a + 4$ **53.** $\dfrac{1}{2}$,
$\dfrac{3}{5}$; $\dfrac{3}{7}$, $\dfrac{1}{2}$; $\dfrac{1}{3}$, $\dfrac{1}{3}$; $\dfrac{1}{5}$, 0; no **55.** $-\dfrac{1}{4}$, $-\dfrac{1}{4}$; $-\dfrac{1}{3}$, $-\dfrac{1}{3}$;
-1, -1; 1, 1; yes **57.** $\dfrac{s}{6}$ **59.** $\dfrac{hs}{2(s + 2h)}$ **61.** $\dfrac{r}{3}$
63. $\dfrac{n^2 - 1}{n - 1} = n + 1$ **65a.** -3 **65b.** 0 **65c.** $\dfrac{a}{3}$
67a. -4 **67b.** 4 **67c.** $\dfrac{a - 4}{a + 4}$ **69a.** $-\dfrac{3}{2}$, 2
69b. -5, $\dfrac{3}{2}$ **69c.** As is

Page 364, Review Exercises

1. $\dfrac{4}{9}$ **2.** 1 **3.** 15 **4.** 15 **5.** -2 **6.** -2

Page 368, Classroom Exercises

1. 1 **2.** 11 **3.** $\frac{3}{10}$ **4.** $\frac{3}{2}$ **5.** $\frac{9x^3}{y^2}$ **6.** 1

Pages 368–370, Written Exercises

1. $\frac{1}{a}$ **3.** $2x$ **5.** $3a$ **7.** $\frac{6}{5}$ **9.** $\frac{5x^2}{y}$ **11.** $4a+3b$
13. $6+5x$ **15.** $3xy+x$ **17.** $\frac{2}{3}+\frac{2}{x}$ **19.** $\frac{a}{2}$
21. $\frac{2a}{3b}$ **23.** $\frac{a}{2}$ **25.** $\frac{x^4}{y^2}$ **27.** $\frac{5c^2}{6d^2}$ **29.** $\frac{a^2}{a+b}$
31. $\frac{x+3}{2(x-4)}$ **33.** $\frac{2}{a}$ **35.** $6xy$ **37.** $\frac{6}{y^2}$ **39.** $\frac{y}{4}$
41. $3(t+2)$ **43.** $\frac{x+3}{x+4}$ **45.** $a+4$ **47.** $\frac{3(x+2)}{x+6}$
49. $2(x-4)$ **51.** $\frac{10}{r+1}$ **53.** $\frac{2(a-b)}{a+b}$ **55.** $\frac{1}{3}$
57. $\frac{2(a+1)}{3(a+2)}$ **59.** $\frac{3a+2}{3a-2}$ **61.** $\frac{a-4}{2a+3}$ **63.** $\frac{x+2}{x-3}$

Page 370, Review Exercises

1. $\frac{3}{5}$ **2.** $\frac{15}{16}$ **3.** $\frac{31}{20}$ **4.** $-\frac{1}{20}$ **5.** 3 **6.** 27

Page 372, Classroom Exercises

1. b **2.** $\frac{x}{3}$ **3.** $\frac{5}{a}$ **4.** 1 **5.** $\frac{1}{x+1}$ **6.** $\frac{1}{a-1}$

Pages 372–373, Written Exercises

1. $6x$ **3.** $\frac{6}{a}$ **5.** $3x$ **7.** $\frac{3}{2y}$ **9.** $\frac{b+1}{b+2}$ **11.** $\frac{c-3}{c-7}$
13. 2 **15.** 4 **17.** $\frac{a+4}{a+3}$ **19.** $\frac{4}{3}$ **21.** $\frac{5}{c-4}$
23. $\frac{s+1}{s-3}$ **25.** 2; -3 **27.** $\frac{6}{h+1}$; $-1, 1$ **29.** 0; 4
31. $\frac{2q+3}{4}$; 0 **33.** $\frac{x-3}{x-2}$; $-2, 2$ **35.** $a+2$; -1
37. $\frac{x-4}{x-3}$; 2, 3 **39.** $\frac{(a+2)^2}{(2a-3)(a+4)}$; $\frac{3}{2}, -4$
41. $\frac{2(2a+3)}{2a-3}$; $\frac{3}{2}$

Pages 373–374, Review Exercises

1. 7 **2.** 1 **3.** 24 **4.** $\left\{17\frac{1}{4}\right\}$ **5.** {80} **6.** -2
7. -2 **8.** $-1\frac{8}{11}$

Page 374, Self-Quiz 1

1. $a\neq0, b\neq0$ **2.** $x\neq-2, 2$ **3.** $\frac{1}{4x}$ **4.** $\frac{5}{a+2b}$
5. $\frac{20a}{b}$ **6.** 1 **7.** $\frac{3}{2mn}$ **8.** $12r$ **9.** $\frac{5}{x}$ **10.** 1
11. $\frac{b+2}{b+4}$ **12.** $\frac{3}{x+3}$

Page 374, Extension

1. No **2.** Yes **3.** Yes **4.** 2813 **5.** 8214 **6.** 53

Page 378, Classroom Exercises

1. $2x$ **2.** $6a^2$ **3.** $12ab$ **4.** $2x^2y^2$ **5.** $(x+1)(x-1)$
6. $\frac{3+4y}{2x}$ **7.** $\frac{3x+2}{x^2}$ **8.** $\frac{2x^2+2y^2}{xy}$ **9.** $\frac{18x-15y}{10x^2y^2}$
10. $\frac{4a}{(a+3)^2}$

Pages 378–380, Written Exercises

1. 18 **3.** $3x$ **5.** ab **7.** a^2 **9.** a^2b^2 **11.** $6a^3b^2$
13. $(x+2)(x-2)$ **15.** $2x(x+3)$ **17.** $\frac{19}{18}$ **19.** $\frac{13}{3a}$
21. $\frac{15+2a}{3a}$ **23.** $\frac{4x-3}{x^2}$ **25.** $\frac{5x}{6}$ **27.** $\frac{x^2-y^2}{xy}$
29. $\frac{3b-2ab}{a^2}$ **31.** $\frac{12xy+7y}{10x^2}$ **33.** $\frac{5}{x}$ **35.** $\frac{5y+x}{xy}$
37. $\frac{3xy+5y+3}{xy}$ **39.** $\frac{5x+4}{x(x+2)}$; 0, -2
41. $\frac{x^2+6x+6}{(x+2)(x+3)}$; $-2, -3$ **43.** $\frac{2x+30}{x(x+6)}$; 0, -6
45. $\frac{4x+1}{(x-5)(x-2)}$; 2, 5 **47.** $\frac{7x+3}{2x(x+1)}$; 0, -1
49. $\frac{x^2+6x-3}{x(2x-1)}$; 0, $\frac{1}{2}$ **51.** $\frac{4x-5}{(x+2)(x-2)}$; 2, -2
53. $\frac{2}{(x+3)(x+1)}$; $-3, -2, -1$ **55.** $\frac{-1}{(x+3)(x+2)}$;
$-4, -3, -2$ **57.** $\frac{30x-5}{(x+2)(x-2)}$ **59.** $2x+2y$

Page 380, Review Exercises

1. 1, -1 **2.** {2} **3.** {14} **4.** $w=\frac{p-2l}{2}$ **5.** $h=\frac{2A}{b}$
6. 12 \$5, 8 \$10 **7.** 20 boys, 10 girls

Page 384, Classroom Exercises

1. $\left\{\frac{19}{2}\right\}$ **2.** {60} **3.** {-7} **4.** {75} **5.** {$-1, 4$}
6. {10} **7.** {3}

Pages 384–385, Written Exercises

1. 12 **3.** $3x$ **5.** $14x$ **7.** $6x$ **9.** {16} **11.** {25}
13. {8} **15.** $\left\{\frac{57}{2}\right\}$ **17.** $\left\{\frac{85}{2}\right\}$ **19.** {8} **21.** {-10}
23. {18} **25.** {-24} **27.** {4} **29.** $\left\{\frac{1}{2}\right\}$ **31.** {-2}
33. {0.5} **35.** {1.4} **37.** {2, 3} **39.** {$-4, 2$}
41. {$-5, 2$} **43.** $\emptyset$ **45.** {4, 6} **47.** {$-1, 3$}
49. {1, 5} **51.** $\left\{-1,\frac{3}{2}\right\}$ **53.** $\left\{\frac{1}{3}, \frac{3}{2}\right\}$ **55.** $\left\{\frac{3}{5}, 4\right\}$
57. $\left\{-\frac{1}{2}, \frac{2}{3}\right\}$ **59.** $\left\{\frac{5}{6}, 5\right\}$

Page 385, Review Exercises

1. $2x^2-x-8$ **2.** $-6a^2b^3$ **3.** $6xy^3$ **4.** $9x^6$
5. $-8a^3b^6$ **6.** $\frac{1}{3x^2}$ **7.** $2x$ **8.** 4,320,000 **9.** $4\cdot10^4$
10. $\frac{1}{4}$ h; Genevieve: 3 km, Helen: 2 km

Page 389, Classroom Exercises

1. 5 Ω **2.** Cindy: 6 h, her brother: 12 h **3.** $3\frac{1}{2}$ Ω
4. 150 Ω **5.** $b=\frac{ac}{a+c}$

Pages 389–392, Written Exercises

1. 3 Ω **3.** $1\frac{1}{3}$ Ω **5.** 4 Ω **7.** 5 Ω **9.** 18 Ω
11. 21 Ω **13.** $\frac{1}{4}+\frac{1}{6}=\frac{1}{t}$, $2\frac{2}{5}$ h **15.** $\frac{1}{6}+\frac{1}{t}=\frac{1}{4}$,
12 h **17.** $\frac{2}{t}+\frac{1}{t}=\frac{1}{3}$, 9 h, $4\frac{1}{2}$ h **19.** $r_1=\frac{r_2R}{r_2-R}$

21. $1\,\Omega$ **23.** $18\,\Omega$ **25.** 5 h **27.** 15 cm **29.** 3 cm
31. $d_i = \dfrac{d_0 f}{d_0 - f}$ **33.** 48 km/h **35.** 1.6 km/h

Pages 392–393, Review Exercises
1a. y-axis **1b.** IV **1c.** II **1d.** I **2a.** $\{0, 1, 2\}$
2b. $\{-2, -1, 0, 1, 2\}$
2c.

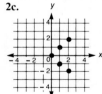

3a. No
3b. Yes
3c. Yes

4.
a	0	5	2	7
b	-10	0	-6	4

5.

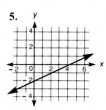

6. $5;\ 0;\ -3;\ -\dfrac{15}{4};\ -4;\ -\dfrac{15}{4};\ -3;\ 0;\ 5$

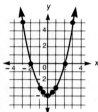

Page 393, Self-Quiz 2
1. $\dfrac{1}{3x}$ **2.** $\dfrac{32 - y^2}{4y}$ **3.** $\dfrac{6 + 7x}{x^2}$ **4.** $\dfrac{4x - 2}{x(x - 2)}$ **5.** $\{10\}$
6. $\{-3, 6\}$ **7.** $\{1\}$ **8.** $6\,\Omega$ **9.** $2\dfrac{1}{17}$ h0

Page 397, Classroom Exercises
1. $\dfrac{1}{12}$ **2.** $\dfrac{52}{5}$ **3.** $\dfrac{1}{100}$ **4.** T **5.** F **6.** T **7.** $\{4\}$
8. $\{2\}$ **9.** $\left\{20\dfrac{1}{4}\right\}$ **10.** $\dfrac{1}{2}$ **11.** $\dfrac{3}{2}$

Pages 397–399, Written Exercises
1. $\dfrac{2}{3}$ **3.** $\dfrac{1000}{1}$ **5.** $\dfrac{1}{6}$ **7.** $\dfrac{1}{4}$ **9.** $\dfrac{10}{17}$ **11.** $\dfrac{34}{99}$ **13.** $\dfrac{3}{4}$
15. $\dfrac{7}{8}$ **17.** $\dfrac{4}{5}$ **19.** $\dfrac{2}{3}$ **21.** T **23.** T **25.** F **27.** F
29. $\{36\}$ **31.** $\{49\}$ **33.** $\{12\}$ **35.** $\left\{13\dfrac{1}{2}\right\}$ **37.** $\dfrac{1}{4}$
39. Length: 24 cm, width: 18 cm **41.** $35,000;
$15,000 **43.** $\{8\}$ **45.** $\{15\}$ **47.** $\{7\}$ **49.** $(4, 5)$
51. $\dfrac{9}{1}$ **53.** $\dfrac{100}{1}$ **55.** $\dfrac{8}{27}$ **57.** Adult: 180,
student: 120 **59.** 60, 80, 100 **61.** $ad = bc$
63. $ad = bc$

Page 399, Review Exercises
1a.–1c.

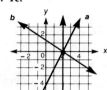

2a. 0 **2b.** $-\dfrac{1}{2}$
2c. 2 **2d.** No slope
3. $2, -5$
4. $y = -\dfrac{3}{5}x + 3$

Page 402, Classroom Exercises
1. 50% **2.** 70% **3.** $62\dfrac{1}{2}$% **4.** $33\dfrac{1}{3}$% **5.** 125%
6. $\dfrac{9}{20}$ **7.** $\dfrac{11}{200}$ **8.** $\dfrac{3}{2}$ **9.** $\{14.25\}$ **10.** $\{58.5\}$
11. $\{160\}$ **12.** $\{50\}$ **13.** $\{75\}$ **14.** $\{116\dfrac{2}{3}\}$ **15.** Yes
16. $2643.75

Pages 403–404, Written Exercises
1. $37\dfrac{1}{2}$% **3.** 120% **5.** $\dfrac{1}{4}$% **7.** 5% **9.** $\dfrac{8}{25}$ **11.** $\dfrac{5}{4}$
13. $\dfrac{1}{200}$ **15.** $\dfrac{3}{200}$ **17.** $\{3\}$ **19.** $\{200\}$ **21.** $\{80\}$
23. $\{4\}$ **25.** $\{6\}$ **27.** $\{8\}$ **29.** $\{75\}$ **31.** $\{80\}$
33. $\{14\}$ **35.** $760 **37.** $2160 **39.** 80% **41.** 220
43. $\{1.5\}$ **45.** $\{195\}$ **47.** $\{8000\}$ **49.** 20%
51. $19,000 **53.** $\left\{-1\dfrac{1}{2}\right\}$ **55.** $\{26\}$ **57.** $\left\{\dfrac{5}{7}\right\}$

Page 405, Review Exercises
1. $\{-2, -1\}$ **2.** $\{2, 2\}$ **3.** $\{4, 10\}$ **4.** Many
solutions **5.** No solution **6.** $x + 3y = 4,\ x + y = 100,$
24 and 76 **7.** $n + d = 50,\ 5n + 10d = 400$, 20 nickels,
30 dimes

Page 405, Extension
regular; diet

Page 407, Classroom Exercises
1. $\dfrac{8}{96} = \dfrac{12}{x}$, 144 **2.** $\dfrac{1.95}{2} = \dfrac{8}{x}$, about 8 mo

Pages 408–410, Written Exercises
1. $\dfrac{3}{10} = \dfrac{n}{60}$ **3.** $\dfrac{3.20}{16} = \dfrac{n}{6}$ **5.** $\dfrac{54}{60} = \dfrac{n}{20}$
7. $\dfrac{45}{100} = \dfrac{n}{1460}$ **9.** $\dfrac{12}{16} = \dfrac{n}{100}$ **11.** 1800 mi
13. 13.5 km/L **15.** $1.14 **17.** 15 **19.** 12%
21. $26.56 **23.** $1.95 **25.** About 22 **27.** 85 cm
29. Alberts: 2286, Barlow: 1905 **31.** $2360 **33.** Bart
Starr **35.** $1125.50 **37.** 8% **39.** 58 **41.** $32

Page 410, Review Exercises
1. $\{1, 3\}$ **2.** $\{-10, 8\}$ **3.** $4/lb: 7.5; $5/lb: 2.5
4. 1, 2, 3, 6, 9, 18 **5.** $2^2 \cdot 3 \cdot 5$ **6.** 20

Page 413, Classroom Exercises
1. Yes   2. No   3. No   4. Yes   5. Yes   6. No
7. No   8. Yes   9. 25   10. Yes

Pages 414–416, Written Exercises
1. No   3. Yes   5. Yes   7. Yes   9. No   11. No
13. Yes   15. Yes   17. Yes   19. No   21. 4   23. $\frac{1}{4}$
25. $\frac{13}{7}$ 27. 16   29. 36   31. 50   33. 32
35. $\frac{5}{18.75} = \frac{8}{x}$, \$30   37. $\frac{25,000}{40,000,000} = \frac{35,000}{x}$,
56,000,000 gal   39. $\frac{1}{450} = \frac{x}{2000}$, 4   41. Yes
43. Yes   45. Yes   47. 10.5 m³   49. 216 cm²
51. 117.850 cm³

Page 417, Review Exercises
1. $x^2 - 2x - 15$ 2. $2x^2 + 11x + 12$ 3. $2a(a - 3)$
4. $(x + 7)(x - 7)$ 5. $(x + 5)^2$ 6. $(x - 4)(x - 3)$
7. $(x + 4)(x - 2)$ 8. $(2x + 3)(x + 1)$ 9. $-\frac{11}{6}$
10. $-\frac{2}{3}$ 11. $\frac{2}{3}$ 12. $\frac{2}{3}$

Page 417, Self-Quiz 3
1. {12}   2. {3}   3. {300}   4. {8}   5. 14%   6. 81 cm
7. 80   8. 112

Page 421, Classroom Exercises
1. Yes, 4   2. Yes, 4   3. Yes, 10   4. No   5. Yes, 6
6. Yes, 2   7. Yes, 18   8. No   9. Yes, 6   10. No
11. $12\frac{1}{2}$ 12. See Classroom Exercise 9.

Pages 421–424, Written Exercises
1. Inversely   3. Directly   5. Inversely   7. Other
9. Inversely   11. Inversely   13. 20   15. 0.5   17. 7

19.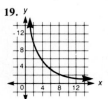

21.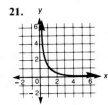

23.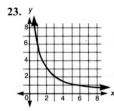

25. $2\frac{1}{2}$ 27. 3   29. 1
31. $(50)(4) = (x)(5)$; 40 mph
33. $(6)(5) = 3x$; 10 h
35. $4x = (5)(12)$; 15
37. $(9)(50) = 10x$; 45 ft
39. $(6)(12) = 8x$; 9 cm

41. $(50)(2) = 40x$; 2.5 m   43. $2x = (2.2)(74)$; 81.4 kg
45. $174x = (114)(12 - x)$; $4\frac{3}{4}$ ft   47. 25 m³
49. 21.22 cm   51. 18 cm

Page 424, Review Exercises
1. $\{0, -5\}$ 2. $\left\{-4, \frac{1}{2}\right\}$ 3. $\{0, -3\}$ 4. $\{0, -3\}$
5. $\{-6\}$ 6. $\{-8\}$ 7. $\left\{-4, \frac{1}{2}\right\}$ 8. $\left\{-2, \frac{2}{3}\right\}$

Pages 426–428, Chapter Review
1. $x = 1$ 3. $x \neq 0$, $y \neq 0$ 5. $\frac{4}{x}$ 7. $\frac{4a}{3}$ 9. 2
11. $\frac{b-4}{b+2}$; $b \neq -2$ 13. $12a^2b^2$ 15. $\frac{n+3m}{mn^2}$
17. {16}   19. {1, 5}   21. $2\frac{2}{5}$ h   23. $\frac{9}{20}$ 25. Length:
7.5 cm, width: 3 cm   27. $\frac{13}{20}$ 29. {250}   31. 10 km
33. No   35. \$17.50   37. 4

Pages 428–429, Chapter 8 Self-Test
1. $\frac{2x}{y}$; x, $y \neq 0$ 2. $\frac{y+3}{5}$; $y \neq 2$ 3. $\frac{6}{b}$ 4. $\frac{3x^2}{5}$
5. $\frac{3}{m}$ 6. $\frac{3(x+y)}{x+2y}$ 7. 2   8. $\frac{9}{a^2}$ 9. $\frac{4+3y}{4y}$
10. $\frac{y-2x}{x^2y}$ 11. $\frac{x+8}{x(x-2)}$ 12. {24}   13. {3}
14. $\{-2, 2\}$ 15. $\{-3\}$ 16. {12}   17. {28}   18. 4 Ω
19. $\frac{4}{25}$ 20. 168   21. 24   22. 6%   23. 12 m
24. 720 mi   25. 16

Page 430, Practice for College Entrance Tests
1. C   2. B   3. A   4. C   5. B   6. A   7. D   8. A
9. D   10. C   11. A

Pages 431–433, Cumulative Review (Chapters 7–8)
1. 1, 3, 13, 39   3. $2^4 \cdot 3$ 5. 8   7. $3xy$ 9. $\{0, 2\}$
11. $\left\{\frac{1}{2}, -\frac{1}{3}\right\}$ 13. $7(2y - x)$ 15. $\{0, -4\}$
17. $a^2 + 9a + 14$ 19. $2n^2 - 2n - 12$ 21. $(7b)^2$
23. $(2x + 1)(2x - 1)$ 25. $\{-9, 9\}$ 27. $4b^2 - 4b + 1$
29. $(x - 5)^2$ 31. $(a + 4)(a + 3)$ 33. $\{-3\}$
35. $(y + 6)(y - 3)$ 37. $(3n + 1)(2n - 1)$ 39. $\{-2, 3\}$
41. No   43. $\frac{1}{3}$ 45. $\frac{1}{b+3}$ 47. 12   49. 3
51. $\frac{12(a+2)}{a}$ 53. $\frac{7}{a-2}$ 55. $\frac{7x}{12}$ 57. $\frac{4x+1}{x(x-1)}$
59. {5}   61. $\left\{\frac{3}{2}\right\}$ 63. $\frac{x}{5} + \frac{x}{6} = 1$; $\frac{30}{11}$ hours   65. $\frac{2}{3}$
67. $\frac{12}{25}$ 69. {60}   71. 105 min or 1 h 45 min
73. 32   75. No   77. 18   79. 55 mph

Chapter 9

Page 437, Classroom Exercises
1. T   2. F   3. T   4. F   5. T   6. T   7. F   8. T
9. T   10. $x > 15$ 11. $x \geq 20$ 12. $x \geq -1$
13. $x < 0$
14.

15.

16.

Pages 437–439, Written Exercises
1. T **3.** T **5.** F **7.** F **9.** T **11.** T **13.** F
15. F **17.** $x < 100$ **19.** $x \geq 70$ **21.** $x > -90$
23. $x \neq 2$

25.

27.

29.

31.

33.

35.

37. $x > 4$ **39.** $x < -2$ **41.** $x \neq 0$ **43.** $x \geq \frac{1}{2}$
45. 20, 0, -20, -100 **47.** 100, 20, 0, -100 **49.** 100,
20, 0 **51.** 0, -20, -100 **53.** 5.5, 0.55, 0.505
55. -0.55, -5 **57.** All **59.** 0.505, -0.55, -5

61. $l \leq 90$

63. $t \geq 10$

65. $c \geq 7$

67. $v < 50$

69. -2, -3 **71.** 3, 2, 1 **73.** 0, -1, -2, -3
75. $b > 1$ **77.** $x > 1$

79.

81.

Page 439, Review Exercises
1. -4 **2.** 2 **3.** 23 **4.** 14 **5.** $\left\{-\frac{3}{5}\right\}$ **6.** $\left\{-\frac{17}{2}\right\}$
7. $r = \frac{P}{2\pi}$ **8.** $y = \frac{c - ax}{b}$

Page 439, Extension
6¢

Page 443, Classroom Exercises
1. T **2.** T **3.** F **4.** T **5.** F **6.** T
7.

8.

9.

10.

11.

12.

13. $x > -2$ and $x \leq 3$

Pages 443–445, Written Exercises
1. T **3.** T **5.** T **7.** F **9.** F **11.** F **13.** T
15. T **17.** T
19.

21.

23.

25.

27.

29.

31.

33.

35.

37. $x < 0$ or $x > 2$ **39.** $x \leq -8$ or $x \geq -5$
41. $x > 33$ and $x < 37$ **43.** $x \geq 0$ and $x \neq 2$

Selected Answers **675**

45. $3, 3\frac{1}{2}$ **47.** $2, 2\frac{1}{2}, 4$ **49.** $3, 4$ **51.** $3\frac{1}{2}, 4$

53. $h > 157$ and $h < 173$

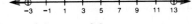

55. $l \geq 6.5$ and $l \leq 6.7$

57. $m \geq 10$ and $m \leq 14$

59. $n < -5$ or $n > 5$

61. $|v| < 4$

63. $-2, 1, 2$ **65.** $-2, 1, 2$ **67.** $-2, 1, 2$ **69.** $-1, 1$

Pages 445–446, Review Exercises

1. $4x - 7$ **2.** $2x^3 y^2$ **3.** $8x^6$ **4.** $2x^3 - 6x^2 + 8x$

5. $\frac{1}{10^2}$ **6.** $4.63 \cdot 10^5$ **7.** $4,650,000$ **8.** 5:30 P.M.

Page 446, Self-Quiz 1

1. F **2.** T

3.

4.

5. $x \geq 6$ **6.** False **7.** True

8.

9.

10. $x \geq -5$ and $x < -2$ **11.** $h \leq 135$ **12.** $t > 3$ and $t < 5$ **13.** $4, 4\frac{1}{2}, 5$ **14.** $3, 3\frac{1}{2}, 4\frac{1}{2}, 5$

Page 449, Classroom Exercises

1. $a > -8$ **2.** $x \leq -1$ **3.** $x < -2$ **4.** $x \geq -2$ **5.** $x < 2$ **6.** $x \neq -8$

Pages 449–451, Written Exercises

1. $a > -6$ **3.** $x \geq 9$ **5.** $x \leq 4$ **7.** $x < 4$ **9.** $x \geq 1$ **11.** $x < -10$

13. $x \leq -2$

15. $x < 1$

17. $x < -3$

19. $x < 3$ or $x > 5$

21. $x > 3$ and $x < 5$

23. $x \geq 5$ or $x < 1$

25. $x \leq -5$ and $x \geq -6$

27. $d \geq 170$ **29.** $k < 8$ **31.** $m > 875$

33. Less than 3 **35.** Less than 5 **37.** $x < -7$

39. $x < -16$ **41.** $x \leq 1$ **43.** $x \leq -4$ **45.** $x < -3$

47. $x < -3$ or $x > 13$

49. $x \leq 2.5$ or $x \geq 3.5$

51. $-2, -1, 2$ **53.** $-2, 1, 2$

Page 451, Review Exercises

1a. x-axis **1b.** IV **1c.** II **2a.** $\{0, 1, 2\}$ **2b.** $\{4, 5\}$

2c.

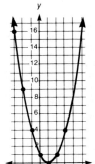

3.

4. 16; 9; 4; 1; 0; 1; 4 **5a.–5c.**

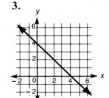

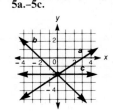

6. $a: -\frac{1}{2}$; $b: 2$; $c: 0$ **7.** $\frac{2}{3}, -2$

Page 454, Classroom Exercises

1. $<$ **2.** $>$ **3.** $<$ **4.** $>$ **5.** $x < 18$

6. $y < -1$

7. $a \geq 2$

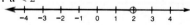

8. $x \geq -\frac{8}{3}$

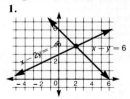

Pages 454–456, Written Exercises

1. < **3.** > **5.** > **7.** < **9.** T **11.** F **13.** T **15.** F
17. $a < 2$

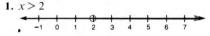

19. $b > 12$ **21.** $c > -3$ **23.** $d < -72$ **25.** $r \geq -\frac{5}{2}$
27. $s \geq 18$ **29.** $c > \frac{7.20}{3}$; cost > \$2.40/lb
31. $c < 2(4.00)$; cost < \$8/lb **33.** $p < \frac{-10,000}{5}$;
profit < $-$\$2000/mo **35.** $\frac{p}{8} < \frac{1.00}{6}$; price < \$1.34
37. $\frac{n}{20} > \frac{12}{25}$; number of field goals > 9 **39.** $x \leq -18$
41. $x \leq 64$ **43.** $y \leq 16$ **45.** $y \geq 32$ **47.** $x > 50$
49. $x < 180$ **51.** $x \geq 3$ and $x \leq 6$ **53.** $x \leq -8$ or
$x \geq 10$ **55.** $x \geq 64$ or $x \leq 16$ **57.** $y \geq \frac{1}{3}$ or $y < 0$
59. $x < 8$ and $x > 0$

Pages 456–457, Review Exercises

1.

2. $\{6, 3\}$ **3.** Many **4.** $\{2, -1\}$ **5.** 30, 70
6. 40%: 15L, 80%: 5L

Page 457, Extension

1. 10.75 ohms **2.** 0.215 ohm **3.** 0.0215 ohm
4. 43 ohms **5.** 465.1 m **6.** 20.93 km **7.** 46.51 m
8. 13.95 m

Page 460, Classroom Exercises

1. $x > 2$

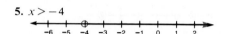

2. $m < 6$

3. $z \leq 4$

4. $y \geq -1$

5. $x > -4$

6. $a > 5$

Pages 460–462, Written Exercises

1. $x > 3$ **3.** $y < 6$ **5.** $n \leq 2$ **7.** $a \geq 2$ **9.** $t > 9$
11. $x > 3$ **13.** $a \geq -\frac{5}{2}$ **15.** $c < -4$ **17.** $x \leq 12$
19. $2w + 2(w + 2) < 28$; width < 6 m **21.** $t + 2t \geq 45$;
Tina's age ≥ 15 **23.** $5x + 6x + 7x \leq 54$;
length ≤ 21 cm **25.** $x + (x + 1) > 16$; integer > $7\frac{1}{2}$
27. $x + (x + 2) < -13$; even integer < $-7\frac{1}{2}$
29. $x \leq -9$

31. $x \geq -42$ **33.** $x < 2$ and $x \neq 0$ **35.** $x > -0.75$
37. $500 + 0.02s \geq 1600$; sales $\geq$ \$55,000
39. $\frac{n}{1.1} \geq 4500$; number of parts ≥ 4546
41. $200 + 0.25s \geq 500$; sales $\geq$ \$1200 **43.** $x > 1$ and
$x < 2$ **45.** $x < -6$ or $x > 9$

Page 462, Review Exercises

1. $2^2 \cdot 7$ **2.** 6 **3.** $3b$ **4.** $2x^2 - 3x - 2$

Page 462, Self-Quiz 2

1. T **2.** T
3. $x < 4$ or $x > 5$

4. $x > 6$ and $x < 9$

5.

6.

7. $x \leq 3$ **8.** $x < -\frac{5}{2}$ **9.** Time < 63 minutes
10. Price > \$400 **11.** Length ≤ 20 cm

Page 466, Classroom Exercises

1. $-5 < x < 5$ **2.** $x > 3$ or $x < -3$
3. $-3 \leq d \leq 11$ **4.** $b \geq -2$ or $b \leq -4$
5.

6.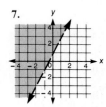

A number line from -4 to 4 with open circles at -1 and 1, shaded between.

7. A number line from -4 to 4 with closed dots at -4 and 4, shaded outward.

8. A number line from -4 to 4 with closed dot at 0.

9. A number line from -5 to 3 with open circles at -5 and 3, shaded outward.

10. A number line from -2 to 6 with open circles at 1 and 4, shaded between.

11. A number line from -4 to 4 with open circles at -2 and 2, shaded outward.

12. A number line from -4 to 4.

13. $|x + 1| > 2$ **14.** $|x - 3| \leq 2$

Pages 466–467, Written Exercises
1. $-10 < x < 10$ **3.** $x > 15$ or $x < -15$
5. $-11 \leq b \leq 5$ **7.** $a \geq 14$ or $a \leq -2$
9. A number line from -4 to 4 with open circles at -3 and 3, shaded between.

11. A number line from -9 to 9 with open circles at -7 and 7, shaded outward.

13. A number line from -12 to 4 with closed dots at -10 and 2, shaded between.

15. A number line from -3 to 5 with closed dots at -2 and 4, shaded outward.

17. A number line from -4 to 4 with closed dots at $-\frac{5}{2}$ and $\frac{5}{2}$, shaded between.

19. A number line from -4 to 4 with open circles at -3 and 3, shaded outward.

21. A number line from -24 to 24 with open circles at -18 and 12, shaded between.

23. A number line from -5 to 5 with closed dots at -4 and 4, shaded between.

25. $|x| \leq 2$ **27.** $|x| > 1$ **29.** $|x - 1| \geq 1$
31. $|x + 2| < 1$
33. $|x| < 5$

A number line from -5 to 5 with open circles at -5 and 5, shaded between.

35. $|x - 2500| < 100$

A number line from 2300 to 2700 with open circles at 2400 and 2600, shaded between.

37. $|x - 25| \leq 0.5$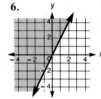

A number line from 24 to 26 with closed dots at 24.5 and 25.5, shaded between.

39. $x > \frac{3}{2}$ or $x < -\frac{5}{2}$ **41.** $-\frac{7}{3} \leq x \leq 1$ **43.** $x \leq -4$
or $x \geq 7$ **45.** $-2 \leq x \leq 3$ **47.** $-1 \leq x \leq 7$

49. A number line from -8 to 10 with closed dots at -6 and 6, open circles at -5 and 5.

51. A number line from 2 to 6 with closed dot at $\frac{5}{2}$, shaded left.

53. A number line from -7 to 1 with closed dot at -5.

55. A number line from -3 to 5 with closed dots at -1 and 1.

Page 467, Review Exercises
1. $\{-3, 0\}$ **2.** $\{-2, 5\}$ **3.** $\{0, 4\}$ **4.** $\left\{0, \frac{5}{2}\right\}$
5. $\{-9, 9\}$ **6.** $\left\{-\frac{5}{4}, \frac{5}{4}\right\}$ **7.** $(x + 5)^2$
8. $(x - 6)(x - 3)$ **9.** $(x + 6)(x - 1)$
10. $(2x - 1)(x + 1)$

Pages 471–472, Classroom Exercises
1. Solid   **2.** Dashed   **3.** Yes   **4.** Yes
5. 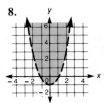   **6.**

7.   **8.**

9. $y < \frac{1}{2}x$

Pages 472–473, Written Exercises
1. Solid   **3.** Dashed   **5.** Yes   **7.** No   **9.** Yes
11. No   **13.** Yes
15. 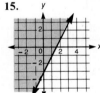   **17.**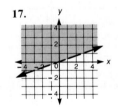

19.

21.

23.

25.

27. $x + y < 4$ **29.** $y < 2x$

31.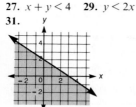

33.

35.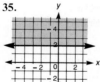

37.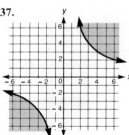

39.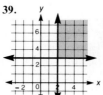

41.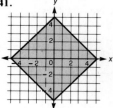

43.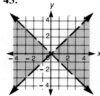

Pages 473–474, Review Exercises

1. 0, 5   **2.** $\frac{2}{3}$ **3.** $\frac{9x}{10}$ **4.** $\frac{3}{2x}$ **5.** {4}   **6.** $\left\{1\frac{7}{8}\right\}$

7. {30}   **8.** {60}   **9.** $4\frac{4}{9}$ h   **10.** $6\frac{2}{3}\Omega$

Page 474, Self-Quiz 3

1.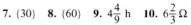

2.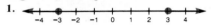

3.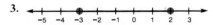

4. $|x| < 1$ **5.** $|x + 1| \geq 1$ **6.** No   **7.** Yes   **8.** Yes

9.   **10.**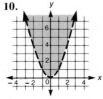

11. $y \geq 315 + 5x$ **12.** $3x + 4y \leq 196$

Pages 479–480, Classroom Exercises

1. Dashed   **2.** Solid   **3.** Solid   **4.** Dashed   **5.** d

6. b   **7.** c   **8.** a

9. 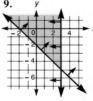   **10.**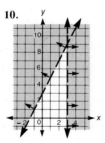

Pages 480–483, Written Exercises

1. d   **3.** a   **5.** g   **7.** f

9. 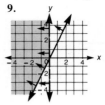   **11.**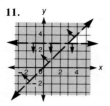

13.   **15.**

17.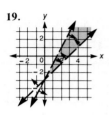

19.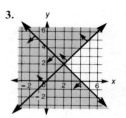

21. $x + y < 5$, $x \geq y + 1$ **23.** $x \geq 2y$, $y < 3$
25. $x + y > 6$, $x - y < 3$ **27.** No   **29.** No
31a. $x + y \leq 100$ **31b.** $x + y \geq 50$ **31c.** $x \geq 40$
31d. $x \leq 70$ **31e.** $y \leq 50$ **31f.** $y \geq 10$ **33.** b
35. f   **37a.** $x + y \leq 200$ **37b.** $x \geq 80$ **37c.** $y \geq 60$
37d. $x + y \geq 160$ **37e.** $y \leq x + 20$ **39.** b, c, e, f, g, h, i   **41.** h   **43.** $94

Page 483, Review Exercises
1. $\frac{2}{5}$ **2.** $90   **3.** 250 mi   **4.** 9   **5.** $450   **6.** 240
7. 4

8.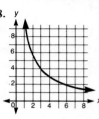

Pages 484–485, Chapter Review
1. $x \geq 21$ **3.** $x \leq 1$
5.

7. $x < 7$
9.

11.

13. $x > 4$

15. $14.90
17.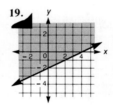

19. 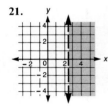   **21.**

Pages 485–486, Chapter 9 Self-Test
1. $n \geq -8$ **2.** $\frac{1}{4}n < 17$
3.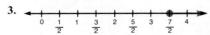

4.

5.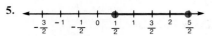

6. $x \leq -4$ or $x \geq -1$ **7.** $3 < x < 8$ **8.** $w > 8$
9. $x \geq -35$
10.

11.

12.

13.

14. $3x \geq 42$; $x \geq 14$ **15.** $2x + 2x + x \leq 25$; $x \leq 5$
16. T   **17.** F   **18.** $s > 4$ or $s < -2$ **19.** $-3 < w < 3$
20. No   **21.** Yes
22. 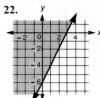   **23.**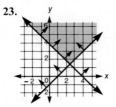

Page 487, Practice for College Entrance Tests
1. D   **2.** B   **3.** B   **4.** C   **5.** A   **6.** D   **7.** A   **8.** B
9. D   **10.** A

Chapter 10

Pages 491, Classroom Exercises
1. positive   **2.** negative   **3.** positive   **4.** 2   **5.** -3

6. 1 **7.** 0 **8.** 14 **9.** 1.4 **10.** -21 **11.** -2.1
12. $\frac{4}{5}$ **13.** $\frac{4}{5}$ **14.** $\frac{7}{6}$ **15.** $\frac{7}{6}$ **16a.** 10 **16b.** 10 and
-10 **17a.** $\{8, -8\}$ Equation **a** has two solutions,
equation **b** has only one solution. **17b.** $\{8\}$

Pages 491–493, Written Exercises
1. Negative **3.** Positive **5.** Not a real **7.** Positive
9. Positive **11.** Negative **13.** 100 **15.** 50
17. -200 **19.** -1.2 **21.** $\frac{3}{4}$ **23.** $\frac{3}{4}$ **25.** 20, -20
27. 40, -40 **29.** $\{4, -4\}$ **31.** $\{5\}$ **33.** $\{-8\}$
35. $\{8, -8\}$ **37.** $\emptyset$ **39.** $\{16\}$ **41.** 10 cm **43.** 6 km
45. 10 km **47.** 0.36 **49.** 1.6 **51.** 12 **53.** 6
55. 0 **57.** $\frac{9}{100}$ **59.** 0.64 **61.** 0.0001
63. 6.75 knots **65.** 9.45 knots **67.** 1 s **69.** 4 s
71. 6.28 s **73.** 12.56 s **75.** 24 **77.** 30
79. 42 ft by 42 ft

Page 493, Review Exercises
1. Distributive property of multiplication over addition
2. Associative property of multiplication **3.** $\{12\}$
4. $\{-5\}$ **5.** $r = \frac{L}{2\pi h}$ **6.** $r = \frac{A - p}{pt}$ **7.** $9.3 \cdot 10^7$

Page 493, Extension
If $x = 1$, $\sqrt{x} = 1$. If $x > 1$, $\sqrt{x} < x$. If $x > 0$ and
$x < 1$, $\sqrt{x} > x$.

Page 496, Classroom Exercises
1. 6, 7 **2.** 8, 9 **3.** 3, 4 **4.** 4.80 **5.** 8.25 **6.** 9.22
7. $\{6.71, -6.71\}$ **8.** $\{7.68, -7.68\}$ **9.** $\{8.72, -8.72\}$

Pages 496–498, Written Exercises
1. 7, 8 **3.** 4, 5 **5.** 1.41 **7.** 5.48 **9.** -4.24
11. -8.49 **13.** 13 **15.** 61 **17.** 5625 **19.** 1936
21. 91 **23.** 38 **25.** $\{\sqrt{10}, -\sqrt{10}\}$ **27.** $\{\sqrt{19}, -\sqrt{19}\}$
29. $\{3.16, -3.16\}$ **31.** $\{4.36, -4.36\}$ **33.** 7.55 cm
35. 2.2 s **37.** 4.5 s **39.** 0.5 amperes **41.** 6.325 cm
43. 6.7 **45.** 3.6 **47.** 13.6 **49.** 20.42 **51.** 26.87
53. 12.49 **55.** False $\sqrt{a - b} \neq \sqrt{a} - \sqrt{b}$ **57.** True
$\sqrt{\frac{a}{b}} = \frac{\sqrt{a}}{\sqrt{b}}$

Page 498, Review Exercises
1.

t	0	2	1	6	-2	0.2
d	40	60	50	100	20	42

2.
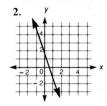

3. 27; 8; 1; $\frac{1}{8}$; 0; $-\frac{1}{8}$;
-1; -8; -27

4.
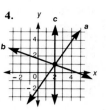

5.

6. $y = \frac{3}{2}x - 3$ **7.** No **8.** No

Page 499, Extension
1. 5.4 **2.** 7.3 **3.** 9.1 **4.** 3.7 **5.** 5.39 **6.** 7.28
7. 9.06 **8.** 3.74 **9.** 4 **10.** 5 **11.** 2 **12.** 3

Page 502, Classroom Exercises
1. Yes **2.** No **3.** Yes **4.** 13 **5.** $\sqrt{5}$ **6.** 6

Pages 502–505, Written Exercises
1. Yes **3.** No **5.** Yes **7.** Yes **9.** 20 **11.** 8
13. 20 **15.** 60 **17.** 80 **19.** 80 **21.** 2.2 **23.** 3.6
25. 7.1 **27.** 8.2 cm **29.** 36.1 km **31.** 60 mm
33. 50 mi **35.** 18 **37.** 61 m **39.** 23 cm
41. $\approx$ 586 m **43.** $\approx$ 47.32 cm, $\approx$ 86.6 cm^2
45. 13 m **47.** $a^2 + b^2 > c^2$

Page 505, Review Exercises
1. $\{0, 2\}$ **2.** $\{4, 20\}$ **3.** 1

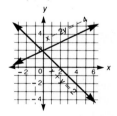

Page 505, Self-Quiz 1
1. $\frac{7}{5}$ **2.** -90 **3.** ± 0.6 **4.** $\{12, -12\}$
5. $\{15, -15\}$ **6.** $\emptyset$ **7.** 3.5 **8.** 140 **9.** 13 **10.** 30
11. 39 **12.** 9 **13.** 48 ft

Page 508, Classroom Exercises
1. $2 \cdot 3^2$ **2.** $2 \cdot 3 \cdot 5^2$ **3.** $2\sqrt{2}$ **4.** $2\sqrt{3}$ **5.** 4
6. $2\sqrt{6}$ **7.** $2\sqrt{10}$ **8.** $4\sqrt{10}$ **9.** $4 - 2\sqrt{6}$ **10.** $5\sqrt{2} + 10$

Pages 508–509, Written Exercises

1. $2^2 \cdot 5$ 3. $2^2 \cdot 3 \cdot 5^2$ 5. $3 \cdot 5^2$ 7. $2^3 \cdot 5^2$ 9. $5\sqrt{2}$
11. $-15\sqrt{2}$ 13. $3\sqrt{5}$ 15. $-6\sqrt{2}$ 17. $2+\sqrt{6}$
19. $2\sqrt{5}+5$ 21. $3-\sqrt{3}$ 23. $4+2\sqrt{3}$ 25. 25
27. 1 29. $21+8\sqrt{5}$ 31. $2\sqrt{3}$ 33. $3\sqrt{2}$ 35. 27
37. 6 39. $12-6\sqrt{3}$ 41. $3\sqrt{10}$ 43. $6\sqrt{3}$ 45. $4\sqrt{3}$
47. 6 49. 17.78 51. 22.36 53. $\sqrt{-4}$ is not a real number.

Page 509, Review Exercises

1. 1, 2, 3, 6, 9, 18 2. $2 \cdot 3 \cdot 7$ 3. 18 4. $4a$
5. $\{0, -5\}$ 6. $\{5, -2\}$ 7. $2a(2a-3)$ 8. 0 or 9

Page 512, Classroom Exercises

1. $\frac{2\sqrt{2}}{5}$ 2. $\frac{\sqrt{15}}{5}$ 3. $\frac{\sqrt{10}}{2}$ 4. $\frac{\sqrt{3}}{2}$ 5. $\frac{\sqrt{15}}{6}$ 6. 2.517
7. $\frac{\sqrt{14}}{2}$ 8. $\frac{\sqrt{10}}{2}$ 9. $\frac{\sqrt{30}}{5}$ 10. $\frac{\sqrt{15}}{3}$ 11. $\frac{2\sqrt{15}}{3}$

Pages 513–514, Written Exercises

1. $\frac{\sqrt{6}}{5}$ 3. $\frac{\sqrt{7}}{3}$ 5. $\frac{2\sqrt{7}}{7}$ 7. $\frac{\sqrt{10}}{5}$ 9. $\frac{\sqrt{14}}{2}$ 11. $\frac{\sqrt{21}}{3}$
13. $\frac{\sqrt{30}}{5}$ 15. $\frac{2\sqrt{6}}{9}$ 17. $\frac{2}{3}$ 19. $\frac{5}{6}$ 21. 0.58
23. 0.32 25. 3.24 27. 1.47 29. 2.65 31. 12.91
33. 2.89 35. 2.53 37. $\frac{7\sqrt{2}}{2}$ cm 39. $\frac{\sqrt{30}}{2}$ m
41. $2\sqrt{2}$ km 43. $\frac{\sqrt{6}}{3}$ 45. $\frac{2\sqrt{3}}{3}$ 47. $\frac{\sqrt{6}}{3}$ 49. $\frac{\sqrt{30}}{3}$
51. $\frac{\sqrt{89}}{10}$ 53. $\frac{1}{2}$ 56. $\frac{18+6\sqrt{2}}{7}$ 57. $4-\sqrt{6}$
59. 3.41 61. -2.41

Page 514, Review Exercises

1. $(x+10)(x-10)$ 2. $(x+6)^2$ 3. $(x-3)(x-4)$
4. $(x-7)(x+6)$ 5. $\{5, -5\}$ 6. $\{-5\}$
7. $\{-2, -10\}$ 8. $\{2, -6\}$

Page 517, Classroom Exercises

1. $7\sqrt{3}$ 2. $-5\sqrt{2}$ 3. Simplest form 4. $12\sqrt{3}-2\sqrt{6}$
5. $-3\sqrt{2}+\sqrt{3}$ 6. $\frac{8\sqrt{5}}{15}$

Pages 517–519, Written Exercises

1. $10\sqrt{2}$ 3. $7\sqrt{3}$ 5. $-2\sqrt{7}$ 7. $7\sqrt{5}$ 9. $7\sqrt{3}+3\sqrt{2}$
11. $\sqrt{21}-2$ 13. $5\sqrt{2}$ 15. Simplest form 17. $\sqrt{5}$
19. $-\sqrt{3}$ 21. $15\sqrt{2}$ 23. $15\sqrt{2}$ 25. $2\sqrt{5}+2\sqrt{3}$, $\sqrt{15}$
27. $24\sqrt{5}$, 135 29. $\frac{5\sqrt{2}}{2}$ 31. $\sqrt{6}$ 33. $\frac{5\sqrt{5}-3\sqrt{15}}{30}$
35. $4\sqrt{3}$ 37. $6\sqrt{3}+6$ 39. $12\sqrt{5}$ 41. $30\sqrt{2}$
43. $4+2\sqrt{2}$ 45. No, $\sqrt{5^2-4^2} \neq 5-4$ 47. No, $(\sqrt{25}-\sqrt{16})^2 \neq 25-16$

Page 519, Review Exercises

1. No 2. $1, -2$ 3. $\frac{9a^2}{2c}$ 4. $\frac{2}{x+2}$ 5. $\frac{2x+3}{x^2}$

6. $\frac{7}{6x}$ 7. $\{27\}$ 8. $\{30\}$ 9. 6 h 10. $6\frac{2}{3}\Omega$
11. $3x^3-15x^2+12x$ 12. $8x^4y^6+8x^3y^7$ 13. 11

Page 520, Self-Quiz 2

1. $6\sqrt{2}$ 2. $5\sqrt{2}-5\sqrt{3}$ 3. $28\sqrt{3}$ 4. $\frac{2\sqrt{7}}{7}$ 5. $\frac{4\sqrt{3}}{3}$
6. $\frac{\sqrt{6}}{3}$ 7. $7\sqrt{2}$ 8. $\frac{2\sqrt{3}}{3}$ 9. 24 10. $\frac{\sqrt{6}}{18}$ 11. 4
12. -2 13. $\frac{\sqrt{6}}{6}$ 14. $12-6\sqrt{3}$ 15. $12\sqrt{2}$
16. $8\sqrt{2}+10$

Page 523, Classroom Exercises

1. $5\sqrt{x}$ 2. $7\sqrt{y}$ 3. $2x\sqrt{3y}$ 4. $4ab\sqrt{3a}$ 5. $\frac{\sqrt{x}}{x}$
6. $x \geq 0$ 7. None 8. $c \geq 0$ 9. $b \geq 0$ 10. $a \geq 0$, $b \neq 0$ 11. $|a|$ 12. $\frac{1}{|a|}$, $a \neq 0$ 13. $b\sqrt{b}$, $b > 0$
14. c^2 15. $2x^2|y|\sqrt{6}$

Pages 523–525, Written Exercises

1. $6\sqrt{y}$ 3. $3b\sqrt{2a}$ 5. $\frac{\sqrt{5x}}{x}$ 7. $\frac{3c^2\sqrt{5a}}{a^2}$ 9. $r \geq 0$
11. None 13. $x > 0$ 15. $s \leq 0$ 17. x 19. $-x$
21. $-|x|$ 23. $|x|$ 25. $|x^3|$ 27. $x\sqrt{x}$, $x \geq 0$ 29. $\frac{1}{|x^5|}$, $x \neq 0$ 31. $\sqrt{5}|x|$ 33. $\frac{2\sqrt{x}}{x}$, $x > 0$ 35. $\frac{\sqrt{2x}}{2}$, $x \geq 0$
37. $\frac{|a|\sqrt{b}}{b}$, $b > 0$ 39. $\frac{a\sqrt{a}}{b^2}$, $a \geq 0$, $b \neq 0$
41. $2\sqrt{a}$, $a > 0$ 43. $a\sqrt{5a}$, $a > 0$ 45. $13a$, $a > 0$
47. F, If $x = -1$, $\sqrt{16} \neq -4$. 49. T 51. F, If $x = 0$, $\sqrt{0} \neq 4$ 53. F, If $x = 0$, $\frac{3}{\sqrt{x}}$ is undefined. 56. $\frac{5\sqrt{x}}{12}$, $x \geq 0$ 57. $\frac{2\sqrt{x}}{x}$ $x > 0$ 59. $c \leq 4$
61. None 63. $h = \frac{\sqrt{3e}}{2}$

Page 525, Review Exercises

1. $\{40\}$ 2. $\{80\}$ 3. \$128 4. 32 5. \$45 6. 90
7. No 8. 10 9. 3

Page 528, Classroom Exercises

1. $\{16\}$ 2. $\{81\}$ 3. $\{7\}$ 4. $\{10\}$ 5. $\{-4\}$ 6. $\{15\}$
7. $\{7\}$ 8. $\{7\}$ 9. $\{1, 5\}$ 10. $\{6\}$

Pages 528–529, Written Exercises

1. $\{25\}$ 3. $\{6\}$ 5. $\{9\}$ 7. $\{3\}$ 9. $\{3, 4\}$ 11. $\{2\}$
13. $\{9\}$ 15. $\{12.5\}$ 17. $\{1\}$ 19. $\{8\}$ 21. $\{4-\sqrt{5}\}$
23. $\{0, 6\}$ 25. $\emptyset$ 27. $\{0\}$ 29. $\{2\}$ 31. $\emptyset$
33. $\{1, 2\}$ 35. $\{5\}$ 37. $\left\{1, \frac{3}{2}\right\}$ 39. $\left\{\frac{3}{5}, -1\right\}$
41. $\{1, 5\}$ 43. $\{4, 8\}$ 45. $\emptyset$

Page 529, Review Exercises

1.

2.

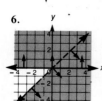

3.

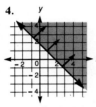

Page 529, Self-Quiz 3

1. $11yz\sqrt{y}$ **2.** $\dfrac{a^2\sqrt{b}}{b}$ **3.** $\dfrac{y^2\sqrt{xv}}{x^2}$ **4.** $xy^2\sqrt{x}$, $x \geq 0$
5. $\dfrac{6\sqrt{z}}{z}$, $z > 0$ **6.** $x^2\sqrt{-2x}$, $x \leq 0$ **7.** $\emptyset$ **8.** $\{4\}$
9. $\{5\}$

Page 532, Classroom Exercises

1. 4, 5, 2, 1   **2.** -3, 2, 4, 8   **3.** -1, -5, -3, -6
4. 8, 6   **5.** 5, 12   **6.** 2, 16   **7.** 9   **8.** 2   **9.** 10

Pages 532–534, Written Exercises

1. 4, 6   **3.** 6, 3   **5.** 6, 13   **7.** 2, 10   **9.** 10   **11.** 5
13. 5   **15.** 13   **17.** 20   **19.** 7   **21.** 1   **23.** 50   **25.** 1
or 13   **27.** 0 or 10   **29.** 2 or 4   **31.** 0 or 10
33. 5.4   **35.** 9.2   **37.** 10   **39.** 20
41. $AB = BC = \sqrt{85}$ **43.** 30   **45.** $17\sqrt{2}$, $17\sqrt{2}$
47. 13, 37   **49.** 41 square units   **51.** (0, 0), (0, 8), or
$(-6, 0)$

Page 534, Review Exercises

1. $|x| > 1$
2.

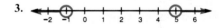

3.

4.

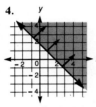

5.

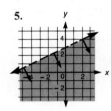

6.

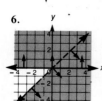

7.

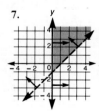

Page 535, Extension

2. The graph is a circle with radius 4 centered at (1, 2).
3. Illegal quantity error in 30. If $x < -3$ or $x > 5$, then
$(16 - (1 - x)^2) < 0$ and has no real-number square root.
4. 10 PRINT "ENTER A NUMBER BETWEEN -6 AND 2."

30 Y = 3 + SQR(16 − (−2 − X) ↑ 2)
60 Y = 3 − SQR(16 − (−2 − X) ↑ 2)

Pages 537–538, Chapter Review

1. -1.2 **3.** $\{9, -9\}$ **5.** $(6.32, -6.32)$ **7.** 13
9. 4.5 cm   **11.** $\sqrt{10} + 2$ **13.** $\dfrac{\sqrt{5}}{3}$ **15.** $4\sqrt{5}$ cm
17. $3\sqrt{3} - 3\sqrt{2}$ **19.** $2|x|$ **21.** $\dfrac{a\sqrt{a}}{|b|}$, $a \geq 0$, $b \neq 0$
23. $\{2\}$ **25.** 5   **27.** -1 or 7

Pages 539–540, Chapter 10 Self-Test

1. NR   **2.** -0.6 **3.** $\dfrac{3}{2}$ **4.** $(0.4, -0.4)$ **5.** $\{-9\}$
6. 7, 8   **7.** $\{2\sqrt{19}, -2\sqrt{19}\}$ **8.** $\{-6\sqrt{5}\}$ **9.** 40
10. 11   **11.** 16 ft   **12.** $3\sqrt{6}$ **13.** $7\sqrt{5} - 14$
14. $5 + 2\sqrt{6}$ **15.** $\dfrac{1}{2}$ **16.** $\dfrac{\sqrt{3}}{2}$ **17.** $\dfrac{3\sqrt{5}}{5}$ **18.** $\dfrac{\sqrt{21}}{3}$
19. $2\sqrt{2}$ **20.** $3\sqrt{2} + \sqrt{3}$ **21.** $\dfrac{\sqrt{5}}{|x|}$, $x \neq 0$ **22.** $\dfrac{b^2\sqrt{a}}{a}$,
$a > 0$ **23.** $\{-24\}$ **24.** $\{3\}$ **25.** $5\sqrt{5}$ **26.** -3 or -7

Pages 540–541, Practice for College Entrance Tests

1. A   **2.** C   **3.** B   **4.** C   **5.** C   **6.** A   **7.** B   **8.** C
9. D   **10.** B   **11.** B   **12.** A   **13.** B   **14.** B

Chapter 11

Page 547, Classroom Exercises

1. 3, 2, -4 **2.** -2, -1, 5   **3.** -1, 1, 0   **4.** 0.5, 0,
-6 **5.** 1, 11, 30   **6.** 2, -14, -2 **7.** $x = 3$,
minimum: $(3, -4)$ **8.** $x = -\dfrac{3}{2}$, minimum:
$\left(-\dfrac{3}{2}, -\dfrac{13}{4}\right)$ **9.** $x = 2$, minimum: $(2, -1)$
10. $x = -4$, minimum: $(-4, 0)$ **11.** Linear

Pages 547–550, Written Exercises

1. No   **3.** Yes   **5.** Yes   **7.** Yes   **9.** Yes   **11.** No
13. 1, 2, 3   **15.** -3, 2, 0   **17.** $y = x^2 + 6x + 8$; 1, 6, 8
19. 0; -3; -4; -3; 0   **21.** 12; 2; -4; -6; -4

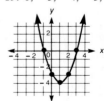

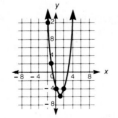

23. 8; 2; 0; 2; 8

25. 5; 0; −3; −4; −3

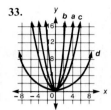

27. $x = 1$, minimum: $(1, -4)$ **29.** $x = -\dfrac{3}{4}$, minimum:

$\left(-\dfrac{3}{4}, -\dfrac{25}{8}\right)$

31.

33.

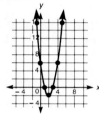

35. The greater c is, the higher the graph is. **37.** The greater a is, the narrower the graph is.

39. 16; 6; 0; −2; 0; 6; 16 axis: $x = 2$; min.: $(2, -2)$; y-int.: 6; x-int.: 1, 3

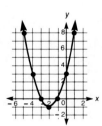

41. 0; $-\dfrac{5}{2}$; −4; $-\dfrac{9}{2}$; −4; $-\dfrac{5}{2}$; 0 axis: $x = 2$; min.: $\left(2, -\dfrac{9}{2}\right)$; y-int.: $-\dfrac{5}{2}$; x int.: −1, 5

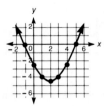

43. 8; 3; 0; −1; 0; 3; 8 axis: $x = -2$; min.: $(-2, -1)$; y-int.: 3; x-int.: −3, −1

45. No, yes, yes, yes
47. No, yes, yes **49.** p must be a whole number.
51. $\left(-\dfrac{b}{2a}, -\dfrac{b^2 + 4ac}{4a}\right)$
53. (5, 0)

1. $-10x$ **2.** $y = \dfrac{2}{3}x - 2$
3.

4. {2, 2}
5. $4/lb:5. $5/lb:15
6. $2(x + 5)(x - 5)$
7. $2(x + 5)(x - 3)$
8. {9, −2}

Pages 553–554, Classroom Exercises

1. {1, 2}

2. {2, −2}

3. {2}

4. 2 **5.** 1 **6.** 0 **7.** −3 and −4, 0 and −1

Pages 554–556, Written Exercises

1.

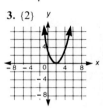

1a. {1, −3}
1b. {0, −2}
1c. {−1}

3.

3a. {2, 4}
3b. {1, 5}
3c. {3}

5.

5a. {3, −1}
5b. {0, 2}
5c. {1}

684 Selected Answers

7. {2, 3}

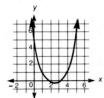

9. {2, 4}

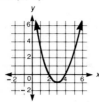

27. h(m)

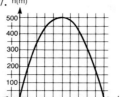

29. 10

31. 8 s, 12 s

33. 3.5 s, 16.5 s

11. ∅

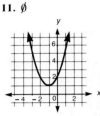

13. 2, 3; −1, 0

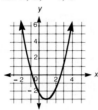

35. 1.8 m

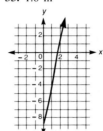

37. By about 0.7 m

39. Very narrow graph

41. Irrational *y*-intercepts

15. −1, 0; −3, −2

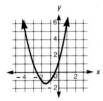

17. 3, 4; 0, 1

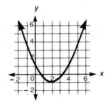

43. {2,4, −0.4}

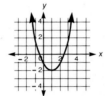

45. {3.4, 0.6}

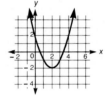

19. {2, 3}

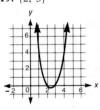

21. $\left\{\frac{4}{3}, -1\right\}$

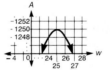

47. Length: 50 ft, width: 25 ft

23. $\left\{\frac{3}{2}, -\frac{1}{2}\right\}$

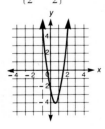

25. ≈ 1.5, ≈ 6.5

Page 556, Review Exercises

1. 0, − 1   **2.** $\frac{4a}{3c}$ **3.** $\frac{18}{a-b}$ **4.** {10}   **5.** {25}

6. {500}   **7.** $\left\{88\frac{8}{9}\right\}$ **8.** $8\frac{1}{3}\Omega$ **9.** $4\frac{4}{9}$ h   **10.** $\frac{7}{3}$

11. 10   **12.** 5

Page 557, Self-Quiz 1

1. $y = x^2 + 3x − 7$; 1, 3, − 7

2.

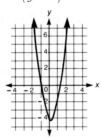

3. $x = − 2$

4. Minimum: $(− 2, − 3)$

5. 6   **6.** 2

7. − 1   **8.** 1

9. $\left\{\frac{1}{2}, 3\right\}$ **10.** $\emptyset$

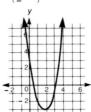 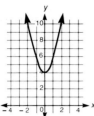

11. 12   **12.** 3 m

Page 560, Classroom Exercises
1. $\{\pm 5\}$ **2.** $\{\pm\sqrt{6}\}$ **3.** $\{1, -7\}$ **4.** $\{-1 \pm \sqrt{5}\}$
5. $\{3 \pm 2\sqrt{2}\}$ **6.** $\emptyset$

Pages 560–562, Written Exercises
1. $\{\pm 7\}$ **3.** $\{12, -8\}$ **5.** $\emptyset$ **7.** $\{-3, -7\}$ **9.** $\emptyset$
11. $\{9\}$ **13.** $\{\pm 5\}$ **15.** $\{\pm 10\}$ **17.** $\{\pm 8\}$
19. $\{\pm 6\}$ **21.** $\{0, 6\}$ **23.** $\left\{\pm\frac{3}{2}\right\}$ **25.** $\{7, -13\}$
27. $\{4, 8\}$ **29.** $\{10\}$ **31.** $\{\pm 3\sqrt{2}\}$ **33.** $\{\pm\sqrt{13}\}$
35. $\{5 \pm \sqrt{7}\}$ **37.** $\{1 \pm \sqrt{3}\}$ **39.** $\{-4 \pm 2\sqrt{2}\}$
41. $\left\{\frac{1 \pm \sqrt{6}}{2}\right\}$ **43.** $\{4.236, -0.236\}$
45. $\{5.367, -11.367\}$ **47.** $\{4.262, -2.062\}$ **49.** 2 m
51. $(10 - 5\sqrt{2})$ cm   **53.** $12\sqrt{10}$ m   **55.** 7.906 s
57. 40 or 60   **59.** $\{\pm 3\sqrt{2}\}$ **61.** $\{\pm 2\sqrt{7}\}$ **63.** $k > 0$
65. $k < 0$ **67.** Any real number

Pages 562, Review Exercises
1. 100   **2.** 84%   **3.** 30 km   **4.** Yes   **5.** 45 min
6. 8   **7.** 2 amperes   **8a.** 1   **8b.** 1   **9a.** 7   **9b.** 7

Page 566, Classroom Exercises
1. 9   **2.** 9   **3.** 36   **4.** 16   **5.** $\frac{9}{4}$ **6.** $\frac{49}{4}$ **7.** $\{2, -6\}$
8. $\{7, -1\}$ **9.** $\left\{-2 \pm \frac{2\sqrt{15}}{3}\right\}$

Pages 566–568, Written Exercises
1. 25   **3.** 9   **5.** $\frac{25}{4}$ **7.** 4   **9.** $\frac{4}{25}$ **11.** 16
13. $\{1, -11\}$ **15.** $\{8, -2\}$ **17.** $\{2, 4\}$
19. $\{-2, -3\}$ **21.** $\{4, -8\}$ **23.** $\{7, -1\}$
25. $\{-2 \pm 3\sqrt{2}\}$ **27.** $\{5 \pm 2\sqrt{10}\}$ **29.** $\{10 \pm 5\sqrt{3}\}$
31. $\{8.099, -2.099\}$ **33.** $\{5.220, -13.220\}$
35. $\{8.599, -1.599\}$ **37.** $-1 + 2\sqrt{3}$ or $-1 - 2\sqrt{3}$
39. $(-6 + 2\sqrt{15})$ cm   **41.** $\left\{\frac{5}{2}, -3\right\}$ **43.** $\left\{\frac{1}{3}, -\frac{7}{3}\right\}$
45. $\left\{\frac{-5 \pm \sqrt{5}}{10}\right\}$ **47.** $\left\{\frac{4 \pm \sqrt{10}}{3}\right\}$
49. $(2 + 2\sqrt{51})$ cm, $(-2 + 2\sqrt{51})$ cm
51. $\frac{-b \pm \sqrt{b^2 - 8}}{2}$ **53.** $\frac{-b \pm \sqrt{b^2 - 4ac}}{2a}$ **55.** 2.3 s

Page 568, Review Exercises
1. $x > -2$
2.
3.
4.
5.
6. $x \leq -2$ **7.** $x < 9$ **8i.** 3   **8ii.** -1 **9i.** $\emptyset$ **9ii.** $\emptyset$
10i. 3   **10ii.** 3

Page 568, Self-Quiz 2
1. $\{\pm 5\sqrt{5}\}$ **2.** $\{-4, -8\}$ **3.** $\emptyset$ **4.** $\{17, -3\}$
5. $\{-4 \pm 2\sqrt{5}\}$ **6.** $\left\{\frac{3 \pm \sqrt{5}}{2}\right\}$ **7.** $\left\{\frac{-4 \pm \sqrt{26}}{2}\right\}$
8. $(1 + 4\sqrt{3})$ cm

Page 571, Classroom Exercises
1. $2x^2 - 3x - 7 = 0$ **2.** $2x^2 - 6x + 4 = 0$
3. $x^2 - 6x - 10 = 0$ **4.** 5, -3, 4   **5.** 6, -2, 0
6. 3, -2, -1 **7.** $\left\{\frac{-1 \pm \sqrt{13}}{3}\right\}$ **8.** $\emptyset$ **9.** $\left\{\pm\frac{1}{2}\right\}$

Pages 572–573, Written Exercises
1. $x^2 + 5x - 2 = 0$ **3.** $2x^2 - x + 6 = 0$
5. $3x^2 + 6x - 6 = 0$ **7.** 5, 7, 2   **9.** 1, 1, -1 **11.** 2,
0, -11 **13.** 1, 3, 2   **15.** $\frac{1}{2}$, -1, $-\frac{1}{3}$ **17.** 1, -5,
-2 **19.** $\{3, -5\}$ **21.** $\left\{-\frac{2}{5}, -1\right\}$ **23.** $\{1, 2\}$
25. $\left\{\frac{1}{2}, -\frac{2}{3}\right\}$ **27.** $\left\{\frac{3}{2}, -\frac{1}{2}\right\}$ **29.** $\{5\}$ **31.** $\{0, 5\}$
33. $\{-0.438, -4.562\}$ **35.** $\{1.449, -3.449\}$
37. $x^2 + x - 1 = 0$; $\left\{\frac{-1 \pm \sqrt{5}}{2}\right\}$ **39.** $x^2 + 20x - 8 = 0$;
$\{-10 \pm 6\sqrt{3}\}$ **41.** $\emptyset$ **43.** $\emptyset$ **45.** $x(x + 2) = 3$; 1, 3 or
-1, -3 **47.** $l^2 + (l - 10)^2 = 11^2$; $\left(\frac{10 + \sqrt{142}}{2}\right)$ cm
49. $-10 + 10\sqrt{2}$ **51.** $\frac{-3 - \sqrt{3}}{3}$; $\frac{-3 + \sqrt{3}}{3}$; -2; $\frac{2}{3}$
53. $\frac{3}{2}$; 4; $\frac{11}{2}$; 6

Page 573, Review Exercises
1. $x < 3$
2. $x < -4$
3. $x \geq 3$

4. $x < -3$

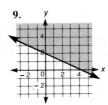

5. $x < \dfrac{3}{2}$

6. $x \leq 4$

7. $x > 2$ or $x < -4$

8. $1 \leq x \leq 3$

9.

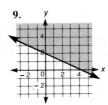

10.

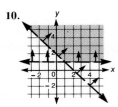

Pages 576–577, Classroom Exercises
1. $b^2 - 4ac$　**2.** 2, rational　**3.** 2, irrational　**4.** None
5. 1, rational　**6.** 2, irrational　**7.** 2, rational
8. 2, rational　**9.** 2, irrational　**10.** None　**11.** None
12. None　**13.** 2, irrational　**14.** 2, irrational
15. 2, irrational　**16.** 1, rational

Pages 577–578, Written Exercises
1. -8　**3.** 25　**5.** 0　**7.** 16　**9.** 2, rational
11. 2, irrational　**13.** None　**15.** 1, rational
17. 2, irrational　**19.** 2, rational　**21.** 1, rational
23. 2, irrational　**25.** 2, rational　**27.** $\{-4 \pm \sqrt{10}\}$
29. $\{-3 \pm \sqrt{13}\}$　**31.** $\emptyset$　**33.** $\{3, -2\}$　**35.** $\emptyset$
37. $\left\{\dfrac{3}{2}\right\}$　**39.** None　**41.** 2　**43.** 2　**45.** $\dfrac{4}{3}$　**47.** $\dfrac{1}{5}$
49. $\dfrac{4}{3}$　**51.** $k < \dfrac{25}{12}$　**53.** $k < \dfrac{4}{3}$　**55.** $|k| > 8$　**57.** True
59. False　**61.** True

Page 578, Review Exercises
1. $\dfrac{3}{5}$　**2.** -8　**3.** $5\sqrt{3}$　**4.** $2 + 2\sqrt{3}$

Page 578, Self-Quiz 3
1. $\{3 \pm \sqrt{6}\}$　**2.** $\left\{\dfrac{-5 \pm \sqrt{29}}{2}\right\}$　**3.** $\left\{-\dfrac{1}{4}, -2\right\}$
4. $\left\{-1, -\dfrac{3}{2}\right\}$　**5.** 2, rational　**6.** None　**7.** 4, -4
8. 9

Page 581, Classroom Exercises
1. $\{1, -8\}$　**2.** $\{-12 \pm 3\sqrt{15}\}$　**3.** $\left\{\dfrac{-2 \pm \sqrt{13}}{3}\right\}$

4. $\left\{\dfrac{-5 \pm \sqrt{57}}{4}\right\}$　**5.** $\{-4 \pm \sqrt{14}\}$　**6.** $\{-4 \pm 4\sqrt{2}\}$

Pages 581–582, Written Exercises
1. $\{7, -4\}$　**3.** $\{3, 6\}$　**5.** $\{-1, -14\}$　**7.** $\{-2 \pm \sqrt{10}\}$
9. $\{4 \pm \sqrt{21}\}$　**11.** $\{6 \pm 2\sqrt{6}\}$　**13.** $\left\{\dfrac{-5 \pm \sqrt{41}}{2}\right\}$
15. $\left\{\dfrac{3 \pm \sqrt{69}}{6}\right\}$　**17.** $\left\{\dfrac{-5 \pm \sqrt{15}}{5}\right\}$　**19.** $\{0, 3\}$
21. $\left\{\dfrac{1 \pm \sqrt{5}}{2}\right\}$　**23.** $\{-2, -4\}$　**25.** $\left\{\dfrac{3}{2}, \dfrac{5}{2}\right\}$
27. $\left\{\dfrac{-3 \pm \sqrt{17}}{4}\right\}$　**29.** $\{11, -3\}$　**31.** $l(l - 3) = 3$;
$\dfrac{3 + \sqrt{21}}{2}$ m　**33.** $n^2 = n + 6$; 3 or -2　**35.** $n^2 = n + 3$;
$\dfrac{1 \pm \sqrt{13}}{2}$　**37.** $2556 = \dfrac{1}{2}n^2 + \dfrac{1}{2}n$; 71
39a. $\dfrac{12 \pm \sqrt{94}}{5}$　**39b.** $\dfrac{12 \pm 2\sqrt{11}}{5}$　**39c.** $\dfrac{12 \pm \sqrt{19}}{5}$
39d. Never　**41.** $(55t)^2 + (45t)^2 = 100^2$; $\dfrac{10\sqrt{202}}{101}$ h
43. $n = 4 + \dfrac{1}{n}$; $2 + \sqrt{5}$ or $2 - \sqrt{5}$　**45.** $\{-3 \pm \sqrt{10}\}$
47. $\left\{2, 3, \dfrac{5 \pm \sqrt{5}}{2}\right\}$

Page 583, Review Exercises
1. $\sqrt{2}$　**2.** $3\sqrt{2}$　**3.** $|x|$　**4.** $2|a|b\sqrt{3b}$, $b \geq 0$　**5.** $\{9\}$
6. $\{8\}$　**7.** 5　**8.** $\sqrt{5}$　**9.** $(x + 7)(x - 2)$
10. $(2x + 1)(x + 2)$

Pages 584–585, Chapter Review
1. $y = x^2 + 2x - 8$; 1, 2, -8
3.

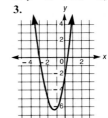

5. $\emptyset$

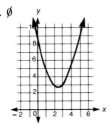

7. $\{10, -4\}$　**9.** $\{-2, \pm \sqrt{5}\}$　**11.** $\{-2 \pm \sqrt{3}\}$
13. $\left\{\dfrac{3}{2}, 2\right\}$　**15.** $(3 + \sqrt{39})$ cm, $(-3 + \sqrt{39})$ cm
17. None　**19.** $\{0, 3\}$　**21.** $1 + \sqrt{10}$ or $1 - \sqrt{10}$

Pages 585–586, Chapter 11 Self-Test
1. No　**2.** $y = -4x^2 + 8x + 3$
3. $x = -3$, $(-3, -4)$　**4.** x-int.: 5, $-\dfrac{1}{2}$; y-int.: -5

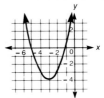

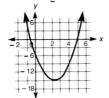

5. $2, 3; -6, -5$

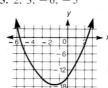

6. $3, -4$ **7.** $\emptyset$

8. $\{\pm 10\}$ **9.** $\{\pm 4\}$

10. $\{3 \pm 2\sqrt{2}\}$

11. $\{6, -10\}$

12. $\left\{\dfrac{5 \pm 2\sqrt{6}}{2}\right\}$

13. $\{1, -11\}$

14. $\{2, 4\}$

15. $\left\{\dfrac{-3 \pm \sqrt{17}}{4}\right\}$ **16.** $\{3 \pm \sqrt{7}\}$ **17.** $\left\{1, -\dfrac{4}{5}\right\}$

18. $\{4\}$ **19.** Positive, perfect square; positive, not a perfect square; 0; negative **20.** $\{2, -22\}$

21. $\left\{\dfrac{5}{2}, -\dfrac{1}{2}\right\}$ **22.** $\left\{\dfrac{9}{5}, -2\right\}$ **23.** $\{-1 \pm 2\sqrt{2}\}$

24. 6 m by 12 m **25.** 8 s and 16 s

Page 587, Practice for College Entrance Tests
1. C **2.** B **3.** D **4.** A **5.** B **6.** D **7.** B **8.** D **9.** C

Pages 588–591, Cumulative Review (Chapters 7–11)
1. $2^2 \cdot 3 \cdot 7$ **3.** $\left\{-4, \dfrac{1}{2}\right\}$ **5.** $c^2 - 10c + 25$

7. $(y + 8)(y - 8)$ **9.** $(x - 5)(x - 3)$

11. $(2x - 1)(x + 4)$ **13.** $\{-5, 3\}$ **15.** $5a$ **17.** $12y$

19. $\dfrac{x}{12}$ **21.** $\left\{\dfrac{1}{2}\right\}$ **23.** $\{60\}$ **25.** \$1.45 **27.** 12

29.

31.

33. $b > -5$ **35.** $w + 15 \geq 128$; 113 lb or more **37.** $a > -4$

39. $y > -2$

41. $x > 11$ or $x < -11$ **43.** No

45.
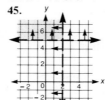

47. 9 **49.** 7 cm **51.** 69 **53.** 20 **55.** $-4\sqrt{3}$

57. $\dfrac{3}{4}$ **59.** $7\sqrt{7}$

61. $10\sqrt{5}$ **63.** $4x^2\sqrt{2x}$

65. $\{4\}$ **67.** -2 and 6

69.
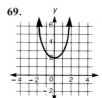

71. $\emptyset$ **73.** 16

75. $a = 6, b = -1, c = 1$

77. $\left\{-2, \dfrac{5}{2}\right\}$ **79.** 48

81. $\dfrac{-5 + \sqrt{30}}{2}$ in.

Supplementary Topic: Probability and Statistics

Page 594, Classroom Exercises
1. $\dfrac{1}{3}$ **2.** $\dfrac{2}{3}$ **3.** $\dfrac{4}{15}$ **4.** $\dfrac{1}{4}$

Page 594, Written Exercises
1. $\dfrac{1}{20}$ **3.** $\dfrac{3}{10}$ **5.** $\dfrac{2}{5}$ **7.** $\dfrac{1}{6}$ **9.** 0 **11.** $\dfrac{1}{2}$ **13.** $\dfrac{1}{4}$

15. $\dfrac{2}{13}$ **17.** $\dfrac{3}{4}$ **19.** $\dfrac{1}{8}$ **21.** $\dfrac{3}{8}$ **23.** $\dfrac{7}{8}$ **25.** $\dfrac{5}{16}$

Page 596, Classroom Exercises
1. $\dfrac{3}{40}$ **2.** $\dfrac{18}{95}$

Pages 596–597, Written Exercises
1. $\dfrac{1}{4}$ **3.** $\dfrac{1}{2}$ **5.** $\dfrac{1}{4}$ **7.** $\dfrac{1}{16}$ **9.** $\dfrac{8}{169}$ **11.** $\dfrac{13}{51}$ **13.** 0

15. $\dfrac{11}{12}$ **17.** $\dfrac{2}{3}$ **19.** $\dfrac{4}{9}$ **21.** 0 **23.** $\dfrac{1}{15}$ **25.** $\dfrac{7}{15}$

27. $\dfrac{1}{3}$ **29.** $\dfrac{10}{13}$ **31.** $\dfrac{3}{8}$ **33.** $\dfrac{7}{20}$ **35.** $\dfrac{3}{20}$

Page 599, Classroom Exercises
1. The median, because all but one of the salaries are close to it. **2.** 177.2, 175, 175

Page 600, Written Exercises
1. 4, 3, 2 **3.** 26.8, 27, 27 **5.** 17, 18, 20 **7.** 70.5, 70, 70 **9.** 52.5 **11.** The Marriage of Figaro; average attendances are 613, 847, and 796. **13.** 158.7, 159.95, 162.45

15.

Points	Frequency
1	1
3	5
6	4

Page 602, Classroom Exercises
1. 1.6 **2.** 1.1 **3.** 5

Pages 602–603, Written Exercises
1. Greatest: A, least: C **3.** 10, 3.9 **5.** 8, 2 **7.** 24, 6 **9.** -23, 18, 5 **11.** F **13.** F **15.** F

Supplementary Topic: Trigonometry

Page 606, Classroom Exercises

1. $85°$ **2.** $a = 5$, $b = 7$ **3.** $a = 6\frac{6}{13}$, $b = 16\frac{5}{7}$

Pages 606–607, Written Exercises

1. $55°$ **3.** $108°$ **5.** $a = 16$, $b = 14$ **7.** $a = 13\frac{5}{7}$, $b = 12\frac{5}{6}$ **9.** 10 m **11.** $\angle A = 66°$, $DF = 22\frac{2}{3}$, $EF = 21\frac{1}{3}$ **13.** $108°$ **15.** $135°$ **17.** $162°$

Page 609, Classroom Exercises

1. $\tan A = 0.63$, $\tan B = 1.60$ **2.** 8.4 **3.** 11.9 **4.** 18.7

Pages 609–610, Written Exercises

1. 0.55, 1.80 **3.** 1.43, 0.70 **5.** 50.1 **7.** $\angle A = 20°$, $\angle B = 70°$ **9.** $\angle A = 35°$, $\angle B = 55°$ **11.** 1371.6 m **13.** $x = 57.2$, $y = 50$ **15.** $\alpha = 45°$, $\beta = 20°$ **17.** $\alpha = 20°$, $\beta = 20°$

Page 612, Classroom Exercises

1. 0.38, 0.92, 0.92, 0.38 **2.** 24.9 **3.** 20.9 **4.** 14.1

Pages 612–613, Written Exercises

1. 0.80, 0.60, 0.60, 0.80 **3.** 0.98, 0.22, 0.22, 0.98 **5.** 47.0 **7.** $\angle A = 30°$, $\angle B = 60°$ **9.** $\angle A = 65°$, $\angle B = 25°$ **11.** 5.1 ft **13.** $x = 76.6$, $y = 81.5$ **15.** $x = 57.4$, $y = 33.8$ **17.** 39.3 mi **19.** $\alpha = 35°$, $\beta = 35°$

Page 615, Classroom Exercises

1. $\angle B = 65°$, $AC = 21.4$, $AB = 23.7$ **2.** ≈ 52.0 m

Pages 615–616, Written Exercises

1. 0.60, 0.80, 0.75 **3.** 0.69, 0.72, 0.95 **5.** $35°$; 45.9, 65.5 **7.** $80°$; 7.1, 40.6 **9.** $55°$; 10.5, 18.3 **11.** 42 ft **13.** $50°$ **15.** 445 cm^2 **17.** T **19.** T **21.** 12, 0.32 **23.** $53°$ **25.** $28°$

■ INDEX

Boldfaced numerals indicate the pages that contain formal or informal definitions.

Inequality *(continued)*
  multiplication property, **452**–454
  sense of, **452**
  solving, 447–448, 452–454,
    458–460
  in two variables, 468–471
Integers
  consecutive, 185
  even and odd, 185
  negative, 39, 306
  positive, 39, 305–306
  in problem solving, 185–186
Intercept, **223**–224, 551–553
  x-intercept, **229**, 551–553
  y-intercept, **223**, 552–553
Interest
  compound, 352
  simple, **402**
Inverse operation(s) in equations,
  108
Inverse property
  for addition, **66**
  for multiplication, **66**
Inverse variation, **418**–420
Irrational number(s), **494**

Least common denominator (LCD),
  376–378
Least common multiple (LCM),
  376
Legs, **500**
Like terms, **24**, 143
Linear equations, 235
  graphs of, 196, 211–212
  systems of, 256
  in two variables, 206–208, **235**
Linear function, **235**
  graph of, 196
  in problem solving, 241–244
Literal equation(s), **122**–123

Maximum (of a quadratic function),
  545–546
Mean
  arithmetic, **598**
  geometric, **493**
Measures of central tendency, **598**
Median, **598**
Minimum (of a quadratic function),
  544–546
Mixture problems, 289–291
Mode, **598**
Monomial(s), **143**
  degree, **148**
  dividing, 175–176

Monomial(s) *(continued)*
  factoring from polynomials,
    318–319
  multiplying, 153–154
    by a polynomial, 165–166
  power of, 162
Multiple, least common, **376**
Multiplication
  on a computer, 28
  cross, 395–396
  of fractions, 365–367
  of a polynomial by a monomial,
    165–166
  of polynomials, 304
  of real numbers, 52–53
  in solving equations, 103–105
  in solving inequalities, 458–460
Multiplication property
  cross-multiplication, **395**
  of equations, **103**–105
  for equivalent fractions, **360**
  of exponents, **162**
  of inequalities, **452**–454
  negative one (−1) of
    multiplication, **67**
  of opposites, **67**
  for radicals, **506**

Negative number(s), **39**
Negative one (−1) property of
  multiplication, **67**
Number(s)
  composite, **305**
  irrational, **494**
  negative, **39**
  positive, **39**
  prime, **305**
  rational, **494**
  real, **38**
Number line, **39**
Numerical coefficient, **10**

Opposite of an opposite property, **67**
Opposites, **40**
Order of operations, 5–8, 11
Order of real numbers, **39**
Ordered pair(s), **197**
  equal, **198**
Ordinate, **198**
Origin, **197**

Parabola, **542**–546
Parallel lines, slope of, 225
Percent, **400**–402

Perfect-square property, **332**
Perfect-trinomial squares, 331–334
Perpendicular lines, slope of, 226
Polynomial(s), **142**–143
  addition, 143–145
  degree, 148–149
  division, 175–177
  evaluating on a computer, 152
  factoring, 304
  multiplication, 153–154,
    165–166, 304, 318–319,
    326–328, 331–334, 337–339,
    342–345, 348–349
  subtraction, 143–145
  types, 148
Positive number(s), **39**
Power, **11**
  of a monomial, **162**
Prime factorization, **305**–306
Prime number, **305**
  on a computer, 309
Principal, **402**
Principal square root, **489**
Probability, **592**
Problem solving and applications
  age, 29
  digit, 274–276
  distance, 274–276
  drawing figures, 131–132
  with factoring, 346
  with fractional equations,
    386–389
  integers, 185–186
  with inverse variation, 418–420
  with linear functions, 241–243
  mixture, 289–291
  money, 179–181
  percents, 402–404
  Pythagorean theorem, 504
  scientific notation, 159–160
  with systems of equations, 261
  with trigonometric ratios,
    610, 613, 616
  uniform motion, 169–171
  using formulas, 13, 215
  using proportions, 406–407
  using tables, 76–78
  work, 76–78, 381, 387–388
  writing equations, 125–127
Product, **10**
  expanding, of binomials, 322–324
  zero, **314**
Properties
  absolute value of inequalities,
    464–465

INDEX

■ CREDITS

Design   *Cover:* Barbara Tonnesen   *Text:* Volney Croswell

Illustrations   *Cover:* Jeff Stock   *Text:* RoseDesign

Photography   *Research:* Linda Finigan

x: Guy Sauvage (Photo Researchers). *5:* Paul Johnson. *15:* Clive Russ. *28:* Bill Gallery (Stock, Boston). *30:* Arthur Grace (Stock, Boston). *30:* Scott Ransom (Taurus Photos). *38:* Ed Robinson (Tom Stack & Associates). *41:* Peter Menzel (Stock, Boston). *47:* James Holland (Stock, Boston). *61:* Joseph Martin/Scala (Art Resource). *66:* The Bettmann Archive. *76:* Gabor Demjen (Stock, Boston). *88:* Porterfield/Chickering (Photo Researchers). *92:* Clive Russ. *103:* Heinz Klutemeir/Sports Illustrated (Time, Inc.). *107:* P.J. Crowley. *118:* Owen Franken. *124:* Edward Jones (Photo Researchers). *131:* Culver Pictures. *142:* John Sutton (Alberta Color Productions). *157:* National Optical Astronomy Observatories. *167:* Richard Paisley. *172:* Dan McCoy (Rainbow). *174:* P.J. Crowley. *174, 178:* Z. Leszczynski (Animals, Animals/Earth Scenes). *179:* Mike Mazzaschi (Stock, Boston). *181:* Edith Haun (Stock, Boston) *184:* Will McIntyre (Photo Researchers). *185:* Tom Stack (Tom Stack & Associates). *189:* Ralph Mercer. *196:* Doug Lee (Tom Stack & Associates) *206:* The Granger Collection. *229:* Steve Allen (Peter Arnold). *234:* Ralph Mercer. 256: Brian Parker (Tom Stack & Associates). *274:* Dave Schaefer (The Picture Cube). *289:* Michael Heron. *304:* John Zoiner (Peter Arnold). *316:* Sissac (Nawrocki Stock Photo). *341:* C.T. Seymour (Photo Researchers). *342:* P.J. Crowley. *351:* Les Van (Nawrocki Stock Photo). *358:* Peter Menzel (Stock, Boston). *381:* Clive Russ. *394:* Sheryl McNee (Tom Stack & Associates). *403:* Tom Hannon (The Picture Cube). *406:* Dave Schaefer (The Picture Cube). *409:* Tom Stack (Tom Stack & Associates). *423:* Ralph Mercer. *434:* Larry Lefever (Grant Heilman). *447:* Clive Russ. *488:* Runk/Schoenberger (Grant Heilman). *492:* Hugh Patrick Brown (Photo Researchers). *494:* The University Museum, University of Pennsylvania. *497:* Robert Dowling (Nawrocki Stock Photo). *504:* H. Silvester (Photo Researchers). *542:* Du Puy (Monkmeyer Press). *582:* NASA (Grant Heilman).